Student Solutions Manual

to accompany

INTRODUCTION TO ORDINARY DIFFERENTIAL EQUATIONS

Fourth Edition

Shepley L. Ross
University of New Hampshire

With the assistance of

Shepley L. Ross, II
Bates College

JOHN WILEY & SONS
New York   Chichester   Brisbane   Toronto   Singapore

20 19 18 17 16 15 14 13 12

Preface

This manual is a supplement to the author's text, *Introduction to Ordinary Differential Equations*, Fourth Edition. It contains the answers to the even-numbered exercises and detailed solutions to approximately one half of both the even- and odd-numbered exercises in the regular exercise sets.

The following abbreviations have been used:

- D.E.   differential equation
- G.S.   general solution
- I.F.   integrating factor
- I.C.   initial condition
- I.V.P.   initial value problem.

The author expresses his thanks to his son Shepley L. Ross, II, of Bates College for his many contributions, especially the solutions and graphs of Chapter 8. The author also thanks his colleague Ellen O'Keefe and his graduate students Rita Fairbrother and Chris McDevitt for providing solutions.

— *Shepley L. Ross*

Contents

Chapter 1

1. (d) We must show that $f(x) = (1 + x^2)^{-1}$ satisfies the D.E.
    $(1 + x^2)y'' + 4xy' + 2y = 0$. Differentiating $f(x)$, we
    find $f'(x) = -(1 + x^2)^{-2}(2x)$ and $f''(x) = (6x^2 - 2)$
    $(1 + x^2)^{-3}$. We now substitute $f(x)$ for y, $f'(x)$ for
    y', and $f''(x)$ for y'' in the stated D.E. We obtain

    $$(1 + x^2)(6x^2 - 2)(1 + x^2)^{-3} - 4x(1 + x^2)^{-2}(2x)$$
    $$+ 2(1+x^2)^{-1} = 0.$$

    This reduces to

    $$(6x^2 - 2 - 8x^2)(1 + x^2)^2 + 2(1 + x^2)^{-1} = 0,$$

    and hence to

    $$[(-2x^2 - 2) + (2 + 2x^2)](1 + x^2)^{-2} = 0,$$

    that is, $0(1 + x^2)^{-2} = 0$ or $0 = 0$. Hence the given D.E.
    is satisfied by $f(x) = (1 + x^2)^{-1}$.

2. (a) We must show that the relation $x^3 + 3xy^2 = 1$ defines
    at least one real function which is an explicit
    solution of the given D.E. on $0 < x < 1$. Solving the
    given relation for y, we obtain

    $$y^2 = \frac{1 - x^3}{3x} \quad , \quad y = \pm\left[\frac{1 - x^3}{3x}\right]^{1/2} \tag{1}$$

1

We choose the plus sign and consider the function f
defined by

$$f(x) = \left[\frac{1 - x^3}{3x}\right]^{1/2} , \quad 0 < x < 1.$$

We note that $(1 - x^3)/3x > 0$ for $0 < x < 1$, and hence
$f(x)$ is indeed defined on $0 < x < 1$. Now differentiating and simplifying, we find

$$f'(x) = -\frac{1}{6}\left[\frac{1 - x^3}{3x}\right]^{-1/2}\left[\frac{2x^3 + 1}{x^2}\right].$$

Substituting $f(x)$ for y and $f'(x)$ for y' in the given
D.E., we find

$$2x\left[\frac{1 - x^3}{3x}\right]^{1/2}\left\{-\frac{1}{6}\left[\frac{1 - x^3}{3x}\right]^{-1/2}\left[\frac{2x^3 + 1}{x^2}\right]\right\}$$

$$+ x^2 + \frac{1 - x^3}{3x} = 0.$$

This simplifies to $-\frac{2x^3 + 1}{3x} + x^2 + \frac{1 - x^3}{3x} = 0$ and

thence to $\frac{0}{3x} = 0$. Thus $f(x)$ is an explicit solution

of the D.E. on $0 < x < 1$, and so the given relation
$x^3 + 3xy^2 = 1$ is an implicit solution on $0 < x < 1$.
We note that choosing the minus sign in (1) would also
have led to an explicit solution of the D.E. on
$0 < x < 1$.

3. (a) We must show that $f(x) = (x^3 + c)e^{-3x}$ satisfies the
D.E. $y' + 3y = 3x^2 e^{-3x}$. Differentiating $f(x)$, we find
$f'(x) = (x^3 + c)(-3e^{-3x}) + e^{-3x}(3x^2)$. We not
substitute $f(x)$ for y and $f'(x)$ for y' in the stated
D.E. We obtain

$$(x^3 + c)(-3e^{-3x}) + e^{-3x}(3x^2) + 3(x^3 + c)e^{-3x} = 3x^2 e^{-3x}.$$

The left member reduces to $3x^2 e^{-3x}$, so we have
$3x^2 e^{-3x} = 3x^2 e^{-3x}$; and thus the D.E. is satisfied by
$f(x) = (x^3 + c)e^{-3x}$.

4. (b) We must show that $g(x) = c_1 e^{2x} + c_2 x e^{2x} + c_3 e^{-2x}$
satisfies the D.E. $y''' - 2y'' - 4y' + 8y = 0$.
Differentiating $g(x)$, we find $g'(x) = 2c_1 e^{2x} + 2c_2 x e^{2x}$
$+ c_2 e^{2x} - 2c_3 e^{-2x}$, $g''(x) = 4c_1 e^{2x} + 4c_2 x e^{2x} + 4c_2 e^{2x} +$
$4c_3 e^{-2x}$, and $g'''(x) = 8c_1 e^{2x} + 8c_2 x e^{2x} + 12c_2 e^{2x} -$
$8c_3 e^{2x}$. We now substitute $g(x)$ for y, $g'(x)$ for y'.
$g''(x)$ for y'', and $g'''(x)$ for y''' in the stated D.E. We
obtain

$$8c_1 e^{2x} + 8c_2 x e^{2x} + 12c_2 e^{2x} - 8c_3 e^{-2x}$$
$$- 2(4c_1 e^{2x} + 4c_2 x e^{2x} + 4c_2 e^{2x} + 4c_3 e^{-2x})$$
$$- 4(2c_1 e^{2x} + 2c_2 x e^{2x} + c_2 e^{2x} - 2c_3 e^{-2x})$$
$$+ 8(c_1 e^{2x} + c_2 x e^{2x} + c_3 e^{-2x}) = 0.$$

This reduces to

$$8c_1e^{2x} + 8c_2xe^{2x} + 12c_2e^{2x} - 8c_3e^{-2x}$$
$$- 8c_1e^{2x} - 8c_2xe^{2x} - 8c_2e^{2x} - 8c_3e^{-2x}$$
$$- 8c_1e^{2x} - 8c_2xe^{2x} - 4c_2e^{2x} + 8c_3e^{-2x}$$
$$+ 8c_1e^{2x} + 8c_2xe^{2x} + 8c_3e^{-2x} = 0$$

and hence to

$$(8c_1 + 12c_2 - 8c_1 - 8c_2 - 8c_1 - 4c_2 + 8c_1)e^{2x}$$
$$+ (8c_2 - 8c_2 - 8c_2 + 8c_2)\, xe^{2x}$$
$$+ (-8c_3 - 8c_3 + 8c_3 + 8c_3)e^{-2x} = 0,$$

that is, $0 \cdot e^{2x} + 0 \cdot xe^{2x} + 0 \cdot e^{-2x} = 0$, or $0 = 0$. Hence the given D.E. is satisfied by $g(x) = c_1e^{2x} + c_2xe^{2x} + c_3e^{-2x}$.

5.   (a) To determine the values of m for which $f(x) = e^{mx}$ satisfies the given D.E., we differentiate $f(x)$ the required number of times and substitute into the D.E. We have $f(x) = e^{mx}$, $f'(x) = me^{mx}$, $f''(x) = m^2e^{mx}$, $f'''(x) = m^3e^{mx}$. Then substituting $f(x)$ for y, $f'(x)$ for y', $f''(x)$ for y'', and $f'''(x)$ for y''', we find $m^3e^{mx} - 3m^2e^{mx} - 4me^{mx} + 12e^{mx} = 0$ or $e^{mx}(m^3 - 3m^2 - 4m + 12) = 0$. Then since $e^{mx} \neq 0$ for all m and x, we must have $m^3 - 3m^2 - 4m + 12 = 0$. That is, if $f(x) = e^{mx}$ is a solution of the given D.E., then the constant m must satisfy the cubic equation $m^3 - 3m^2 - 4m + 12 = 0$. By inspection we observe that $m = 2$ is a root, and so

(m – 2) is a factor of the left member. We now use synthetic division to find the other factor. We have

2	1	–3	–4	12
		2	–2	–12
	1	–1	–6	0

From this we see that the reduced quadratic factor is $m^2 - m - 6$. Hence the cubic equation may be written $(m - 2)(m^2 - m - 6) = 0$ and so $(m - 2)(m - 3)(m + 2) = 0$. Thus we see that its roots are m = 2, 3, -2. These then are the values of m for which $f(x) = e^{mx}$ is a solution of the given D.E.

6. (b) From $f(x) = 3e^{2x} - 2xe^{2x} - \cos 2x$, we find $f'(x) = 4e^{2x} - 4xe^{2x} + 2 \sin 2x$, $f''(x) = 4e^{2x} - 8xe^{2x} + 4 \cos 2x$. We substitute f(x) for y, f'(x) for y', f''(x) for y'' in the given D.E., obtaining

$$(4e^{2x} - 8xe^{2x} + 4 \cos 2x)$$
$$- 4(4e^{2x} - 4xe^{2x} + 2 \sin 2x)$$
$$+ 4(3e^{2x} - 2xe^{2x} - \cos 2x) = -8 \sin 2x .$$

Collecting like terms in the left member, we find

$$(4 - 16 + 12)e^{2x} + (-8 + 16 - 8)xe^{2x}$$
$$+ (4 - 4) \cos 2x - 8 \sin 2x = -8 \sin 2x$$

or $-8 \sin 2x = -8 \sin 2x .$

Thus $f(x)$ satisfies the D.E. Now note that $f(0) = 3e^0 - 2(0)e^0 - \cos 0 = 3 - 0 - 1 = 2$ and $f'(0) = 4e^0 - 4(0)e^0 + 2 \sin 0 = 4 - 0 + 0 = 4$. Hence $f(x)$ also satisfies the stated conditions.

7. (a) We must show that the D.E. $|y'| + |y| + 1 = 0$ has no (real) solutions. Assume, to the contrary, that this D.E. has a real differentiable function f as a solution on some interval I. Since f is a solution on I, we must have $|f'(x)| + |f(x)| + 1 = 0$ for all $x \in$ I. But for a real function f, both $|f'(x)| \geq 0$ and $|f(x)| \geq 0$, so $|f'(x)| + |f(x)| + 1 \geq 1$ for all $x \in$ I. Thus we obtain two contradictory statements about $|f'(x)| + |f(x)| + 1$, and so our original assumption that the D.E. has a real solution is incorrect. So it has no real solutions.

Section 1.3, Page 21

2. (a) We apply the I.C. $y(0) = 2$ to the given family of solutions. That is, we let $x = 0$, $y = 2$ in $y = (x^2 + c)e^{-x}$. We obtain $2 = (0 + c)e^0$ and hence $c = 2$. We thus obtain the particular solution $y = (x^2 + 2)e^{-x}$ satisfying the stated I.V.P.

3. (a) We first apply the I.C. $y(0) = 5$ to the given family of solutions. That is, we let $x = 0$, $y = 5$ in $y = c_1e^{4x} + c_2e^{-3x}$. We obtain $c_1 + c_2 = 5$. We next differentiate the given family to obtain $y' = 4c_1e^{4x} - 3c_2e^{-3x}$. We apply the I.C. $y'(0) = 6$ to this derived

relation. That is, we let $x = 0$, $y' = 6$ in $y' = 4c_1e^{4x} - 3c_2e^{-3x}$. We obtain $4c_1 - 3c_2 = 6$. The two equations

$$\begin{cases} c_1 + c_2 = 5, \\ 4c_1 - 3c_2 = 6 \end{cases} \quad \text{determine } c_1 \text{ and } c_2 \text{ uniquely.}$$

Solving this system, we find $c_1 = 3$, $c_2 = 2$. Substituting these values back into $y = c_1e^{4x} + c_2e^{-3x}$ we obtain the particular solution $y = 3e^{4x} + 2e^{-3x}$ satisfying the stated I.V.P.

4. (a) We first apply the B.C. $y(0) = 0$ to the given family of solutions. That is, we let $x = 0$, $y = 0$, $y = c_1 \sin x + c_2 \cos x$. We obtain $c_2 = 0$. We next apply the B.C. $y(\pi/2) = 1$ to the given family of solutions. That is, we let $x = \pi/2$, $y = 1$ in $y = c_1 \sin x + c_2 \cos x$. We obtain $c_1 = 1$. Substituting the values $c_1 = 1$, $c_2 = 0$ back into $y = c_1 \sin x + c_2 \cos x$, we obtain the particular solution $y = \sin x$ satisfying the stated boundary-value problem.

(c) We first apply the B.C. $y(0) = 0$ to the given family of solutions. That is, we let $x = 0$, $y = 0$ in $y = c_1 \sin x + c_2 \cos x$. We obtain $c_2 = 0$. We next apply the B.C. $y(\pi) = 1$ to the given family of

solutions.   That is, we let $x = \pi$, $y = 1$ in
$y = c_1 \sin x + c_2 \cos x$. We obtain $-c_2 = 1$, so $c_2 = -1$.
At this point we have both $c_2 = 0$ and, at the same
time, $c_2 = -1$.   This is impossible! So the given
boundary-value problem has no solution.

5. We are given that every solution of the stated D.E. may be
written in the form

$$y = c_1 x + c_2 x^2 + c_3 x^3 \tag{1}$$

for some choice of the constants c_1, c_2, c_3. We must
determine these constants so that (1) will satisfy the
three stated conditions. We differentiate (1) twice to
obtain

$$y' = c_1 + 2c_2 x + 3c_3 x^2 \tag{2}$$

and

$$y'' = 2c_2 x + 6c_3 x \tag{3}$$

We now apply the condition $y(2) = 0$ to (1), letting $x = 2$,
$y = 0$ in (1). We obtain $2c_1 + 4c_2 + 8c_3 = 0$. Similarly,
we apply the condition $y'(2) = 2$ to (2), thereby obtaining
$c_1 + 4c_2 + 12c_3 = 2$. Finally, we apply the condition
$y''(2) = 6$ to (3), obtaining $2c_2 + 12c_3 = 6$. Thus we have
the three equations

$$\begin{cases} 2c_1 + 4c_2 + 8c_3 = 0, \\ c_1 + 4c_2 + 12c_3 = 2, \\ 2c_2 + 12c_3 = 6, \end{cases}$$

in the three unknowns. These can be solved in various ways. One easy way is to eliminate c_1 from the first two equations, obtaining the equivalent of $c_2 + 4c_3 = 1$. Combining this last with $c_2 + 6c_3 = 3$, which is equivalent to the third equation of the system, we readily find $c_2 = -3$, $c_3 = 1$. Then from the second equation one finds $c_1 = 2$. Thus $c_1 = 2$, $c_2 = -3$, $c_3 = 1$. Substituting these values back into (1), we find the solution of the stated I.V.P. is

$$y = 2x - 3x^2 + x^3.$$

(a) Let us apply Theorem 1.1. We first check the hypothesis. Here $f(x,y) = x^2 \sin y$ and
$\dfrac{\partial f(x,y)}{\partial y} = x^2 \cos y$. Both f and $\dfrac{\partial f}{\partial y}$ are continuous in every domain D in the xy plane. The initial condition $y(1) = -2$ means that $x_0 = 1$ and $y_0 = -2$, and the point $(1, -2)$ certainly lies in some such domain D. Thus all hypothesis are satisfied and the conclusion holds. That is, there is a unique solution ϕ of the D.E. $y' = x^2 \sin y$, defined on some interval $|x - 1| \le h$ about $x_0 = 1$, which satisfies the initial condition, that is, which is such that $\phi(1) = -2$.

Chapter 2

In these solutions we denote derivatives by primes and partial derivatives by subscripts. The solutions of Exercises 3, 5, 6, 7, and 9 follow the pattern of Example 2.5 on page 31.

3. Here $M(x,y) = 2xy + 1$, $N(x,y) = x^2 + 4y$. From these we find $M_y(x,y) = 2x = N_x(x,y)$, so the D.E. is exact. We seek $F(x,y)$ such that

$$F_x(x,y) = M(x,y) = 2xy + 1 \text{ and } F_y(x,y) = N(x,y) = x^2 + 4y.$$

From the first of these, we find

$$F(x,y) = \int M(x,y)\,\partial x + \phi(y) = \int (2xy + 1)\,\partial x + \phi(y)$$

$$= x^2 y + x + \phi(y). \text{ From this,}$$

$$F(x,y) = x^2 + \phi'(y).$$

But we must have $F_y(x,y) = N(x,y) = x^2 + 4y$. Therefore

$$x^2 + \phi'(y) = x^2 + 4y$$

or $\dfrac{d\phi(y)}{dy} = 4y$. Then $\phi(y) = 2y^2 + c_0$. Thus $F(x,y) = x^2 y + x + 2y^2 + c_0$. The one-parameter family of solutions $F(x,y) = c_1$ is $x^2 y + x + 2y^2 = c$, where $c = c_1 - c_0$.

Alternatively, by the method of grouping, we first write the D.E. in the form $2xy\,dx + x^2\,dy + dx + 4y\,dy = 0$. We recognize this as $d(x^2y) + d(x) + d(2y^2) = d(c)$ or $d(x^2y + x + 2y^2) = d(c)$. Hence we obtain the solution $x^2y + x + 2y^2 = c$.

4. Here $M(x,y) = 3x^2y + 2$, $N(x,y) = -(x^3 + y)$. From these we find $M_y(x,y) = 3x^2 \neq -3x^2 = N_x(x,y)$. Since $M_y(x,y) \neq N_x(x,y)$, the D.E. is not exact.

5. Here $M(x,y) = 6xy + 2y^2 - 5$, $N(x,y) = 3x^2 + 4xy - 6$. From these we find $M_y(x,y) = 6x + 4y = N_x(x,y)$, so the D.E. is exact. We seek $F(x,y)$ such that $F_x(x,y) = M(x,y) = 6xy + 2y^2 - 5$ and $F_y(x,y) = N(x,y) = 3x^2 + 4xy - 6$. From the first of these, we find $F(x,y) = \int M(x,y)\,\partial x + \phi(y)$ $\int (6xy + 2y^2 - 5)\,\partial x + \phi(y) = 3x^2y + 2y^2x - 5x + \phi(y)$. From this, $F_y(x,y) = 3x^2 + 4xy + \phi(y)$. But we must have $F_y(x,y) = N(x,y) = 3x^2 + 4xy - 6$. Therefore

$$3x^2 + 4xy + \phi'(y) = 3x^2 + 4xy - 6,$$

or $\phi'(y) = -6$. Then $\phi(y) = -6y + c_0$. The one-parameter family of solutions $F(x,y) = c_1$ is $3x^2y + 2y^2x - 5x - 6y = c$, where $c = c_1 - c_0$.

Alternatively, by the method of grouping, we first write the D.E. in the form $(6xy\,dx + 3x^2dy) + (2y^2dx + 4xy\,dy) - 5dx - 6dy = 0$. We recognize this as $d(3x^2y) + d(2y^2x) - d(5x) - d(6y) = d(c)$ or $d(3x^2y + 2y^2x - 5x - 6y) = d(c)$. Hence we obtain the solution $3x^2y + 2y^2x - 5x - 6y = c$.

6.   Here $M(r,\theta) = (\theta^2 + 1)\cos r$, $N(r,\theta) = 2\theta \sin r$. From these we find $M_\theta(r,\theta) = 2\theta \cos r = N_r(r,\theta)$, so the D.E. is exact. We seek $F(r,\theta)$ such that $F_r(r,\theta) = M(r,\theta) = (\theta^2 + 1)\cos r$ and $F_\theta(r,\theta) = N(r,\theta) = 2\theta \sin r$. From the first of these, we find $F(r,\theta) = \int M(r,\theta)\,\partial r + \phi(\theta) = \int (\theta^2 + 1)\cos r\,\partial r + \phi(\theta) = (\theta^2 + 1)\sin r + \phi(\theta)$. From this, $F_\theta(r,\theta) = 2\theta \sin r + \phi'(\theta)$. But we must have $F_\theta(r,\theta) = N(r,\theta) = 2\theta \sin r$. Therefore

$$2\theta \sin r + \phi'(\theta) = 2\theta \sin r$$

or $\phi'(\theta) = 0$. Then $\phi'(\theta) = c_0$. Thus $F(r,\theta) = (\theta^2 + 1)\sin r + c_0$. The one-parameter family of solutions $F(r,\theta) = c_1$ is $(\theta^2 + 1)\sin r = c_1$ where $c = c_1 - c_0$.

Alternately, by the method of grouping, we first write the D.E. in the form $(\theta^2 \cos r\,dr + 2\theta \sin r\,d\theta) + \cos r\,dr = 0$. We recognize this as $d(\theta^2 \sin r) + d(\sin r) = d(c)$ or $d(\theta^2 \sin r + \sin r) = d(c)$. Hence we obtain the solution $\theta^2\sin r + \sin r = c$, or $(\theta^2 + 1)\sin r = c$.

7.   Here $M(x,y) = y \sec^2 x + \sec x \tan x$, $N(x,y) = \tan x + 2y$.
From these we find $M_y(x,y) = \sec^2 x = N_x(x,y)$, so the D.E.
is exact.   We seek $F(x,y)$ such that $F_x(x,y) = M(x,y) = y$
$\sec^2 x + \sec x \tan x$ and $F_y(x,y) = N(x,y) = \tan x + 2y$.
From the first of these, we find $F(x,y) = \int M(x,y) \partial x +$
$\phi(y) = \int (y \sec^2 x + \sec x \tan x) \partial x + \phi(y) = y \tan x +$
$\sec x + \phi(y)$.   From this, $F_y(x,y) = \tan x + \phi'(y)$.   But we
must have $F_y(x,y) = N(x,y) = \tan x + 2y$.   Therefore, $\tan x$
$+ \phi'(y) = \tan x + 2y$ or $\phi'(y) = 2y$.   Then $\phi(y) = y^2 + c_0$.
Thus $F(x,y) = y \tan x + \sec x + y^2 + c_0$.   The one-
parameter family of solutions $F(x,y) = c_1$ is $y \tan x +$
$\sec x + y^2 = c$, where $c = c_1 - c_0$.

Alternatively, by the method of grouping, we first
write D.E. in the form $(y \sec^2 x \, dx + \tan x \, dy) +$
$\sec x \tan x \, dx + 2y \, dy = 0$.   We recognize this as $d(y \tan x)$
$+ d(\sec x) + d(y^2) = d(c)$, and hence obtain the solution
$y \tan x + \sec x + y^2 = c$.

8.   Here $M(x,y) = \dfrac{x}{y^2} + x$. $N(x,y) = \dfrac{x^2}{y^3} + y$.   From these we find
$M_y(x,y) = \dfrac{-2x}{y^3} \neq \dfrac{2x}{y^3} = N_x(x,y)$.   Since $M_y(x,y) \neq N_x(x,y)$,
the D.E. is not exact.

9. Here $M(s,t) = \dfrac{2s - 1}{t}$, $N(s,t) = \dfrac{s - s^2}{t^2}$. From these we find

$M_t(s,t) = \dfrac{1 - 2s}{t^2} = N_s(s,t)$, so the D.E. is exact. We seek

$F(s,t)$ such that $F_s(s,t) = M(s,t) = \dfrac{2s - 1}{t}$ and $F_t(s,t) =$

$N(s,t) = \dfrac{s - s^2}{t^2}$. From the first of these, we find $F(s,t)$

$= \displaystyle\int M(s,t)\,\partial s + \phi(t) = \int \dfrac{2s - 1}{t}\,\partial s + \phi(t) = \dfrac{s^2 - s}{t} + \phi(t)$.

From this, $F_t(s,t) = \dfrac{s - s^2}{t^2} + \phi'(t)$. Therefore, $\dfrac{s - s^2}{t^2} +$

$\phi'(t) = \dfrac{s - s^2}{t^2}$, or $\phi'(t) = 0$. Then $\phi(t) = c_0$. Thus

$F(s,t) = \dfrac{s - s^2}{t} + c_0$. The one-parameter family of

solutions $F(s,t) = c_1$ is $\dfrac{s^2 - s}{t} + c_0 = c_1$ or $s^2 - s = ct$,

where $c = c_1 - c_0$.

The solutions of Exercises 12, 13, 14, and 16 follow
the pattern of Example 2.6 on page 32.

12. Here $M(x,y) = 3x^2y^2 - y^3 + 2x$, $N(x,y) = 2x^3y - 3xy^2 + 1$.
From these we find $M_y(x,y) = 6x^2y - 3y^2 = N_x(x,y)$, so D.E.
is exact. We seek $F(x,y)$ such that $F_x(x,y) = M(x,y) =$
$3x^2y^2 - y^3 + 2x$ and $F_y(x,y) = N(x,y) = 2x^3y - 3xy^2 + 1$.
From the first of these, we have

$F(x,y) = \displaystyle\int M(x,y)\,\partial x + \phi(y) = \int (3x^2y^2 - y^3 + 2x)\,\partial x +$
$\phi(y) = x^3y^2 - xy^3 + x^2 + \phi(y)$

From this, $F_y(x,y) = 2x^3y - 3xy^2 + \phi'(y)$. But we must have $F_y(x,y) = N(x,y) = 2x^3y - 3xy^2 + 1$. Therefore, $2x^3y - 3xy^2 + \phi'(y) = 2x^3y - 3xy^2 + 1$, or $\phi'(y) = 1$. Then $\phi(y) = y + c_0$. Thus $F(x,y) = x^3y^2 - xy^3 + x^2 + y + c_0$. The one-parameter family of solutions $F(x,y) = c_1$ is

$$x^3y^2 - xy^3 + x^2 + y = c, \qquad (*)$$

where $c = c_1 - c_0$.

Applying the I.C. $y(-2) = 1$, we let $x = -2$, $y = 1$ in $(*)$, obtaining $-8 + 2 + 4 + 1 = c$, from which $c = -1$. Thus the particular solution of the stated I.V. problem is $x^3y^2 - xy^3 + x^2 + y = -1$ or

$$x^3y^2 - xy^3 + x^2 + y + 1 = 0. \qquad (**)$$

Alternately, the one-parameter family of solutions $(*)$ could also be found by the method of grouping. We write the D.E. in the form $(3x^2y^2dx + 2x^3y\,dy) - (y^3dx + 3xy^2dy) + 2x\,dx + dy = 0$. We recognize this as $d(x^3y^2) - d(xy^3) + d(x^2) + d(y) = d(c)$, and so again obtain the one-parameter family of solutions

$$x^3y^2 - xy^3 + x^2 + y = c. \qquad (*)$$

The I.C. again yields the particular solution $(**)$.

13.   Here $M(x,y) = 2y \sin x \cos x + y^2 \sin x$, $N(x,y) = \sin^2 x -$
2y cos x .   From these we find $M_y(x,y) = 2 \sin x \cos x +$
$2y \sin x = N_x(x,y)$, so D.E. is exact.   We first seek $F(x,y)$
such that $F_x(x,y) = M(x,y) = 2y \sin x \cos x + y^2 \sin x$ and
$F_y(x,y) = N(x,y) = \sin^2 x - 2y \cos x$.   From the first of
these, we have $F(x,y) = \int M(x,y) \, \partial x + \phi(y) =$
$\int (2y \sin x \cos x + y^2 \sin x) \, \partial x + \phi(y) = y \sin^2 x - y^2 \cos x$
$+ \phi(y)$.   From this, $F_y(x,y) = \sin^2 x - 2y \cos x + \phi'(y)$.
But we must have $F_y(x,y) = N(x,y) = \sin^2 x - 2y \cos x$.
Therefore, $\sin^2 x - 2y \cos x + \phi'(y) = \sin^2 x - 2y \cos x$ or
$\phi'(y) = 0$.   Then $\phi(y) = c_0$.   Thus $F(x,y) = y \sin^2 x - y^2$
$\cos x + c_0$.   The one-parameter family of solutions $F(x,y)$
$= c_1$ is

$$y \sin^2 x - y^2 \cos x = c, \qquad\qquad (*)$$

where $c = c_1 - c_0$.

Applying the I.C. $y(0) = 3$, we let $x = 0$, $y = 3$ in
$(*)$, obtaining $3 \sin^2 0 - 9 \cos 0 = c$, from which $c = -9$.
Thus the particular solution of the stated I.V. problem is
$y \sin^2 x - y^2 \cos x = -9$ or

$$y^2 \cos x - y \sin^2 x = 9. \qquad\qquad (**)$$

Alternatively, the one-parameter family of solutions
$(*)$ could also be found by the method of grouping.   To do

so, we first write the D.E. in the form $(2y \sin x \cos x \, dx + \sin^2 x \, dy) + (y^2 \sin x \, dx - 2y \cos x \, dy) = 0$. We recognize this as $d(y \sin^2 x) + d(-y^2 \cos x) = d(c)$ and hence again obtain the solution (*) in the form $y \sin^2 x - y^2 \cos x = c$. The I.C. again yields the particular solution (**).

14. Here $M(x,y) = ye^x + 2e^x + y^2$, $N(x,y) = e^x + 2xy$. From these we find $M_y(x,y) = e^x + 2y = N_x(x,y)$, so the D.E. is exact. We first seek $F(x,y)$ such that $F_x(x,y) = M(x,y) = ye^x + 2e^x + y^2$ and $F_y(x,y) = N(x,y) = e^x + 2xy$. From the first of these, we have $F(x,y) = \int M(x,y) \, \partial x + \phi(y) = \int (ye^x + 2e^x + y^2) \, \partial x + \phi(y) = ye^x + 2e^x + xy^2 + \phi(y)$. From this, $F_y(x,y) = e^x + 2xy + \phi'(y)$. But we must have $F_y(x,y) = N(x,y) = e^x + 2xy$. Therefore, $e^x + 2xy + \phi'(y) = e^x + 2xy$ or $\phi'(y) = 0$. Then $\phi(y) = c_0$. Thus $F(x,y) = ye^x + 2e^x + xy^2 + c_0$. The one-parameter family of solutions $F(x,y) = c_1$ is

$$ye^x + 2e^x + xy^2 = c, \qquad (*)$$

where $c = c_1 - c_0$.

Applying the I.C. $y(0) = 6$, we let $x = 0$, $y = 6$ in (*), obtaining $6e^0 + 2e^0 + 0(6^2) = c$, from which $c = 8$. Thus the particular solution of the stated I.V. problem is $ye^x + 2e^x + xy^2 = 8$ or

$$e^x y + xy^2 + 2e^x = 8. \qquad (**)$$

Alternatively, the one-parameter family of solutions (*) could also be found by the method of grouping. To do so, we first write the D.E. in the form $(ye^x dx + e^x dy) + (y^2 dx + 2xy\, dy) + 2e^x dx = 0$. We recognize this as $d(e^x y) + d(xy^2) + d(2e^x) = d(c)$ and hence once again obtain the solution (*) in the form $e^x y + xy^2 + 2e^x = c$. The I.C. again yields the particular solution (**).

16. Here $M(x,y) = x^{-2/3}y^{-1/3} + 8x^{1/3}y^{1/3}$, $N(x,y) = 2x^{4/3}y^{-2/3} - x^{1/3}y^{-4/3}$. From these we find $M_y(x,y) = -\frac{1}{3}x^{-2/3}y^{-4/3} + \frac{8}{3}x^{1/3}y^{-2/3} = N_x(x,y)$ so the D.E. is exact. We first seek $F(x,y)$ such that $F_x(x,y) = M(x,y) = x^{-2/3}y^{-1/3} + 8x^{1/3}y^{1/3}$ and $F_y(x,y) = N(x,y) = 2x^{4/3}y^{-2/3} - x^{1/3}y^{-4/3}$. From the first of these, we have $F(x,y) = \int M(x,y)\,\partial x + \phi(y) = \int (x^{-2/3}y^{-1/3} + 8x^{1/3}y^{1/3})\,\partial x + \phi(y) = 3x^{1/3}y^{-1/3} + 6x^{4/3}y^{1/3} + \phi(y)$. From this, $F_y(x,y) = -x^{1/3}y^{-4/3} + 2x^{4/3}y^{-2/3} + \phi'(y)$. But we must have $F_y(x,y) = N(x,y) = 2x^{4/3}y^{-2/3} - x^{1/3}y^{-4/3}$. Thus $-x^{1/3}y^{-4/3} + 2x^{4/3}y^{-2/3} + \phi'(y) = 2x^{4/3}y^{-2/3} - x^{1/3}y^{-4/3}$ or $\phi'(y) = 0$. Then $\phi(y) = c_0$. Thus $F(x,y) = 3x^{1/3}y^{-1/3} + 6x^{4/3}y^{1/3}$. The one-parameter family of solutions $F(x,y) = c_1$ is $3x^{1/3}y^{-1/3} + 6x^{4/3}y^{1/3} = c_2$, where $c_2 = c_1 - c_0$. We can simplify this slightly by dividing through by 3 and replacing $c_2/3$ by c, thus obtaining

$$x^{1/3}y^{-1/3} + 2x^{4/3}y^{1/3} = c. \qquad (*)$$

Applying the I.C. $y(1) = 8$, we let $x = 1$, $y = 8$ in (*) to obtain $\frac{1}{2} + 4 = c$, from which $c = 9/2$. Thus the particular solution of the stated I.V. problem is $x^{1/3}y^{-1/3} + 2x^{4/3}y^{1/3} = 9/2$ or

$$2x^{1/3}y^{-1/3} + 4x^{4/3}y^{1/3} = 9. \qquad (**)$$

Alternatively, the one-parameter family of solutions (*) could also be found by the method of grouping. To do so, we first write the D.E. in the form $(x^{-2/3}y^{-1/3}dx - x^{1/3}y^{-4/3}dy) + (8x^{1/3}y^{1/3}dx + 2x^{4/3}y^{-2/3}dy) = 0$. We recognize this as $d(3x^{1/3}y^{-1/3}) + d(6x^{4/3}y^{1/3}) = d(c_2)$ and hence obtain the solution $3x^{1/3}y^{-1/3} + 6x^{4/3}y^{1/3} = c_2$. Once again, this quickly reduces to (*), and the I.C. again yields the particular solutions (**).

18. (a) Here $M(x,y) = Ax^2y + 2y^2$, $N(x,y) = x^3 + 4xy$. From these we find $M_y(x,y) = Ax^2 + 4y$ and $N_x(x,y) = 3x^2 + 4y$. The given D.E. is exact if and only if $M_y(x,y) = N_x(x,y)$, i.e., if and only if $Ax^2 + 4y = 3x^2 + 4y$. So the given D.E. is exact if and only if $A = 3$. Substituting 3 for A in the given D.E. yields the exact D.E. $(3x^2y + 2y^2)dx + (x^3 + 4xy)dy = 0$. We now proceed to solve this D.E.

Here $M(x,y) = 3x^2y + 2y^2$, $N(x,y) = x^3 + 4xy$. From these we find $M_y(x,y) = 3x^2 + 4y = N_x(x,y)$, so the D.E. is exact. We seek $F(x,y)$ such that $F_x(x,y)$ = $M(x,y)$ = $3x^2y + 2y^2$ and $F_y(x,y)$ = $N(x,y)$ = $x^3 + 4xy$. From the first of these, we find $F(x,y)$ = $\int M(x,y)\,\partial x$ + $\phi(y)$ = $\int (3x^2y + 2y^2)\,\partial x + \phi(y)$ = $x^3y + 2xy^2 + \phi(y)$. From this, $F_y(x,y)$ = $x^3 + 4xy + \phi'(y)$. Therefore x^3 + $4xy + \phi'(y)$ = $x^3 + 4xy$, or $\phi'(y)$ = 0. Then $\phi(y)$ = c_0. Thus $F(x,y)$ = $x^3y + 2xy^2 + c_0$. The one- parameter family of solutions $F(x,y)$ = c_1 is $x^3y + 2xy^2$ = c, where c = $c_1 - c_0$.

Alternatively, by the method of grouping, we first write the D.E. in the form $(3x^2y\,dx + x^3dy) + (2y^2dx + 4xydy)$ = 0. We recognize this as $d(x^3y) + d(2xy^2)$ = $d(c)$ or $d(x^3y + 2xy^2)$ = $d(c)$. Hence we obtain the solution $x^3y + 2xy^2$ = c.

20. (b) Here $N(x,y)$ = $2ye^x + y^2e^{3x}$, so $N_x(x,y)$ = $2ye^x$ + $3y^2e^{3x}$. For the stated D.E. to be exact, we need $M_y(x,y)$ = $N_x(x,y)$ = $2ye^x + 3y^2e^{3x}$. From this, we find $M(x,y)$ = $\int M_y(x,y)\,\partial y + \phi(x)$ = $\int (2ye^x + 3y^2e^{3x})\,\partial y$ + $\phi(x)$ = $y^2e^x + y^3e^{3x} + \phi(x)$. Hence the most general function $M(x,y)$ such that the stated D.E. is exact is $M(x,y)$ = $y^2e^x + y^3e^{3x} + \phi(x)$, where $\phi(x)$ is an arbitrary function of x.

21. Here $M(x,y) = 4x + 3y^2$, $N(x,y) = 2xy$.

(a) Since $M_y(x,y) = 6y \neq 2y = N_x(x,y)$, the D.E. is not exact.

(b) We multiply the given equation through by x^n to obtain $(4x^{n+1} + 3x^n y^2)dx + 2x^{n+1} y\, dy = 0$. For this equation, we have $M(x,y) = 4x^{n+1} + 3x^n y^2$, $N(x,y) = 2x^{n+1}y$. For this equation to be exact, we must have $M_y(x,y) = 6x^n y = 2(n + 1)x^n y = N_x(x,y)$, and hence $6 = 2(n + 1)$, from which $n = 2$. Thus an I.F. of the form x^n is x^2.

(c) We multiply the given equation through by the I.F. x^2, obtaining $(4x^3 + 3x^2 y^2)dx + 2x^3 y\, dy = 0$. Here $M(x,y) = 4x^3 + 3x^2 y^2$, $N(x,y) = 2x^3 y$. Since $M_y(x,y) = 6x^2 y = N_x(x,y)$, this D.E. is indeed exact. We seek $F(x,y)$ such that $F_x(x,y) = M(x,y) = 4x^3 + 3x^2 y^2$ and $F_y(x,y) = N(x,y) = 2x^3 y$. From the first of these, $F(x,y) = \int M(x,y)\,\partial x + \phi(y) = \int (4x^3 + 3x^2 y^2)\,\partial x + \phi(y) = x^4 + x^3 y^2 + \phi(y)$. From this $F_y(x,y) = 2x^3 y + \phi'(y)$. But we must have $F_y(x,y) = N(x,y) = 2x^3 y$, so $\phi'(y) = 0$. Then $\phi(y) = c_0$. Thus $F(x,y) = x^4 + x^3 y^2 + c_0$. The one-parameter family of solutions $F(x,y) = c_1$ is $x^4 + x^3 y^2 = c$, where $c = c_1 - c_0$. Alternatively, by the **method of grouping**, we first write the D.E. $(4x^3 + 3x^2 y^2)dx + 2x^3 y\, dy = 0$ in the form $4x^3 dx + (3x^2 y^2 dx + 2x^3 y\, dy) = 0$. We recognize this as $d(x^4) + d(x^3 y^2) = d(c)$, and hence we obtain the solutions $x^4 + x^3 y^2 = c$.

23.   (a) Here $M(x,y) = y + x f(x^2 + y^2)$ and $N(x,y) = y f(x^2 + y^2) - x$.   Since $M_y(x,y) = 1 + 2xy f'(x^2 + y^2) \neq 2xy$ $f'(x^2 + y^2) - 1 = N_x(x,y)$, the given D.E. is not exact.

(b) We multiply the given D.E. through by $1/(x^2 + y^2)$ to obtain

$$\left[\frac{y}{x^2 + y^2} + \frac{xf(x^2 + y^2)}{x^2 + y^2} \right] dx + \left[\frac{yf(x^2 + y^2)}{x^2 + y^2} - \frac{x}{x^2 + y^2} \right] dy = 0.$$

For this equation, we have

$$M(x,y) = \frac{y}{x^2 + y^2} + \frac{xf(x^2 + y^2)}{x^2 + y^2} \quad \text{and}$$

$$N(x,y) = \frac{yf(x^2 + y^2)}{x^2 + y^2} - \frac{x}{x^2 + y^2}$$

From these we find

$$M_y(x,y) = \frac{2xy(x^2 + y^2)f'(x^2 + y^2) - 2xyf(x^2 + y^2) + x^2 - y^2}{(x^2 + y^2)^2}$$

$$= N_x(x,y)$$

So the D.E. of part (b) is exact, and hence $1/(x^2 + y^2)$ is an I.F. of the given D.E.

24. Applying Exercise 23(a) with $f(x^2 + y^2) = (x^2 + y^2)^2$, we see that the given D.E. is not exact. By Exercise 23(b), we know that $1/(x^2 + y^2)$ is an I.F. of the given D.E. Hence we multiply the given D.E. through by $1/(x^2 + y^2)$ to obtain the equivalent D.E.

$$\left[\frac{y}{x^2 + y^2} + x(x^2 + y^2)\right]dx + \left[y(x^2 + y^2) - \frac{x}{x^2 + y^2}\right]dy = 0,$$

which is therefore exact. Here $M(x,y) = y/(x^2 + y^2) + x(x^2 + y^2)$, $N(x,y) = y(x^2 + y^2) - x/(x^2 + y^2)$, $M_y(x,y) = (x^2 - y^2)/(x^2 + y^2)^2 + 2xy = N_x(x,y)$. We seek $F(x,y)$ such that $F_x(x,y) = M(x,y) = y/(x^2 + y^2) + x(x^2 + y^2)$ and $F_y(x,y) = N(x,y) = y(x^2 + y^2) - x/(x^2 + y^2)$. From the first of these, we find $F(x,y) = \int M(x,y)\,\partial x + \phi(y) = \int [y/(x^2 + y^2) + x(x^2 + y^2)]\,\partial x + \phi(y) = $ arc tan $(x/y) + x^4/4 + x^2y^2/2 + \phi(y)$. From this, $F_y(x,y) = -x/(x^2 + y^2) + x^2y + \phi'(y)$. But we must have $F_y(x,y) = N(x,y) = x^2y + y^3 - x/(x^2 + y^2)$. Therefore, $\phi'(y) = y^3$ or $\frac{d\phi(y)}{dy} = y^3$. Then, $\phi(y) = y^4/4 + c_0$. Thus $F(x,y) = $ arc tan $(x/y) + x^4/4 + x^2y^2/2 + y^4/4 + c_0$, or more simply, $F(x,y) = $ arc tan $(x/y) + (x^2 + y^2)^2/4 + c_0$. The one-parameter family of solutions $F(x,y) = c$ is arc tan$(x/y) + (x^2 + y^2)^2/4 = c$, where $c = c_1 - c_0$.

Section 2.2, Page 46

The equations in Exercises 1-7 and 15-17 are separable, and those in Exercises 8-14 and 18-20 are homogeneous.

2. Since $xy + 2x + y + 2 = x(y + 2) + 1(y + 2) = (x + 1)$ $(y + 2)$, the D.E. can be rewritten as $(x + 1)(y + 2)dx + (x^2 + 2x)dy = 0$. The D.E. is separable. We first separate variables to obtain $\dfrac{(x + 1)dx}{x^2 + 2x} + \dfrac{dy}{y + 2} = 0$. Next we integrate:

$$\int \frac{(x + 1)dx}{x^2 + 2x} = \frac{1}{2} \ln|x^2 + 2x| \text{ and } \int \frac{dy}{y + 2} = \ln|y + 2|.$$

Hence we find the one-parameter family of solutions in the form $\frac{1}{2} \ln|x^2 + 2x| + \ln|y + 2| = \ln c_1$ (where we write $\ln c_1$ for the arbitrary constant, since each term on the left is an ln term). We multiply by 2 and simplify to obtain $\ln|x^2 + 2x| + \ln(y + 2)^2 = \ln c_1{}^2$ or $\ln\left[|x^2 + 2x|(y + 2)^2\right] = \ln c_1{}^2$. From this, we have $|x^2 + 2x|(y + 2)^2 = c$. If $x \geq 0$ (or $x \leq -2$), this may be expressed somewhat more simply as $(x^2 + 2x)(y + 2)^2 = c$ or $x(x + 2)(y + 2)^2 = c$.

3. The D.E. is separable. We first separate variables to obtain $2r \, dr/(r^4 + 1) + ds/(s^2 + 1) = 0$. Next we integrate. By a well-known formula, $\int ds/(s^2 + 1) = $ arc tan s. We next apply the same formula with $s = r^2$, $ds = 2r \, dr$. Thus we obtain $\int 2r \, dr/(r^4 + 1) = $ arc tan r^2.

Hence we find the one-parameter family of solutions in the form

$$\text{arc tan } r^2 + \text{arc tan } s = \text{arc tan } c, \qquad (*)$$

(where we write arc tan c for the arbitrary constant, since each term on the left is an arc tan). We could leave the solutions in this form, but they are unweildy. We take the tangent of each side of (*), applying the formula $\tan(A + B) = \dfrac{\tan A + \tan B}{1 - \tan A \tan B}$ with $A = \text{arc tan } r^2$ and $B = \text{arc tan } s$, to the left member. We obtain

$$\frac{\tan(\text{arc tan } r^2) + \tan(\text{arc tan } s)}{1 - \tan(\text{arc tan } r^2)\tan(\text{arc tan } s)} = \tan(\text{arc tan } c),$$

which reduces to

$$\frac{r^2 + s}{1 - r^2 s} = c \text{ or } r^2 + s = c(1 - r^2 s).$$

6. The D.E. is separable. We first separate variables to obtain $\dfrac{\cos u \ du}{\sin u + 1} + \dfrac{e^v dv}{e^v + 1} = 0$. Next we integrate:

$$\int \frac{\cos u \ du}{\sin u + 1} = \ln(\sin u + 1) \text{ and } \int \frac{e^v dv}{e^v + 1} = \ln(e^v + 1).$$

Hence we find the one-parameter family of solutions in the form $\ln(\sin u + 1) + \ln(e^v + 1) = \ln c$ (where we write ln c for the arbitrary constant, since each term on the left is a ln term). We simplify to obtain $\ln[(\sin u + 1)$ $(e^v + 1)] = \ln c$. From this, we have

$$(\sin u + 1)(e^v + 1) = c.$$

7. This equation is separable. We first separate variables to obtain $(x + 4) dx /(x^2 + 3x + 2) + y\, dy /(y^2 + 1) = 0$. Next we integrate: $\int \dfrac{y\, dy}{y^2 + 1} = \dfrac{1}{2} \ln (y^2 + 1)$. To integrate the dx term, we use partial fractions. We set

$$\frac{x + 4}{x^2 + 3x + 2} = \frac{x + 4}{(x + 1)(x + 2)} = \frac{A}{x + 1} + \frac{B}{x + 2}, \quad x + 4 =$$

$A(x + 2) + B(x + 1)$. Then $x = -1$ gives $A = 3$ and $x = -2$ gives $B = -2$. Thus we find $\int \dfrac{(x + 4)\, dx}{x^2 + 3x + 2} = 3\int \dfrac{dx}{x + 1} -$

$2\int \dfrac{dx}{x + 2} = 3 \ln|x + 1| - 2 \ln|x + 2| = \ln \dfrac{|x + 1|^3}{(x + 2)^2}$. Hence

we obtain solutions in the form $\ln \dfrac{|x + 1|^3}{(x + 2)^2} + \dfrac{1}{2} \ln (y^2 + 1)$

$= \ln c_1$ (where we write $\ln c_1$ for the arbitrary constant, since each term on the left is a ln term). We multiply by 2 and simplify to obtain $\ln \dfrac{(x + 1)^6}{(x + 2)^4} + \ln (y^2 + 1) +$

$\ln c_1^2$ or $\ln \dfrac{(x + 1)^6 (y^2 + 1)}{(x + 2)^4} = \ln c_1^2$. From this, we

have $(x + 1)^6 (y^2 + 1) = c(x + 2)^4$.

9. We first write the D.E. in the form $\dfrac{dy}{dx} = \dfrac{2xy + 3y^2}{2xy + x^2}$ and

thence $\dfrac{dy}{dx} = \dfrac{2(y/x) + 3(y/x)^2}{2(y/x) + 1}$. In this form we recognize

the D.E. is homogeneous. We let $y = vx$. Then $\frac{dy}{dx} = v +$
$x\frac{dv}{dx}$ and $v = \frac{y}{x}$. We make these substitutions to obtain

$$v + x\frac{dv}{dx} = \frac{2v + 3v^2}{2v + 1} \text{ or } x\frac{dv}{dx} = \frac{2v + 3v^2}{2v + 1} - v =$$

$$\frac{2v + 3v^2 - 2v^2 - v}{2v + 1} \text{ or } x\frac{dv}{dx} = \frac{v^2 + v}{2v + 1}. \text{ We now separate}$$

variables to obtain $\frac{(2v + 1)\,dv}{v^2 + v} = \frac{dx}{x}$. We integrate to

obtain $\ln|v^2 + v| = \ln|x| + \ln|c|$, or $\ln|v^2 + v| = \ln|cx|$.
From this, we have $|v^2 + v| = |cx|$. We now resubstitute

$v = \frac{y}{x}$ to obtain $\left| \frac{y^2}{x^2} + \frac{y}{x} \right| = |cx|$. We simplify to obtain

$|y^2 + xy| = |cx|x^2$, from which we find $y^2 + xy = cx^3$.

10. We first write the D.E. in the form $\frac{du}{dv} = \frac{uv^2 - u^3}{v^3}$ from

which we at once obtain $\frac{du}{dv} = \frac{u}{v} - \left(\frac{u}{v}\right)^3$. In this form we
recognize that the D.E. is homogeneous in u and v. If a
D.E. $\frac{dy}{dx} = f\left(\frac{y}{x}\right)$ is homogeneous in x and y, we introduce a
new variable v by letting $y = vx$. Here u and v are the
original variables, so we introduce a new variable w by
letting $u = wv$. Then $\frac{du}{dv} = w + v\frac{dw}{dv}$ and $w = \frac{u}{v}$.

Substituting into the D.E. $\frac{du}{dv} = \frac{u}{v} - \left(\frac{u}{v}\right)^3$, we obtain $w +$
$v\frac{dw}{dv} = w - w^3$. Hence $v\frac{dw}{dv} = -w^3$ and separating variables,

we have $-w^{-3}dw = \frac{dv}{v}$. Integrating we obtain $-\frac{w^{-2}}{-2} = \ln|v|$
$+ c_1$ or $\frac{1}{w^2} = 2\ln|v| + 2c_1$. Resubstituting $w = \frac{u}{v}$, we have

$\frac{v^2}{u^2} = \ln v^2 + c$ or $v^2 = u^2(\ln v^2 + c)$.

11. We first write the D.E. in the form $\frac{dy}{dx} = \tan\left(\frac{y}{x}\right) + \frac{y}{x}$. In this form we recognize that the D.E. is homogeneous. We let $y = vx$. Then $\frac{dy}{dx} = v + x\frac{dv}{dx}$ and $v = \frac{y}{x}$. We make these substitutions in the D.E. to obtain $v + x\frac{dv}{dx} = \tan v + v$ or $x\frac{dv}{dx} = \tan v$. We now separate variables to obtain $\cot v\, dv = \frac{dx}{x}$. We integrate to obtain $\ln|\sin v| = \ln|x| + \ln|c|$, or $\ln|\sin v| = \ln|cx|$. From this we have $|\sin v| = |cx|$. We now resubstitute $v = \frac{y}{x}$ to obtain $\left|\sin\frac{y}{x}\right| = |cx|$. For suitable x and y, this may be expressed somewhat more simply as $\sin\frac{y}{x} = cx$.

13. We first write the D.E. in the form $\frac{dy}{dx} = \frac{x^3 + y^2\sqrt{x^2 + y^2}}{xy\sqrt{x^2 + y^2}}$.

We then divide numerator and denominator by x^3 to obtain

$$\frac{dy}{dx} = \frac{1 + \frac{y^2}{x^3}\sqrt{x^2 + y^2}}{\frac{y}{x^2}\sqrt{x^2 + y^2}} = \frac{1 + \left[\frac{y}{x}\right]^2\left[\frac{1}{x}\right]\sqrt{x^2 + y^2}}{\left[\frac{y}{x}\right]\left[\frac{1}{x}\right]\sqrt{x^2 + y^2}}.$$ Assuming

$x > 0$ so that $x = \sqrt{x^2}$, this becomes

$$\frac{dy}{dx} = \frac{1 + \left[\frac{y}{x}\right]^2\left[\frac{1}{\sqrt{x^2}}\right]\left(\sqrt{x^2 + y^2}\right)}{\left[\frac{y}{x}\right]\left[\frac{1}{\sqrt{x^2}}\right]\sqrt{x^2 + y^2}} = \frac{1 + \left[\frac{y}{x}\right]^2\sqrt{1 + y^2/x^2}}{\left[\frac{y}{x}\right]\sqrt{1 + y^2/x^2}}$$ or

$$\frac{dy}{dx} = \frac{1 + \left[\frac{y}{x}\right]^2\sqrt{1 + (y/x)^2}}{\left[\frac{y}{x}\right]\sqrt{1 + (y/x^2)}}.$$ In this form we recognize the

D.E. is homogeneous. We let $y = vx$. Then $\frac{dy}{dx} = v + x\frac{dv}{dx}$

and $v = \frac{y}{x}$. We make these substitutions to obtain

$$v + x\frac{dv}{dx} = \frac{1 + v^2\sqrt{1 + v^2}}{v\sqrt{1 + v^2}} \quad \text{or} \quad x\frac{dv}{dx} = \frac{1 + v^2\sqrt{1 + v^2}}{v\sqrt{1 + v^2}} - v$$

$$= \frac{1 + v^2\sqrt{1 + v^2} - v^2\sqrt{1 + v^2}}{v\sqrt{1 + v^2}} \quad \text{or} \quad x\frac{dv}{dx} = \frac{1}{v\sqrt{1 + v^2}}. \quad \text{We}$$

now separate variables to obtain $v\sqrt{1 + v^2}\, dv = \frac{dx}{x}$. We

integrate to obtain $1/3\,(1 + v^2)^{3/2} = \ln|x| + \ln|c_0|$

(where we have chosen to write $\ln|c_0|$ for the arbitrary

constant). We multiply by 3 and simplify to obtain

$(1 + v^2)^{3/2} = 3\ln|c_0 x|$ or $(1 + v^2)^{3/2} = \ln|c_0|^3|x|^3$.

Since we have assumed $x > 0$, this may be expressed

somewhat more simply as $(1 + v^2)^{3/2} = \ln c x^3$, where $c =$

$|c_0|^3$. We now resubstitute $v = \frac{y}{x}$ to obtain $\left(1 + \left(\frac{y}{x}\right)^2\right)^{3/2}$

$= \ln c x^3$. We simplify to obtain $\left(\frac{x^2 + y^2}{x^2}\right)^{3/2} = \ln c x^3$ or

$\dfrac{\left[x^2 + y^2\right]^{3/2}}{x^3} = \ln c x^3$, from which we find $\left(x^2 + y^2\right)^{3/2} =$

$x^3 \ln c x^3$.

14. This D.E. is homogeneous. Recognizing this, we could let
    $y = vx$ and substitute in order to separate the variables.
    However, the resulting separable equation is not readily
    tractable. This being the case, we assume $x > y > 0$ and

divide the entire equation through by $\sqrt{y}$. Then we solve for dx/dy, putting the D.E. in the form

$$\frac{dx}{dy} = \frac{\sqrt{x/y + 1} - \sqrt{x/y - 1}}{\sqrt{x/y + 1} + \sqrt{x/y - 1}}.$$

We now let x = uy (see Exercise 24 concerning this). Then $\frac{dx}{dy} = u + y\frac{du}{dy}$ and x/y = u. Substituting in the D.E. it takes the form

$$u + y\frac{du}{dy} = \frac{\sqrt{u + 1} - \sqrt{u - 1}}{\sqrt{u + 1} + \sqrt{u - 1}}.$$

We simplify the right member by multiplying both its numerator and denominator by $\sqrt{u + 1} - \sqrt{u - 1}$ and then simplifying. As a result of this, the D.E. takes the form $u + y\frac{du}{dy} = u - \sqrt{u^2 - 1}$ which readily simplifies to the separable equation $\frac{du}{\sqrt{u^2 - 1}} = -\frac{dy}{y}$. Integrating (tables are useful!), we obtain $\ln\left|u + \sqrt{u^2 - 1}\right| = -\ln|y| + \ln|c|$ or $\ln\left|u + \sqrt{u^2 - 1}\right| = \ln\left|\frac{c}{y}\right|$. Since x > y > 0 and u = x/y, we can write this more simply as $\ln\left(u + \sqrt{u^2 - 1}\right) = \ln(c/y)$, from which we at once have $u + \sqrt{u^2 - 1} = c/y$. Now substituting u = x/y, we have $x/y + \sqrt{(x/y)^2 - 1} = c/y$ which readily simplifies to $x + \sqrt{x^2 - y^2} = c$.

17. The D.E. is separable. We separate variables to obtain

$$\frac{(3x + 8)}{x^2 + 5x + 6} - \frac{4y\,dy}{y^2 + 4} = 0.$$ tegrate the dx term, we

use partial fractions. We write $\dfrac{3x + 8}{x^2 + 5x + 6} =$

$\dfrac{3x + 8}{(x + 2)(x + 3)} = \dfrac{A}{x + 2} + \dfrac{B}{x + 3}$, and so $3x + 8 = A(x + 3) +$

$B(x + 2)$. Then $x = -2$ gives $A = 2$; and $x = -3$ gives $B =$

1. Thus we find

$$\int \frac{3x + 8}{x^2 + 5x + 6}\,dx = 2\int \frac{dx}{x + 2} + \int \frac{dx}{x + 3} = 2\ln|x + 2| +$$

$\ln|x + 3| = \ln(x + 2)^2(x + 3)$. Using this, we obtain

solutions in the form $\ln(x + 2)^2(x + 3) - 2\ln(y^2 + 4) =$

$\ln|c|$ or $\ln\dfrac{(x + 2)^2(x + 3)}{(y^2 + 4)^2} = \ln|c|$. Take $|x + 3| = x + 3$

≥ 0 and $|c| = c \geq 0$, and we have $(x + 2)^2(x + 3) = c(y^2 +$

$4)^2$. We now apply the I.C. $y(1) = 2$ to this, obtaining 36

$= 64c$, $c = \dfrac{9}{16}$. Thus we find the particular solution

$$16(x + 2)^2(x + 3) = 9(y^2 + 4).$$

19. We first write the D.E. in the form $\dfrac{dy}{dx} = \dfrac{5y - 2x}{4x - y}$ or

$\dfrac{dy}{dx} = \dfrac{5(y/x) - 2}{4 - y/x}$. We recognize that this D.E. is

homogeneous. We let $y = vx$. Then $\dfrac{dy}{dx} = v + x\dfrac{dv}{dx}$ and

$v = y/x$. Making these substitutions, the D.E. becomes

$v + x\dfrac{dv}{dx} = \dfrac{5v - 2}{4 - v}$ or $x\dfrac{dv}{dx} = \dfrac{v^2 + v - 2}{4 - v}$. We now separate

variables to obtain $\dfrac{(4 - v)\,dv}{v^2 + v - 2} = \dfrac{dx}{x}$. Next we integrate:

$\int \frac{dx}{x} = \ln|x|$. To integrate the dv term, we use partial fractions. We set $\frac{4 - v}{v^2 + v - 2} = \frac{4 - v}{(v + 2)(v - 1)} = \frac{A}{v + 2} + \frac{B}{v - 1}$, so $4 - v = A(v - 1) + B(v + 2)$. Then $v = -2$ gives $A = -2$ and $v = 1$ gives $B = 1$. Thus we find $\int \frac{(4 - v)\,dv}{v^2 + v - 2} = -2\int \frac{dv}{v + 2} + \int \frac{dv}{v - 1} = -2\ln|v + 2| + \ln|v - 1| = \ln \frac{|v - 1|}{(v + 2)^2}$. Hence we obtain solutions in the form

$\ln \frac{|v - 1|}{(v + 2)^2} = \ln|x| + \ln c$ or $\ln \frac{|v - 1|}{(v + 2)^2} = \ln c |x|$. From this we have $\frac{|v - 1|}{(v + 2)^2} = c|x|$ or $|v - 1| = c|x|(v + 2)^2$.

We resubstitute $v = y/x$ to obtain $|y/x - 1| = c|x|(y/x + 2)^2$. We multiply both sides by x^2 to obtain $|yx - x^2| = c|x|(y + 2x)^2$ or $|x||y - x| = c|x|(y + 2x)^2$ from which we obtain $|y - x| = c(y + 2x)^2$. Taking $y \geq x$, this may be expressed somewhat more simply as $y - x = c(y + 2x)^2$.

Applying the I.C. $y(1) = 4$ we let $x = 1$, $y = 4$ and find $3 = c(6^2)$, from which we find $c = 1/12$. We thus obtain the particular solution $y - x = (1/12)(y + 2x)^2$, or $(2x + y)^2 = 12(y - x)$.

20. We first write the D.E. in the form

$$\frac{dy}{dx} = \frac{3x^2 + 9xy + 5y^2}{6x^2 + 4xy}$$

or

$$\frac{dy}{dx} = \frac{3 + 9(y/x) + 5(y/x)^2}{6 + 4 \ (y/x)} \ .$$

We recognize that this D.E. is homogeneous. We let $y =$ vx, and then $\frac{dy}{dx} = v + x\frac{dv}{dx}$ and $v = \frac{y}{x}$. Making these substitutions, the D.E. becomes

$$v + x\frac{dv}{dx} = \frac{3 + 9v + 5v^2}{6 + 4v}$$

or

$$x\frac{dv}{dx} = \frac{v^2 + 3v + 3}{4v + 6}.$$

We now separate variables to obtain

$$\frac{4v + 6}{v^2 + 3v + 3} \ dv = \frac{dx}{x}.$$

Integrating, we find $2 \ \ln|v^2 + 3v + 3| = \ln|x| + \ln|c|$ or $\ln(v^2 + 3v + 3)^2 = \ln|cx|$. From this, $(v^2 + 3v + 3)^2 = \ln|cx|$. We resubstitute $v = y/x$ to obtain

$$\left[\frac{y^2 + 3xy + 3x^2}{x^2}\right]^2 = |cx| \ .$$

We simplify, taking $|x| = x > 0$, thereby obtaining

$$(y^2 + 3xy + 3x^2)^2 = cx^5.$$

Applying the I.C. $y(2) = -6$, we let $x = 2$, $y = -6$ and find $144 = 32c$ or $c = 9/2$. We thus obtain the particular solution $2(y^2 + 3xy + 3x^2)^2 = 9x^5$.

21. (b) Here $M(x,y) = Ax^2 + Bxy + Cy^2$, $N(x,y) = Dx^2 + Exy + Fy^2$. From these we find $M_y(x,y) = Bx + 2Cy$ and $N_x(x,y) = 2Dx + Ey$. Now the given homogeneous D.E. is exact if and only if $M_y(x,y) = N_x(x,y)$, i.e., if and only if $Bx + 2Cy = 2Dx + Ey$. But $Bx + 2Cy = 2Dx + Ey$ if and only if $B = 2D$ and $2C = E$. Therefore, we have it that the given homogeneous D.E. is exact if and only if $B = 2D$ and $E = 2C$.

23. (a) The given D.E. is of the form $(Ax^2 + Bxy + Cy^2)dx + (Dx^2 + Exy + Fy^2)dy = 0$, where $A = 1$, $B = 0$, $C = 2$, $D = 0$, $E = 4$, $F = -1$. From exercise 21(b), we know the D.E. is homogeneous. Also, since $B = 0 = 2D$ and $E = 4 = 2C$, we know from Exercise 21(b) that the given D.E. is also exact.

First, let us solve the given D.E. as a homogeneous D.E. We first write the D.E. in the form $\dfrac{dy}{dx} = \dfrac{x^2 + 2y^2}{y^2 - 4xy}$ or $\dfrac{dy}{dx} = \dfrac{1 + 2(y/x)^2}{(y/x)^2 - 4(y/x)}$. In this form we recognize that the D.E. is homogeneous. We let $y = vx$. Then $\dfrac{dy}{dx} = v + x\dfrac{dv}{dx}$ and $v = y/x$. We make these substitutions to obtain $v + x\dfrac{dv}{dx} = \dfrac{1 + 2v^2}{v^2 - 4v}$ or

$x\dfrac{dv}{dx} = \dfrac{1 + 6v^2 - v^3}{v^2 - 4v}$. We now separate variables to

obtain $\dfrac{(v^2 - 4v)\,dv}{1 + 6v^2 - v^3} = \dfrac{dx}{x}$. We integrate to obtain

$-\dfrac{1}{3}\ln|1 + 6v^2 - v^3| = \ln|x| + \ln c_0$ or

$-\dfrac{1}{3}\ln|1 + 6v^2 - v^3| = \ln c_0|x|$. We multiply by -3

and simplify to obtain $\ln|1 + 6v^2 - v^3| = -3 \ln c_0|x|$

or $\ln|1 + 6v^2 - v^3| = c_0^{-3}|x|^{-3}$ or $|1 + 6v^2 - v^3| =$

$c|x|^{-3}$, where $c = c_0^{-3}$. We now resubstitute $v = y/x$

to obtain $|1 + 6(y/x)^2 - (y/x)^3| = c|x|^{-3}$. We

multiply by $|x|^3$ to obtain $|x^3 + 6xy^2 - y^3| = c$. For

suitable values of x and y, this may be expressed

somewhat more simply as $x^3 + 6xy^2 - y^3 = c$, or

$x^3 - y^3 + 6xy^2 = c$.

Now let us solve the given D.E. as an exact D.E.

Here $M(x,y) = x^2 + 2y^2$, $N(x,y) = 4xy - y^2$. From these

we find $M_y(x,y) = 4y = N_x(x,y)$, so the D.E. is exact.

We seek $F(x,y)$ such that $F_x(x,y) = M(x,y) = x^2 + 2y^2$

and $F_y(x,y) = N(x,y) = 4xy - y^2$. From the first of

these, $F(x,y) = \int M(x,y)\,\partial x + \phi(y) = \int (x^2 + y^2)\,\partial x +$

$\phi(y) = \dfrac{x^3}{3} + 2xy^2 + \phi(y)$. From this, $F_y(x,y) = 4xy +$

$\phi'(y)$. But we must have $F_y(x,y) = N(x,y) = 4xy - y^2$.

Therefore $4xy + \phi'(y) = 4xy - y^2$ or $\phi'(y) = -y^2$. So

$\phi(y) = -\dfrac{y^3}{3} + c_0$. Thus $F(x,y) = \dfrac{x^3}{3} + 2xy^2 - \dfrac{y^3}{3} + c_0$.

The one-parameter family of solutions $F(x,y) = c_1$ is

$\dfrac{x^3}{3} + 2xy^2 - \dfrac{y^3}{3} = c_2$, where $c_2 = c_1 - c_0$. We multiply

by 3 to obtain $x^3 + 6xy^2 - y^3 = c$, where $c = 3c_2$. So

the one-parameter family of solutions of the given

D.E. is $x^3 - y^3 + 6xy^2 = c$.

25. Since the D.E. is homogeneous, it can be expressed in the

form $\frac{dy}{dx} = g\left(\frac{y}{x}\right)$. Let x = r cos θ, y = r sin θ. Then

$$\frac{dy}{dx} = \frac{r \cos \theta \, d\theta + \sin \theta \, dr}{-r \sin \theta \, d\theta + \cos \theta \, dr} = \frac{r \cos \theta + \sin \theta \, dr/d\theta}{-r \sin \theta + \cos \theta \, dr/d\theta}$$

Substituting into $\frac{dy}{dx} = g\left(\frac{y}{x}\right)$. The D.E. reduces

successively to $\frac{r \cos \theta + \sin \theta \, dr/d\theta}{-r \sin \theta + \cos \theta \, dr/d\theta} = g(\tan \theta)$,

$r \cos \theta + \sin \theta \frac{dr}{d\theta} = g(\tan \theta)[-r \sin \theta + \cos \theta \frac{dr}{d\theta}]$,

$[\sin \theta - \cos \theta \, g(\tan \theta)] \frac{dr}{d\theta} = -[g(\tan \theta) \sin \theta + \cos \theta]r$,

$\frac{dr}{r} = \frac{\sin \theta \, g(\tan \theta) + \cos \theta}{\cos \theta \, g(\tan \theta) - \sin \theta} \, d\theta$, which is separable in r

and θ.

26. (a) The D.E. of Exercise 8 is (x + y) dx $-$ x dy = 0. This

can be written $\frac{dy}{dx} = 1 + y/x$, and so is homogeneous.

Using the method of Exercise 25, we let x = r cos θ,

y = r sin θ. Then $\frac{dy}{dx} = \frac{r \cos \theta \, d\theta + \sin \theta \, dr}{-r \sin \theta \, d\theta + \cos \theta \, dr}$. We

substitute this into $\frac{dy}{dx} = 1 + y/x$. The D.E. reduces

successively to

$$\frac{r \cos \theta \, d\theta + \sin \theta \, dr}{-r \sin \theta \, d\theta + \cos \theta \, dr} = 1 + \tan \theta,$$

$r \cos \theta \, d\theta + \sin \theta dr = (1 + \tan \theta)(-r \sin \theta \, d\theta +$

$\cos \theta \, dr)$, $[\sin \theta - (1 + \tan \theta) \cos \theta] \, dr = -r[\cos \theta +$

$(1 + \tan \theta) \sin \theta] \, d\theta$,

$$\frac{dr}{r} = \frac{\cos\theta + (1 + \tan\theta)\sin\theta}{(1 + \tan\theta)\cos\theta - \sin\theta}\,d\theta\ .$$

$$= \left[\frac{\cos\theta + \sin\theta + \tan\theta\sin\theta}{\cos\theta}\right]d\theta$$

$$= (1 + \tan\theta + \tan^2\theta)\,d\theta\ ,$$

$$= (\sec^2\theta + \tan\theta)\,d\theta\ ,$$

or finally

$$\frac{dr}{r} = (\sec^2\theta + \tan\theta)\,d\theta\ .$$

This is separable. Integrating, assuming $r > 0$, $\cos\theta > 0$, we obtain $\ln r = \tan\theta - \ln\cos\theta + \ln c$, where $c > 0$. Now resubstitute, according to $x = r\cos\theta$, $y = r\sin\theta$; that is, let $r = \sqrt{x^2 + y^2}$, $\tan\theta = y/x$. We obtain successively

$$\ln\sqrt{x^2 + y^2} = y/x - \ln(x/\sqrt{x^2 + y^2}) - \ln c\ ,$$

$$\ln\sqrt{x^2 + y^2} + \ln x/\sqrt{x^2 + y^2} + \ln c = y/x,$$

$$\ln(cx) = y/x, \text{ or finally } y = x\,\ln(cx).$$

Section 2.3, Page 56

The equations of Exercises 1 through 14 are linear. In solving, first express the equation in the standard form of equation (2.26) and then follow the procedure of Example 2.14.

1. The D.E. is already in the standard form (2.26), with $P(x)$
    $= 3/x$, $Q(x) = 6x^2$. An I.F. is $e^{\int P(x)dx} = e^{\int (3/x)dx} =$
    $e^{3\ln|x|} = e^{\ln|x|^3} = |x|^3 = \pm x^3$ (+ if $x \geq 0$, - if $x < 0$).
    We multiply the D.E. through by this I.F. to obtain $x^3 \dfrac{dy}{dx}$
    $+ 3x^2 = 6x^5$ or $\dfrac{d}{dx}(x^3y) = 6x^5$. Integrating, we obtain x^3y
    $= x^6 + c$ or $y = x^3 + cx^{-3}$.

3. The D.E. is already in the standard form (2.26), wih $P(x)$
    $= 3$, $Q(x) = 3x^2e^{-3x}$. An I.F. is $e^{\int P(x) \, dx} = e^{\int 3 \, dx} = e^{3x}$.
    We multiply the D.E. through by the I.F. to obtain

    $$e^{3x} \frac{dy}{dx} + 3e^{3x}y = 3x^2$$

    or $\dfrac{d}{dx}\left(e^{3x}y\right) = 3x^2$. Integrating we obtain $e^{3x}y = x^3 + c$

    or $y = \left(x^3 + c\right)e^{-3x}$.

6. This equation is linear in v. We divide through by $u^2 + 1$
    to put it in the standard form $\dfrac{dv}{du} + \dfrac{4u}{u^2 + 1} v = \dfrac{3u}{u^2 + 1}$,

    with $P(u) = \dfrac{4u}{u^2 + 1}$ and $Q(u) = \dfrac{3u}{u^2 + 1}$. An I.F. is

    $e^{\int P(u)du} = \exp \int \dfrac{4u}{u^2 + 1} \, du = e^{2\ln(u^2+1)} = (u^2 + 1)^2$.

    We multiply the standard form equation through by this to
    obtain $(u^2 + 1)^2 \dfrac{dv}{du} + 4u(u^2 + 1)v = 3u(u^2 + 1)$ or

    $\dfrac{d}{du}[(u^2 + 1)^2 v] = 3u^3 + 3u$. Integrating, we obtain

    $(u^2 + 1)^2 v = \dfrac{3u^4}{4} + \dfrac{3u^2}{2} + c$.

7. We first divide through by x to obtain $\frac{dy}{dx} + \frac{2x + 1}{x^2 + x} y =$

$\frac{x - 1}{x}$, where $P(x) = \frac{2x + 1}{x^2 + x}$ and $Q(x) = \frac{x - 1}{x}$. An I.F. is

$e^{\int P(x)dx} = \exp\left(\int \frac{2x + 1}{x^2 + x} dx\right) = e^{\ln|x^2 + x|} = |x^2 + x| = \pm (x^2$

$+ x)$ (+ if $x < -1$ or $x > 0$, $-$ if $-1 < x < 0$). In any
case, upon multiplying the standard form equation through
by the I.F., we obtain $(x^2 + x) \frac{dy}{dx} + (2x + 1) y = (x +$

$1)(x - 1)$ or $(x^2 + x) \frac{dy}{dx} + (2x + 1) y = x^2 - 1$, that is,

$\frac{d}{dx} [(x^2 + x)y] = x^2 - 1$. Integrating, we find $(x^2 + x)y =$

$x^3/3 - x + c/3$ or $3(x^2 + x)y = x^3 - 3x + c$.

8. We first divide through by $x^2 + x - 2$ to put the equation
in the standard form $\frac{dy}{dx} + \frac{3(x + 1)}{(x + 2)(x - 1)} y = \frac{1}{x + 2}$, where

$P(x) = \frac{3(x + 1)}{(x + 2)(x - 1)}$ and $Q(x) = \frac{1}{x + 2}$. An I.F. is

$e^{\int P(x)dx} = \exp \int \left[\frac{3(x + 1)}{(x + 2)(x - 1)}\right] dx =$

$e^{\ln|x+2|+2\ln|x-1|} = |x + 2|(x - 1)^2 = \pm(x + 2)(x - 1)^2,$
(+ if $x \geq -2$, $-$ if $x \leq -2$), where partial fractions have
been used to perform the integration. In either case ($x >$
2 or $x \leq -2$), upon multiplying the standard form equation
through by the I.F., we obtain $(x + 2)(x - 1)^2 \frac{dy}{dx} +$

$3(x + 1)(x - 1)y = (x - 1)^2$, that is, $\frac{d}{dx} [(x + 2)(x -1)^2 y]$

$= (x - 1)^2$. Integrating, we find $(x + 2)(x - 1)^2 y =$

$\frac{(x - 1)^3}{3} + \frac{c}{3}$ or $3(x + 2)y = x - 1 + c(x - 1)^{-2}$.

9.   We first put the equation in the form

$$\frac{dy}{dx} + \frac{xy + y - 1}{x} = 0,$$

and then in the standard form

$$\frac{dy}{dx} + \left(1 + \frac{1}{x}\right)y = \frac{1}{x},$$

where $P(x) = 1 + 1/x$ and $Q(x) = 1/x$. An I.F. is $e^{\int P(x)dx}$
$= e^{(1+1/x) \, dx} = e^{x+\ln|x|} = |x|e^x = \pm \, x\,e^x$, ($+$ if $x \geq 0$,
$-$ if $x < 0$). In either case, we multiply the standard
form equation through by the I.F. and obtain

$$xe^x \frac{dy}{dy} + (x + 1)e^x y = e^x$$

or $\frac{d}{dx} [x \, e^x y] = e^x$. Integrating, we find $x \, e^x y = e^x + c$
or $y = x^{-1}(1 + c \, e^{-x})$.

10.   This D.E. is linear in x (like Example 2.16). We divide
through by y and dy to put it in the standard form

$$\frac{dx}{dy} + \frac{xy^2 + x - y}{y} = 0,$$

or $\frac{dx}{dy} + \left(y + \frac{1}{y}\right) x = 1$, where $P(y) = y + \frac{1}{y}$, $Q(y) = 1$. An
I.F. is

$$e^{\int P(y)dy} = e^{\int (y+1/y)dy} = e^{y^2/2+\ln|y|} = |y|e^{y^2/2} =$$

$\pm \, ye^{y^2/2}(+$ if $y \geq 0$; $-$ if $y < 0$). Multiplying the
standard form equation through by this, we obtain

$$ye^{y^2/2} \frac{dx}{dy} + (y^2 + 1)e^{y^2/2} x = ye^{y^2/2}$$

or $\frac{d}{dy}(ye^{y^2/2}x) = ye^{y^2/2}$.   Integrating, we find $ye^{y^2/2}x = e^{y^2/2} + c$ or $xy = 1 + ce^{-y^2/2}$.

12.   This D.E. is linear in r. We first divide through by $d\theta$ to obtain $\cos\theta \, \frac{dr}{d\theta} + r\sin\theta - \cos^4\theta = 0$ or $\cos\theta\frac{dr}{d\theta} + r\sin\theta = \cos^4\theta$. Next we divide through by $\cos\theta$ to put the equation in the standard form $\frac{dr}{d\theta} + (\tan\theta)\,r = \cos^3\theta$, where $P(\theta) = \tan\theta$ and $Q(\theta) = \cos^3\theta$. An I.F. is $e^{\int P(\theta)d\theta} = e^{\int \tan\theta \, d\theta} = e^{\ln|\sec\theta|} = |\sec\theta| = \pm\sec\theta$ (+ if $\frac{4k-1}{2}\pi < \theta < \frac{4k+1}{2}\pi$ for some integer k, and $-$ if $\frac{4k+1}{2}\pi < \theta < \frac{4k+3}{2}\pi$ for some integer k). In any case, upon multiplying the standard form equation through by the I.F., we obtain $\sec\theta\frac{dr}{d\theta} + (\sec\theta\tan\theta)r = \cos^2\theta$, that is, $\frac{d}{d\theta}[r\sec\theta] = \cos^2\theta$. Upon integrating, we find $r\sec\theta = \frac{1}{2}\theta + \frac{1}{4}\sin 2\theta + c_0$, where the right-side was integrated as follows: $\int\cos^2\theta d\theta = \int \frac{1 + \cos 2\theta}{2}d\theta = \frac{1}{2}\theta + \frac{1}{4}\sin 2\theta$. Substituting $2\sin\theta\cos\theta$ for $\sin 2\theta$, the solutions may be expressed as $r\sec\theta = \frac{1}{2}\theta + \frac{1}{2}\sin\theta\cos\theta + c_0$. Finally, we multiply through by $2\cos\theta$ to obtain

$$2r = (\theta + \sin\theta\cos\theta + c)\cos\theta, \text{ where } c = 2c_0.$$

13. We first put the D.E. in the form

$$(1 + \sin x)\frac{dy}{dx} + (\cos x)y = \cos^2 x,$$

and then in the standard form

$$\frac{dy}{dx} + \frac{\cos x}{1 + \sin x}\ y = \frac{\cos^2 x}{1 + \sin x},$$

where $P(x) = \dfrac{\cos x}{1 + \sin x}$, $Q(x) = \dfrac{\cos^2 x}{1 + \sin x}$.   An I.F. is

$$e^{\int P(x)dx} = e^{\int \frac{\cos x \, dx}{1 + \sin x}} = e^{\ln|1+\sin x|} = |1 + \sin x| = 1$$

$+ \sin x$, since $1 + \sin x \geqslant 0$. Multiply the standard form
equation through by this, to obtain

$$(1 + \sin x)\frac{dy}{dx} + (\cos x)y = \cos^2 x,$$

which is in fact the form preceding the standard form.   We
observe that this is $\frac{d}{dx}[(1 + \sin x)y] = \cos^2 x$.
Integrating we obtain $(1 + \sin x)y = \frac{x}{2} + \frac{\sin 2x}{4} + c_1$ or
$2(1 + \sin x)y = x + \sin x \cos x + c$, where $c = 2c_1$ and we
have employed double-angle formulas.

14. This D.E. is linear in y.   We first divide through by dy
to obtain $y \sin 2x - \cos x + (1 + \sin^2 x)\frac{dy}{dx} = 0$ or
$(1 + \sin^2 x)\frac{dy}{dx} + (\sin 2x)y = \cos x$.   Next we divide
through by $1 + \sin^2 x$ to put the equation in the standard

form $\dfrac{dy}{dx} + \dfrac{\sin 2x}{1 + \sin^2 x} y = \dfrac{\cos x}{1 + \sin^2 x}$, where $P(x) = \dfrac{\sin 2x}{1 + \sin^2 x}$

and $Q(x) = \dfrac{\cos x}{1 + \sin^2 x}$. An I.F. is

$e^{\int P(x)dx} = \exp\left[\int \dfrac{\sin 2x}{1 + \sin^2 x} dx\right] = \exp\left[\int \dfrac{2 \sin x \cos x}{1 + \sin^2 x} dx\right]$

$= e^{\ln(1+\sin^2 x)} = 1 + \sin^2 x$. Upon multiplying

the standard form equation through by the I.F., we obtain

$(1 + \sin^2 x)\dfrac{dy}{dx} + (\sin 2x)y = \cos x$, that is,

$\dfrac{d}{dx}[(1 + \sin^2 x)y] = \cos x$. Integrating, we find

$(1 + \sin^2 x)y = \sin x + c$.

15. This is a Bernoulli D.E., where n = 2. We multiply

through by y^{-2} to obtain $y^{-2}\dfrac{dy}{dx} - \dfrac{1}{x}y^{-1} = -\dfrac{1}{x}$. Let

$v = y^{1-n} = y^{-1}$; then $\dfrac{dv}{dx} = -y^{-2}\dfrac{dy}{dx}$. The preceeding D.E.

readily transforms into the linear equation $\dfrac{dv}{dx} + \dfrac{1}{x}v = \dfrac{1}{x}$.

An I.F. is $e^{\int dx/x} = e^{\ln|x|} = |x| = \pm x$. Multiplying

through by this, we find $x\dfrac{dv}{dx} + v = 1$ or $\dfrac{d}{dx}(xv) = 1$.

Integrating, we find $xv = x + c$, from which $v = 1 + cx^{-1}$.

But $v = 1/y$. Thus we obtain the solution in the form $1/y$

$= 1 + cx^{-1}$.

18. This is a Bernoulli D.E. in the dependent variable x and

independent variable t, where n = -1. We multiply through

by x to obtain $x\dfrac{dx}{dt} + \dfrac{t + 1}{2t}x^2 = \dfrac{t + 1}{t}$. Let $v = x^{1-n} = $

x^2; then $\dfrac{dv}{dt} = 2x\dfrac{dx}{dt}$. The preceeding D.E. readily transforms into the linear equation $\dfrac{dv}{dt} + \dfrac{t + 1}{t} v =$

$2\left(\dfrac{t + 1}{t}\right)$.   An I.F. is $e^{\int \frac{t + 1}{1} dt} = e^{t + \ln|t|} = |t| e^t$

$= \pm t e^t$.   Multiplying through by this, we find $t e^t \dfrac{dv}{dt} +$

$(t + 1)e^t v = 2(t + 1)e^t$ or $\dfrac{d}{dt}[t e^t v] = 2(t + 1)e^t$.

Integrating, we obtain $t e^t v = 2 t e^t + c$ or $v = 2 +$

$c t^{-1} e^{-t}$.   But $v = x^2$.   Thus we obtain the solution in

the form $x^2 = 2 + c t^{-1} e^{-t}$.

The equations of Exercises 19 through 24 are linear.

21.   We divide thru by $(e^x + 1)$ and dx to put in the form

$e^x[y(e^x + 1)^{-1} - 3(e^x + 1)] + \dfrac{dy}{dx} = 0$ and thence in the

standard form

$$\dfrac{dy}{dx} + \dfrac{e^x}{e^x + 1} y = 3e^x(e^x + 1),$$

where $P(x) = \dfrac{e^x}{e^x + 1}$, $Q(x) = 3e^x(e^x + 1)$.   An I.F. is

$e^{\int P(x)dx} = \exp\int \dfrac{e^x}{e^x + 1} dx = e^{\ln(e^x + 1)} = e^x + 1$.   We

multiply the standard form equation through by this,

obtaining $(e^x + 1)\dfrac{dy}{dx} + e^x y = 3e^x(e^x + 1)^2$ or

$\dfrac{d}{dx}[(e^x + 1)y] = 3e^x(e^x + 1)^2$.   Integrating, we obtain

$(e^x + 1)y = (e^x + 1)^3 + c$ or

$$y = (e^x + 1)^2 + c(e^x + 1)^{-1}. \qquad (*)$$

We apply the I.C. $y(0) = 4$: Let $x = 0, y = 4$ in (*) to obtain $4 = 4 + c/2$. Thus $c = 0$, and the particular solution of the stated I.V.P. is $y = (e^x + 1)^2$.

24. This equation is linear in the dependent variable x with t as the independent variable, and is already in the standard form, with $P(t) = -1$, $Q(t) = \sin 2t$. An I.F. is $e^{\int P(t)dt} = e^{\int (-1)dt} = e^{-t}$. We multiply the equation through by this to obtain $e^{-t}\frac{dx}{dt} - e^{-t}x = e^{-t}\sin 2t$ or $\frac{d}{dt}(e^{-t}x) = e^{-t}\sin 2t$. We next integrate, using integration by parts twice, or an integral table, on the right. We obtain

$$e^{-t}x = e^{-t}\frac{-\sin 2t - 2\cos 2t}{5} + c$$

or

$$x = \frac{-(\sin 2t + 2\cos 2t)}{5} + ce^t. \qquad (*)$$

We apply the I.C. $x(0) = 0$: Let $t = x = 0$ in (*). We obtain $0 = -1/5(0 + 2) + c$ from which $c = 2/5$. Thus the general solution of the stated I.V.P. is

$$x = \frac{2c^t - \sin 2t - 2\cos 2t}{5}.$$

25. This is a Bernoulli D.E. with $n = -3$. We first multiply through by y^3 to obtain $y^3\frac{dy}{dx} + \frac{y^4}{2x} = x$. Now let $v = y^{1-n} = y^4$. Then $\frac{dv}{dx} = 4 y^3\frac{dy}{dx}$ and the preceeding D.E.

transforms into $\frac{1}{4}\frac{dv}{dx} + \frac{v}{2x} = x$, which is linear in v. In

the standard form this linear equation is $\frac{dv}{dx} + \frac{2}{x}v = 4x$,

with $P(x) = 2/x$, $Q(x) = 4x$. An I.F. is

$$e^{\int P(x)dx} = e^{\int (2/x)dx} = e^{2\ln|x|} = x^2.$$

We multiply the standard form through by this to obtain
$x^2\frac{dv}{dx} + 2xv = 4x^3$ or $\frac{d}{dx}(x^2) = 4x^3$. We integrate to

obtain $x^2v = x^4 + c$. But $v = y^4$. Hence we obtain the
one-parameter family of solutions of the given Bernoulli
Equation in the form

$$x^2y^4 = x^4 + c. \qquad (*)$$

We apply the I.C. $y(1) = 2$. Let $x = 1$, $y = 2$ in $(*)$
to obtain $16 = 1 + c$, so $c = 15$. Thus the particular
solution of the stated I.V.P. is $x^2y^4 = x^4 + 15$.

26. We first write the D.E. in the form $x\frac{dy}{dx} + y = x^{3/2}y^{3/2}$

and multiply through by $1/x$ to obtain $\frac{dy}{dx} + \frac{1}{x}y = x^{1/2}y^{3/2}$.

We recognize this D.E. as a Bernoulli D.E. with $n = 3/2$.
We now multiply through by $y^{-3/2}$ to obtain $y^{-3/2}\frac{dy}{dx} + $

$\frac{1}{x}y^{-1/2} = x^{1/2}$. Now let $v = y^{1-n} = y^{-1/2}$. Then

$\frac{dv}{dx} = -\frac{1}{2}y^{-3/2}\frac{dy}{dx}$ and the preceeding D.E. transforms into

$-2\frac{dv}{dx} + \frac{1}{x}v = x^{1/2}$, which is linear in v. In the standard form this linear D.E. is $\dfrac{dv}{dx} + \left(\dfrac{-1}{2x}\right)v = -\dfrac{x^{1/2}}{2}$, with $P(x) =$

$-\dfrac{1}{2x}$, $Q(x) = -\dfrac{x^{1/2}}{2}$. An I.F. is $e^{\int P(x)dx} = e^{\int(-1/2x)dx} =$

$e^{-1/2\ln|x|} = e^{\ln|x|^{-1/2}} = |x|^{-1/2} = x^{-1/2}$ (assuming $x >$

0). We multiply the standard form through by this I.F. to

obtain $x^{-1/2}\dfrac{dv}{dx} - \dfrac{1}{2x^{3/2}}v = -\dfrac{1}{2}$ or $\dfrac{d}{dx}(x^{-1/2}v) = -\dfrac{1}{2}$. We

integrate to obtain $x^{-1/2}v = -\dfrac{1}{2}x + c$. But $v = y^{-1/2}$.

Hence we obtain the one-parameter family of solutions of

the given Bernoulli equation in the form $x^{-1/2}y^{-1/2} =$

$-\dfrac{1}{2}x + c$ or

$$\frac{1}{\sqrt{xy}} = -\frac{1}{2}x + c \qquad (*)$$

We apply the I.C. $y(1) = 4$. Let $x = 1$, $y = 4$ in $(*)$

to obtain $\dfrac{1}{2} = -\dfrac{1}{2} + c$, so $c = 1$. Thus the particular

solution of the stated I.V.P. is $\dfrac{1}{\sqrt{xy}} = -\dfrac{1}{2}x + 1$.

27. Here we actually have two I.V. problems:

$$(I)\begin{cases} \text{For } 0 \le x < 1, \\ \dfrac{dy}{dx} + y = 2, \\ y(0) = 0, \end{cases} \qquad (II)\begin{cases} \text{For } x \ge 1, \\ \dfrac{dy}{dx} + y = 0, \\ y(1) = a, \end{cases}$$

where a is the value of $\lim\limits_{x \to 1-} \phi(x)$ and ϕ denotes the

solution of (I). This is prescribed so that the solution

of the entire problem will be continuous at $x = 1$.

We first solve (I). The D.E. of this problem is linear in standard form with $P(x) = 1$, $Q(x) = 2$. An I.F. is $e^{\int P(x)dx} = e^{\int (1)dx} = e^x$. We multiply the D.E. of (I) through by this to obtain $e^x \frac{dy}{dx} + e^x y = 2e^x$ or $\frac{d}{dx}[e^x y] = 2e^x$. We integrate to obtain $e^x y = 2e^x + c$ or $y = 2 + ce^{-x}$. We apply the I.C. of (I), $y(0) = 0$, to this, obtaining $0 = 2 + c$ or $c = -2$. Thus the particular solution of problem (I) is $y = 2 - 2e^{-x}$. This is valid for $0 \leq x < 1$.

Letting ϕ denote the solution just obtained, we note that $\lim_{x \to 1^-} \phi(x) = \lim_{x \to 1^-} (2 - 2e^{-x}) = 2 - 2e^{-1}$. This is the a of Problem (II); that is, the I.C. of (II) is $y(1) = 2 - 2e^{-1}$.

Now solve (II). The D.E. of this problem is linear is standard form with $P(x) = 1$, $Q(x) = 0$. An I.F., as in problem (I), is e^x. We multiply through by this to obtain $e^x \frac{dy}{dx} + e^x y = 0$ or $\frac{d}{dx}(e^x y) = 0$. We integrate to obtain $e^x y = c$ or $y = ce^{-x}$. Now apply the I.C. of (II), $y(1) = 2 - 2e^{-1}$, to this. Let $x = 1$, $y = 2 - 2e^{-1}$. We obtain $2 - 2e^{-1} = ce^{-1}$, from which $c = 2e - 2$. Thus the particular solution of problem (II) is $y = (2e - 2)e^{-x}$. This is valid for $x \geq 1$.

We write the solution of the entire problem, showing intervals where each part is valid, as

$$
y = \begin{cases}
2(1 - e^{-x}), & 0 \leq x < 1, \\[2ex]
2(e - 1)e^{-x}, & x \geq 1 .
\end{cases}
$$

29. As in Exercise 27, we actually have two I.V. problems:

(I) $\begin{cases} \text{For } 0 \le x < 2, \\[6pt] \dfrac{dy}{dx} + y = e^{-x}, \\[6pt] y(0) = 1, \end{cases}$ (II) $\begin{cases} \text{For } x \ge 2, \\[6pt] \dfrac{dy}{dx} + y = e^{-2}, \\[6pt] y(2) = a, \end{cases}$

where a is the value of $\lim\limits_{x \to 2^-} \phi(x)$ and ϕ denotes the

solution of (I). This is prescribed so that the solution of the entire problem will be continuous at x = 2.

We first solve (I). The D.E. of this problem is linear in standard form with $P(x) = 1$, $Q(x) = e^{-x}$. An I.F. is $e^{\int P(x)dx} = e^{\int (1)dx} = e^{x}$. We multiply the D.E. of (I) through by this to obtain $e^{x}\dfrac{dy}{dx} + e^{x}y = 1$ or $\dfrac{d}{dx}(e^{x}y)$ = 1. Integrating we obtain $e^{x}y = x + c$ or $y = e^{-x}(x + c)$.

We apply the I.C. of (I), $y(0) = 1$ to this, obtaining 1 = 0 + c, or c = 1. Thus the particular solution of problem (I) is $y = e^{-x}(x + 1)$. This is valid for $0 \le x < 2$.

Letting ϕ denote the solution just obtained, we find $\lim\limits_{x \to 2^-} \phi(x) = \lim\limits_{x \to 2^-} e^{-x}(x + 1) = 3e^{-2}$. This is the a of Problem (II); that is, the I.C. of (II) is $y(2) = 3e^{-2}$.

Now solve (II). The D.E. of this problem is linear in standard form with $P(x) = 1$, $Q(x) = e^{-2}$. An I.F., as in Problem (I), is e^{x}. We multiply through by this to obtain $e^{x}\dfrac{dy}{dx} + e^{x}y = e^{x-2}$ or $\dfrac{d}{dx}[e^{x}y] = e^{x-2}$. We integrate to obtain $e^x y = e^{x-2} + c$ or $y = e^{-x}(e^{x-2} + c)$ or $y = ce^{-x} + e^{-2}$. Now apply the I.C. of (II), $y(2) = 3e^{-2}$, to this. Let $x = 2$, $y = 3e^{-2}$. We obtain $3e^{-2} = ce^{-2} + e^{-2}$, from which $c = 2$. Thus the particular solution of problem (II) is $y = 2e^{-x} + e^{-2}$. This is valid for $x \geq 2$.

We write the solution of the entire problem, showing intervals where each part is valid, as

$$
y = \begin{cases}
e^{-x}(x + 1), & 0 \leq x < 2 , \\
\\
2e^{-x} + e^{-2}, & x \geq 2
\end{cases} .
$$

30. As in Exercises 27 and 29, we actually have two I.V. problems:

(I) $\begin{cases} \text{For } 0 \leq x < 3, \\ \\ (x + 1)\dfrac{dy}{dx} + y = x, \\ \\ y(0) = 1/2. \end{cases}$ (II) $\begin{cases} \text{For } x \geq 3, \\ \\ (x + 1)\dfrac{dy}{dx} + y = 3, \\ \\ y(3) = a, \end{cases}$

where a is the value of $\lim\limits_{x \to 3^-} \phi(x)$ and ϕ denotes the solution of (I). This is prescribed so that the solution of the entire problem will be continuous at $x = 3$.

We first solve (I). We first express the D.E. in the standard form $\dfrac{dy}{dx} + \dfrac{1}{x + 1} y = \dfrac{x}{x + 1}$, where $P(x)$

$= 1/(x + 1)$, $Q(x) = x/(x + 1)$. An I.F. is $e^{\int P(x)dx} =$ $e^{\int dx/x+1} = e^{\ln|x+1|} = |x + 1|$. Since $x + 1 \geq 1 > 0$ in (I), we have $|x + 1| = x + 1$. We multiply the standard

form equation through by this to obtain $(x + 1)\dfrac{dy}{dx} + y = x$,
which turns out to be exactly the equation that we started
with. We note that this is $\dfrac{d}{dx}[(x + 1)y] = x$.

Integrating we obtain $(x + 1)y = x^2/2 + c$. Applying the
I.C. $y(0) = 1/2$, we obtain $c = 1/2$. Thus the particular
solution of Problem (I) is $(x + 1)y = \dfrac{1}{2}(x^2 + 1)$ or

$y = \dfrac{x^2 + 1}{2(x + 1)}$. This is valid for $0 \le x < 3$.

Letting ϕ denote the solution just obtained, we find

$\lim\limits_{x \to 3^-} \phi(x) = \lim\limits_{x \to 3^-}\left[\dfrac{x^2 + 1}{2(x + 1)}\right] = \dfrac{5}{4}$. This is the a of Problem

(II); that is, the I.C. of (II) is $y(3) = 5/4$.

Now solve (II). We put the D.E. in standard form
$\dfrac{dy}{dx} + \dfrac{1}{x + 1}\, y = \dfrac{3}{x + 1}$ and find, as in Problem (I), that an
I.F. is $(x + 1)$. We multiply the standard form equation
through by this to obtain $(x + 1)\dfrac{dy}{dx} + y = 3$, which is
exactly what we started with here in Problem (II). We
note that this is $\dfrac{d}{dx}[(x + 1)y] = 3$. Integrating we
obtain $(x + 1)y = 3x + c$. Applying the I.C. $y(3) = 5/4$,
we obtain $5 = 9 + c$ or $c = -4$. Thus the particular
solution of Problem (II) is $(x + 1)y = 3x - 4$ or
$y = \dfrac{3x - 4}{x + 1}$. This is valid for $x \ge 3$.

The solution of the entire problem is

$$y = \begin{cases} \dfrac{x^2 + 1}{2(x + 1)}, & 0 \le x < 3, \\\\ \dfrac{3x - 4}{x + 1}, & x \ge 3 \end{cases} .$$

31. (a) The D.E. is linear. In the standard form it is $\dfrac{dy}{dx} +$
$\dfrac{b}{a} y = \dfrac{k}{a} e^{-\lambda x}$, with $P(x) = \dfrac{b}{a}$, $Q(x) = \dfrac{k}{a} e^{-\lambda x}$. An I.F.
is $e^{\int P(x)dx} = e^{\int (b/a)dx} = e^{(b/a)x}$. Multiplying the
standard form equation through by this, we obtain

$$e^{(b/a)x} \frac{dy}{dx} + \frac{b}{a} e^{(b/a)x} y = \frac{k}{a} e^{(b/a-\lambda)x}$$

$$\text{or } \frac{d}{dx} [e^{(b/a)x} y] = \frac{k}{a} e^{(b/a-\lambda)x} . \qquad (1)$$

We now consider two cases: (i) $\lambda \ne \dfrac{b}{a}$; and (ii) $\lambda = \dfrac{b}{a}$.

In case (i), where $\lambda \ne \dfrac{b}{a}$, integrating (1) we obtain
$$e^{(b/a)x} y = \frac{k}{a}\left(\frac{b}{a} - \lambda\right)^{-1} e^{(b/a-\lambda)x} + c \text{ or}$$

$$y = \frac{ke^{-\lambda x}}{b - a\lambda} + ce^{-bx/a} . \qquad (2)$$

In case (ii), where $\lambda = \dfrac{b}{a}$, (1) becomes $\dfrac{d}{dx} [e^{(b/a)x} y] =$
$\dfrac{k}{a}$. Integrating, we obtain $e^{(b/a)x} y = \dfrac{kx}{a} + c$ or

$$y = \frac{k x e^{-bx/a}}{a} + ce^{-bx/a} . \qquad (3)$$

(b) Suppose $\lambda = 0$. Then since $\dfrac{b}{a} > 0$, we can have only

case (i), $\lambda \neq \dfrac{b}{a}$, here. Then with $\lambda = 0$, (2) becomes $y = \dfrac{k}{b} + ce^{-bx/a}$. As $x \to \infty$, $e^{-bx/a} \to 0$ and $y \to \dfrac{k}{b}$.

Suppose $\lambda > 0$. Then either case (i) or case (ii) can occur. In case (i), y is given by (2); and since $e^{-\lambda x} \to 0$ and $e^{-bx/a} \to 0$ as $x \to \infty$, we have $y \to 0$ as $x \to \infty$. In case (ii), y is given by (3); and since $xe^{-bx/a} \to 0$ and $e^{-bx/a} \to 0$ as $x \to \infty$, we again have $y \to 0$ as $x \to \infty$.

33. (a) The function f such that $f(x) = 0$ for all $x \in I$ has derivative $f'(x) = 0$ for all $x \in I$. We substitute this $f(x)$ for you and $f'(x)$ for $\dfrac{dy}{dx}$ in the D.E. We obtain $f'(x) + P(x)f(x) = 0 + P(x)0 = 0$ for all $x \in I$, so $f(x)$ is a solution.

(b) By the Theorem 1.1, there is a unique solution f of the D.E. satisfying the I.C. $f(x_0) = 0$, where $x_0 \in I$. By (a), $\phi(x) = 0$ for all $x \in I$ is a solution of the D.E. and obviously it satisfies the I.C. Thus by the uniqueness of Theorem 1.1, we must have $f(x) = \phi(x)$, that is, $f(x) = 0$ for all $x \in I$.

(c) Since f and g are solutions of the D.E., so is their difference $h = f - g$ (by Exercise 32(a)). Also $h(x_0) = f(x_0) - g(x_0) = 0$. So h is a solution such that $h(x_0) = 0$ for some $x_0 \in I$. By part (b), $h(x) = 0$ for all $x \in I$. But this means $f(x) - g(x) = 0$ and hence $f(x) = g(x)$ for all $x \in I$.

34.   (a) Since f is a solution of (A),

$$f'(x) + P(x)f(x) = Q(x). \qquad (1)$$

Since g is a solution of (A),

$$g'(x) + P(x)g(x) = Q(x). \qquad (2)$$

Then, subtracting (2) from (1), we have

$$f'(x) + P(x)f(x) - g'(x) - P(x)g(x) = Q(x) - Q(x)$$

or

$$f'(x) - g'(x) + P(x)[f(x) - g(x)] = 0$$

or, finally,

$$[f(x) - g(x)]' + P(x)[f(x) - g(x)] = 0.$$

This shows that f − g is a solution of the D.E.

$$\frac{dy}{dx} + P(x)y = 0.$$

(b) Since f is a solution of Equation (A), we have

$$f'(x) + P(x)f(x) = Q(x). \qquad (3)$$

By part (a), since f and g are solutions of (A), their difference f − g is a solution of dy/dx + P(x)y = 0. Thus we have

$$[f(x) - g(x)]' + P(x)[f(x) - g(x)] = 0.$$

Multiplying this by c, where c is an arbitrary constant, we have

$$c[f(x) - g(x)]' + cP(x)[f(x) - g(x)] = 0. \quad (4)$$

Adding (3) and (4), we have

$$\{f'(x) + P(x)f(x)\} + \{c[f(x) - g(x)]'$$
$$+ cP(x)[f(x) - g(x)]\} = Q(x).$$

Rearranging terms, this takes the form

$$\{c[f(x) - g(x)]' + f'(x)\}$$
$$+ P(x)\{c[f(x) - g(x)] + f(x)\} = Q(x).$$

or

$$\{c[f(x) - g(x)] + f(x)\}'$$
$$+ P(x)\{c[f(x) - g(x)] + f(x)\} = P(x).$$

We see from this that $c(f - g) + f$ satisfies equation (A). That is, since c is an arbitrary constant, $c(f - g) + f$ is a one-parameter family of solutions of (A).

37. (a) This is of the form (2.41) with $f(y) = \sin y$, $P(x) = \frac{1}{x}$, and $Q(x) = 1$. We let $v = f(y) = \sin y$, from which $\frac{dv}{dx} = (\cos y)\frac{dy}{dx}$. Substituting this into the stated D.E., it becomes $\frac{dv}{dx} + \frac{1}{x}v = 1$, which is linear in v. An I.F. is $e^{\int P(x)dx} = e^{\int (1/x)dx} = e^{\ln|x|} = |x|$.

Multiplying the linear equation through by this, we have $x\frac{dv}{dx} + v = x$ or $\frac{d}{dx}(xv) = x$. Integrating, we find $xv = \frac{x^2}{2} + c_0$, or $2xv = x^2 + c$, where $c = 2c_0$.

Now replacing v by $\sin y$, we obtain the solution in the form $2x\sin y - x^2 = c$.

40. This is a Riccati Equation of the form (A) of Exercise 38, with $A(x) = -1$, $B(x) = x$, $C(x) = 1$. The solution $f(x) = x$ is given. By Exercise 38(b), we make the transformation $y = f(x) + \frac{1}{v} = x + \frac{1}{v}$, from which $\frac{dy}{dx} = 1 - \left(\frac{1}{v^2}\right)\frac{dv}{dx}$.

Substituting in the given D.E., it takes the form

$$1 - \frac{1}{v^2}\frac{dv}{dx} = -\left(x + \frac{1}{v}\right)^2 + x\left(x + \frac{1}{v}\right) + 1.$$

This reduces to $-\frac{1}{v^2}\frac{dv}{dx} = -\frac{1}{v^2} - \frac{x}{v}$ or $\frac{dv}{dx} - xv = 1$, which is linear in v, with $P(x) = -x$, $Q(x) = 1$. An I.F. is $e^{\int P(x)dx} = e^{\int(-x)dx} = e^{-x^2/2}$. Multiplying the linear D.E. through by this, we obtain

$$e^{-x^2/2}\frac{dv}{dx} - e^{-x^2/2}xv = e^{-x^2/2}$$

or

$$\frac{d}{dx}(e^{-x^2/2}v) = e^{-x^2/2}.$$

Integrating, we obtain $e^{-x^2/2}v = \int e^{-x^2/2}dx + c$. Since the integral on the right cannot be expressed as a finite sum of known elementary functions, we leave it as indicated. But from the transformation $y = x + \dfrac{1}{v}$, we see that $v = \dfrac{1}{(y - x)}$. We replace v accordingly and thus obtain the solution,

$$\frac{e^{-x^2/2}}{y - x} = \int e^{-x^2/2}\, dx + c.$$

Section 2.3, Miscellaneous Review Exercises, Page 59

The solution of each of these review problems is given, but most of these solutions are presented in abbreviated form with many details omitted.

1. The D.E. is both separable and linear. Upon separating variables and integrating, we obtain $2\ln\ (x^3 + 1) + \ln|c| = \ln|y|$, from which we find $y = c(x^3 + 1)^2$.

    The solution as a linear equation is almost as easy. The I.F. is $(x^3 + 1)^{-2}$.

2. The D.E. is exact. We seek $F(x,y)$ such that $F_x(x,y) = M(x,y) = 2xy^3 - y$ and $F_y(x,y) = N(x,y) = 3x^2y^2 - x$. From the first of these, we find $F(x,y) = x^2y^3 - xy + \phi(y)$. From this $F_y(x,y) = 3x^2y^2 - x + \phi'(y)$. But since $F_y(x,y) = N(x,y) = 3x^2y^2 - x$, we find $\phi'(y) = 0$, $\phi(y) = c_0$. Thus $F(x,y) = x^2y^3 - xy + c_0$, and the family of solutions is $x^2y^3 - xy = c$.

3. The D.E. is both separable and linear. First consider it
    as a separable equation. Upon separating variables and
    using partial fractions (preparatory to integrating), we
    find $\left[\frac{1}{x} - \frac{1}{x + 1}\right]dx + \frac{dy}{y - 1} = 0$. Integration gives $\ln|x|$ -
    $\ln|x + 1| + \ln|y - 1| = \ln c_1$. Upon simplifying and
    taking antilogs, we find $|x(y - 1)| = |c_1(x + 1)|$.
    Assuming $x > 0$, $y > 1$ and simplifying, we can write this
    as $xy = cx + (c - 1)$, or $xy + 1 = c(x + 1)$, where
    $c = 1 + c_1$.

    Alternately, consider the D.E. as a linear equation.
    The I.F. is $\exp\left[\frac{dx}{x(x + 1)}\right] = \pm\frac{x}{x + 1}$. Multiplying the
    "standard form" of the D.E. through by this we have
    $\left[\frac{xy}{x + 1}\right]' = (x + 1)^{-2}$. Integrating, we find $\frac{xy}{x + 1} = $
    $-(x + 1)^{-1} + c$. From this we find $xy + 1 = c(x + 1)$, as
    before.

4. The D.E. is linear. The I.F. is x. Multiplying the
    "standard form" of the equation through by this, we find
    $x^2 y' + 2xy = x^3$ and hence $(x^2 y)' = x^3$. Integration and
    simplification at once give $y = \frac{x^2}{4} + \frac{c}{x^2}$.

5. Writing the D.E. in the form $\frac{dy}{dx} = \dfrac{5\left[\frac{y}{x}\right] - 3}{1 + \frac{y}{x}}$, we recognize
    that it is homogeneous. We let $y = vx$. Then the D.E.
    becomes $v + x\frac{dv}{dx} = \frac{5v - 3}{1 + v}$. Simplification yields

$\dfrac{(v + 1)dv}{v^2 - 4v + 3} = \dfrac{-dx}{x}$. Then upon applying partial fractions

to the left member, we find

$$\left[-\frac{1}{v - 1} + \frac{2}{v - 3} \right] dv = -\frac{dx}{x}.$$

Integration gives $-\ln|v - 1| + 2 \ln|v - 3| = -\ln|x| +$

$\ln|c|$. Simplification and taking of antilogarthims

results in $\dfrac{(v - 3)^2}{|v - 1|} = \left|\dfrac{c}{x}\right|$. But $v = \dfrac{y}{x}$. Upon

resubstituting accordingly and simplifying, we find

$(3x - y)^2 = |c(y - x)|$.

6.   The D.E. is exact, separable, and linear. Considered as

an exact equation, we seek $F(x,y)$ such that $F_x(x,y) =$

$M(x,y) = e^{2x}y^2$ and $F_y(x,y) = N(x,y) = e^{2x}y - 2y$. From the

first of these, we find $F(x,y) = e^{2x}\dfrac{y^2}{2} + \phi(y)$. From

this, $F_y(x,y) = e^{2x}y^2 + \phi'(y)$. But since $F_y(x,y) = N(x,y)$

$= e^{2x}y - 2y$, we find $\phi'(y) = -2y$, $\phi(y) = -y^2 + c_0$. Thus

$F(x,y) = \dfrac{e^{2x}y^2}{2} - y^2 + c_0$, and the family of solutions is

$e^{2x}y^2 - 2y^2 = c$, [where we have set $F(x,y) = c_1$,

multiplied through by 2, and let $c = 2(c_1 - c_0)$].

Assuming $e^{2x} - 2 \geq 0$, this can also be written as $y =$

$c_2(e^{2x} - 2)^{-1/2}$ where $c_2 = \sqrt{c}$.

We also consider the D.E. as a linear equation. We

rewrite it as $(e^{2x} - 2)y' + e^{2x}y = 0$ and then put it in

the "standard form" $y' + e^{2x}y/(e^{2x} - 2) = 0$. From this we find the I.F. is $(e^{2x} - 2)^{1/2}$, where we assume $e^{2x} - 2 \geq 0$. Multiplying the standard form through by the I.F., we find $(e^{2x} - 2)^{1/2}y' + e^{2x}y(e^{2x} - 2)^{-1/2} = 0$ or $[(e^{2x} - 2)^{1/2}y]' = 0$. Integration and simplification readily yield $y = c(e^{2x} - 2)^{-1/2}$.

7. The D.E. is both separable and linear. We first solve this as a separable equation. Separating variables, we have $\dfrac{4x^3dx}{x^4 + 1} + \dfrac{dy}{2y - 3} = 0$. Integration yields $\ln(x^4 + 1) + \dfrac{1}{2}\ln|2y - 3| = \ln|c_1|$. Simplify, assuming $2y - 3 \geq 0$, to obtain $\ln(x^4 + 1)(2y - 3)^{1/2} = \ln|c_1|$. From this we at once have $(x^4 + 1)^2(2y - 3) = c_1^2$. Then $y = 3/2 + c(x^4 + 1)^{-2}$, where $c = \dfrac{c_1^2}{2}$.

Consider the D.E. as a linear equation. In standard form the D.E. is $y' + \dfrac{8x^3y}{x^4 + 1} = \dfrac{12x^3}{x^4 + 1}$. The I.F. is $(x^4 + 1)^2$. Multiplying the standard form equation through by this we find $[(x^4 + 1)^2y]' = (x^4 + 1)(12x^3)$. Integration and division by $(x^4 + 1)^2$ then give $y = \dfrac{3}{2} + c(x^4 + 1)^{-2}$.

8. Writing the D.E. in the form $\dfrac{dy}{dx} = -\dfrac{\left[2 + \dfrac{y}{x} + \left[\dfrac{y}{x}\right]^2\right]}{2}$, we see that the D.E. is homogeneous. Let $y = vx$. Then $\dfrac{dy}{dx} = v + x\dfrac{dv}{dx}$, and the D.E. becomes $v + x\dfrac{dv}{dx} = -1 - \dfrac{v}{2} - \dfrac{v^2}{2}$.

Simplification yields $x\dfrac{dv}{dx} = -\dfrac{(v^2 + 3v + 2)}{2}$, and hence

$\dfrac{2\,dv}{v^2 + 3v + 2} = -\dfrac{dx}{x}$. Use of partial fractions gives

$2\left[\dfrac{1}{v + 1} - \dfrac{1}{v + 2}\right]dv = -\dfrac{dx}{x}$. Integration and immediate

simplification give $2\ln\left|\dfrac{v + 1}{v + 2}\right| = \ln\left|\dfrac{c}{x}\right|$. Further

simplification gives $\left|\dfrac{v + 1}{v + 2}\right|^2 = \left|\dfrac{c}{x}\right|$. But $v = \dfrac{y}{x}$.

Resubstituting accordingly and again simplifying, we

obtain $\left[\dfrac{y + x}{y + 2x}\right]^2 = \left|\dfrac{c}{x}\right|$.

9. Rewriting the D.E. in the form $(4x^3y^2 - 3x^2y)dx +$
$(2x^4y - x^3)dy = 0$, we find that it is exact. We seek
$F(x,y)$ such that $F_x(x,y) = M(x,y) = 4x^3y^2 - 3x^2y$ and
$F_y(x,y) = N(x,y) = 2x^4y - x^3$. From the first of these,
$F(x,y) = x^4y^2 - x^3y + \phi(y)$. From this, $F_y(x,y) = 2x^4y -$
$x^3 + \phi'(y)$. But since $F_y(x,y) = N(x,y) = 2x^4y - x^3$, we
must have $\phi'(y) = 0$, $\phi(y) = c_0$. Thus $F(x,y) = x^4y^2 - x^3y$
$+ c_0$, and the solution is $x^4y^2 - x^3y = c$.

10. The D.E. is linear. The standard form is $y' + \dfrac{xy}{x + 1} =$
$e^{-x}(x + 1)$, and the I.F. is $\pm e^x(x + 1)$. Multiplying the
standard form through by this, we obtain the equivalent of
$\left[\dfrac{e^x y}{x + 1}\right]' = (x + 1)^{-2}$. Integration and multiplication by
$(x + 1)e^{-x}$ result in $y = e^{-x}[-1 + c(x + 1)]$.

11.   Writing the D.E. in the form $\dfrac{dy}{dx} = \dfrac{2 - 7\left[\dfrac{y}{x}\right]}{3\left[\dfrac{y}{x}\right] - 8}$, we recognize

that it is homogeneous. We let $y = vx$. Then $\dfrac{dy}{dx} = v +$

$x\dfrac{dv}{dx}$. The D.E. becomes $v + x\dfrac{dv}{dx} = \dfrac{2 - 7v}{3v - 8}$. Simplification

gives $\dfrac{(3v - 8)dv}{3v^2 - v - 2} = -\dfrac{dx}{x}$. Partial fractions decomposition

puts this in the form $\left[\dfrac{6}{3v + 2} - \dfrac{1}{v - 1}\right] dv = -\dfrac{dx}{x}$.

Integration gives $2 \ln (3v + 2) - \ln|v - 1| = -\ln|x| +$

$\ln|c|$. Simplification and taking of antilogarithms

results in $\dfrac{(3v + 2)^2}{|v - 1|} = \left|\dfrac{c}{x}\right|$. But $v = \dfrac{y}{x}$. Resubstituting

accordingly and simplifying gives $\dfrac{(3y + 2x)^2}{x^2} \cdot \dfrac{|x|}{|y - x|} =$

$\left|\dfrac{c}{x}\right|$ and hence $(2x + 3y)^2 = |c(y - x)|$.

12.   The D.E. is a Bernoulli Equation with $n = 3$. We multiply

through by y^{-3} to obtain $x^2 y^{-3}\dfrac{dy}{dx} + xy^{-2} = x$. Let $v =$

$y^{1-n} = y^{-2}$; then $\dfrac{dv}{dx} = -2y^{-3}\dfrac{dy}{dx}$. The preceeding D.E.

readily transforms into the linear equation $\dfrac{dv}{dx} - \dfrac{2}{x}v$

$= -\dfrac{2}{x}$. An I.F. is $e^{-\int 2/x\,dx} = \dfrac{1}{x^2}$. Multiplying through by

this, we obtain $\dfrac{d}{dx}\left[\dfrac{v}{x^2}\right] = -\dfrac{2}{x^3}$. Integrating, we find $\dfrac{v}{x^2} =$

$\dfrac{1}{x^2} + c$. Replacing v by y^{-2} and simplifying, we find $y^2 =$

$\dfrac{1}{1 + cx^2}$.

13. This is a linear equation. The standard form is

$$y' + \frac{6x^2 y}{x^3 + 1} = \frac{6x^2}{x^3 + 1}.$$ An I.F. is $(x^3 + 1)^2$. Multiplying

through by this we obtain $[(x^3 + 1)y]' = (x^3 + 1)(6x^2)$.
Integration and division by $(x^3 + 1)^2$ then give the
solutions $y = 1 + c(x^3 + 1)^{-2}$.

14. Writing the D.E. in the form $\dfrac{dy}{dx} = \dfrac{2 + \left[\dfrac{y}{x}\right]^2}{2\left[\dfrac{y}{x}\right] - 1}$, we recognize

that it is homogeneous. We let $y = vx$. Then $\dfrac{dy}{dx} = v +$

$x \dfrac{dv}{dx}$. The D.E. becomes $v + x \dfrac{dv}{dx} = \dfrac{2 + v^2}{2v - 1}$. Simplification

gives $\dfrac{(2v - 1)dv}{v^2 - v - 2} = - \dfrac{dx}{x}$. Integration gives $\ln|v^2 - v - 2|$

$= \ln\left|\dfrac{c}{x}\right|$. Taking antilogarithms and simplifying slightly,

we obtain $v^2 - v - 2 = c/x$. But $v = y/x$. Resubstituting
accordingly and simplifying, we find $y^2 - xy - 2x^2 = cx$ or
$(y - 2x)(y + x) = cx$.

15. The D.E. is both homogeneous and Bernoulli. We first
consider it as a homogeneous equation, which form we

recognize by writing it as $\dfrac{dy}{dx} = \dfrac{1 + \left[\dfrac{y}{x}\right]^2}{2\left[\dfrac{y}{x}\right]}$. We let $y = vx$.

Then $\dfrac{dy}{dx} = v + x \dfrac{dv}{dx}$. The D.E. becomes $v + x \dfrac{dv}{dx} = \dfrac{1 + v^2}{2v}$.

Simplification quickly leads to $\dfrac{2vdv}{v^2 - 1} = - \dfrac{dx}{x}$, and then

integration gives $\ln|v^2 - 1| = \ln\left|\dfrac{c}{x}\right|$. From this, we

readily find $v^2 - 1 = \dfrac{c}{x}$. Resubstituting $v = \dfrac{y}{x}$ and

simplifying, we find $y^2 - x^2 = cx$.

Now apply the initial condition $y(1) = 2$. Letting $x = 1$, $y = 2$, we get $3 = c$, and hence obtain the solution in the form $y^2 = x^2 + 3x$ or $y = \sqrt{x^2 + 3x}$.

Alternatively, we recognize that the given D.E. is a Bernoulli equation, with $n = -1$, by writing in the equivalent form $2x\frac{dy}{dx} - y = x^2 y^{-1}$. We rewrite this as $2xy\frac{dy}{dx} - y^2 = x^2$; let $v = y^2$ and $\frac{dv}{dx} = 2y\frac{dy}{dx}$; and transform the D.E. into the linear equation $x\frac{dy}{dx} - v = x^2$. The standard form of this is $\frac{dv}{dx} - \frac{1}{x}v = x$, and an I.F. is x^{-1}. Multiplying the standard form equation through by this, we obtain $(x^{-1}v)' = 1$. Integration then gives $x^{-1}v = x + c$. Now resubstituting $v = y^2$ and simplifying slightly, we find $y^2 = x^2 + cx$. Application of the I.C. again gives the particular solution $y^2 = x^2 + 3x$ or $y = \sqrt{x^2 + 3x}$.

16.   The D.E. is both a separate equation and a Bernoulli Equation. We first solve it as a separable equation. Separating variables, we get $\dfrac{2\,dx}{1 - x^2} + \dfrac{y\,dy}{y^2 + 4} = 0$. We can use partial fractions or an integral table to integrate the dx term. Integrating the equation and simplifying slightly, we find

$$\ln\left|\frac{1 + x}{1 - x}\right| + \ln(y^2 + 4)^{1/2} = \ln|c|.$$

Further simplification results in the one-parameter family of solutions

$$|(1 + x)(y^2 + 4)^{1/2}| = |c(1 - x)|.$$

Now apply the I.C. $y(3) = 0$. Letting $x = 3$, $y = 0$ in the preceeding gives $8 = 2|c|$, so $|c| = 4$. Substituting this value of $|c|$ back into the equation of the family of solutions, we get

$$\left| (1 + x)(y^2 + 4)^{1/2} \right| = 4|1 - x|.$$

Since the initial x value is $3 > 1$, we take $x > 1$ and $1 - x < 0$, so $|1 - x| = x - 1$. Thus we obtain the particular solution

$$(x + 1)(y^2 + 4)^{1/2} = 4(x - 1).$$

Alternately, we now solve the given D.E. as a Bernoulli Equation. We first rewrite it as $\dfrac{dy}{dx} + \dfrac{2}{1 - x^2}y = -\dfrac{8}{1 - x^2}y^{-1}$, in which we recognize that it is indeed a Bernoulli D.E. with $n = -1$. We thus let $v = y^{1-n} = y^2$, find $\dfrac{dv}{dx} = 2y\dfrac{dy}{dx}$, and transform the D.E. into

$$\frac{dv}{dx} + \frac{4}{1 - x^2}v = -\frac{16}{1 - x^2},$$

which is linear in v, with $P(x) = \dfrac{4}{1 - x^2}$, $Q(x) = \dfrac{-16}{1 - x^2}$.

An I.F. is $e^{\int P(x)dx} = e^{\int 4dx/(1-x^2)} = e^{2\ln|(1+x)/(1-x)|} = \left(\dfrac{1 + x}{1 - x}\right)^2$. Multiplying the linear D.E. through by this, we obtain

$$\frac{d}{dx}\left[\left(\frac{1 + x}{1 - x}\right)^2 v\right] = -\frac{16(1 + x)}{(1 - x)^3}.$$

To integrate the right member, we note that

$$-\frac{16(1 + x)}{(1 - x)^3} = -16\left[-\frac{1}{(1 - x)^2} + \frac{2}{(1 - x)^3}\right].$$

Integrating the D.E., we thus find

$$\left(\frac{1 + x}{1 - x}\right)^2 v = -16\left[-\frac{1}{1 - x} + \frac{1}{(1 - x)^2}\right] + c,$$

which readily simplifies into

$$\left(\frac{1 + x}{1 - x}\right)^2 v = \frac{-16x}{(1 - x)^2} + c.$$

Application of the I.C. $x = 3$ at $y = 0$ gives $c = 12$.
Substituting this back and multiplying by $(1 - x)^2$, we
obtain $(1 + x)^2 y^2 = -16x + 12(1-x)$. Adding $4(1 + x)^2$ to
both sides of this and simplifying, we find $(1 + x)^2$
$(y^2 + 4) = 16(x - 1)^2$, from which the solution previously
found is readily obtained.

17. This D.E., with $M(x,y) = e^{2x}y^2 - 2x$ and $N(x,y) = e^{2x}y$, is
exact; for $M_y(x,y) = 2e^{2x}y = N_x(x,y)$. We seek $F(x,y)$ such
that $F_x(x,y) = M(x,y) = e^{2x}y^2 - 2x$ and $F_y(x,y) = N(x,y) =$
$e^{2x}y$. From the first of these, $F(x,y) = e^{2x}\frac{y^2}{2} - x^2 +$
$\phi(y)$. From this, $F_y(x,y) = e^{2x}y + \phi'(y)$. But this must

equal $N(x,y) = e^{2x}y$. Thus we must have $\phi'(y) = 0$ and

hence $\phi(y) = c_0$. Thus $F(x,y) = e^{2x}\frac{y^2}{2} - x^2 + c_0$; and from

this the solution $F(x,y) = c_1$ take the form $e^{2x}\frac{y^2}{2} - x^2 = $

c, where $c = c_1 - c_0$. Now apply the given I.C. $y(0) = 2$

to this. Thus letting $x = 0$, $y = 2$, we find $c = 2$. Thus

we obtain the solution $e^{2x}\frac{y^2}{2} - x^2 = 2$ or $y^2 e^{2x} = 2x^2 + 4$

or, finally, $y = e^{-x}(2x^2 + 4)^{1/2}$.

18. This D.E., with $M(x,y) = 3x^2 + 2xy^2$ and $N(x,y) = 2x^2 y$
$+ 6y^2$, is exact; for $M_y(x,y) = 4xy = N_x(x,y)$. We seek
$F(x,y)$ such that $F_x(x,y) = M(x,y) = 3x^2 + 2xy^2$ and $F_y(x,y)$
$= N(x,y) = 2x^2 y + 6y^2$. From the first of these, $F(x,y) = $
$x^3 + x^2 y^2 + \phi(y)$. From this, $F_y(x,y) = 2x^2 y + \phi(y)$. But
this must equal $N(x,y) = 2x^2 y + 6y^2$. Thus we must have
$\phi'(y) = 6y^2$ and hence $\phi(y) = 2y^3 + c_0$. Thus $F(x,y) = $
$x^3 + x^2 y^2 + 2y^3 + c_0$; and from this the solutions $F(x,y) = $
c_1 take the form $x^3 + x^2 y^2 + 2y^3 = c$, where $c = c_1 - c_0$.
Applying the I.C. $y(1) = 2$, we let $x = 1$, $y = 2$, to obtain
$c = 21$. Thus we obtain the solution in the form $x^3 + x^2 y^2$
$+ 2y^3 = 21$.

19. This D.E. is both a separable equation and a Bernoulli
Equation. We first solve it as a separable D.E., writing
it as $\frac{4\,y\,dy}{y^2 + 1} = \frac{dx}{x}$. Integration then gives $2 \ln (y^2 + 1) = $
$\ln|x| + \ln|c|$. Simplifying we find $\ln(y^2 + 1)^2 = \ln|cx|$
and hence $(y^2 + 1)^2 = |cx|$. Applying the I.C. $y(2) = 1$,

we set x = 2, y = 1 in this to obtain $4 = 2|c|$ and hence $|c| = 2$. Thus we obtain the solution $(y^2 + 1)^2 = 2|x|$. Since the initial x value is 2 > 0, we take x > 0 and write the solution as $(y^2 + 1)^2 = 2x$. Alternately, writing the D.E. as $4x\frac{dy}{dx} - y = y^{-1}$, we recognize that it is a Bernoulli Equation with n = -1. Thus we let $v = y^{1-n} = y^2$, $\frac{dv}{dx} = 2y\frac{dy}{dx}$, and substitute into the stated equation $4xy\frac{dy}{dx} = y^2 + 1$, obtaining $2x\frac{dv}{dx} = v + 1$ or $\frac{dv}{dx} - \frac{1}{2x}v = \frac{1}{2x}$. This is linear in v, with I.F. $e^{-\int(1/2x)dx} = x^{-1/2}$. Multiplying the standard form linear equation through by this, we find $\frac{d}{dx}(x^{-1/2}v) = \frac{x^{-3/2}}{2}$. Integrating, we obtain $x^{-1/2}v = -x^{-1/2} + c$ or $v = -1 + cx^{1/2}$. But $v = y^2$, so we have $y^2 = -1 + cx^{1/2}$. Applying the I.C. $y(2) = 1$ gives $c = \sqrt{2}$. Thus we obtain the solution $y^2 = -1 + \sqrt{2}\,x^{1/2}$, from which we readily obtain the solution $(y^2 + 1)^2 = 2x$ previously found.

20. Writing the D.E. in the form $\frac{dy}{dx} = \dfrac{2 + 7\left[\frac{y}{x}\right]}{2 - 2\left[\frac{y}{x}\right]}$, we see that

it is homogeneous. Thus we let $y = vx$, $\frac{dy}{dx} = v + x\frac{dv}{dx}$. The D.E. successively reduces to $v + x\frac{dv}{dx} = \frac{2 + 7v}{2 - 2v}$, $x\frac{dv}{dx} = \frac{2v^2 + 5v + 2}{2 - 2v}$, and $\frac{(2v - 2)dv}{2v^2 + 5v + 2} = -\frac{dx}{x}$. This last equation is separable, with the variables separated. Using partial fractions, we find $\frac{2v - 2}{2v^2 + 5v + 2} = -\frac{2}{2v + 1} + \frac{2}{v + 2}$. We thus have $\left[\dfrac{-2}{2v + 1} + \dfrac{2}{v + 2}\right]dv = -\dfrac{dx}{x}$. Integrating, we

obtain $-\ln|2v + 1| + 2\ln|v + 2| = -\ln|x| + \ln|c|$, which

reduces to

$$\ln\dfrac{(v + 2)^2}{|2v + 1|} = \ln\left|\dfrac{c}{x}\right|.$$

From this we find $\dfrac{(v + 2)^2}{|2v + 1|} = \left|\dfrac{c}{x}\right|$. But $v = \dfrac{y}{x}$; and

resubstituting and simplifying, we find

$$\dfrac{(y + 2x)^2}{x^2} \cdot \dfrac{|x|}{|2y + x|} = \left|\dfrac{c}{x}\right|$$

or $(2x + y)^2 = |c(2y + x)|$. Applying the I.C. $y(1) = 2$,

we let $x = 1$, $y = 2$, to obtain $16 = |c|5$, from which $|c| = \dfrac{16}{5}$. Replacing $|c|$ with this value, and taking $x + 2y \geq 0$,

we express the solution as $5(2x + y)^2 = 16(x + 2y)$.

21. This equation is both separable and linear. Treating it
    as a separable equation, we have $\dfrac{dy}{y} = \dfrac{x\,dx}{x^2 + 1}$. Integrating
    then gives $2\ln|y| = \ln(x^2 + 1) + \ln|c|$, which quickly
    simplifies to $y^2 = c(x^2 + 1)$, with $c > 0$. Applying the
    I.C. $y(\sqrt{15}) = 2$, we let $x = \sqrt{15}$, $y = 2$, to obtain $16c = 4$,
    $c = \dfrac{1}{4}$. Thus we find the solution $y^2 = \dfrac{(x^2 + 1)}{4}$ or $y = \dfrac{(x^2 + 1)^{1/2}}{2}$. Now treating the equation as a linear
    equation, we have $\dfrac{dy}{dx} - \dfrac{x}{x^2 + 1}\,y = 0$. with $P(x) = \dfrac{-x}{x^2 + 1}$,

$Q(x) = 0$.   An I.F. is $e^{\int P(x)dx} = e^{-\int [x/(x^2+1)]dx} =$

$e^{-(1/2)\ln(x^2+1)} = (x^2 + 1)^{-1/2}$.   Multiplying the D.E.

through by this, we obtain $\frac{d}{dx} [(x^2 + 1)^{-1/2}y] = 0$.

Integrating, we find $(x^2 + 1)^{-1/2}y = c$ or $y =$

$c(x^2 + 1)^{1/2}$.   Applying the I.C. $y(\sqrt{15}) = 2$ gives $c = \frac{1}{2}$.

Thus we again obtain $y = \dfrac{(x^2 + 1)^{1/2}}{2}$.

22.   Here we actually have two I.V. problems

$\text{(I)} \begin{cases} \text{For } 0 \le x < 2, \\[6pt] \dfrac{dy}{dx} + y = 1, \\[6pt] y(0) = 0; \end{cases}$
$\qquad$
$\text{(II)} \begin{cases} \text{For } x \ge 2, \\[6pt] \dfrac{dy}{dx} + y = 0, \\[6pt] y(2) = a; \end{cases}$

where a is the value of $\lim\limits_{x \to 2^-} \phi(x)$ and ϕ denotes the

solution of (I).   This is prescribed so that the solution

of the entire problem will be continuous at $x = 2$.

We first solve (I).   The D.E. is linear in standard

form, with $P(x) = 1$, $Q(x) = 1$.   An I.F. is $e^{\int P(x)dx} = e^{\int dx}$

$= e^x$.   Multiplying the D.E. through by this we obtain

$e^x \frac{dy}{dx} + e^x y = e^x$ or $\frac{d}{dx} (e^x y) = e^x$.   Integrating, we find

$e^x y = e^x + c$ or $y = 1 + ce^{-x}$.   Applying the I.C. $y(0) = 0$

gives $0 = 1 + ce^0$, $c = -1$.   Hence the solution of Problem

(I) is $y = 1 - e^{-x}$, valid for $0 \le x < 2$.

Letting ϕ denote the solution just obtained, we note that $\lim\limits_{x \to 2^-} \phi(x) = \lim\limits_{x \to 2^-} (1 - e^{-x}) = 1 - e^{-2}$. This is the a of problem (II); that is, the I.C. of problem (II) is $y(2) = 1 - e^{-2}$.

We now solve problem (II). The D.E. is linear in standard form, with $P(x) = 1$, $Q(x) = 0$. An I.F., as in Problem (I), is e^x. Multiplying the D.E. through by this and integrating, we obtain $e^x \frac{dy}{dx} + e^x y = 0$ or $\frac{d}{dx}(e^x y) = 0$. Integrating, we find $e^x y = c$ or $y = ce^{-x}$. Now apply the I.C. of problem (II), namely, $y(2) = 1 - e^{-2}$. We have $1 - e^{-2} = ce^{-2}$, from which we find $c = e^2 - 1$. Thus the solution of problem (II) is $y = (e^2 - 1)e^{-x}$, valid for $x > 2$.

We write the solution of the entire problem, showing intervals where each part is valid, as

$$y = \begin{cases} 1 - e^{-x}, & 0 \leq x < 2, \\ (e^2 - 1)e^{-x}, & x > 2. \end{cases}$$

23. Here, as in problem 22, we actually have two I.V. problems:

$$(I) \begin{cases} \text{For } 0 \leq x < 2, \\ (x + 2) \dfrac{dy}{dx} + y = 2x, \\ y(0) = 4; \end{cases} \qquad (II) \begin{cases} \text{For } x > 2, \\ (x + 2) \dfrac{dy}{dx} + y = 4, \\ y(2) = a; \end{cases}$$

where a is the value of $\lim\limits_{x \to 2^-} \phi(x)$ and ϕ denotes the

solution of (I).

We first solve (I). The D.E. of this problem is
linear. In standard form it is $\frac{dy}{dx} + \left(\frac{1}{x + 2}\right)y = \frac{2x}{x + 2}$ with

$P(x) = \frac{1}{x + 2}$, $Q(x) = \frac{2x}{x + 2}$. An I.F. is $e^{\int P(x)dx} =$

$e^{\int [1/(x+2)]dx} = e^{\ln|x+2|} = |x + 2|$. Multiplying the
standard form equation through by this, we obtain the
originally stated D.E., $(x + 2)\frac{dy}{dx} + y = 2x$, or

$\frac{d}{dx} [(x + 2)y] = 2x$. Integrating, we find $(x + 2)y$

$= x^2 + c$. Thus the particular solution of Problem (I) is
$(x + 2)y = x^2 + 8$, valid for $0 \le x < 2$.

We can write this solution as $y = \frac{x^2 + 8}{x + 2}$.

Denoting this solution by ϕ, we note that $\lim\limits_{x \to 2^-} \phi(x) =$

$\lim\limits_{x \to 2^-} \left[\frac{x^2 + 8}{x + 2}\right] = 3$. This is the a of problem (II); that is

the I.C. of problem (II) is $y(2) = 3$.

We now solve (II). The D.E. is linear. In standard

form it is $\frac{dy}{dx} + \left(\frac{1}{x + 2}\right)y = 4$. Just as in (I), an I.F. is

$|x + 2|$. Multiplying the standard form equation through
by this, we obtain the originally stated D.E.,
$(x + 2)\frac{dy}{dx} + y = 4$ or $\frac{d}{dx} [(x + 2)y] = 4$. Integrating, we

find $(x + 2)y = 4x + c$. Applying the I.C. of (II), $y(2) =$
3, we find $c = 4$. Thus the particular solution of problem
(I) is $(x + 2)y = 4x + 4$, valid for $x > 2$.

We write the solution of the entire problem as

$$\begin{cases} (x + 2)y = x^2 + 8, \ 0 \le x \le 2, \\ (x + 2)y = 4x + 4, \ x > 2. \end{cases}$$

24. This D.E. is both a Bernoulli D.E. and a homogeneous D.E. We first solve it as a Bernoulli Equation, with $n = 3$. We thus multiply through by y^{-3}, expressing the D.E. in the equivalent form $x^2 y^{-3} \dfrac{dy}{dx} + xy^{-2} = x^{-1}$ or $-2y^{-3} \dfrac{dy}{dx} - \dfrac{2}{x} y^{-2}$ $= -\dfrac{2}{x^3}$. We let $v = y^{1-n} = y^{-2}$, $\dfrac{dv}{dx} = -2y^{-3} \dfrac{dy}{dx}$. The D.E. then transforms into the linear D.E. in v, $\dfrac{dv}{dx} - \dfrac{2}{x} v =$ $-\dfrac{2}{x^3}$. This linear equation is in standard form with $P(x)$ $= -\dfrac{2}{x}$, $Q(x) = -\dfrac{2}{x^3}$. An I.F. is $e^{\int P(x)dx} = e^{-\int (2/x)dx} =$ $e^{-2\ln|x|} = x^{-2}$. Multiplying the linear equation through by this we get $x^{-2} \dfrac{dv}{dx} - 2x^{-3}v = -2x^{-5}$ or $\dfrac{d}{dx}(x^{-2}v) =$ $-2x^{-5}$. Integrating, we find $x^{-2}v = \dfrac{x^{-4}}{2} + c$ or $v = \dfrac{x^2}{2} +$ cx^2. But $v = y^{-2}$, thus we obtain $y^{-2} = \dfrac{x^{-2}}{2} + cx^2$. Applying the I.C. $y(1) = 1$ to this, we find $c = 1/2$. Thus we obtain the solution $y^{-2} = \dfrac{x^{-2}}{2} + \dfrac{x^2}{2}$, which can be written as $y^2 = \dfrac{2x^2}{x^4 + 1}$, or $y = \dfrac{\sqrt{2} \, x}{(x^4 + 1)^{1/2}}$.

Now we express the D.E. in the form $\frac{dy}{dx} = -\frac{y}{x} + \left(\frac{y}{x}\right)^3$ and solve it as a homogeneous equation. We let $y = vx$, $\frac{dy}{dx} = v + x\frac{dv}{dx}$ and obtain the transformed separable equation $v + x\frac{dv}{dx} = -v + v^3$, which reduces to $\frac{dv}{v(v^2 - 2)} = \frac{dx}{x}$. We rewrite the left member, using partial fractions, and thus have $\left(-\frac{1}{v} + \frac{v}{v^2 - 2}\right)dv = 2\frac{dx}{x}$.

Integration gives $-\ln|v| + \frac{\ln|v^2 - 2|}{2} = 2\ln|x| + \ln|c_1|$. This simplifies to

$$-\ln v^2 + \ln|v^2 - 2| = \ln x^4 + \ln c,$$

$\ln \frac{|v^2 - 2|}{v^2} = \ln cx^4$. From this, $|v^2 - 2| = cv^2x^4$.

Resubstituting $v = \frac{y}{x}$, we find $|y^2 - 2x^2| = cy^2x^4$.

Applying the I.C. $y(1) = 1$, we find $c = 1$. Thus we find the solution $|y^2 - 2x^2| = y^2x^4$. Since $y^2 - 2x^2$ is initially -1, we take $|y^2 - 2x^2| = 2x^2 - y^2$, and write the solution as $2x^2 - y^2 = x^4y^2$ or $(x^4 + 1)y^2 = 2x^2$ or $y^2 = \frac{2x^2}{x^4 + 1}$, as we obtained before.

Section 2.4, Page 67

1. This equation is of the form (2.42) of Theorem 2.6 with $M(x,y) = 5xy + 4y^2 + 1$, $N(x,y) = x^2 + 2xy$. To apply that theorem, we first find

$$\frac{M_y(x,y) - N_x(x,y)}{N(x,y)} = \frac{5x + 8y - 2x - 2y}{x^2 + 2xy} =$$

$$\frac{3(x + 2y)}{x(x + 2y)} = \frac{3}{x}.$$

This depends on x only, so

$$e^{\int (3/x)dx} = e^{3\ln|x|} = e^{\ln|x|^3} = |x|^3$$

is an I.F. of the given equation. Multiplying the given equation through by this and simplifying, we find $(5x^4y + 4x^3y^2 + x^3)dx + (x^5 + 2x^4y)dy = 0$. For this equation $M(x,y) = 5x^4y + 4x^3y^2 + x^3$, $N(x,y) = x^5 + 2x^4y$, and $M_y(x,y) = 5x^4 + 8x^3y = N_x(x,y)$. Thus this equation is exact. We seek $F(x,y)$ such that $F_x(x,y) = M(x,y)$ and $F_y(x,y) = N(x,y)$. From the first of these, $F(x,y)$

$$\int M(x,y)\,\partial x = \int (5x^4y + 4x^3y^2 + x^3)\,\partial x = x^5y + x^4y^2 + \frac{x^4}{4} +$$

$\phi(y)$. From this, $F_y(x,y) = x^5 + 2x^4y + \phi'(y)$. But we must have $F_y(x,y) = N(x,y) = x^5 + 2x^4y$. Thus $\phi'(y) = 0$, $\phi(y) = c_0$. Hence $F(x,y) = x^5y + x^4y^2 + \frac{x^4}{4} + c_0$. The family of solutions is then $F(x,y) = c_1$, and multiplying through by 4 we express this as $4x^5y + 4x^4y^2 + x^4 = c$, where $c = 4(c_1 - c_0)$.

2.   This equation is of the form (2.42) of Theorem 2.6 with
$M(x,y) = 2x + \tan y$, $N(x,y) = x - x^2\tan y$. To apply that
theorem, we use $1 + \tan^2 y = \sec^2 y$ and find

$$\frac{N_x(x,y) - M_y(x,y)}{M(x,y)} = \frac{1 - 2x\tan y - \sec^2 y}{2x + \tan y}$$

$$= \frac{-2x\tan y - \tan^2 y}{2x + \tan y} = -\tan y.$$

This depends on y only, so $e^{-\int \tan y\, dy} = e^{\ln|\cos y|}$
$= |\cos y|$ is an I.F. of the given equation. We assume
$\cos y \geq 0$, and take $|\cos y| = \cos y$. Multiplying the
given equation through by this and simplifying, we find
$(2x\cos y + \sin y)dx + (x\cos y - x^2\sin y)dy = 0$. For this
equation $M(x,y) = 2x\cos y + \sin y$, $N(x,y) = \cos y -$
$x^2\sin y$, and $M_y(x,y) = -2x\sin y + \cos y = N_x(x,y)$. Thus
this equation is exact. We seek $F(x,y)$ such that $F_x(x,y)$
$= M(x,y)$ and $F_y(x,y) = N(x,y)$. From the first of these
$F(x,y) = \int M(x,y)\,\partial x = \int (2x\cos y + \sin y)\,\partial x = x^2\cos y +$
$x\sin y + \phi(y)$. From this, $F_y(x,y) = -x^2\sin y + x\cos y +$
$\phi'(y)$. But we must have $F_y(x,y) = N(x,y) = x\cos y -$
$x^2\sin y$. Thus $\phi'(y) = 0$, $\phi(y) = c_0$. The family of
solutions is then $F(x,y) = c_1$. We express this as
$x^2\cos y + x\sin y = c$, when $c = c_1 - c_0$.

4.   This equation is of the form (2.42) of Theorem 2.6 with
$M(x,y) = 2xy^2 + y$, $N(x,y) = 2y^3 - x$. To apply that
theorem, we first find $\dfrac{M_y(x,y) - N_x(x,y)}{N(x,y)} = \dfrac{2(2xy + 1)}{2y^3 - x}$; but

since this depends on y as well as x, we cannot proceed

using this. We next find $\dfrac{N_x(x,y) - M_y(x,y)}{M(x,y)} = \dfrac{-2(1 + 2xy)}{y(1 + 2xy)}$

$= -\dfrac{2}{y}$. This depends on y only, so

$$e^{-\int(2/y)dy} = e^{-2\ln|y|} = e^{\ln|y|^{-2}} = |y|^{-2} = \dfrac{1}{y^2}$$

is an I.F. of the given equation. Multiplying the given equation through by this and simplifying, we find

$$(2x + y^{-1})dx + (2y - xy^{-2})dy = 0.$$

For this equation $M(x,y) = 2x + y^{-1}$, $N(x,y) = 2y - xy^{-2}$, $M_y(x,y) = -y^{-2} = N_x(x,y)$. Thus this equation is exact. We seek $F(x,y)$ such that $F_x(x,y) = M(x,y)$ and $F_y(x,y) = N(x,y)$. From the first of these, $F(x,y) = \int M(x,y)\partial x = \int (2x + y^{-1})\partial x = x^2 + xy^{-1} + \phi(y)$. From this, $F_y(x,y) = -xy^{-2} + \phi'(y)$. But we must have $F_y(x,y) = N(x,y) = 2y - xy^{-2}$. Thus $\phi'(y) = 2y$, $\phi(y) = y^2 + c_0$. Hence $F(x,y) = x^2 + xy^{-1} + y^2 + c_0$. The family of solutions $F(x,y) = c_1$ is then expressed as $x^2 + xy^{-1} + y^2 = c$, where $c = c_1 - c_0$.

5.   We first multiply the stated equation through by $x^p y^q$ to

obtain $(4x^{p+1}y^{q+2} + 6x^p y^{q+1})dx + (5x^{p+2}y^{q+1} + 8x^{p+1}y^q)dy = 0$. For this equation, we have $M(x,y) = 4x^{p+1}y^{q+2}$
$+ 6x^p y^{q+1}$, $N(x,y) = 5x^{p+2}y^{q+1} + 8x^{p+1}y^q$, $M_y(x,y) =$
$4(q + 2)x^{p+1}y^{q+1} + 6(q + 1)x^p y^q$, and $N_x(x,y) =$
$5(p + 2)x^{p+1}y^{q+1} + 8(p + 1)x^p y^q$. In order for the
equation to be exact, we must have $M_y(x,y) = N_x(x,y)$, and

hence

$$\begin{cases} 4(q + 2) = 5(p + 2) \\ 6(q + 1) = 8(p + 1), \text{ that is,} \end{cases} \qquad \begin{cases} 5p - 4q = -2 \\ 8p - 6q = -2. \end{cases}$$

Solving these, we find $p = 2$, $q = 3$. Thus the desired
I.F. of the form $x^p y^q$ is $x^2 y^3$.

Now multiplying the original equation through by this
I.F. $x^2 y^3$, we obtain the equivalent exact equation
$(4x^3 y^5 + 6x^2 y^4)dx + (5x^4 y^4 + 8x^3 y^3)dy = 0$. It is easy to
check that this is indeed exact. We seek $F(x,y)$ such that
$F_x(x,y) = M(x,y) = 4x^3 y^5 + 6x^2 y^4$ and $F_y(x,y) = N(x,y) =$
$5x^4 y^4 + 8x^3 y^3$. From the first of these,

$$F(x,y) = \int M(x,y)\,\partial x = \int (4x^3 y^5 + 6x^2 y^4)\,\partial x$$

$$= x^4 y^5 + 2x^3 y^4 + \phi(y).$$

From this, $F_y(x,y) = 5x^4 y^4 + 8x^3 y^3 + \phi'(y)$. But we must
have $F_y(x,y) = N(x,y) = 5x^4 y^4 + 8x^3 y^3$. Thus $\phi'(y) = 0$,
$\phi(y) = c_0$. Thus $F(x,y) = x^4 y^5 + 2x^3 y^4 + c_0$, and the

solution $F(x,y) = c_1$ is $x^4y^5 + 2x^3y^4 = c$ or $x^3y^4(xy + 2)$
$= c$, where $c = c_1 - c_0$.

7. For this equation $a_1 = 5$, $b_1 = 2$, $a_2 = 2$, $b_2 = 1$, so

$$\frac{a_2}{a_1} = \frac{2}{5} \neq \frac{1}{2} = \frac{b_2}{b_1}.$$

Therefore this is Case 1 of Theorem 2.7. We make the transformation

$$\begin{cases} x = X + h \\ y = Y + k, \end{cases}$$

where (h,k) is the solution of the system

$$\begin{cases} 5h + 2k + 1 = 0 \\ 2h + k + 1 = 0. \end{cases}$$

The solution of this system is $h = 1$, $k = -3$, and so the transformation is

$$\begin{cases} x = X + 1 \\ y = Y - 3. \end{cases}$$

This reduces the given equation to the homogeneous equation

$$(5X + 2Y)dX + (2X + Y)dY = 0.$$

We write this in the form

$$\frac{dY}{dX} = - \frac{5 + 2\left[\frac{Y}{X}\right]}{2 + \left[\frac{Y}{X}\right]}$$

and let $Y = vX$ to obtain

$$v + X\frac{dv}{dX} = - \frac{5 + 2v}{2 + v}.$$

This reduces to

$$X\frac{dv}{dX} = \frac{5 - v^2}{2 + v} \quad \text{or} \quad \frac{(v + 2)dv}{v^2 + 4v + 5} = - \frac{dX}{X}.$$

Integrating we find

$$\ln|v^2 + 4v + 5| = -2 \ln|X| + \ln|c_1|.$$

or

$$\ln|v^2 + 4v + 5| = \ln(|c_1|X^{-2}).$$

From this, $|v^2 + 4v + 5| = |c_1|X^{-2}$.   Now replacing v by $\frac{Y}{X}$, and simplifying, we obtain

$$|5X^2 + 4XY + Y^2| = |c_1|.$$

Finally replacing X by $x - 1$ and Y by $y + 3$, we obtain the solutions of the original D.E. in the form

$$\left| 5(x - 1)^2 + 4(x - 1)(y + 3) + (y + 3)^2 \right| = \left| c_1 \right|$$

or

$$\left| 5x^2 + 4xy + y^2 + 2x + 2y + 14 \right| = \left| c_1 \right|.$$

Assuming $x > 0$, $y > 0$, and writing $c = c_1 - 14$, this takes the form

$$5x^2 + 4xy + y^2 + 2x + 2y = c.$$

8. Here $a_1 = 3$, $b_1 = -1$, $a_2 = -6$, $b_2 = 2$, and $\dfrac{a_2}{a_1} = -2 = \dfrac{b_2}{b_1}$.

Therefore this is Case 2 of Theorem 2.7. We therefore let $z = 3x - y$. Then $dy = 3dx - dz$, and the given D.E. transforms into $(z + 1)dx - (2z - 3)(3dx - dz) = 0$ or $(-5z + 10)dx + (2z - 3)dz = 0$, which is separable. Separating variables, we obtain

$$dx + \frac{2z - 3}{-5z + 10}\,dz = 0$$

or

$$dx + \left(-\frac{2}{5} - \frac{1}{5(z - 2)} \right) dz = 0.$$

Integrating, we have

$$x - \frac{2}{5}z - \frac{1}{5}\ln|z - 2| = c_1.$$

or

$$5x - 2z - \ln|z - 2| = 5c_1.$$

Replacing z by 3x - y and simplifying, we obtain the
solution of the given D.E. in the form x - 2y +
$\ln|3x - y - 2| = c$, where $c = -5c_1$.

10.   For this equation $a_1 = 10$, $b_1 = -4$, $a_2 = -1$, $b_2 = -5$, so
$\dfrac{a_2}{a_1} = -\dfrac{1}{10} \neq \dfrac{5}{4} = \dfrac{b_2}{b_1}$.   Therefore this is Case 1 of Theorem
2.7.   We make the transformation

$$
\begin{cases}
x = X + h, \\
y = Y + k,
\end{cases}
$$

where (h,k) is the solution of the system

$$
\begin{cases}
10h - 4k + 12 = 0, \\
h + 5k + 3 = 0.
\end{cases}
$$

The solution of this system is $h = -\dfrac{4}{3}$, $k = -\dfrac{1}{3}$, and so
the transformation is

$$
\begin{cases}
x = X - \dfrac{4}{3}, \\
y = Y - \dfrac{1}{3}.
\end{cases}
$$

This reduces the given D.E. to the homogeneous equation

$$(10X - 4Y)dX - (X + 5Y)dY = 0.$$

We write this in the form

$$\frac{dY}{dX} = \frac{10 - 4(Y/X)}{1 + 5(Y/X)}$$

and let $Y = vX$ to obtain

$$v + X \frac{dv}{dX} = \frac{10 - 4v}{1 + 5v}.$$

This reduces to

$$X \frac{dv}{dX} = - \frac{5v^2 + 5v - 10}{5v + 1}$$

or

$$\frac{(5v + 1)dv}{v^2 + v - 2} = -5 \frac{dX}{X}.$$

We write $v^2 + v - 2 = (v + 2)(v - 1)$ and use partial fractions to integrate the left member. We find

$$3 \ln |v + 2| + 2 \ln |v - 1| = -5 \ln |X| + \ln|c|$$

or

$$\ln|v + 2|^3 (v - 1)^2 = \ln \frac{|c|}{|X|^5}$$

and hence

$$|v + 2|^3 (v - 1)^2 = \frac{|c|}{|X|^5}.$$

Now replacing v by Y/X and simplifying, we find

$$|Y + 2X|^3 (Y - X)^2 = |c|.$$

Finally, replacing X by $x + 4/3$ and Y by $y + 1/3$ and simplifying, we obtain

$$|2x + y + 3|^3 (x - y + 1) = |c|.$$

We assume $2x + y + 3 \geq 0$, take $c \geq 0$, and express this in the form

$$(2x + y + 3)^3 (x - y + 1)^2 = c.$$

12. The D.E. of this problem is of the form (2.52) of Theorem 2.7, within $a_1 = 3$, $b_1 = -1$, $a_2 = 1$, $b_2 = 1$, and hence $\dfrac{a_2}{a_1}$

$= \dfrac{1}{3} \neq -1 = \dfrac{b_2}{b_1}$. Therefore this is Case 1 of the theorem.

We let $x = X + h$, $y = Y + k$, where (h,k) is the solution of the system $\begin{cases} 3h - k - 6 = 0 \\ h + k + 2 = 0 \end{cases}$. The solution of this is h $= 1$, $k = -3$, and so the transformation is $\begin{cases} x = X + 1 \\ y = Y - 3 \end{cases}$.

This reduces the given equation to the homogeneous equation $(3X - Y)dX + (X + Y)dY = 0$. We write this in the form

$$\frac{dY}{dX} = \frac{(Y/X) - 3}{(Y/X) + 1}$$

and let $Y = vX$ to obtain

$$v + X \frac{dv}{dX} = \frac{v - 3}{v + 1}.$$

This reduces to

$$X \frac{dv}{dX} = -\frac{v^2 + 3}{v + 1} \quad \text{or} \quad \frac{(v + 1)dv}{v^2 + 3} = -\frac{dX}{X}.$$

Integrating (tables may be needed!), we find

$$\frac{1}{2} \ln (v^2 + 3) + \frac{1}{\sqrt{3}} \arctan \frac{v}{\sqrt{3}} = -\ln |X| + c$$

or

$$\ln (v^2 + 3) + \frac{2}{\sqrt{3}} \arctan \frac{v}{\sqrt{3}} = -\ln X^2 + c_1 .$$

Now replacing v by $\frac{Y}{X}$, we obtain

$$\ln [Y^2/X^2 + 3] + \frac{2}{\sqrt{3}} \arctan \frac{Y}{\sqrt{3X}} = -\ln X^2 + c_1 .$$

Combining the ln terms, this takes the form

$$\ln (Y^2 + 3X^2) + \frac{2}{\sqrt{3}} \arctan \frac{Y}{\sqrt{3X}} = C_1 .$$

Finally replacing X by $x - 1$ and Y by $y + 3$, we obtain the solutions of the original D.E. in the form

$$\ln [3(x - 1)^2 + (y + 3)^2] + \frac{2}{\sqrt{3}} \arctan \frac{y + 3}{\sqrt{3}(x - 1)} = c_1 .$$

We now apply the I.C. $y(2) = -2$ to this, obtaining

$$\ln 4 + \frac{2}{\sqrt{3}} \arctan \frac{1}{\sqrt{3}} = c_1$$

or $\ln 4 + (2/\sqrt{3})(\pi/6) = c_1$. Thus we obtain the solution of the stated I.V. problem as

$$\ln[3(x - 1)^2 + (y + 3)^2] + \frac{2}{\sqrt{3}} \text{ arc tan } \frac{y + 3}{\sqrt{3}(x - 1)} =$$

$$\ln 4 + \frac{\pi}{3\sqrt{3}}.$$

13. Here $a_1 = 2$, $b_1 = 3$, $a_2 = 4$, $b_2 = 6$, and $a_2/a_1 = 2 =$
     b_2/b_1. Thus this is Case 2 of Theorem 2.7. We let
     $z = 2x + 3y$. Then $dy = (dz - 2dx)/3$, and the given D.E.
     transforms into

$$(z + 1)dx + (2z + 1)\left(\frac{dz - 2\,dx}{3}\right) = 0$$

or

$$-(z - 1)dx + (2z + 1)dz = 0,$$

which is separable in x and z. Separating variables, we
obtain

$$dx - \frac{2z + 1}{z - 1}\,dz = 0$$

or

$$dx - \left(2 + \frac{3}{z - 1}\right)dz = 0.$$

Integrating, we find

$$x - 2z + 3\ln|z - 1| = c_1.$$

Applying the I.C. $y(-2) = 2$, we find $c_2 = 2$ and hence obtain

$$x + 2y - 2 - \ln|2x + 3y - 1| = 0.$$

21. (a) The D.E. is of the form (A) of Exercise 20, with $f(p) = p^2$.

Following the procedure outlined there, we differentiate $y = px + p^2$ with respect to x, to obtain $\frac{dy}{dx} = p + x\frac{dp}{dx} + 2p\frac{dp}{dx}$, since $p = \frac{dy}{dx}$, this is $[x + 2p]\frac{dp}{dx} = 0$, which is of form (B) of Exercise 20.

Assume $x + 2p \neq 0$, divide through by it, and we are left with just $\frac{dp}{dx} = 0$, from which $p = c$ (where c is an arbitrary constant). Then, returning to the given D.E., we find the one-parameter family of solutions $y = cx + c^2$.

(b) Now assume $x + 2p = 0$, and eliminate p between this and the given D.E. From this, $p = -x/2$. Substituting this into the given D.E. and simplifying, we obtain $y = -x^2/4$. This is the desired "extra" solution.

Chapter 3

1. **Step 1.** We first find the D.E. of the given family $y = cx^3$. Differentiating, we obtain $\frac{dy}{dx} = 3cx^2$.

   Eliminating the parameter c between the equation of the given family and its derived equation, we obtain the D.E. of the given family in the form $\frac{dy}{dx} = 3(y/x^3)x^2$ or $\frac{dy}{dx} = \frac{3y}{x}$.

   Step 2. We now find the D.E. of the orthogonal trajectories by replacing $\frac{3y}{x}$ in the D.E. (of Step 1) by its negative reciprocal $-\frac{x}{3y}$, thus obtaining

   $$\frac{dy}{dx} = -\frac{x}{3y}.$$

   Step 3. We solve this last D.E. Separating variables, we have $3y\,dy = -x\,dx$. Integrating, we find $\frac{3y^2}{2} = -(x^2/2) + k_1$ or $x^2 + 3y^2 = k^2$, where $k^2 = -2k_1 \geq 0$. This is the family of orthogonal trajectories of the given family of cubics. It is a family of ellipses with centers at the origin and major axes along the x axis.

4. **Step 1.** We first find the D.E. of the given family $y = e^{cx}$. Differentiating, we obtain $\frac{dy}{dx} = ce^{cx}$. We eliminate the parameter c between the equation of the given family and its derived equation. From the given equation, we have $y > 0$ and $\ln y = cx$. Thus $c = \frac{\ln y}{x}$. Substituting in

the derived equation we find $\frac{dy}{dx} = \frac{y \ln y}{x}$. This is the D.E. of the given family.

Step 2. We now find the D.E. of the orthogonal trajectories by replacing $\frac{y \ln y}{x}$ by its negative reciprocal $- \frac{x}{y \ln y}$, thereby obtaining the D.E.

$$\frac{dy}{dx} = - \frac{x}{y \ln y}.$$

Step 3. We solve this last D.E. Separating variables, we have $y \ln y \, dy = -x \, dx$. Integrating (by parts), we find $\frac{y^2 \ln y}{2} - \frac{y^2}{4} = - \frac{x^2}{2} + k_1$ or $y^2 (\ln y - 1/2) = -x^2 + k$.

This is the family of orthogonal trajectories of the given family of exponentials.

5. **Step 1.** We first find the D.E. of the given family $y = x - 1 + c e^{-x}$. Differentiating, we obtain $\frac{dy}{dx} = 1 - c e^{-x}$. Eliminating the parameter c between the equation of the given family and its derived equation, we obtain $\frac{dy}{dx} = 1 - (y - x + 1)$ or $\frac{dy}{dx} = x - y$.

Step 2. We now find the D.E. of the orthogonal trajectories by replacing $x - y$ by its negative reciprocal $\frac{1}{y - x}$, thereby obtaining the D.E. $\frac{dy}{dx} = \frac{1}{y - x}$.

Step 3. We solve this last D.E. We regard x as the dependent variable and write the equation in the form $\frac{dx}{dy} = y - x$ or $\frac{dx}{dy} + x = y$. This is linear in x, with I.F. $e^{\int 1 dy} = e^y$. Multiplying $\frac{dx}{dy} + x = y$ through by this, we have $e^y \frac{dx}{dy} + e^y x = y e^y$ or $\frac{d}{dy}[x e^y] = y e^y$. Integrating, we obtain $x e^y = e^y(y - 1) + k$ or $x = y - 1 + k e^{-y}$. This is the family of orthogonal trajectories of the given family.

6. **Step 1.** We first find the D.E. of the given family $y = cx^2/(x + 1)$. Differentiating we obtain, $\frac{dy}{dx} = c(x^2 + 2x)/(x + 1)^2$. From the given equation, $c = (x + 1)y/x^2$; and substituting this into the derived equation and simplifying, we obtain $\frac{dy}{dx} = \frac{(x + 2)y}{x(x + 1)}$. This is the D.E. of the given family.

Step 2. We now find the D.E. of the orthogonal trajectories by replacing $\frac{(x + 2)y}{x(x + 1)}$ in the D.E. of step 1 by its negative reciprocal $- \frac{x(x + 1)}{(x + 2)y}$, thereby obtaining the D.E. $\frac{dy}{dx} = - \frac{x(x + 1)}{(x + 2)y}$.

Step 3. We solve this last D.E. Separating variables, we have $y \, dy = - \frac{x(x + 1)}{x + 2} dx$, which can be rewritten as $y \, dy = \left(-x + 1 - \frac{2}{x + 2}\right) dx$. Integrating this, we have $\frac{y^2}{2} = - \frac{x^2}{2} + x - 2 \ln |x + 2| + k_1$ or $x^2 + y^2 - 2x + \ln(x + 2)^4 = k$.

8. **Step 1.** We first find the D.E. of the given family $x^2 = 2y - 1 + ce^{-2y}$. Differentiating, we obtain $2x = 2y \frac{dy}{dx} - 2ce^{-2y} \frac{dy}{dx}$. We eliminate the parameter c between these two equations as follows. From the given equation, $c = (x^2 - 2y + 1)e^{2y}$. Substituting this into the second equation, we get

$$x = \frac{dy}{dx} - (x^2 - 2y + 1) \frac{dy}{dx} \quad \text{or} \quad \frac{dy}{dx} = \frac{x}{2y - x^2}$$

This is the D.E. of the given family.

Step 2. We now find the D.E. of the orthogonal trajectories by replacing $\frac{x}{2y - x^2}$ by its negative reciprocal $\frac{x^2 - 2y}{x}$, thereby obtaining the D.E.

$$\frac{dy}{dx} = \frac{x^2 - 2y}{x}.$$

Step 3. We solve this last D.E. Writing it as $\frac{dy}{dx} + \frac{2}{x}y = x$, we recognize that it is a linear equation in standard form with $P(x) = \frac{2}{x}$ and $Q(x) = x$. An I.F. is $e^{\int P(x)dx} = e^{\int (2/x)dx} = e^{2\ln|x|} = x^2$. Multiplying through by this, we have $x^2 \frac{dy}{dx} + 2xy = x^3$ or $\frac{d}{dx}[x^2y] = x^3$. Integrating, we find $x^2y = \frac{x^4}{4} + k$. This is the equation of the one-parameter family of orthogonal trajectories.

10. **Step 1.** We first find the D.E. of the given family $x^2 - y^2 = cx^3$. Differentiating, we obtain $2x - 2y\frac{dy}{dx} = 3cx^2$. From the given equation, $c = (x^2 - y^2)/x^3$. Substituting this in the derived equation and simplifying, we obtain successively $2x - 2y\frac{dy}{dx} = 3(x^2 - y^2)/x$ or

$$2y\frac{dy}{dx} = -x + 3\frac{y^2}{x} \text{ or } \frac{dy}{dx} = \frac{3y^2 - x^2}{2xy}.$$

Step 2. We now find the D.E. of the orthogonal trajectories by replacing $\frac{3y^2 - x^2}{2xy}$ by its negative reciprocal $\frac{2xy}{x^2 - 3y^2}$, thereby obtaining the D.E.

$$\frac{dy}{dx} = \frac{2xy}{x^2 - 3y^2}$$

Step 3. We solve this last D.E. Writing it in the form $\frac{dy}{dx} = \frac{2(y/x)}{1 - 3(y/x)^2}$, we recognize that it is homogeneous. We let $y = vx$. Then $\frac{dy}{dx} = v + x\frac{dv}{dx}$, and the D.E. becomes $v + $

$$x\frac{dv}{dx} = \frac{2v}{1 - 3v^2} \text{ or } x\frac{dv}{dx} = \frac{3v^3 + v}{1 - 3v^2}. \text{ Separating variables,}$$

we have $\left[\frac{1 - 3v^2}{3v^3 + v}\right] dv = \frac{dx}{x}$. Using partial fractions, this

becomes $\left[\frac{1}{v} - \frac{6v}{3v^2 + 1}\right] dv = \frac{dx}{x}$. Integrating, we find $\ln|v|$

$- \ln(3v^2 + 1) = \ln|x| + \ln k$,

where $k > 0$, or $\ln\frac{|v|}{3v^2 + 1} = \ln k|x|$. From this,

$\dfrac{|v|}{3v^2 + 1} = k|x|$.   Now replacing v by y/x and taking

x > 0 and y > 0, we have $\dfrac{y/x}{3y^2/x^2 + 1} = kx$, from which

$y = k(x^2 + 3y^2)$.

11.   **Step 1.**   The given family of ellipse has the equation $\dfrac{x^2}{4c^2}$

$+ \dfrac{y^2}{3c^2} = 1$ or $\dfrac{x^2}{4} + \dfrac{y^2}{3} = c^2$.   We first find the D.E. of this

given family.   Differentiating $\dfrac{x^2}{4} + \dfrac{y^2}{3} = c^2$, we find $\dfrac{x}{2} +$

$\dfrac{2y}{3}\dfrac{dy}{dx} = 0$.   From this, we have $\dfrac{dy}{dx} = -\dfrac{3x}{4y}$, which is the

D.E. of the given family.

Step 2.   The D.E. of the orthogonal trajectories is
obtained from this by replacing -3x/4y by its negative
reciprocal 4y/3x, and so it is $\dfrac{dy}{dx} = \dfrac{4y}{3x}$.

Step 3.   We solve this lost D.E.   Separating variables, we
have $\dfrac{dy}{y} = \dfrac{4}{3}\dfrac{dx}{x}$.   Integrating, we find $\ln|y| = \dfrac{4}{3}\ln|x| +$

$\ln|k|$; and simplifying, we obtain $y = k\,x^{4/3}$.

13.   We first find the D.E. of the given family of parabolas
$y = c_1 x^2 + K$.   Differentiating, we obtain $\dfrac{dy}{dx} = 2c_1 x$.

Eliminating the parameter c between the given equation and
the derived equation, we obtain the D.E. of the family of
parabolas in the form $\dfrac{dy}{dx} = \dfrac{2y - 2K}{x}$.   It will be convenient

to rewrite this slightly, as follows: $\dfrac{dy}{dx} = \dfrac{4y - 4K}{2x}$.   Then

the D.E. of the family of orthogonal trajectories of the
given family of parabolas is

$$\frac{dy}{dx} = \frac{2x}{4K - 4y}. \tag{*}$$

We now find the D.E. of the given family of ellipses $x^2 + 2y^2 - y = c_2$. Differentiating, we obtain $2x + 4y\frac{dy}{dx} - \frac{dy}{dx} = 0$, from which we at once obtain the D.E. of the family of ellipses, that is,

$$\frac{dy}{dx} = \frac{2x}{1 - 4y}. \tag{**}$$

Comparing (*) amd (**) we see that the two given families are orthogonal provided $K = \frac{1}{4}$.

14. **Step 1.** We first find the D.E. of the given family $y = \frac{x}{1 - c_2 x}$. Differentiating, we obtain $\frac{dy}{dx} = 1/(1 - c_2 x)^2$. From the given D.E., $1 - c_2 x = \frac{x}{y}$. Substituting this into the derived equation, we have $\frac{dy}{dx} = \frac{y^2}{x^2}$. This is the D.E. of the given family.

Step 2. The D.E. of the orthogonal trajectories is obtained from this by replacing y^2/x^2 by its negative reciprocal $-x^2/y^2$, and so it is $\frac{dy}{dx} = -\frac{x^2}{y^2}$.

Step 3. We solve this last D.E. Separating variables, we have $y^2 dy = -x^2 dx$. Then integrating and simplifying, we find $x^3 + y^3 = c_1$. Thus the value of n is 3.

18. **Step 1.** We first find the D.E. of the given family $x + y = cx^2$. Differentiating, we obtain $1 + \dfrac{dy}{dx} = 2cx$.

Eliminating the parameter c between the given equation and the derived equation, we obtain the D.E. of the given family in the form $\dfrac{dy}{dx} = \dfrac{x + 2y}{x}$.

Step 2. We replace $f(x,y) = \dfrac{x + 2y}{x}$ in this D.E. by

$$\frac{f(x,y) + \tan \alpha}{1 - f(x,y)\tan \alpha} = \frac{f(x,y) + 2}{1 - 2f(x,y)} = \frac{\dfrac{x + 2y}{x} + 2}{\dfrac{1 - 2(x + 2y)}{x}} = -\frac{3x + 2y}{x + 4y}.$$

We have the D.E. $\dfrac{dy}{dx} = -\dfrac{3x + 2y}{x + 4y}$ of the desired oblique trajectories.

Step 3. We now solve this D.E. Writing it in the form $\dfrac{dy}{dx}$ $= -\dfrac{3 + 2(y/x)}{1 + 4(y/x)}$, we see that it is homogeneous. We let $y = vx$ to obtain $v + x\dfrac{dv}{dx} = -\dfrac{3 + 2v}{1 + 4v}$ or $\dfrac{(4v + 1)dv}{4v^2 + 3v + 3} = -\dfrac{dx}{x}$.

Integrating this (we recommend tables), we find $(1/2)\ln|4v^2 + 3v + 3| - (1/2)(2/\sqrt{39}) \text{ arc tan } \dfrac{8v + 3}{\sqrt{39}} =$

$\ln x + (1/2)c$, where the final 1/2 is for convenience. We multiply through by 2 and replace v by y/x to obtain

$$\ln\left|\frac{4y^2 + 3xy + 3x^2}{x^2}\right| - \left(\frac{2}{\sqrt{39}}\right) \text{ arc tan }\left[\frac{8y + 3x}{\sqrt{39}\, x}\right] = -\ln x^2 +$$

c. Thus we find the desired family of oblique trajectories in the form $\ln|3x^2 + 3xy + 4y^2| =$

$-\left(\dfrac{2}{\sqrt{39}}\right) \text{ arc tan }\left[\dfrac{3x + 8y}{\sqrt{39}\, x}\right] = c.$

Section 3.2, Page 88

1. We choose the positive x axis vertically downward along
    the path of the stone and the origin at the point from
    which the body fell. The forces acting on the stone are:
    (1) F_1, its weight, 4 lbs., which acts downward and so is
    positive, and (2) F_2, the air resistance, numerically
    equally to $v/2$, which acts upward and so is the negative
    quantity $-v/2$. Newton's Second Law gives $m\dfrac{dv}{dt} = F_1 + F_2$.
    Using $g = 32$ and $m = \dfrac{w}{g} = \dfrac{4}{32} = \dfrac{1}{8}$, this becomes
    $\dfrac{1}{8}\dfrac{dv}{dt} = 4 - \dfrac{v}{2}$.

    The initial condition is $v(0) = 0$.

    The D.E. is separable. We write it as

    $$\frac{dv}{4 - v/2} = 8\, dt\ .$$

    Integrating we find $-2\ln\left|4 - v/2\right| = 8t + c_0$ which
    reduces to $\left|4 - v/2\right| = c_1 e^{-4t}$. Applying the I.C. $v(0) = 0$
    to this, we find $c_1 = 4$. Thus the velocity at time t is
    given by $v = 8 - 8e^{-4t}$. Writing this as $\dfrac{dx}{dt} = 8 - 8e^{-4t}$
    and integrating, we find $x = 8t + 2e^{-4t} + c_2$. Applying
    the I.C. $x(0) = 0$, we find $c_2 = -2$. Thus the distance
    fallen is $x = 8t + 2e^{-4t} - 2$. Thus the answers to part
    (a) are: $v = 8(1 - e^{-4t})$ and $x = 2(4t + e^{-4t} - 1)$.

To answer part (b), we simply let t = 5 in these expressions for v and x. Thus $v(5) = 8(1 - e^{-20}) \approx 8$ ft/sec and $x(5) = 2(19 + e^{-20}) \approx 38$ feet.

3. We choose the positive x axis vertically upward along the path of the ball and the origin at the ground level. The forces acting on the ball are: (1) F_1, its weight, (3/4) lb, which acts downward and so is negative and (2) F_2, the air resistance, numerically equal to v/64, which also acts downward and so is the negative quantity - v/64. Newton's Second Law gives $m\dfrac{dv}{dt} = F_1 + F_2$. Using g = 32, and m = w/g $= \dfrac{3/4}{32} = \dfrac{3}{128}$, this becomes $\dfrac{3}{128}\dfrac{dv}{dt} = -\dfrac{3}{4} - \dfrac{v}{64}$. The initial condition is v(0) = 20.

The D.E. is separable. We write it as

$$\frac{3dv}{2v + 96} = -dt.$$

Integrating we find $(3/2) \ln|2v + 96| = -t + c_0$ which reduces to $|v + 48| = c_1 e^{-2t/3}$. Applying the I.C. v(0) = 20 to this, we find $c_1 = 68$. Thus the velocity at time t is given by $v = 68e^{-2t/3} - 48$.

From this we obtain $x = -102e^{-2t/3} - 48t + c_2$. Applying the I.C. x(0) = 6, we obtain $c_2 = 108$. Thus we have the distance $x = -102e^{-2t/3} - 48t + 108$.

The question asks how high the ball will rise. It will stop rising and start falling when the velocity v = 0. Thus we set $68e^{-2t/3} - 48 = 0$, obtaining $e^{-2t/3} = \frac{12}{17}$ and $t \approx 0.5225$ (seconds). For this value of t, we find x ≈ 10.92, which is the height above the ground that the ball will rise.

4. The forces acting on the ship are: (1) F_1, the constant propeller thrust of 100,000 lb., which moves the ship forward and so is positive, and (2) F_2, the resistence, numerically equal to 8000v, which acts against the forward motion of the ship and so is the negative quantity $-8000v$. Newton's Second Law gives $m\frac{dv}{dt} = F_1 + F_2$. Using g = 32 and $m = w/g = (32000 \text{ tons}) \frac{(2000 \text{ lbs/ton})}{32} = 2,000,000$, this becomes

$$2,000,000 \frac{dv}{dt} = 100,000 - 8000v.$$

The initial condition is $v(0) = 0$.

The D.E. is separable. We write it as

$$\frac{dv}{dt} = \frac{1}{20} - \frac{v}{250} \text{ or } \frac{dv}{25 - 2v} = \frac{dt}{500}.$$

Integrating we find $\ln|25 - 2v| = -\frac{t}{250} + c_0$ which reduces to $|25 - 2v| = c_1 e^{-t/250}$. Applying the I.C. $v(0) = 0$ to this, we find $c_1 = 25$. Thus the velocity at time t is given by

$$v = \frac{25}{2}(1 - e^{-t/250}) \qquad (*)$$

This is the answer to part (a). The answer to part (b) is then $\lim\limits_{t \to \infty} v = \lim\limits_{t \to \infty} \frac{25}{2}(1 - e^{-t/250}) = 12.5 (\text{ft/sec})$. To answer part (c), let $v = (0.80)(12.5)$ in $(*)$ and solve for t. We obtain $0.80 = 1 - e^{-t/250}$ and hence $e^{-t/250} = 1/5$. Thus $-\frac{t}{250} = \ln(1/5)$, from which $t = 402$ (sec.).

6. We break the problem into two parts: (1) while the body is rising, and (2) while the body is falling back toward earth.

We consider part (1). We choose the positive x-axis upward along the path of the object. The forces acting on the object are: $(1)F_1$, its weight, $- 98000$ dynes; and (2) F_2, air resistance, $- 200 v$ dynes. Each is negative, since each acts in the downward direction. Newton's Second Law gives $m\frac{dv}{dt} = F_1 + F_2$ or $100\frac{dv}{dt} = - 98000 - 200v$. The I.C. is $v(0) = 150$.

We simplify the D.E. to read $\frac{dv}{dt} = - (2v + 980)$. It is separable; and separating variables, we have $\frac{dv}{2v + 980} = -dt$. Integrating, we find $\ln|2v + 980| = - 2t + c_0$. Since $v \geq 0$ in part (1), this reduces to $2v + 980 = c_1 e^{-2t}$ or $v = - 490 + c e^{-2t}$. Applying the I.C., $v(0) = 150$ to this, we find $c = 640$. Hence

$$v = - 490 + 640 e^{-2t}. \qquad (*)$$

The object stops rising and starts falling when v = 0.
Thus this happens at t such that $e^{-2t} = \frac{49}{64}$. We find

t ≈ 0.1336 (sec.).

In (a) we seek the velocity 0.1 second after the
object is thrown. Since 0.1 < 0.1336, we let t = 0.1 in
(*), and find v = - 490 + 640 $e^{-1.2}$ ≈ 33.99 cm/sec.

Now consider part (2). We choose the positive x-axis
vertically downward from the highest point reached. Now
the weight is + 98000, since it acts in this downward
direction. The D.E. is now 100 $\frac{dv}{dt}$ = 98000 - 200v, and the
I.C. is v(0) = 0. Simplifying the D.E. and separating
variables, we have $\frac{dv}{980 - 2v}$ = dt. Integrating, we find
ln|980 - 2v| = - 2t + c_0. Since 980 - 2v ≥ 0, upon
simplifying we have v = 490 - c e^{-2t} . The I.C. gives
c = 490, and so

$$v = 490(1 - e^{-2t}). \qquad (**)$$

In (b) we seek the velocity 0.1 second after the
object stops rising and starts falling. So we let t = 0.1
in (**) and obtain v = 490(1 - $e^{-0.2}$) ≈ 88.82 cm./sec.

8. We choose the positive x-axis horizontally along the given
    direction of motion. The forces acting on the boat and
    rider are: (1) F_1, the constant force of 12 lb., which
    acts in the given direction and so is positive, and (2)
    F_2, the resistance force, numerically equal to 2v, which

acts opposite to the given direction and so is the negative quantity $-2v$.

Newton's Second Law gives $m \dfrac{dv}{dt} = F_1 + F_2$. Using $g = 32$ and $m = \dfrac{w}{g} = \dfrac{150 + 170}{32} = 10$, this becomes

$$10 \frac{dv}{dt} = 12 - 2v.$$

The initial velocity is 20 m.p.h. $= \dfrac{88}{3}$ ft/sec. Thus the I.C. is $v(0) = \dfrac{88}{3}$.

The D.E. is separable. We write it as

$$\frac{5 \, dv}{6 - v} = dt.$$

Integrating we find $5 \ln|v - 6| = -t + c_0$ which reduces to $|v - 6| = c \, e^{-t/5}$. Applying the I.C. $v(0) = \dfrac{88}{3}$ to this, we find $c = \dfrac{70}{3}$. Thus the velocity at time t is given by

$$v = 6 + \left(\frac{70}{3}\right) e^{-t/5}. \qquad\qquad (*)$$

To answer part (a), we let $t = 15$ in $(*)$ to obtain $v(15) = 6 + \left(\dfrac{70}{3}\right) e^{-3} \approx 7.16$ ft/sec. To answer part (b), let $v = \left(\dfrac{1}{2}\right)\left(\dfrac{88}{3}\right) = \dfrac{44}{3}$ in $(*)$ and solve for t. We have $6 + \left(\dfrac{70}{3}\right) e^{-t/5} = (44)(3)$, from which $e^{-t/5} = \dfrac{13}{35}$. From this, $-\dfrac{t}{5} \approx -0.99$ and $t \approx 4.95$ (seconds).

10. We choose the positive x-axis vertically upward along the
     path of the shell with the origin at the earth's surface.
     The forces acting on the shell are: (1) F_1, its weight, 1
     lb., which acts downward and so is the negative quantity
     -1; and (2) F_2, the air resistance numerically equal to
     $10^{-4}v^2$, which also acts downward (against the rising
     shell) and so is the negative quantity $-10^{-4}v^2$.   Newton's
     Second Law gives $m\dfrac{dv}{dt} = F_1 + F_2$.   Using g = 32 and $m = \dfrac{w}{g} =$
     $\dfrac{1}{32}$, this becomes

$$\frac{1}{32}\frac{dv}{dt} = -1 - 10^{-4}v^2.$$

The initial condition is v(0) = 1000.

   The D.E. is separable.   We write it as

$$\frac{1}{32}\frac{dv}{dt} = -\frac{10^4 + v^2}{10^4} \quad \text{or} \quad \frac{dv}{v^2 + 10^4} = -\frac{32}{10^4}dt \ .$$

Integrating we find $\left(\dfrac{1}{100}\right)$ arc tan $\left(\dfrac{v}{100}\right) = -\dfrac{32t}{10^4} + c$ or

arc tan $\left(\dfrac{v}{100}\right) = -\dfrac{32t}{100} + c$.   Application of the initial

condition at once gives c = arc tan 10.   Thus we obtain the
solution in the form

$$\text{arc tan}\left(\frac{v}{100}\right) = \text{arc tan } 10 - \left(\frac{32t}{100}\right).$$

Taking the tan of each side and multiplying by 100 gives

$$v = 100 \tan(\text{arc tan } 10 - 0.32t).$$

This is the answer to part (a). To answer part (b), note that the shell will stop rising when v = 0. Setting v = 0, we at once have arc tan 10 - 0.32t = 0, and thus t = arc tan 10/(0.32) $\approx$ 4.60 (sec).

13. We choose the positive x-axis horizontally along the given direction of motion and the origin at the point at which the man stops pushing. The forces acting on the loaded sled as it continues are: (1) F_1, the air resistance, numerically equal to $\frac{3v}{4}$, which acts opposite to the direction of motion and so is given by $- \frac{3v}{4}$, and (2) F_2, the frictional force, having numerical value μN = (0.04) (80), which also acts opposite to the direction of motion and so is given by $-(0.04)(80) = - \frac{16}{5}$. Newton's Second Law gives $m \frac{dv}{dt} = F_1 + F_2$. Using g = 32 and m = $\frac{w}{g} = \frac{80}{32} = \frac{5}{2}$, this becomes

$$\frac{5}{2} \frac{dv}{dt} = - \frac{3v}{4} - \frac{16}{5}.$$

The initial velocity is 10 ft/sec., so the I.C. is v(0) = 10.   The D.E. is separable. We write it as

$$\frac{dv}{15v + 64} = - \frac{dt}{50}.$$

Integrating, we find $\ln|15v + 64| = - \frac{3t}{10} + c_0$ which reduces to $|15v + 64| = c e^{-3t/10}$. Applying the I.C. v(0) = 10 to this, we find c = 214. Thus the velocity is given by

$$v = \left(\frac{214}{15}\right)e^{-3t/10} - \frac{64}{15}.$$

From this, integration and simplification give x =

$-\left(\frac{428}{9}\right)e^{-3t/10} - \frac{64t}{15} + c_1$. Application of the condition

$x(0) = 0$ then gives $c_1 = \frac{428}{9}$. Hence the distance x is

given by

$$x = \left(\frac{428}{9}\right)(1 - e^{-3t/10}) - \frac{64t}{15}.$$

To answer the stated question, note that the sled will continue until the velocity v = 0. Thus we set v = 0 and solve for t, finding $e^{-3t/10} = \frac{32}{107}$ from which t ≈ 4.02.

We now evaluate x at t ≈ 4.02 to determine the distance which the sled will continue. We find

$$x(4.02) \approx \left(\frac{428}{9}\right)\left(1 - \frac{32}{107}\right) - \frac{64(4.02)}{15} \approx 16.18 \text{ feet.}$$

15. We choose the positive x direction down the slide with the origin at the top. The forces acting on the case are: (1) F_1, the component of its weight parallel to the slide, (2) F_2, the frictional force; and (3) F_3, the air resistance. The case weighs 24 lbs., and the component parallel to the slide has numerical value $24 \sin 45^0 = \frac{24}{\sqrt{2}}$.

Since this acts in the positive (downward) direction along the slide, $F_1 = \frac{24}{\sqrt{2}}$. The frictional force F_2, has

numerical value μN, where μ is the coefficient of the friction and N is the normal force. Here $\mu = 0.4$ and the magnitude of N is $24 \cos 45^0 = \dfrac{24}{\sqrt{2}}$.   Since the force F_2

acts in the negative (upward) direction along the slide,

we have $F_2 = -(0.4)\left(\dfrac{24}{\sqrt{2}}\right)$.   Finally, the air resistance F_3

has numerical value $\dfrac{v}{3}$,   Since this also acts in the

negative direction, we thus have $F_3 = -\dfrac{v}{3}$.   Newton's

Second Law now gives $m\dfrac{dv}{dt} = F_1 + F_2 + F_3$.   With $g = 32$,

$m = \dfrac{w}{g} = \dfrac{24}{32} = \dfrac{3}{4}$ and the above forces, this becomes

$$\left(\frac{3}{4}\right)\left(\frac{dv}{dt}\right) = \left(\frac{24}{\sqrt{2}}\right) - (0.4)\left(\frac{24}{\sqrt{2}}\right) - \frac{v}{3}$$

The initial condition is $v(0) = 0$.

The D.E. is separable.   We write it as $\left(\dfrac{3}{4}\right)\left(\dfrac{dv}{dt}\right)$

$= \dfrac{(-5v + 108\sqrt{2})}{15}$ or $\dfrac{dv}{5v - 108\sqrt{2}} = -\dfrac{4}{45} dt$.   Integrating we

find $\ln|5v - 108\sqrt{2}| = -\dfrac{4t}{9} + c_0$ or $|5v - 108\sqrt{2}| = $

$c e^{-4t/9}$.   Applying the initial condition, we at once find

$c = 108\sqrt{2}$.   Thus the velocity is given by

$$v = \frac{108\sqrt{2}}{5} (1 - e^{-4t/9}). \qquad (*)$$

To answer (a), simply let $t = 1$ in $(*)$.   We obtain

$v(1) = \left(\dfrac{108\sqrt{2}}{5}\right)\left(1 - \dfrac{e^{-4}}{9}\right) \approx 10.96 (\text{ft/sec})$.   To answer (b),

more work is required.   We first integrate $(*)$ to find the

distance x from the top of the slide.   We have

$$x = \left(\frac{108\sqrt{2}}{5}\right)t + \left(\frac{243\sqrt{2}}{5}\right)e^{-4t/9} + c.$$

Since x = 0 at t = 0, we find $c = -\frac{243\sqrt{2}}{5}$, and hence

$$x = \left(\frac{108\sqrt{2}}{5}\right)t + \left(\frac{243\sqrt{2}}{5}\right)e^{-4t/9} - \frac{243\sqrt{2}}{5}.$$

The slide being 30 ft. long, we let x = 30 in this to
determine the time t at which the case reaches the bottom.
That is, we must find t such that

$$30 = \left(\frac{108\sqrt{2}}{5}\right)t + \left(\frac{243\sqrt{2}}{5}\right)e^{-4t/9} - \frac{243\sqrt{2}}{5}.$$

We simplify this so that it takes the form

$$68.72\, e^{-4t/9} = -30.54t + 98.72.$$

A little trial-and-error calculation with a hand
calculator shows that the two sides of this are
approximately equal for t = 2.49. This is the time at
which the case reaches the bottom. Letting t = 2.49 in
(*), we find the velocity at that time to be approximately
20.46 (ft/sec).

16. We choose the positive x direction down the hill with the
origin at the starting point. The forces acting on the
boy and sled are: (1) F_1, the component of their weight
parallel to the hill; (2) F_2, the frictional force; and
F_3, the air resistance. The boy and sled weigh 72 lbs.,

and the component parallel to the hill has numerical value
$72 \sin 30^0 = 36$.   Since this acts in the positive
(downward) direction on the hill, $F_1 = 36$.   The frictional

force F_2 has numerical value μN, where $\mu > 0$ is the

coefficient of friction and N is the normal force.   The
magnitude of N is $72 \cos 30^0 = 36\sqrt{3}$; the μ is an unknown
which will be determined in due course from the given data
of the problem.   Since the force F_2 acts in the negative

(upward) direction on the hill, we have $F_2 = -\mu(36\sqrt{3})$.

Finally, the air resistance F_3 has numerical value 2v.

Since this also acts in the negative direction, we thus
have $F_3 = -2v$.   Newton's Second Law now gives $m\frac{dv}{dt} =$
$F_1 + F_2 + F_3$.   With $g = 32$, $m = \frac{w}{g} = \frac{72}{32} = \frac{9}{4}$, and the above
forces, this becomes

$$\left(\frac{9}{4}\right)\left(\frac{dv}{dt}\right) = 36 - \mu\, 36\sqrt{3} - 2v.$$

The initial condition is $v(0) = 0$.   Another condition is
also given, that is $v(5) = 10$; and this extra condition
will be sufficient for us to eventually determine μ.

The D.E. is separable.   We write it as

$$\frac{dv}{dt} = \left(\frac{8}{9}\right)(18 - \mu\, 18\sqrt{3} - v)$$

or

$$\frac{dv}{v + 18(\mu\sqrt{3} - 1)} = -\frac{8}{9}\, dt .$$

Integrating we find $\ln|v + 18(\mu\sqrt{3} - 1)| = -\dfrac{8t}{9} + c_0$ or

$|v + 18(\mu\sqrt{3} - 1)| = c\,e^{-8t/9}$. Applying the initial

condition, we at once find that $c = |18(\mu\sqrt{3} - 1)|$. Thus

we obtain the solution in the form $v + 18(\mu\sqrt{3} - 1) =$

$18(\mu\sqrt{3} - 1)e^{-8t/9}$ or $v = 18(1 - \mu\sqrt{3})(1 - e^{-8t/9})$. Now

we apply the extra condition $v(5) = 10$ to this to

determine μ. We have $10 = 18(1 - \mu\sqrt{3})(1 - e^{-40/9})$ and

hence $1 - \mu\sqrt{3} = \left(\dfrac{5}{9}\right)(0.988)^{-1}$, from which we obtain $\mu =$

0.25.

17. This problem has two parts: (A), *before* the object

reaches the surface of the lake; and (B), *after* it passes

beneath the surface. We consider (A) first. We take the

positive x-axis vertically downward with the origin at the

point of release of the object. The forces acting on the

body are: (1) F_1, its weight, 32 lbs., which acts

downward and so is positive; and (2) F_2, the air

resistance, numerically equal to 2v, which acts upward and

so is the negative quantity -2v. Applying Newton's Second

Law, with $m\dfrac{w}{g} = \dfrac{32}{32} = 1$, we at once obtain the D.E.

$$\frac{dv}{dt} = 32 - 2v.$$

The initial condition is $v(0) = 0$. The D.E. is separable.

We write it as $\dfrac{dv}{32 - 2v} = dt$. Integrating we find

$\ln|32 - 2v| = -2t + c_0$ or $|32 - 2v| = c_1 e^{-2t}$. Applying

the initial condition we find $c = 32$. With this, we at

once have

$$v = 16(1 - e^{-2t}) \qquad\qquad (*)$$

This gives the velocity at each time t before the object reaches the surface of the lake. To solve problem (B), we will need to know the velocity at the instant when the object reaches the surface. To find out when this is, we need to know distance fallen as a function of time. This is found by integrating (*). We at once obtain

$$x = 16t + 8e^{-2t} + k.$$

Since $x(0) = 0$, we have $k = -8$, and hence the distance fallen (*before* striking the water) is given by

$$x = 16t + 8e^{-2t} - 8.$$

Since the point of release was 50 feet above the water, if we let x = 50 in this, it will determine the time at which the object hits the surface. Thus we must solve

$$50 = 16t + 8e^{-2t} - 8$$

for t. We write this in the form $58 - 16t = 8e^{-2t}$. A little trial-and-error calculation with a hand calculator leads to the approximate solution t = 3.62. The velocity at this instant is then found by letting t = 3.62 in (*). The result is approximately 15.99 (ft/sec).

We now turn to problem (B). We again take the positive x-axis vertically downward, but now we take the origin at the point where the object hits the surface of the lake. The forces now acting on the body are given by

$F_1 = 32$, $F_2 = -6v$, and $F_3 = -8$. The last two (the water resistance and the buoyancy) have negative signs since they act upward. Newton's Second Law leads to the D.E.

$$\frac{dv}{dt} = 32 - 6v - 8.$$

By part (A), the velocity of the object at the surface of the lake is 15.99 (ft/sec). We take this as the I.C. here: $v(0) = 15.99$.

Separating variables, the D.E. becomes $\frac{dv}{24 - 6v} = dt$. Integrating, we find $\ln|24 - 6v| = -6t + c_1$ or $|24 - 6v| = ce^{-6t}$. Applying the I.C. gives $c = 71.94$. Thus we obtain $-(24 - 6v) = 71.94\, e^{-6t}$ and hence $v = 4 + 11.99\, e^{-6t}$. This is the velocity after the object passes beneath the surface. We want to know what this is 2 sec. after. Hence we let $t = 2$ in this to obtain $v \approx 4.00$ (ft/sec).

Section 3.3, Page 102

1. Let x be the amount of radioactive nuclei present after t years. Then x satisfies the D.E. $\frac{dx}{dt} = -kx$, where $k > 0$. Letting x_0 denote the amount initially present, we have the I.C. $x(0) = x_0$. Also, since 10% of the original number have undergone disintegration in 100 years, 90% remain, and so we have the additional condition $x(100) = \frac{9x_0}{10}$. The D.E. is separable; and separating variables and integrating, we at once obtain $x = c\, e^{-kt}$. Application of the I.C. immediately gives $x_0 = c$. Hence we have

$x = x_0 e^{-kt}$. Now apply the additional condition to

determine k. We have $\dfrac{9x_0}{10} = x_0 e^{-100k}$, which reduces to

$e^{-k} = \left(\dfrac{9}{10}\right)^{1/100}$. Thus the solution takes the form

$$x = x_0 \left(\frac{9}{10}\right)^{t/100} \tag{$*$}$$

To answer question (a), we let t = 1000 in ($*$). We find

$x(1000) = x_0\left(\dfrac{9}{10}\right)^{10} \approx 0.3487\, x_0$. Thus the answer to

34.87%. To answer question (b), we let $x = \dfrac{x_0}{4}$ in ($*$). We

have $\dfrac{x_0}{4} = x_0\left(\dfrac{9}{10}\right)^{t/100}$. From this, $\dfrac{1}{100}\ln\left(\dfrac{9}{10}\right) = \ln\left(\dfrac{1}{4}\right)$,

from which $t = 100\left[\dfrac{\ln\dfrac{1}{4}}{\ln\dfrac{9}{10}}\right] \approx 1315.28$. Thus $t \approx 1315$

years.

4. Let x = the amount of the first chemical present. Then x

satisfies the D.E. $\dfrac{dx}{dt} = -kx$, where k > 0. Two conditions

are given: $x(1) = \dfrac{2}{3}$ and $x(4) = \dfrac{1}{3}$. The D.E. is separable;

and separating variables and integrating, we at once

obtain $x = ce^{-kt}$. Applying the two conditions, we obtain

respectively $ce^{-k} = \dfrac{2}{3}$ and $ce^{-4k} = \dfrac{1}{3}$. These two equations

will determine c and k. Dividing the first by the second

gives $e^{3k} = 2$, from which $e^{k} = 2^{1/3}$. Then from the second

equation for c and k, $c = \left(\dfrac{1}{3}\right)e^{4k} = \dfrac{2^{4/3}}{3}$. Thus the

solution of the D.E. which satisfies the two given

conditions is $x = \left(\dfrac{2^{4/3}}{3}\right)(e^k)^{-t} = \left(\dfrac{2^{4/3}}{3}\right)(2^{1/3})^{-t}$ or

$x = \dfrac{2^{(4-t)/3}}{3}$. $\hspace{4cm}$ (*)

To answer question (a), let $t = 7$ in (*). This gives

$x(7) = \dfrac{2^{-1}}{3} = \dfrac{1}{6}$ (kg). Now compare this with the orignial

amount, which is $x(0) = \dfrac{2^{4/3}}{3}$ (kg). We have $\dfrac{x(7)}{x(0)} = \dfrac{\frac{1}{6}}{\frac{2^{4/3}}{3}} =$

$\dfrac{1}{4^3\sqrt{2}} \approx 0.1984$. Thus 19.8% of the first chemical remains

at the end of seven hours.

To answer question (b), we first note that one tenth

of the first chemical is $\dfrac{x(0)}{10} = \dfrac{2^{4/3}}{30}$ (kg). We thus let x

$= \dfrac{2^{4/3}}{30}$ in (*) and solve for t. We have $\dfrac{2^{(4-t)/3}}{3} = \dfrac{2^{4/3}}{30}$.

From this $\left[\dfrac{4-t}{3}\right][\ln 2] = \ln \dfrac{2^{4/3}}{10}$; and solving for t, we

find $t \approx 9.97$. Thus the answer is 9 hours, 58 minutes.

7. Let x = the temperature of the body at time t. Assuming
    Newton's Law of Cooling, we have the D.E. $\dfrac{dx}{dt} = k(x - 40)$.

    We also have the I.C. $x(0) = 100$ and the additional
    condition $x(10) = 90$. The D.E. is separable. We write it
    in the form $\dfrac{dx}{x - 40} = k\,dt$. Integrating, we obtain
    $\ln|x - 40| = kt + c_0$ and hence $|x - 40| = c\,e^{kt}$. Since
    $x \geq 40$, $|x - 40| = x - 40$, so we have $x = 40 + c\,e^{kt}$.
    Application of I.C. $x(0) = 100$ gives $100 = 40 + c$ or $c =$
    60. Thus $x = 40 + 60\,e^{kt}$. Then application of the

additional condition $x(10) = 90$ gives $90 = 40 + 60 \, e^{10k}$ from which $e^{10k} = \frac{5}{6}$ or $e^{k} = \left(\frac{5}{6}\right)^{1/10}$. Thus $e^{kt} = \left(\frac{5}{6}\right)^{t/10}$,

and we have the solution

$$x = 40 + 60 \left(\frac{5}{6}\right)^{t/10}. \qquad (*)$$

(a) Let $t = 30$ in $(*)$. We find $x(30) = 40 + 60 \left(\frac{5}{6}\right)^{3} \approx$

74.72^{0} F.

(b) Let $x = 50$ in $(*)$. We have $50 = 40 + 60 \left(\frac{5}{6}\right)^{t/10}$, from

which $\left(\frac{5}{6}\right)^{t/10} = \frac{1}{6}$. Then $t = \frac{10 \, \ln(1/6)}{\ln \, (5/6)} \approx 98.29$.

Hence $t \approx 98$ minutes, 17 seconds.

9. Let x = the temperature of the pie at time t. Assuming Newton's Law of Cooling, we have the D.E. $\frac{dx}{dt} = k(x - 80)$.

We also have the I.C. $x(0) = 350$ and the additional condition $x(5) = 300$. Separating variables in the D.E., we have $\frac{dx}{x - 80} = k \, dt$; and integrating, we find $\ln|x - 80|$ $= kt + c_0$, and hence $|x - 80| = c \, e^{kt}$. Since $x \geq 80$, this simplifies to $x - 80 + c \, c^{kt}$. Applying the I.C. $x(0) = 350$, we have $350 = 80 + c$, so $c = 270$. Thus $x = 80 + 270 \, e^{kt}$. Application of the additional condition $x(5) = 300$ gives $300 = 80 + 270 \, e^{5k}$, from which $e^{5k} = \frac{220}{270}$ or $e^{k} = \left(\frac{22}{27}\right)^{1/5}$. Thus we obtain the solution

$$x = 80 + 270 \left(\frac{22}{27}\right)^{t/5}. \qquad (*)$$

(a) Let $t = 10$ in $(*)$. We find $x(10) = 80 + 270\left(\frac{22}{27}\right)^2 \approx$ 259.26^0.

(b) Let $x = 100$ in $(*)$. We have $100 = 80 + 270\left(\frac{22}{27}\right)^{t/5}$, from which $\left(\frac{22}{27}\right)^{t/5} = \frac{2}{27}$. Then $t = 5 \frac{\ln(2/27)}{\ln(22/27)} \approx$ 63.54. Hence $t \approx 63$ minutes, 32 seconds.

10.   Let $x =$ the temperature of the coffee at time t. Assuming Newton's Law of Cooling, we have the D.E. $\frac{dx}{dt} = k(x - 70)$.

We also have the I.C. $x(0) = 180$ and the additional condition $x(10) = 160$. Separating variables in the D.E., we have $\frac{dx}{x - 70} = k\,dt$; and integrating, we find $\ln|x - 70|$ $= kt + c_0$, and hence $|x - 70| = c\,e^{kt}$. Since $x \geq 70$, this simplifies to $x = 70 + c\,e^{kt}$. Applying the I.C. $x(0) =$ 180, we have $180 = 70 + c$, so $c = 110$. Thus $x =$ $70 + 110\,e^{kt}$. Application of the additional condition $x(10) = 160$ gives $160 = 70 + 110\,e^{10k}$, from which $e^{10k} =$ $\frac{90}{110}$ or $e^k = \left(\frac{9}{11}\right)^{1/10}$. Thus we obtain the solution

$$x = 70 + 110\left(\frac{9}{11}\right)^{t/10} \qquad\qquad (*).$$

(a) Let $t = 15$ in $(*)$. We have $x(15) = 70 + 110\left(\frac{9}{11}\right)^{3/2} \approx$ 151.41^0.

(b) Let x = 140 in (∗).   We have $140 = 70 + 110\left(\frac{9}{11}\right)^{t/10}$,

from which $\left(\frac{9}{11}\right)^{t/10} = \frac{7}{11}$.   Then $t = \frac{10\ \ln(7/11)}{\ln(9/11)} \approx$

22.52.   So x = 140 at t ≈ 22 minutes, 31 seconds.   Now

let x = 130 in (∗).   We have $130 = 70 + 110\left(\frac{9}{11}\right)^{t/10}$,

from which $\left(\frac{9}{11}\right)^{t/10} = \frac{6}{11}$.   Then $t = \frac{10\ \ln(6/11)}{\ln(9/11)} \approx$

30.20.   So x = 130 at t ≈ 30 minutes, 12 seconds.

Thus the woman should have drunk the coffee between 22
minutes, 31 seconds past 10 and 30 minutes, 12 seconds
past 10.

11.   Let x = the population at time t.   Then we at once have
the D.E. $\frac{dx}{dt} = kx$.   Letting x_0 denote the population at the
start of the given 40 year period, we have the I.C. x(0) =
x_0.   Since the population doubles in 40 years, we have the
additional condition $x(40) = 2x_0$.   The solution of the
D.E. is $x = c\,e^{kt}$.   Applying the I.C. to this we at once
have $c = x_0$, and hence $x = x_0 e^{kt}$.   Now applying the
additional condition to this, we have $2x_0 = x_0 e^{40k}$, from
which $e^{40k} = 2$ and $k = \ln\frac{2}{40} \approx 0.0173$.   Thus the solution
of the D.E. which satisfies the two given conditions is x
$= x_0 e^{0.0173t}$.   To answer the stated question, we let x =
$3x_0$ in this, from which $e^{0.0173t} = 3$ and $0.0173t = \ln 3$.
From this we find t ≈ 63.5.   Thus the population triples
in approximately 63.5 years.

12.   Let x = the population at time t. We at once have the
      D.E. $\frac{dx}{dt}$ = kx. We take 1970 as the zeroth year of the
      problem and have the I.C. x(0) = 30,000. Then 1980 is the
      tenth year, and we have the additional condition x(10) =
      35,000. The one-parameter family of solutions of the D.E.
      is x = c e^{kt}. Applying the I.C. to this, we have c =
      30,000. Thus x = 30,000 e^{kt}. We apply the additional
      condition to this, obtaining 35,000 = 30,000 e^{10k}. From
      this, e^k = $\left(\frac{7}{6}\right)^{1/10}$. Thus we obtain the solution x =
      30,000$\left(\frac{7}{6}\right)^{t/10}$. We seek the population in 1990, which is
      the twentieth year on or scale. Letting t = 20 in the
      solution, we find x = 30,000$\left(\frac{7}{6}\right)^2$ ≈ 40,833.

15.   The population x satisfies the D.E.

$$\frac{dx}{dt} = \frac{3x}{100} - \frac{3x^2}{10^8}.$$

The I.C. is x(1980) = 200,000. The D.E. is separable. We
write it in the form

$$\frac{10^6 dx}{x(x - 10^6)} = -\frac{3\,dt}{100}.$$

To integrate the left member, we use partial fractions.
Thus we obtain

$$\left[\frac{1}{x - 10^6} - \frac{1}{x}\right] dx = -\frac{3}{100}\,dt .$$

Integrating we find $\ln|x - 10^6| - \ln|x| = -\dfrac{3t}{100} + c_0$ or

$\ln\left|\dfrac{x - 10^6}{x}\right| = -\dfrac{3t}{100} + c_0$. From this we obtain $\dfrac{x - 10^6}{x} =$

$ce^{-3t/100}$. We apply the I.C. $x(1980) = 200{,}000 = (2)(10^5)$

to this. We have $\dfrac{[(2)(10^5) - 10^6]}{2(10)^5} = ce^{-3(198)/10}$, or $c =$

$-4e^{59.4}$. Thus we obtain the solution in the form $\dfrac{x - 10^6}{x}$

$= -4e^{59.4-3t/100}$. We solve this for x, obtaining

$x + 4x\,e^{59.4-3t/100} = 10^6$ and hence

$$x = \frac{10^6}{1 + 4e^{59.4-3t/100}} \qquad (*)$$

This is the answer to part (a). To answer part (b),

we let $t = 2000$ in $(*)$. We find $x(2000) = \dfrac{10^6}{[1 + 4e^{0.6}]} \approx$

312,966. To answer (c), simply find $\lim\limits_{t \to \infty} x$, where x is

given by $(*)$. Since $\lim\limits_{t \to \infty} e^{59.4-3t/100} = 0$, we have

$\lim\limits_{t \to \infty} x = \dfrac{10^6}{1} = 1{,}000{,}000.$

16. (a) We have the initial-value problem $\dfrac{dx}{dt} = kx - \lambda x^2$,

$x(t_0) = x_0$, where we assume $kx - \lambda x^2 > 0$. The D.E.

is separable. We separate variables and use partial

fractions to obtain $\dfrac{1}{k}\left[\dfrac{1}{x} + \dfrac{\lambda}{k - \lambda x}\right]dx = dt$. Then

integration gives $\dfrac{1}{k}[\ln x - \ln(k - \lambda x)] = t + \ln c_0$.

Simplifying, we obtain $\dfrac{x}{k - \lambda x} = c\,e^{kt}$. Applying the

I.C. $x(t_0) = x_0$ to this, we find $c = \dfrac{x_0 e^{-kt_0}}{k - \lambda x_0}$. Before

using this, we solve the one-parameter family of

solutions for x. We have $x = k c e^{kt} - \lambda c x e^{kt}$, from

which $(1 + \lambda c e^{kt})x = k c e^{kt}$ and hence $x = \dfrac{k c e^{kt}}{1 + \lambda c e^{kt}}$

or finally $x = \dfrac{kc}{\lambda c + e^{-kt}}$. We now substitute into this

the value of c already determined from the initial

condition. We find

$$
x = \frac{k \left[\dfrac{x_0 e^{-k t_0}}{k - \lambda x_0} \right]}{\lambda \left[\dfrac{x_0 e^{-k t_0}}{k - \lambda x_0} \right] + e^{-kt}}.
$$

We simplify this by multiplying numerator and

denominator through by $(k - \lambda x_0)e^{k t_0}$. We obtain

$$
x = \frac{k x_0}{\lambda x_0 + (k - \lambda x_0)e^{-k(t - t_0)}}.
$$

17.   Let x = the population at time t. Since 100 people leave

the island every year, the D.E. (3.58) is correctly

modified by subtracting 100 from its right member. Thus

we have the D.E.

$$
\frac{dx}{dt} = \left(\frac{1}{400}\right)x - 10^{-8}x^2 - 100.
$$

We also have the I.C. $x(1980) = 20{,}000$. The D.E. is

separable. We write it in the successive forms

$$\frac{dx}{dt} = \frac{(10)^6 x - 4x^2 - 4(10)^{10}}{4(10)^8},$$

$$\frac{dx}{[x^2 - 25(10)^4 x + (10)^{10}]} = -\frac{dt}{(10)^8},$$

$$\frac{dx}{[x - 5(10)^4][x - 20(10)^4]} = -\frac{dt}{(10)^8}.$$

We now apply partial fractions to the left member and multiply through by $(15)(10)^4$ to obtain

$$\left[\frac{1}{x - 20(10)^4} - \frac{1}{x - 5(10)^4}\right] dx = -\frac{15\ dt}{(10)^4}.$$

Integrating and simplification give

$$\ln\left[\frac{x - 20(10)^4}{x - 5(10)^4}\right] = -\frac{15t}{(10)^4} + c_0$$

or

$$\frac{x - 20(10)^4}{x - 5(10)^4} = c\,e^{-15t/(10)^4}.$$

We apply the I.C. $x(1980) = 20,000 = (20)(10)^3$ to this. With $x = (20)(10)^3$, the left member reduces thus:

$$\frac{(20)(10)^3(1 - 10)}{(5)(10)^3(4 - 10)} = 6.$$

Thus we find $6 = c\,e^{-(15)(1980)/(10)^4}$, and so $c = 6\,e^{(15)(1980)/(10)^4}$. Thus the solution takes the form

$$\frac{x - (20)(10)^4}{x - 5(10)^4} = 6 \, e^{15(1980-t)/(10)^4}.$$

We must solve this for x. After some algebraic
manipulations, we obtain the desired result

$$x = \frac{(10)^5 [3 \, e^{15(1980-t)/(10)^4} - 2]}{6 \, e^{15(1980-t)/(10)^4} - 1}.$$

18.   This falls into two problems: (A) Before the cats
arrived, and (B) after the cats arrived. We let x = the
number of mice on the island at time t and proceed to
solve problem (A). During this time, in the absence of
cats killing mice, the D.E. is simply $\frac{dx}{dt} = kx$, with
solution $x = c \, e^{kt}$. Choosing January 1, 1970 as time
t = 0, the I.C. is x(0) = 50,000. Then, measuring t in
years, Jan. 1, 1980 is time t = 10; and we have the
additional condition x(10) = 100,000. Applying the I.C.
x(0) = 50,000, we at once find c = 50,000 and hence
$x = 50,000 \, e^{kt}$. The additional condition now gives
$e^{10k} = 2$, from which $k = \ln \frac{2}{10}$. With this, the solution
takes the form $x = 50,000 \, e^{t \ln(2/10)}$. But this solution
only holds until 1980 and so will not help us to answer
the stated question. It is the number k, which represents
the natural rate of population increase, which we need
here to solve problem (B) and answer the stated question.

Turning to problem (B), since the cats kill 1000 mice/
month = 12,000 mice/year, the D.E. of problem (A) must be
modified by subtracting 12,000 from its right member.

Thus we have $\frac{dx}{dt}$ = kx - 12,000, where k = ln $\frac{2}{10}$ was found in problem (A). The D.E. is separable, and we write it in the form $\frac{dx}{kx - 12,000}$ = dt. Integrating we obtain kx = 12,000 + c e^{kt}. Now we have an I.C. here, for on Jan. 1, 1980, there were 100,000 mice. Taking this date as time t = 0 for problem (B), we thus have the I.C. x(0) = 100,000. Application of this gives c = 100,000 k - 12,000. With this value of c and k = ln $\frac{2}{10}$, we obtain the solution of problem (B):

$$x = \left(\frac{10}{\ln 2}\right)[12,000 + (10,000 \ln 2 - 12,000)e^{t \ln 2/10}]$$

The answer to the stated question is now found by letting t = 1 in this expression. We have

$$x(1) = (14.427)[12,000 + (6931 - 12,000)(1.0718)]$$
$$= 94,742.$$

19. Let x = the amount of money at time t. We have the D.E. $\frac{dx}{dt}$ = kx. Since the annual rate is 6%, we let k = .06 and so have the D.E. $\frac{dx}{dt}$ = .06x. We also have the I.C. x(0) = 1000. Solving the D.E., we find x = c $e^{.06t}$. The I.C. gives c = 1000, so we have the solution x = 1000 $e^{.06t}$.

(a) We let t = 10 in the solution. We find x = 1000 $e^{0.6}$ ≈ 1822.12. Thus there will be $1822.12.

(b) We let x = 2000 in the solution.   We obtain 2000 =
1000 $e^{.06t}$ or $e^{.06t}$ = 2.   From this .06t = ln 2 and t
$\approx$ 11.55.   Thus the amount will double in approximately
11.55 years.

20.   From exercise 19, we have the D.E. $\frac{dx}{dt}$ = kx, where k gives

the annual rate.   We have the I.C. x(0) = x_0, where x_0

denotes the original amount.   Solving the D.E., we find x
= $c\,e^{kt}$.   Applying the I.C. to this, we find c = x_0 and so

have the solution x = $x_0 e^{kt}$.

(a) If the original amount doubles in two years, then x =
$2x_0$ when t = 2.   Applying this to the solution, we

obtain $2x_0$ = $x_0 e^{2k}$.   From this k = $\frac{\ln 2}{2}$ $\approx$ 0.3466.

Thus the annual rate is a remarkable 34.66%.

(b) Here we have the additional condition x(1/2) = $3x_0/2$.

We apply this to the solution.   We have $3x_0/2$ =
$x_0 e^{k/2}$, from which e^k = 9/4.   Then the solution

becomes x = $x_0\left(\frac{9}{4}\right)^t$.   We ask how long it will take the

original amount to double, so we let x = $2x_0$ in this

solution and solve for t.   We find $\left(\frac{9}{4}\right)^t$ = 2, from

which t = $\frac{\ln 2}{\ln(9/4)}$ $\approx$ 0.8547.   So the amount doubles in

approximately 0.85 years.

21. Let x = the amount of salt at time t. We use the basic
     equation (3.63), $\frac{dx}{dt}$ = IN - OUT. We find IN = (3 lb./gal)

     (4 gal./min.) = 12(lb./min.); and OUT = (C lb./gal.)
     (4 gal./min.) where C lb./gal. is the concentration.
     Since the rates of inflow and outflow are the same, there
     is always 100 gal. in the tank and so this is simply $\frac{x}{100}$.

     Thus we have OUT = $\frac{4x}{100}$ (lb./min.) Hence the D.E. of the

     problem is $\frac{dx}{dt}$ = 12 - $\frac{4x}{100}$. The I.C. is x(0) = 20.

         The D.E. is separable. We write it as $\frac{dx}{dt} = \frac{300 - x}{25}$ or

     $\frac{dx}{300 - x} = \frac{dt}{25}$. Integrating, we find $\ln|300 - x|$ = $-\frac{t}{25}$ +

     c_0 or $|300 - x| = c\,e^{-t/25}$. Applying the I.C., we find

     280 = c; and since 300 - x > 0, we take $|300 - x|$ = 300 -
     x. Thus we obtain the solution in the form 300 - x =
     280 $e^{-t/25}$ or

                    x = 300 - 280 $e^{-t/25}$ (*)

         To answer question (a), let t = 10 in (*). We find x
     ≈ 112.31 (lb.). To answer question (b), let x = 160 in
     (*). We have 280 $e^{-t/25}$ = 140 or $e^{-t/25}$ = 0.5. Then $-\frac{t}{25}$
     = ln 0.5, from which we find t ≈ 17.33 (min.).

22. Let x = the amount of salt at time t. We use the basic
     equation (3.63), $\frac{dx}{dt}$ = IN - OUT. We have IN = 0, since

     pure water flows into the tank; and we find OUT = (C lb./
     gal.)(2gal/min), where c lb./gal. is the concentration.
     Since water flows in at 5 gal/min. and the mixture flows

out at 2 gal/min., there is a gain of 3 gal/min of brine in the tank. Since there was initially 100 gal. there, the amount of brine at time t is 100 + 3t. Thus the concentration is $\frac{x}{100 + 3t}$. Thus OUT = $\frac{2x}{100 + 3t}$ (lb./min.). Hence the D.E. of the problem is $\frac{dx}{dt} = -\frac{2x}{100 + 3t}$. The I.C. is x(0) = 10.

The D.E. is separable. We write it as $\frac{dx}{x} = -\frac{2 \, dt}{3t + 100}$. Integrating, we find $\ln x = -\frac{2}{3} \ln (3t + 100) + \ln c_0$, from which $x = \frac{c}{(3t + 100)^{2/3}}$. We now apply the I.C., obtaining

$10 = \frac{c}{(100)^{2/3}}$ or $c = 10(100)^{2/3}$. Thus we have the

solution $x = \frac{10(100)^{2/3}}{(3t + 100)^{2/3}}$.

(a) We let t = 15 in the solution. We find $x = \frac{10(100)^{2/3}}{(145)^{2/3}}$

≈ 7.81. So there is approximately 7.81 lb. The concentration is $\frac{7.81}{145} \approx 0.0539$ lb./gal.

(b) Since there was initially 100 gallons of brine in the tank and the amount is increasing at the rate of 3 gal./min., the number of minutes t needed to obtain a full tank of 250 gallons is given by 100 + 3t = 250; so t = 50. We let t = 50 in the solution and find

$x = \frac{10(100)^{2/3}}{(250)^{2/3}} \approx 5.4279$. The concentration is then

$\frac{5.4279}{250} \approx 0.0217$ lb./gal.

25. Let x = the amount of salt at time t. We use the basic equation $\frac{dx}{dt}$ = IN - OUT. We find IN = (30 gm/liter)(4 liters/min) = 120 (gm/min); and OUT = (C gm/liter) $\left(\frac{5}{2} \text{ liters/min}\right)$, where C gm/liter is the concentration.

The rate of inflow is 4 liters/min and that of outflow is $\frac{5}{2}$ liters/min, so there is a net gain of $\frac{3}{2}$ liters/min of fluid in the tank. Hence at the end of t minutes the amount of fluid in the tank is $300 + \frac{3t}{2}$ liters. Thus the concentration at time t is $\dfrac{x}{300 + \frac{3t}{2}}$ gm/liter, and so OUT = $\dfrac{5x}{2\left[300 + \frac{3t}{2}\right]} = \dfrac{5x}{3t + 600}$ (gm/min). Hence the D.E. of the problem is $\dfrac{dx}{dt} = 120 - \dfrac{5x}{3t + 600}$. The I.C. is x(0) = 50.

The D.E. is linear. We write it in the standard form

$$\frac{dx}{dt} + \frac{5}{3t + 600} x = 120.$$

An I.F. is $e^{\int [(5/(3t+600)]dt} = e^{(5/3)\ln(3t+600)} = (3t + 600)^{5/3}$. Multiplying through by this we have

$$(3t + 600)^{5/3} \frac{dx}{dt} + 5(3t + 600)^{2/3}x = 120(3t + 600)^{5/3}$$

or

$$\frac{d}{dt} [(3t + 600)^{5/3}x] = 120(3t + 600)^{5/3}.$$

Integration gives $(3t + 600)^{5/3}x = 15(3t + 600)^{8/3} + c$ or

$$x = 15(3t + 600) + c(3t + 600)^{-5/3}.$$

Applying the I.C. x(0) = 50 gives 50 = (15)(600) + $c(600)^{-5/3}$, from which c = $-8950(600)^{5/3}$. Thus the solution is given by

$$x = 15(3t + 600) - (8950)(600)^{5/3}(3t + 600)^{-5/3}. \quad (*)$$

The stated question asks for x at the instant when the tank overflows. Since the amount of fluid increases at the rate of $\frac{3}{2}$ liter/min and the 500 liter tank originally had 300 liters in it, this time t is given by $\frac{3t}{2}$ = 200 or t = $\frac{400}{3}$. We thus let t = $\frac{400}{3}$ in (*) and obtain x $\left(\frac{400}{3}\right)$ =

$1500(1000) - (8950)(600)^{5/3}(1000)^{5/3} = 15000 - 8950\left(\frac{3}{5}\right)^{5/3}$

≈ 11,179.96 (grams).

26. Let x = the amount of salt at time t. We use the basic equation $\frac{dx}{dt}$ = IN - OUT. We find IN = (50 gm/liter) (5 liters/min) = $\frac{250}{gm/min}$; and OUT = (C gm/liter)(7 liter/min), where C gm/liter is the concentration. The rate of inflow is 5 liters/min. and that of outflow is 7 liters/min., so there is a net loss of 2 liters/min of fluid in the tank. Hence at the end of t minutes the amount of fluid in the tank is 200 - 2t liters. Thus the concentration at time t is $\frac{x}{200 - 2t}$ gm/liter, and so OUT = $\frac{7x}{200 - 2t}$ (gm/min). Hence the D.E. is $\frac{dx}{dt}$ = 250 - $\frac{7x}{200 - 2t}$. The I.C. is x(0) = 40.

The D.E. is linear. We write it in the standard form

$$\frac{dx}{dt} + \frac{7}{200 - 2t}x = 250.$$

An I.F. is $e^{\int [7/(200-2t)]dt} = e^{-(7/2)\ln(200-2t)} =$ $(200 - 2t)^{-7/2}$. Multiplying through by this, we have

$$(200 - 2t)^{-7/2}\frac{dx}{dt} + 7(200 - 2t)^{-9/2}x = 250(200 - 2t)^{-7/2}$$

or

$$\frac{d}{dt}[(200 - 2t)^{-7/2}x] = 250(200 - 2t)^{-7/2}.$$

Integration gives $(200 - 2t)^{-7/2}x = 50(200 - 2t)^{-5/2} + c$ or

$$x = 50(200 - 2t) + c(200 - 2t)^{7/2}.$$

Applying the I.C. $x(0) = 40$, we find $40 = 10,000 + c(200)^{7/2}$, from which $c = -9960(200)^{-7/2}$. Thus the solution is given by

$$x = 50(200 - 2t) - 9960\left(1 - \frac{t}{100}\right)^{7/2}. \qquad (*)$$

The tank will be half full when it contains 100 liters. Since it initially contained 200 liters and it loses 2 liters/min., it will contain 100 liters after 50 min. Thus we let $t = 50$ in $(*)$. We find

$$x(50) = 5000 - 9960\left(\frac{1}{2}\right)^{7/2} \approx 4119.65 \text{ (gm)}.$$

30. Let x = the number of people who have the disease after t
     days. Then 10,000 - x people do not have it, and we have
     the D.E. $\frac{dx}{dt}$ = kx(10,000 - x). We also have the I.C. x(0)

     = 1.   The D.E. is separable; and separating variables, we
     have $\frac{dx}{x(10,000 - x)}$ = k dt . Using partial fractions and

     integrating, we find $\frac{1}{10000}$ [ln|x|-ln|10,000 - x|] =

     kt + c_0. Noting that x $\geq$ 0 and 10,000 - x $\geq$ 0, we

     simplify and obtain ln $\frac{x}{10,000 - x}$ = 10,000 kt + c_1.   Then

     we have $\frac{x}{10,000 - x}$ = c e$^{10,000\ kt}$. We apply the I.C. to

     obtain c = 1/9999. Thus we obtain the solution $\frac{x}{10,000 - x}$

     = $\frac{e^{10,000\ kt}}{9999}$.

     We solve this for x in terms of t.   We find

     $\left(1 + \frac{e^{10,000\ kt}}{9999}\right)x = \frac{10,000}{9999} e^{10,000\ kt}$ and hence

$$x = \frac{10,000}{1 + 9999\ e^{-10,000\ kt}}.$$

     Now besides the I.C., we also have the additional
     condition x(5) = 50. We apply this to the preceeding
     solution. We have

$$50 = \frac{10,000}{1 + 9999\ e^{-50,000\ k}}.$$

     Then (50)(9999)e$^{-50,000\ k}$ = 9950, e$^{-50,000\ k}$ = $\frac{199}{9999}$, and

     so e$^{-10,000\ kt}$ = $\left(\frac{199}{9999}\right)^{t/5}$.   Thus the solution takes the

     form

$$x = \frac{10,000}{1 + 9999\left[\frac{199}{9999}\right]^{t/5}}.$$

We are asked how many people have the disease after 10 days. We let $t = 10$ in this to obtain

$$x(10) = \frac{10,000}{1 + 9999\left[\frac{199}{9999}\right]^2} \approx 478.88.$$

Thus about 479 people have the disease at that time.

31. Let x = the number of lbs. of c_3 formed in time t. We have the D.E.

$$\frac{dx}{dt} = k\left(10 - \frac{x}{4}\right)\left(15 - \frac{3x}{4}\right),$$

the I.C. $x(0) = 0$, and the additional condition $x(15) = 5$. Separating variables and simplifying, the D.E. takes the form $\frac{16\,dx}{3(x - 40)(x - 20)} = k\,dt$.

Applying partial fractions, we have $\frac{4}{15}\left[\frac{1}{x - 40} - \frac{1}{x - 20}\right]dx$
$= k\,dt$. Integrating, we obtain $\frac{4}{15}\left[\ln(x - 40)\right.$
$- \ln(x - 20)] = kt + c_0$ or $\ln\frac{x - 40}{x - 20} = \frac{15}{4}kt + c_1$. From this,

$$\frac{x - 40}{x - 20} = c\,e^{\frac{15}{4}kt}. \tag{*}$$

We apply the I.C. $x(0) = 0$ to this, and at once obtain $c = 2$. We set $c = 2$ in (*), and apply the additional condition $x(15) = 5$.

We have $\dfrac{35}{15} = 2\left(e^{\frac{15}{4}k}\right)^{15}$ and hence $e^{\frac{15}{4}k} = \left(\dfrac{7}{6}\right)^{1/15}$.

Thus we obtain the solution

$$\frac{x - 40}{x - 20} = 2\left(\frac{7}{6}\right)^{t/15}.$$

We solve this for x in terms of t, obtaining

$$x = \frac{40\left[\left(\frac{7}{6}\right)^{t/15} - 1\right]}{2\left(\frac{7}{6}\right)^{t/15} - 1}.$$

This is the answer to part (a). To answer (b), let $t = 60$ in this. We have

$$x(60) = \frac{40\left[\left(\frac{7}{6}\right)^{4} - 1\right]}{2\left(\frac{7}{6}\right)^{4} - 1} \approx \frac{40(0.8526)}{2.7052} \approx 12.6068$$

Thus we find $x \approx 12.61$ (lb).

Chapter 4

6. (a) Suppose $c_1 e^x + c_2 e^{3x} = 0$ for all x on a $\leq$ x $\leq$ b.

    Differentiating, we obtain $c_1 e^x + 3c_2 e^{3x} = 0$ for all x

    on a $\leq$ x $\leq$ b. Subtracting the first of these from the

    second, we have $2c_2 e^{3x} = 0$ for all x on a $\leq$ x $\leq$ b.

    Since $2e^{3x} \neq 0$ for all such x, we must have $c_2 = 0$.

    Then $c_1 e^x + c_2 e^{3x} = 0$ reduces to $c_1 e^x = 0$ for all x on

    a $\leq$ x $\leq$ b. Since $e^x \neq 0$ for all such x, we must have

    $c_1 = 0$. Thus $c_1 e^x + c_2 e^{3x} = 0$ for all x on a $\leq$ x $\leq$ b

    implies that $c_1 = c_2 = 0$, so e^x and e^{3x} are linearly

    independent on a $\leq$ x $\leq$ b.

    (b) $W(e^x, e^{3x}) = \begin{vmatrix} e^x & e^{3x} \\ e^x & 3e^{3x} \end{vmatrix} = 2e^{4x} \neq 0$ on a $\leq$ x $\leq$ b, so

    e^x and e^{3x} are linearly independent on this interval.

7. (a) One readily verifies by direct substitution into the

    D.E. that each of these functions is indeed a

    solution. To show that they are linearly independent,

    we apply Theorem 4.4. We have

$$W(e^{2x}, e^{3x}) = \begin{vmatrix} e^{2x} & e^{3x} \\ 2e^{2x} & 3e^{3x} \end{vmatrix} = 3e^{5x} - 2e^{5x} = e^{5x} \neq 0$$

for all x, $-\infty < x < \infty$. Thus the two solutions are
linearly independent on $-\infty < x < \infty$.

(b) $y = c_1 e^{2x} + c_2 e^{3x}$.

(c) To satisfy the condition $y(0) = 2$, we let $x = 0$, $y = 2$
in the general solution of part (b). We have

$$c_1 + c_2 = 2. \qquad\qquad (*)$$

Now we differentiate the general solution, obtaining
$y' = 2c_1 e^{2x} + 3c_2 e^{3x}$. To satisfy the condition $y'(0)$
$= 3$, we let $x = 0$, $y' = 3$ in this derived equation.
We have

$$2c_1 + 3c_2 = 3. \qquad\qquad (**)$$

Solving the two equations $(*)$ and $(**)$ for c_1 and c_2, $-$
we find $c_1 = 3$, $c_2 = -1$. Substituting these values of
c_1 and c_2 into the general solution of part (b), we
have the desired particular solution $y = 3e^{2x} - e^{3x}$.
This is unique by Theorem 4.1; and it is defined on
$-\infty < x < \infty$.

8. (a) One readily verifies by direct substitution into the
D.E. that each of these functions is indeed a
solution. To show that they are linearly independent,
we apply Theorem 4.4. We have

$$W(e^x, xe^x) = \begin{vmatrix} e^x & xe^x \\ e^x & (x+1)e^x \end{vmatrix} = e^{2x} \neq 0$$

for all x, $-\infty < x < \infty$. Thus the two solutions are linearly independent on the interval $-\infty < x < \infty$.

(b) $y = c_1 e^x + c_2 xe^x$.

(c) To satisfy the condition $y(0) = 1$, we let $x = 0$, $y = 1$ in the general solution of part (b). We have $c_1 = 1$.

Now we differentiate the general solution, obtaining $y' = c_1 e^x + c_2(x+1)e^x$. To satisfy the condition $y'(0) = 4$, we let $x = 0$, $y' = 4$ in the derived equation. We have $c_1 + c_2 = 4$. Since $c_1 = 1$, this yields $c_2 = 3$. Substituting these values of c_1 and c_2 into the general solution, we have the desired particular solution $y = e^x + 3xe^x$. This is unique by Theorem 4.1; and it is defined on $-\infty < x < \infty$.

10. (a) One readily verifies by direct substitution into the D.E. that each of these functions is indeed a solution. To show that they are linearly independent, we apply Theorem 4.4. We have

$$W\left(x^2, \frac{1}{x^2}\right) = \begin{vmatrix} x^2 & \dfrac{1}{x^2} \\ 2x & -\dfrac{2}{x^3} \end{vmatrix} = -\frac{2}{x} - \frac{2}{x} = -\frac{4}{x} \neq 0$$

for all x, $0 < x < \infty$. Thus the two solutions are linearly independent on $0 < x < \infty$.

(b) $y = c_1 x^2 + \dfrac{c_2}{x^2}$.

(c) To satisfy the condition $y(2) = 3$, we let $x = 2$, $y = 3$ in the general solution of part (b). We have

$$4c_1 + \frac{c_2}{4} = 3. \qquad\qquad (*)$$

Now we differentiate the general solution, obtaining $y' = c_1 e^x + c_2(x + 1)e^x$. To satisfy the condition $y'(0) = 4$, we let $x = 0$, $y' = 4$ in the derived equation. We have $c_1 + c_2 = 4$. Since $c_1 = 1$, this yields $c_2 = 3$. Substituting these values of c_1 and c_2 into the general solution, we have the desired particular solution $y = e^x + 3xe^x$. This is unique by Theorem 4.1; and it is defined on $-\infty < x < \infty$.

10. (a) One readily verifies by direct substitution into the D.E. that each of these functions is indeed a solution. To show that they are linearly independent, we apply Theorem 4.4. We have

$$\left(W\ x^2,\ \frac{1}{x^2}\right) = \begin{vmatrix} x^2 & \dfrac{1}{x^2} \\[2ex] 2x & -\dfrac{2}{x^3} \end{vmatrix} = -\frac{2}{x} - \frac{2}{x} = -\frac{4}{x} \neq 0$$

for all x, $0 < x < \infty$. Thus the two solutions are linearly independent on $0 < x < \infty$.

(b) $y = c_1 x^2 + \dfrac{c_2}{x^2}$.

(c) To satisfy the condition $y(2) = 3$, we let $x = 2$, $y = 3$ in the general solution of part (b). We have

$$4c_1 + \frac{c_2}{4} = 3 \qquad\qquad (*).$$

Now we differentiate the general solution, obtaining $y' = 2c_1 x - 2c_2 x^{-3}$. To satisfy the condition $y'(2) = -1$, we let $x = 2$, $y' = -1$ in this derived equation. We have

$$4c_1 - \frac{c_2}{4} = -1 \qquad\qquad (**).$$

Solving the two equations (*) and (**) for c_1 and c_2, we find $c_1 = \frac{1}{4}$, $c_2 = 8$. Substituting these values of c_1 and c_2 into the general solution of part (b) we have the desired particular solution $y = \left(\dfrac{1}{4}\right) x^2 + \dfrac{8}{x^2}$.

This is unique by Theorem 4.1; and it is defined on $0 < x < \infty$.

12.   We use Theorem 4.4. We have

$$W(e^{-x}, e^{3x}, e^{4x}) = \begin{vmatrix} e^{-x} & e^{3x} & e^{4x} \\ -e^{-x} & 3e^{3x} & 4e^{4x} \\ e^{-x} & 9e^{3x} & 16e^{4x} \end{vmatrix}$$

$$= e^{-x}e^{3x}e^{4x} = \begin{vmatrix} 1 & 1 & 1 \\ -1 & 3 & 4 \\ 1 & 9 & 16 \end{vmatrix}$$

$$= e^{6x} \left\{ \begin{vmatrix} 3 & 4 \\ 9 & 16 \end{vmatrix} - \begin{vmatrix} -1 & 4 \\ 1 & 16 \end{vmatrix} + \begin{vmatrix} -1 & 3 \\ 1 & 9 \end{vmatrix} \right\}$$

$$= e^{6x}[12 + 20 - 12] = 20e^{6x} \neq 0$$

for all x, $-\infty < x < \infty$. Thus the three solutions are linearly independent on $-\infty < x < \infty$. The general solution is $y = c_1 e^{-x} + c_2 e^{3x} + c_3 e^{4x}$.

13.   We have

$$W(x, x^2, x^4) = \begin{vmatrix} x & x^2 & x^4 \\ 1 & 2x & 4x^3 \\ 0 & 2 & 12x^2 \end{vmatrix}$$

$$= x \begin{vmatrix} 2x & 4x^3 \\ 2 & 12x^2 \end{vmatrix} - \begin{vmatrix} x^2 & x^4 \\ 2 & 12x^2 \end{vmatrix}$$

$$= 16x^4 - 10x^4 = 6x^4 \neq 0$$

on $0 < x < \infty$. Thus by Theorem 4.4, the three given solutions are linearly independent on $0 < x < \infty$. The general solution is $y = c_1 x + c_2 x^2 + c_3 x^4$, where c_1, c_2 and c_3 are arbitrary constants.

Section 4.1.D, Page 132

1. Let $y = vx$. Then $y' = xv' + v$ and $y'' = xv'' + 2v'$. Substituting these into the given D.E., we obtain
$x^2(xv'' + 2v') - 4x(xv' + v) + 4xv = 0$ or $x^3 v'' - 2x^2 v' = 0$
or $xv'' - 2v' = 0$. Letting $w = v'$, we obtain $x \dfrac{dw}{dx} - 2w = 0$
or $\dfrac{dw}{w} = 2 \dfrac{dx}{x}$. Integrating, we find
$\ln|w| = 2 \ln|x| + \ln|c|$ or $w = cx^2$. We choose $c = 1$,
recall $v' = w$, and integrate to obtain $v = \dfrac{x^3}{3}$. Now

forming $y = xv$, we obtain $y = \dfrac{x^4}{3}$. This or any nonzero constant multiple thereof serves as the desired linearly independent solution. Choosing multiple 3, we have $y = x^4$. The general solution is then $y = c_1 x + c_2 x^4$.

4. Let $y = xv$. Then $y' = xv' + v$ and $y'' = xv'' + 2v'$. Substituting these into the given D.E., we obtain
$(x^2 - x + 1)(xv'' + 2v') - (x^2 + x)(xv' + v) + (x + 1)xv = 0$ or $x(x^2 - x + 1)v'' + (-x^3 + x^2 - 2x + 2)v' = 0$ or
$x(x^2 - x + 1)v'' + (x^2 + 2)(1 - x)v' = 0$. Letting $w = v'$,
we obtain $x(x^2 - x + 1) \dfrac{dw}{dx} + (x^2 + 2)(1 - x)w = 0$, or $\dfrac{dw}{w} = \dfrac{(x - 1)(x^2 + 2)}{x(x^2 - x + 1)} dx$. Using long division and partial

fractions, we express this in the form $\dfrac{dw}{w} =$

$\left[1 - \dfrac{2}{x} + \dfrac{2x - 1}{x^2 - x + 1} \right] dx.$ Integration now yields

$$\ln|w| = x - 2\ln|x| + \ln|x^2 - x + 1| + \ln|c|$$

and so

$$\ln|w| = x + \ln \dfrac{|c(x^2 - x + 1)|}{x^2}.$$

We choose $c = 1$, recall $w = v'$, and obtain

$$v' = \left(1 - \dfrac{1}{x} + \dfrac{1}{x^2} \right) e^x.$$

Thus $v = e^x - \int \dfrac{e^x}{x}\, dx + \int \dfrac{1}{x^2} e^x\, dx$.   Integrating

$\int \dfrac{1}{x^2} e^x\, dx$ by parts, we obtain $-\dfrac{e^x}{x} + \int \dfrac{e^x}{x}\, dx$.   Thus

$v = e^x - \dfrac{e^x}{x}$ or $y = e^x\left(1 - \dfrac{1}{x} \right)$.   Now forming $y = xv$, we find
the desired linearly independent solution $y = (x - 1)e^x$.
Thus the general solution is $y = c_1 x + c_2(x - 1)e^x$.

5. Let $y = e^{2x}v$. Then $y' = e^{2x}v' + 2e^{2x}v$ and
    $y'' = e^{2x}v'' + 4e^{2x}v' + 4e^{2x}v$. Substituting this into the
    given D.E., we obtain $(2x + 1)(e^{2x}v'' + 4e^{2x}v' + 4e^{2x}v)$

$$-4(x + 1)(e^{2x}v' + 2e^{2x}v) + 4e^{2x}v = 0$$

or

$$(2x + 1)e^{2x}v'' + 4xe^{2x}v' = 0$$

or

$$(2x + 1)v'' + 4xv' = 0.$$

Letting $w = v'$, we obtain $(2x + 1)\dfrac{dw}{dx} + 4xw = 0$ or

$\dfrac{dw}{w} = -4x\dfrac{dx}{2x + 1}$.   Using long division on the right member,

we rewrite this as $\dfrac{dw}{w} = \left(-2 + \dfrac{2}{2x + 1}\right)dx$.   Integrating, we

have $\ln|w| = -2x + \ln|2x + 1| + \ln|c|$ or

$w = c(2x + 1)e^{-2x}$.   We choose $c = 1$ and recall $w = v'$ to

obtain $\dfrac{dv}{dx} = (2x + 1)e^{-2x}$.   Integrating this, we obtain

$v = -(x + 1)e^{-2x}$.   Now forming $y = e^{2x}v$, we obtain

$y = -(x + 1)$.   This or any nonzero constant multiple

thereof serves as the desired linearly independent

solution.   Choosing multiple -1, we have $y = x + 1$.   The

general solution is then $y = c_1e^{2x} + c_2(x + 1)$.

8.   Let $y = e^x v$.   Then $y' = e^x v' + e^x v$ and

$y'' = e^x v'' + 2e^x v' + e^x v$.   Substituting into the given

D.E., we obtain $(x^2 + x)(e^x v'' + 2e^x v' + e^x v)$

$- (x^2 - 2)(e^x v' + e^x v) - (x + 2)(e^x v) = 0$ or

$(x^2 + x)v'' + (x^2 + 2x + 2)v' = 0$.   Letting $w = v'$, we

obtain $(x^2 + x) + (x^2 + 2x + 2)w = 0$ or

$\dfrac{dw}{w} = -\dfrac{x^2 + 2x + 2}{x^2 + x}\,dx$.   Using long division and partial

fractions, we express this in the form

$\dfrac{dw}{w} = -\left[1 + \dfrac{2}{x} - \dfrac{1}{x + 1}\right]dx$.   Integration now yields

$\ln|w| = -x - 2 \ln|x| + \ln|x + 1| + \ln|c|$ and so

$$\ln|w| = -x + \ln \frac{|c(x + 1)|}{x^2}.$$

We choose $c = 1$, recall $w = v'$, and obtain

$$v' = \left(\frac{1}{x} + \frac{1}{x^2}\right)e^{-x}.$$

Thus

$$v = \int \frac{e^{-x}}{x}\,dx + \int \frac{e^{-x}}{x^2}\,dx.$$

Integrating $\int \frac{1}{x}\,e^{-x}dx$ by parts, we obtain

$-\frac{e^{-x}}{x} - \int \frac{e^{-x}}{x^2}\,dx$. Thus we have $v = -\frac{e^{-x}}{x}$. Now forming

$y = e^x v$, we find the desired linearly independent solution $y = -1/x$. Thus the general solution is $y = c_1 e^x + c_2/x$.

11. We first apply Theorem 4.10 to obtain the solution of the D.E.

$$y'' - 5y' + 6y = 2 - 12x. \qquad\qquad (*)$$

From the first two given D.E.'s, $F_1(x) = 1$, $F_2(x) = x$, with respective given particular integrals $f_1(x) = 1/6$, $f_2(x) = x/6 + 5/36$. The right member of (*) is $2F_1(x) - 12F_2(x)$. Then by the theorem, a particular integral of (*) is $y = 2f_1(x) - 12f_2(x) = 1/3 - 2x - 5/3 = -2x - 4/3$.

We now apply the theorem over again, this time to the third given D.E. and the D.E.(*), with $F_1(x) = 2 - 12x$, $F_2(x) = e^x$, $f_1(x) = -2x - 4/3$, $f_2(x) = e^x/2$, $k_1 = 1$, and $k_2 = 6$. Doing so, we find the particular integral $y = -2x - 4/3 + 3e^x$ of the fourth given D.E.

Section 4.2, Page 143

1. The auxiliary equation is $m^2 - 5m + 6 = 0$. Its roots are 2 and 3 (real and distinct). The G.S. is
   $$y = c_1 e^{2x} + c_2 e^{3x}.$$

4. The auxiliary equation is $3m^2 - 14m - 5 = 0$. Its roots are 5 and $-\frac{1}{3}$ (real and distinct). The G.S is
   $$y = c_1 e^{5x} + c_2 e^{-x/3}.$$

6. The auxiliary equation is $2m^2 + 3m - 2 = 0$. Its roots are $\frac{1}{2}$ and -2 (real and distinct). The G.S. is
   $$y = c_1 e^{x/2} + c_2 e^{-2x}.$$

7. The auxiliary equation is $4m^2 - 4m + 1 = 0$. Its roots are $\frac{1}{2}$ and $\frac{1}{2}$ (a double real root). The G.S. is
   $$y = (c_1 + c_2 x)e^{x/2}.$$

10. The auxiliary equation is $16m^2 + 32m + 25 = 0$. We solve, using the quadratic formula, and obtain

$$m = \frac{-32 \pm \sqrt{(32)^2 - (4)(16)(25)}}{2(32)} = \frac{-32 \pm \sqrt{64(-9)}}{32}$$

$$= \frac{-32 \pm 24i}{32} = -1 \pm \frac{3}{4} i.$$

Thus these conjugate complex members are the roots of the auxiliary equation. The G.S. is
$$y = e^{-x}(c_1 \sin 3x/4 + c_2 \cos 3x/4).$$

11. The auxiliary equation is $m^3 - 3m^2 - m + 3 = 0$, which can be written $m^2(m - 3) - (m - 3) = 0$ or $(m + 1)(m - 1)(m - 3) = 0$. Hence its roots are 1, -1, and 3 (real and distinct). The G.S. is $y = c_1 e^x + c_2 e^{-x} + c_3 e^{3x}$.

14. The auxiliary equation is $4m^3 + 4m^2 - 7m + 2 = 0$. Observe by inspection that -2 is a root. Then by synthetic division,

$$
\begin{array}{r|rrrr}
-2 & 4 & 4 & -7 & 2 \\
   &   & -8 & 8 & -2 \\
\hline
   & 4 & -4 & 1 & \;0 \\
\end{array}
\quad ,
$$

Find the factorization $(m + 2)(4m^2 - 4m + 1) = 0$ and hence $(m + 2)(2m - 1)^2 = 0$. Thus the roots are -2 (real simple root) $\frac{1}{2}, \frac{1}{2}$ (real double root). The G.S. is

$$y = c_1 e^{-2x} + (c_2 + c_3 x) \, e^{(1/2)x}$$

or

$$y = c_1 e^{-2x} + c_2 e^{(1/2)x} + c_3 x e^{(1/2)x}.$$

15. The auxiliary equation is $m^3 - m^2 + m - 1 = 0$, which can be written $m^2(m - 1) + (m - 1) = 0$ or $(m^2 + 1)(m - 1) = 0$. Thus the roots are the real simple root 1 and the conjugate complex pair $\pm i$. The G.S. is
$$y = c_1 e^x + c_2 \sin x + c_3 \cos x.$$

16. The auxiliary equation is $m^3 + 4m^2 + 5m + 6 = 0$. Observe by inspection that -3 is a root. Then by synthetic division,

$$
\begin{array}{r|rrrr}
-3 & 1 & 4 & 5 & 6 \\
   &   & -3 & -3 & -6 \\
\hline
   & 1 & 1 & 2 & 0
\end{array}
$$

find the factorization $(m + 3)(m^2 + m + 2) = 0$. From $m^2 + m + 2 = 0$, obtain

$$m = \frac{-1 \pm \sqrt{1 - 8}}{2} = -\frac{1}{2} \pm \frac{\sqrt{7}}{2} i.$$

Thus the roots are the real simple root -3 and the conjugate complex roots $-\frac{1}{2} \pm \left(\frac{\sqrt{7}}{2}\right)i$. The G.S. is
$$y = c_1 e^{-3x} + e^{-x/2}\left[c_2 \sin\left(\frac{\sqrt{7}}{2}\right)x + c_3 \cos\left(\frac{\sqrt{7}}{2}\right)x\right].$$

17.   The auxiliary equation is $m^2 - 8m + 16 = 0$.   Its roots are 4, 4 (real, double root).   The G.S is

$$y = (c_1 + c_2 x)e^{4x}$$

or

$$y = c_1 e^{4x} + c_2 x e^{4x}.$$

19.   The auxiliary equation is $m^2 - 4m + 13 = 0$.   Solving it, obtain

$$m = \frac{4 \pm \sqrt{16 - 52}}{2} = 2 \pm 3i.$$

So the roots are the conjugate complex numbers $2 \pm 3i$. The G.S. is $y = e^{2x}(c_1 \sin 3x + c_2 \cos 3x)$.

22.   The auxiliary equation is $4m^2 + 1 = 0$.   Its roots are $m = \pm \left(\frac{1}{2}\right)i$.   The G.S. is $y = c_1 \sin\left(\frac{1}{2}\right)x + c_2 \cos\left(\frac{1}{2}\right)x$.

24.   The auxiliary equation is $8m^3 + 12m^2 + 6m + 1 = 0$.   By inspection, we find that $-1/2$ is a root.   Then by synthetic division,

$$
\begin{array}{r|rrrr}
-1/2 & 8 & 12 & 6 & 1 \\
     &   & -4 & -4 & -1 \\
\hline
     & 8 & 8 & 2 & \underline{0} \\
\end{array} \ ,
$$

We find the factorization $(m + 1/2)^2(8m^2 + 8m + 2) = 0$ or $2(m + 1/2)^3 = 0$.   Thus $-1/2$ is a triple real root.   The G.S. is $y = (c_1 + c_2 x + c_3 x^2)e^{-x/2}$.

27. The auxiliary equation is $m^4 + 8m^2 + 16 = 0$ or
    $(m^2 + 4)^2 = 0$. The roots are $\pm 2i$, $\pm 2i$. That is, each of
    the conjugate complex pair $\pm 2i$ is a double root. The G.S.
    is $y = (c_1 + c_2 x)\sin 2x + (c_3 + c_4 x)\cos 2x$.

28. The auxiliary equation is $m^4 - m^3 - 3m^2 + m + 2 = 0$.
    Observe by inspection that -1 is a root. Then by
    synthetic division

$$
\begin{array}{r|rrrrr}
-1 & 1 & -1 & -3 & 1 & 2 \\
   &   & -1 & 2 & 1 & -2 \\
\hline
   & 1 & -2 & -1 & 2 & \,0 \\
\end{array}
$$

we find the factorization $(m + 1)(m^3 - 2m^2 - m + 2) = 0$
which can be written $(m + 1)[m^2(m - 2) - (m - 2)] = 0$ or
$(m + 1)(m^2 - 1)(m - 2) = 0$ or $(m + 1)^2(m - 1)(m - 2) = 0$.
Thus the roots are 1, 2 (real and distinct) and -1, -1
(real double root). The G.S. is

$$y = c_1 e^x + c_2 e^{2x} + (c_3 + c_4 x)e^{-x}$$

or

$$y = c_1 e^x + c_2 e^{2x} + c_3 e^{-x} + c_4 x e^{-x}.$$

30. The auxiliary equation is $m^4 + 6m^3 + 15m^2 + 20m + 12 = 0$.
    Observe by inspection that -2 is a root. Then by
    synthetic division,

$$
\begin{array}{r|rrrrr}
-2 & 1 & 6 & 15 & 20 & 12 \\
   &   & -2 & -8 & -14 & -12 \\
\hline
   & 1 & 4 & 7 & 6   & \,0 \\
\end{array}
$$

find the factorization $(m + 2)(m^3 + 4m^2 + 7m + 6) = 0$.
Now observe by inspection that -2 is also a root of the
reduced cubic $m^3 + 4m^2 + 7m + 6 = 0$. Then by synthetic
division,

$$
\begin{array}{r|rrr|r}
-2 & 1 & 4 & 7 & 6 \\
   &   & -2 & -4 & -6 \\
\hline
   & 1 & 2 & 3 & 0
\end{array}
$$

find the factorization $(m + 2)(m^2 + 2m + 3) = 0$ of the
reduced cubic. Hence the auxiliary equation has the
factored form $(m + 2)^2(m^2 + 2m + 3) = 0$. The factor
$(m + 2)^2$ yields the real double root -2, -2. The factor
$m^2 + 2m + 3 = 0$ gives the conjugate complex roots

$$
m = \frac{-2 \pm \sqrt{4 - 12}}{2} = \frac{-2 \pm 2\sqrt{2}i}{2} = -1 \pm \sqrt{2}i.
$$

Thus the roots are -2, -2, $-1 \pm \sqrt{2}i$. The G.S. is

$$
y = (c_1 + c_2x)e^{-2x} + e^{-x}(c_3\sin\sqrt{2}x + c_4\cos\sqrt{2}x)
$$

or

$$
y = c_1e^{-2x} + c_2xe^{-2x} + e^{-x}(c_3\sin\sqrt{2}x + c_4\cos\sqrt{2}x).
$$

31. The auxiliary equation is $m^5 - 2m^4 + m^3 = 0$. Factoring
gives $m^3(m^2 - 2m + 1) = 0$ or $m^3(m - 1)^2 = 0$. Thus the
roots are 0, 0, 0 [triple root, from factor m^3], 1, 1,
[double root, from factor $(m - 1)^2$]. The G.S. is

$$
y = (c_1 + c_2x + c_3x^2)e^{0x} + (c_4 + c_5x)e^x,
$$

that is,

$$y = c_1 + c_2x + c_3x^2 + (c_4 + c_5x)e^x$$

or

$$y = c_1 + c_2x + c_3x^2 + c_4e^x + c_5xe^x.$$

32. The auxiliary equation is
$m^5 + 5m^4 + 10m^3 + 10m^2 + 5m + 1 = 0$. We recognize this as
a binomial expansion, and write it as $(m + 1)^5 = 0$. Thus
-1 is a five-fold root of the auxiliary equation. The
general solution of the D.E. is
$$y = (c_1 + c_2x + c_3x^2 + c_4x^3 + c_5x^4)e^{-x}.$$

34. The auxiliary equation is $m^6 - 2m^3 + 1 = 0$ or
$(m^3 - 1)^2 = 0$ or $(m - 1)^2(m^2 + m + 1)^2 = 0$. From
$(m - 1)^2 = 0$, we find the double real root 1. Solving the
quadratic $m^2 + m + 1 = 0$, we obtain

$$m = \frac{-1 \pm \sqrt{1 - 4}}{2} = -\frac{1}{2} \pm \frac{\sqrt{3}}{2} i,$$

which is a conjugate complex pair. Thus the roots of the
auxiliary equation are $1, 1, -\frac{1}{2} + \frac{\sqrt{3}}{2} i, -\frac{1}{2} + \frac{\sqrt{3}}{2} i,$

$-\frac{1}{2} - \frac{\sqrt{3}}{2} i, -\frac{1}{2} - \frac{\sqrt{3}}{2} i$. Note that each root is double.
The general solution of the D.E. is $y = (c_1 + c_2x)e^x$

$+ e^{-x/2}\left[(c_3 + c_4x)\sin\frac{\sqrt{3}}{2}x + (c_5 + c_6x)\cos\frac{\sqrt{3}}{2}x\right].$

35. The auxiliary equation is $m^4 + 1 = 0$. Hence we seek m
such that $m^4 = -1$; that is, we seek the four fourth roots
of $z = -1$. We express this in the so-called polar form
$z = 1(\cos \pi + i \sin \pi)$. From complex number theory, we
know that the n nth roots of $z = r(\cos \theta + i \sin \theta)$ are
given by the formula

$$z^{1/n} = \sqrt[n]{r}\left[\cos\left(\frac{\theta + 2k\pi}{n}\right) + i \sin\left(\frac{\theta + 2k\pi}{n}\right)\right],$$

$$k = 0,1,2,\cdots,n - 1. \tag{*}$$

Since $z = 1(\cos \pi + i \sin \pi)$, we have $r = 1$, $\theta = \pi$ here;
and since we want the four fourth roots, we take $n = 4$.
Thus the formula (*) becomes

$$z^{1/4} = \sqrt[4]{1}\left[\cos\left(\frac{\pi + 2k\pi}{4}\right) + i \sin\left(\frac{\pi + 2k\pi}{4}\right)\right], \quad k = 0,1,2,3.$$

Letting $k = 0$, 1, 2, and 3 successively in this, we find

$$\begin{cases} \cos \dfrac{\pi}{4} + i \sin \dfrac{\pi}{4} = \dfrac{\sqrt{2}}{2} + i \dfrac{\sqrt{2}}{2}, \\[2ex] \cos \dfrac{3\pi}{4} + i \sin \dfrac{3\pi}{4} = -\dfrac{\sqrt{2}}{2} + i \dfrac{\sqrt{2}}{2}, \\[2ex] \cos \dfrac{5\pi}{4} + i \sin \dfrac{5\pi}{4} = -\dfrac{\sqrt{2}}{2} - i \dfrac{\sqrt{2}}{2}, \\[2ex] \cos \dfrac{7\pi}{4} + i \sin \dfrac{7\pi}{4} = \dfrac{\sqrt{2}}{2} - i \dfrac{\sqrt{2}}{2}, \end{cases}$$

respectively. These are the four fourth roots of -1.
Thus they are the four roots of the auxiliary equation

$m^4 + 1 = 0$. Observe that they are in fact the two pairs of conjugate complex numbers $\frac{\sqrt{2}}{2} \pm i \frac{\sqrt{2}}{2}$ and $-\frac{\sqrt{2}}{2} \pm i \frac{\sqrt{2}}{2}$.

Alternatively, the auxiliary equation is $m^4 + 1 = 0$, which can be written as $(m^4 + 2m^2 + 1) - 2m^2 = 0$ or $(m^2 + 1)^2 - (\sqrt{2}m)^2 = 0$. Factoring this as a difference of squares gives $(m^2 + 1 - \sqrt{2}m)(m^2 + 1 + \sqrt{2}m) = 0$ or $(m^2 - \sqrt{2}m + 1)(m^2 + \sqrt{2}m + 1) = 0$. The factor $m^2 - \sqrt{2}m + 1$ gives the conjugate complex roots

$$m = \frac{\sqrt{2} \pm \sqrt{2 - 4}}{2} = \frac{\sqrt{2} \pm \sqrt{2}i}{2} = \frac{\sqrt{2}}{2} \pm \frac{\sqrt{2}}{2}i,$$

while the factor $m^2 + \sqrt{2}m + 1$ gives the conjugate complex roots

$$m = \frac{-\sqrt{2} \pm \sqrt{2 - 4}}{2} = \frac{-\sqrt{2} \pm \sqrt{2}i}{2} = -\frac{\sqrt{2}}{2} \pm \frac{\sqrt{2}}{2}i.$$

Thus the roots are $\frac{\sqrt{2}}{2} \pm \frac{\sqrt{2}}{2}i$, $-\frac{\sqrt{2}}{2} \pm \frac{\sqrt{2}}{2}i$.

Thus the G.S. of the D.E. is

$$y = e^{\frac{\sqrt{2}}{2}x}\left(c_1 \sin \frac{\sqrt{2}}{2}x + c_2 \cos \frac{\sqrt{2}}{2}x\right)$$
$$+ e^{-\frac{\sqrt{2}}{2}x}\left(c_3 \sin \frac{\sqrt{2}}{2}x + c_4 \cos \frac{\sqrt{2}}{2}x\right).$$

36. The auxiliary equation is $m^6 + 64 = 0$. So we seek m such
    that $m^6 = -64$; that is, we seek the six sixth roots of
    $z = -64$. In the so-called polar form, this is
    $z = 64(\cos \pi + i \sin \pi)$. We use the formula (*) given in
    the solution of Exercise 35, with $r = 64$, $\theta = \pi$, and
    $n = 6$. The formula becomes

$$z^{1/6} = \sqrt[6]{64}\left[\cos\left(\frac{\pi + 2k\pi}{6}\right) + i \sin\left(\frac{\pi + 2k\pi}{6}\right)\right],$$

$$k = 0,1,2,\cdots,5.$$

Letting $k = 0, 1, 2, 3, 4,$ and 5 successively in this, we
find

$$2\left[\cos \frac{\pi}{6} + i \sin \frac{\pi}{6}\right] = 2\left[\frac{\sqrt{3}}{2} + i\left(\frac{1}{2}\right)\right] = \sqrt{3} + i,$$

$$2\left[\cos \frac{\pi}{2} + i \sin \frac{\pi}{2}\right] = 2[0 + i(1)] = 2i,$$

$$2\left[\cos \frac{5\pi}{6} + i \sin \frac{5\pi}{6}\right] = 2\left[-\frac{\sqrt{3}}{2} + i\left(\frac{1}{2}\right)\right] = -\sqrt{3} + i,$$

$$2\left[\cos \frac{7\pi}{6} + i \sin \frac{7\pi}{6}\right] = 2\left[-\frac{\sqrt{3}}{2} + i\left(-\frac{1}{2}\right)\right] = -\sqrt{3} - i,$$

$$2\left[\cos \frac{3\pi}{2} + i \sin \frac{3\pi}{2}\right] = 2[0 + i(-1)] = -2i,$$

$$2\left[\cos \frac{11\pi}{6} + i \sin \frac{11\pi}{6}\right] = 2\left[\frac{\sqrt{3}}{2} + i\left(-\frac{1}{2}\right)\right] = \sqrt{3} - i,$$

respectively. These are the six sixth roots of -64. Thus
they are the six roots of the auxiliary equation $m^6 + 64 =
0$. Note that they are in fact the three pairs of

conjugate complex numbers $\pm 2i$, $\sqrt{3} \pm i$, and $-\sqrt{3} \pm i$.   Thus the G.S. of the D.E. is

$$y = c_1 \sin 2x + c_2 \cos 2x + e^{\sqrt{3}x}(c_3 \sin x + c_4 \cos x)$$

$$+ e^{-\sqrt{3}x}(c_5 \sin x + c_6 \cos x).$$

37.   The auxiliary equation is $m^2 - m - 12 = 0$.   Its roots are 4, -3 (real and distinct).   The G.S. is

$$y = c_1 e^{4x} + c_2 e^{-3x}. \tag{A}$$

From this,

$$y' = 4c_1 e^{4x} - 3c_2 e^{-3x}. \tag{B}$$

Apply condition $y(0) = 3$ to (A) to obtain $c_1 + c_2 = 3$.

Apply condition $y'(0) = 5$ to (B) to obtain $4c_1 - 3c_2 = 5$.

From these two equations in c_1 and c_2, find $c_1 = 2$,

$c_2 = 1$.   Thus the solution of the stated I.V.P. is

$y = 2e^{4x} + e^{-3x}.$

40.   The auxiliary equation is $3m^2 + 4m - 4 = 0$, which can be written $(3m - 2)(m + 2) = 0$.   Its roots are 2/3, -2 (real and distinct).   The G.S. is

$$y = c_1 e^{\frac{2}{3}x} + c_2 e^{-2x}. \tag{A}$$

From this,

$$y' = \frac{2}{3} c_1 e^{\frac{2}{3}x} - 2c_2 e^{-2x}. \qquad (B)$$

Apply condition $y(0) = 2$ to (A) to obtain $c_1 + c_2 = 2$.

Apply condition $y'(0) = -4$ to (B) to obtain $\frac{2}{3} c_1 - 2c_2 = -4$. From these two equations in c_1 and c_2, find $c_1 = 0$, $c_2 = 2$. Thus the solution of the stated I.V.P. is $y = 2e^{-2x}$.

41. The auxiliary equation is $m^2 + 6m + 9 = 0$. Its roots are -3, -3 (real double root). The G.S. is

$$y = (c_1 + c_2 x)e^{-3x}. \qquad (A).$$

From this,

$$y' = (-3c_1 + c_2 - 3c_2 x)e^{-3x}. \qquad (B)$$

Apply condition $y(0) = 2$ to (A) to obtain $c_1 = 2$. Apply condition $y'(0) = -3$ to (B) to obtain $-3c_1 + c_2 = -3$. From this, find $c_2 = 3$. Thus the solution of the stated I.V.P. is $y = (2 + 3x)e^{-3x}$.

44. The auxiliary equation is $9m^2 - 6m + 1 = 0$, which can be written $(3m - 1)^2 = 0$. Its roots are $1/3$, $1/3$ (real double root). The G.S. is

$$y = (c_1 + c_2 x)e^{\frac{1}{3}x} \qquad\qquad \text{(A)}$$

From this,

$$y' = \left(\frac{1}{3}c_1 + c_2 + \frac{1}{3}c_2 x\right)e^{\frac{1}{3}x} \qquad\qquad \text{(B)}$$

Apply condition $y(0) = 3$ to (A) to obtain $c_1 = 3$. Apply condition $y'(0) = -1$ to (B) to obtain $\frac{1}{3}c_1 + c_2 = -1$. From this, find $c_2 = -2$. Thus the solution of the stated I.V.P is $y = (3 - 2x)e^{\frac{1}{3}x}$.

45. The auxiliary equation is $m^2 - 4m + 29 = 0$. Solving, obtain the conjugate complex roots

$$m = \frac{4 \pm \sqrt{16 - 116}}{2} = \frac{4 \pm 10i}{2} = 2 \pm 5i.$$

The G.S. is

$$y = e^{2x}(c_1 \sin 5x + c_2 \cos 5x) \qquad\qquad \text{(A)}.$$

From this,

$$y' = e^{2x}[(2c_1 - 5c_2)\sin 5x + (5c_1 + 2c_2)\cos 5x]. \qquad \text{(B)}.$$

Apply condition $y(0) = 0$ to (A) to obtain

$$e^0(c_1 \sin 0 + c_2 \cos 0) = 0 \quad\text{or}\quad c_2 = 0.$$

Apply condition $y'(0) = 5$ to (B) to obtain

$$e^0[(2c_1 - 5c_2)\sin 0 + (5c_1 + 2c_2)\cos 0] = 5$$

or

$$5c_1 + 2c_2 = 5.$$

From this, $c_1 = 1$. Thus the solution of the stated I.V.P. is $y = e^{2x}\sin 5x$.

49. The auxiliary equation is $9m^2 + 6m + 5 = 0$. Solving, obtain the conjugate complex roots

$$m = \frac{-6 \pm \sqrt{36 - (36)(5)}}{18} = \frac{-6 \pm 12i}{18} = -\frac{1}{3} \pm \frac{2}{3}i.$$

The G.S. is

$$y = e^{-\frac{1}{3}x}\left(c_1\sin\frac{2}{3}x + c_2\cos\frac{2}{3}x\right). \qquad (A)$$

From this,

$$y' = e^{-\frac{1}{3}x}\left[\left(-\frac{1}{3}c_1 - \frac{2}{3}c_2\right)\sin\frac{2}{3}x\right.$$

$$\left. + \left(\frac{2}{3}c_1 - \frac{1}{3}c_2\right)\cos\frac{2}{3}x\right]. \qquad (B)$$

Apply condition $y(0) = 6$ to (A) to obtain $c_2 = 6$. Apply condition $y'(0) = 0$ to (B) to obtain $\frac{2}{3}c_1 - \frac{1}{3}c_2 = 0$.

From this, $c_1 = 3$. Thus the solution of the stated I.V.P.
is

$$y = e^{-\frac{1}{3}x}\left(3 \sin \frac{2}{3} x + 6 \cos \frac{2}{3} x\right)$$

or

$$y = 3e^{-\frac{1}{3}x}\left(\sin \frac{2}{3} x + 2 \cos \frac{2}{3} x\right).$$

50.   The auxiliary equation is $4m^2 + 4m + 37 = 0$. Solving,
obtain the conjugate complex roots

$$m = \frac{-4 \pm \sqrt{16 - 16(37)}}{8} = \frac{-4 \pm 24i}{8} = -\frac{1}{2} \pm 3i.$$

The G.S. is $y = e^{-x/2}(c_1 \sin 3x + c_2 \cos 3x)$. (A). From

this, $y' = e^{-x/2}\left[\left(-\frac{1}{2} c_1 - 3c_2\right) \sin 3x + \left(3c_1 - \frac{c_2}{2}\right) \cos 3x\right].$

(B).   Apply condition $y(0) = 2$ to (A) to obtain $c_2 = 2$.

Apply condition $y'(0) = -4$ to (B) to obtain $3c_1 - \frac{c_2}{2} = -4$.

From this, $c_1 = -1$. Thus the solution of the stated

I.V.P. is $y = e^{-x/2}(-\sin 3x + 2 \cos 3x)$.

51.   The auxiliary equation is $m^3 - 6m^2 + 11m - 6 = 0$. Observe
by inspection that 1 is a root. Then by synthetic
division,

$$
\begin{array}{r|rrrr}
1 & 1 & -6 & 11 & -6 \\
  &   & 1 & -5 & 6 \\
\hline
  & 1 & -5 & 6 & 0 \\
\end{array}
$$

find the factorization $(m - 1)(m^2 - 5m + 6) = 0$, which can
be written $(m - 1)(m - 2)(m - 3) = 0$. Thus the roots are
1, 2, 3 (real and distinct). The G.S. is

$$y = c_1 e^x + c_2 e^{2x} + c_3 e^{3x} \qquad \text{(A)}$$

From this,

$$y' = c_1 e^x + 2c_2 e^{2x} + 3c_3 e^{3x} \qquad \text{(B)}$$

$$y'' = c_1 e^x + 4c_2 e^{2x} + 9c_3 e^{3x} \qquad \text{(C)}$$

Apply condition $y(0) = 0$ to (A) to obtain $c_1 + c_2 + c_3 = 0$. Apply condition $y'(0) = 0$ to (B) to obtain
$c_1 + 2c_2 + 3c_3 = 0$. Apply condition $y''(0) = 2$ to (C) to
obtain $c_1 + 4c_2 + 9c_3 = 2$. The solution of these three
equations in c_1, c_2, c_3 is $c_1 = 1$, $c_2 = -2$, $c_3 = 1$. Thus
the solution of the stated I.V.P. is
$y = e^x - 2e^{2x} + e^{3x}$.

53.   The auxiliary equation is $m^3 - 3m^2 + 4 = 0$. Observe by
inspection that -1 is a root. Then by synthetic division,

$$
\begin{array}{r|rrrr}
-1 & 1 & -3 & 0 & 4 \\
   &   & -1 & 4 & -4 \\
\hline
   & 1 & -4 & 4 & 0 \\
\end{array}
$$

find the factorization $(m + 1)(m^2 - 4m + 4) = 0$, which can be written $(m + 1)(m - 2)^2 = 0$. Thus the roots are -1 (real simple root) and $2, 2$ (real double root). The G.S. is

$$y = c_1 e^{-x} + (c_2 + c_3 x) e^{2x} \qquad (A)$$

From this,

$$y' = -c_1 e^{-x} + (2c_2 + c_3 + 2c_3 x) e^{2x} \qquad (B)$$
$$y'' = c_1 e^{-x} + (4c_2 + 4c_3 + 4c_3 x) e^{2x} \qquad (C)$$

Apply condition $y(0) = 1$ to (A) to obtain $c_1 + c_2 = 1$. Apply condition $y'(0) = -8$ to (B) to obtain $-c_1 + 2c_2 + c_3 = -8$. Apply condition $y''(0) = -4$ to (C) to obtain $c_1 + 4c_2 + 4c_3 = -4$. The solution of these three equations in c_1, c_2, c_3 is $c_1 = 32/9$, $c_2 = -23/9$, $c_3 = 2/3$. Thus the solution of the stated I.V.P. is

$$y = \frac{32}{9} e^{-x} + \left(-\frac{23}{9} + \frac{2}{3} x\right) e^{2x}$$

or

$$y = \frac{32}{9} e^{-x} - \frac{23}{9} e^{2x} + \frac{2}{3} x e^{2x}.$$

54. The auxiliary equation is $m^3 - 5m^2 + 9m - 5 = 0$. Observe by inspection that 1 is a root. Then by synthetic division,

$$
\begin{array}{r|rrrr}
1 & 1 & -5 & 9 & -5 \\
  &   & 1 & -4 & 5 \\
\hline
  & 1 & -4 & 5 & 0
\end{array}
$$

Find the factorization $(m - 1)(m^2 - 4m + 5) = 0$. From $m^2 - 4m + 5 = 0$, obtain

$$m = \frac{4 \pm \sqrt{16 - 20}}{2} = 2 \pm i.$$

Thus the roots are the real simple root 1 and the conjugate complex roots $2 \pm i$. The G.S. is

$$y = c_1 e^x + e^{2x}(c_2 \sin x + c_3 \cos x). \qquad (A)$$

From this,

$$y' = c_1 e^x + e^{2x}[(2c_2 - c_3)\sin x$$
$$+ (c_2 + 2c_3)\cos x], \qquad (B)$$

$$y'' = c_1 e^x + e^{2x}[(3c_2 - 4c_3)\sin x$$
$$+ (4c_2 + 3c_3)\cos x] \qquad (C)$$

Apply condition $y(0) = 0$ to (A) to obtain $c_1 + c_3 = 0$.
Apply condition $y'(0) = 1$ to (B) to obtain
$c_1 + c_2 + 2c_3 = 1$. Apply condition $y''(0) = 6$ to (C) to
obtain $c_1 + 4c_2 + 3c_3 = 6$. The solution of these three
equations in c_1, c_2, c_3 is $c_1 = 1$, $c_2 = 2$, $c_3 = -1$. Thus
the solution of the stated I.V.P. is $y = e^x + e^{2x}(2 \sin x - \cos x)$.

55.   The auxiliary equation is $m^4 - 3m^3 + 2m^2 = 0$ or $m^2(m - 1)(m - 2)$, and so its roots are $m = 0, 0, 1, 2$. Thus the general solution of the D.E. is

$$y = c_1 + c_2x + c_3e^x + c_4e^{2x}. \qquad (1)$$

Its first three derivatives are

$$y' = c_2 + c_3e^x + 2c_4e^{2x}, \qquad (2)$$

$$y'' = c_3e^x + 4c_4e^{2x}, \qquad (3)$$

and $$y''' = c_3e^x + 8c_4e^{2x}, \qquad (4)$$

respectively. Apply I.C. $y(0) = 2$ to (1), I.C. $y'(0) = 0$ to (2), I.C. $y''(0) = 2$ to (3), and I.C. $y'''(0) = 2$ to (4). We obtain respectively

$$c_1 + c_3 + c_4 = 2, \qquad c_2 + c_3 + 2c_4 = 0,$$

$$c_3 + 4c_4 = 2, \qquad \text{and} \quad c_3 + 8c_4 = 2.$$

From the last two of these four, $c_3 = 2$, $c_4 = 0$. Then the second of the four gives $c_2 = -2 + 0 = -2$. Finally, the first equation gives $c_1 = 2 - 2 + 0 = 0$. Substituting these values into (1), we have the solution $y = -2x + 2e^x$.

59.   Since 4 is the 4-fold real root of the auxiliary equation, the part of the general solution corresponding to it is

$$(c_1 + c_2x + c_3x^2 + c_4x^3)e^{4x}.$$

Since the conjugate complex numbers 2 + 3i and 2 - 3i are each 3-fold roots of the auxiliary equation, the corresponding part of the general solution is

$$e^{2x}[(c_5 + c_6x + c_7x^2)\sin 3x + (c_8 + c_9x + c_{10}x^2)\cos 3x].$$

Thus the general solution is

$$y = (c_1 + c_2x + c_3x^2 + c_4x^3)e^{4x}$$
$$+ e^{2x}[(c_5 + c_6x + c_7x^2)\sin 3x$$
$$+ (c_8 + c_9x + c_{10}x^2)\cos 3x].$$

61.   Since $\sin x$ is a solution of the D.E., $m = \pm i$ must be roots of the auxiliary equation $m^4 + 2m^3 + 6m^2 + 2m + 5 = 0$, and hence $(m - i)(m + i) = m^2 + 1$ must be a factor of $m^4 + 2m^3 + 6m^2 + 2m + 5$. By long division, we find the other factor is $m^2 + 2m + 5$. Hence in factored form the auxiliary equation is

$$(m^2 + 1)(m^2 + 2m + 5) = 0.$$

The factor $m^2 + 2m + 5 = 0$ gives

$$m = \frac{-2 \pm \sqrt{4 - 20}}{2} = -1 \pm 2i.$$

Thus the roots of the auxiliary equation are the two pairs of conjugate complex numbers $\pm i$ and $-1 \pm 2i$. The G.S. of the D.E. is

$$y = c_1\sin x + c_2\cos x + e^{-x}(c_3\sin 2x + c_4\cos 2x).$$

62. Since $e^x \sin 2x$ is a solution of the D.E., $m = 1 \pm 2i$ must
be roots of the auxiliary equation
$m^4 + 3m^3 + m^2 + 13m + 30 = 0$, and hence $[m - (1 + 2i)]$
$[m - (1 - 2i)] = m^2 - 2m + 5$ must be a factor of
$m^4 + 3m^3 + m^2 + 13m + 30$. By long division, we find the
other factor is $m^2 + 5m + 6$. Since $m^2 + 5m + 6 =$
$(m + 2)(m + 3)$, the auxiliary equation in factored form is
$(m + 2)(m + 3)(m^2 - 2m + 5) = 0$; and its roots are $m = -2$,
-3, $1 \pm 2i$. Thus the general solution is
$y = c_1 e^{-2x} + c_2 e^{-3x} + e^x(c_3 \sin 2x + c_4 \cos 2x)$.

Section 4.3, Page 159

Exercises 1 through 24 follow the pattern of Examples
4.36, 4.37 and 4.38 on Pages 154-159 of the text. The
five steps of the method outlined on pg. 152 are indicated
in each solution.

1. The corresponding homogeneous D.E. is $y'' - 3y' + 2y = 0$.
The auxiliary equation is $m^2 - 3m + 2 = 0$, with roots 1,
2. The complementary function is $y_c = c_1 e^x + c_2 e^{2x}$. The
NH term is a constant multiple of the UC function given by
x^2.

Step 1: Form the UC set of x^2. It is $S_1 = \{x^2, x, 1\}$.

Step 2: This step does not apply, since there is only one
UC set present.

Step 3: An examination of the complementary function shows
that none of the functions in S_1 is a solution of

the corresponding homogeneous D.E.   Hence S_1 does not need revision.

Step 4: Thus the original set S_1 remains.   Form a linear combination of its three members.

Step 5: Thus we take

$$y_p = Ax^2 + Bx + C$$

as a particular solution.   Then

$$y_p' = 2Ax + B, \quad y_p'' = 2A.$$

We substitute in the D.E., obtaining

$$2A - 3(2Ax + B) + 2(Ax^2 + Bx + C) = 4x^2 \quad \text{or}$$

$$2Ax^2 + (-6A + 2B)x + (2A - 3B + 2C) = 4x^2.$$

We equate coefficients of like terms on both sides of this to obtain

$$2A = 4, \; -6A + 2B = 0, \; 2A - 3B + 2C = 0.$$

From these, we find $A = 2$, $B = 6$, $C = 7$.   Thus we obtain the particular integral

$$y_p = 2x^2 + 6x + 7.$$

The G.S. of the D.E. is

$$y = c_1 e^x + c_2 e^{2x} + 2x^2 + 6x + 7.$$

4. The corresponding homogeneous D.E. is $y'' + 2y' + 2y = 0$. The auxiliary equation is $m^2 + 2m + 2 = 0$, with roots $-1 \pm i$. The complementary function is $y_c = e^{-x}(c_1 \sin x + c_2 \cos x)$. The NH term is a constant multiple of the UC function given by $\sin 4x$.

Step 1: Form the UC set of $\sin 4x$. It is $S_1 = \{\sin 4x, \cos 4x\}$.

Step 2: This step does not apply.

Step 3: An examination of the complementary function shows that none of the functions in S_1 is a solution of the corresponding homogeneous D.E. Hence S_1 does not need revision.

Step 4: Thus the original set S_1 remains. Form a linear combination of its two members.

Step 5: Thus we take $y_p = A \sin 4x + B \cos 4x$ as a particular solution. Then

$$y_p' = 4A \cos 4x - 4B \sin 4x.$$

$$y_p'' = -16A \sin 4x - 16B \cos 4x.$$

We substitute in the D.E., obtaining

$-16A \sin 4x - 16B \cos 4x + 8A \cos 4x - 8B \sin 4x$

$\qquad + 2A \sin 4x + 2B \cos 4x = 10 \sin 4x$ \qquad\qquad or

$(-14A - 8B) \sin 4x + (8A - 14B) \cos 4x = 10 \sin 4x.$

We equate coefficients of like terms on both sides of this to obtain

$$-14A - 8B = 10, \quad 8A - 14B = 0.$$

From these, we find $A = -\dfrac{7}{13}$, $B = -\dfrac{4}{13}$. Thus we obtain the particular integral

$$y_p = -\left(\frac{7}{13}\right) \sin 4x - \left(\frac{4}{13}\right) \cos 4x.$$

The G.S. of the D.E. is

$$y = e^{-x}(c_1 \sin x + c_2 \cos x) - \left(\frac{7}{13}\right) \sin 4x - \left(\frac{4}{13}\right) \cos 4x.$$

8.   The corresponding homogeneous D.E. is $y'' + 2y' + 10y = 0$. The auxiliary equation is $m^2 + 2m + 10 = 0$, with roots $-1 \pm 3i$. The complementary function is $y_c = e^{-x}(c_1 \sin 3x + c_2 \cos 3x)$. The NH term is a constant multiple of the UC function given by xe^{-2x}

Step 1: Form the UC set of xe^{-2x}. It is
$\qquad S_1 = \{xe^{-2x}, e^{-2x}\}.$

Step 2: This step does not apply.

Step 3: An examination of the complementary function shows that none of the functions in S_1 is a solution of the corresponding homogeneous D.E.   Hence S_1 does not need revision.

Step 4: Thus the original set S_1 remains.   Form a linear combination of its two members.

Step 5: Thus we take $y_p = Axe^{-2x} + Be^{-2x}$ as a particular solution.   Then

$$y_p' = -2Axe^{-2x} + (A - 2B)e^{-2x},$$

$$y_p'' = 4Axe^{-2x} + (-4A + 4B)e^{-2x}.$$

We substitute into the D.E., obtaining
$$4Axe^{-2x} + (-4A + 4B)e^{-2x} - 4Axe^{-2x}$$
$$+ (2A - 4B)e^{-2x} + 10Axe^{-2x} + 10Be^{-2x} = 5xe^{-2x}$$

or

$$10Axe^{-2x} + (-2A + 10B)e^{-2x} = 5xe^{-2x}.$$

We equate coefficients of like terms on both sides of this to obtain $10A = 5$, $-2A + 10B = 0$.   From these, we find $A = \frac{1}{2}$, $B = \frac{1}{10}$.   Thus we obtain the particular integral

$$y_p = \frac{xe^{-2x}}{2} + \frac{e^{-2x}}{10}.$$

The G.S. of the D.E. is

$$y = e^{-x}(c_1\sin 3x + c_2\cos 3x) + \frac{xe^{-2x}}{2} + \frac{e^{-2x}}{10}.$$

11. The corresponding homogeneous D.E. is $y'' + 4y = 0$. The auxiliary equation is $m^2 + 4 = 0$, with conjugate complex roots $m = \pm 2i$. The complementary function is $y_c = c_1 \sin 2x + c_2 \cos 2x$. The NH member is a linear combination of the UC functions $\sin 2x$ and $\cos 2x$.

Step 1: Form the UC set of each of these two UC functions: $S_1 = \{\sin 2x, \cos 2x\}$, $S_2 = \{\cos 2x, \sin 2x\}$.

Step 2: Each set is identical with the other, so we only keep one of them, say S_1.

Step 3: Observe that each member of S_1 is included in the complementary function and so is a solution of the corresponding homogeneous equation. Thus we multiply each member of S_1 by x to obtain the revised set $S_1' = \{x \sin 2x, x \cos 2x\}$, whose members are not solutions of the homogeneous D.E.

Step 4: We now have only the revised set $S_1' = \{x \sin 2x, x \cos 2x\}$. We form a linear combination $Ax \sin 2x + Bx \cos 2x$ of its two elements.

Step 5: Thus we take $y_p = Ax \sin 2x + Bx \cos 2x$ as a particular integral. Then

$$y_p' = 2Ax \cos 2x + A \sin 2x - 2Bx \sin 2x + B \cos 2x,$$

$$y_p'' = -4Ax \sin 2x + 4A \cos 2x - 4Bx \cos 2x - 4B \sin 2x.$$

We substitute into the D.E., obtaining

$$[-4Ax \sin 2x + 4A \cos 2x - 4Bx \cos 2x - 4B \sin 2x]$$
$$+ 4[Ax \sin 2x + Bx \cos 2x] = 4 \sin 2x + 8 \cos 2x$$

or $4A \cos 2x - 4B \sin 2x = 4 \sin 2x + 8 \cos 2x.$

We equate coefficients of like terms on both sides of this to obtain $4A = 8$, $-4B = 4$. From this, $A = 2$, $B = -1$. Thus we obtain the particular integral $y_p = 2x \sin 2x - x \cos 2x$. The G.S. is

$$y = c_1 \sin 2x + c_2 \cos 2x + 2x \sin 2x - x \cos 2x.$$

12. The corresponding homogeneous D.E. is $y'' - 4y = 0$. The auxiliary equation is $m^2 - 4 = 0$ or $(m + 2)(m - 2) = 0$, with roots $m = 2, -2$. The complementary function is $y_c = c_1 e^{2x} + c_2 e^{-2x}$. The NH member is a constant multiple of the UC functions xe^{2x}.

Step 1: Form the UC set of xe^{2x}: $S_1 = \{xe^{2x}, e^{2x}\}$.

Step 2: This step does not apply.

Step 3: Observe that the member of e^{2x} of S_1 is included in the complementary function and so is a solution of the corresponding homogeneous equation. Thus we multiply each member of S_1 by x to obtain the revised set $S_1' = \{x^2 e^{2x}, xe^{2x}\}$, whose members are not solutions of the homogeneous D.E.

Step 4: We now have only the revised set
$S_1' = \{x^2e^{2x}, xe^{2x}\}$. We form a linear combination
$Ax^2e^{2x} + Bxe^{2x}$ of its two elements.

Step 5: Thus we take $y_p = Ax^2e^{2x} + Bxe^{2x}$ as a particular
integral. Then

$$y_p' = 2Ax^2e^{2x} + (2A + 2B)xe^{2x} + Be^{2x},$$

$$y_p'' = 4Ax^2e^{2x} + (8A + 4B)xe^{2x} + (2A + 4B)e^{2x}.$$

We substitute into the D.E., obtaining

$$[4Ax^2e^{2x} + (8A + 4B)xe^{2x} + (2A + 4B)e^{2x}]$$

$$- 4[Ax^2e^{2x} + Bxe^{2x}] = 16xe^{2x} \qquad \text{or}$$

$$8Axe^{2x} + (2A + 4B)e^{2x} = 16xe^{2x}.$$

We equate coefficients of like terms on both sides of this
to obtain $8A = 16$, $2A + 4B = 0$. From this, we find $A = 2$,
$B = -1$. Thus we obtain the particular integral $y_p =$
$2x^2e^{2x} - xe^{2x}$. The G.S. is $y = c_1e^{2x} + c_2e^{-2x} + 2x^2e^{2x} -$
xe^{2x}.

15. The corresponding homogeneous D.E. is $y''' + 4y'' + y' - 6y = 0$. The auxiliary equation is $m^3 + 4m^2 + m - 6 = 0$. By
inspection note that $m = 1$ is a root. From this (by
synthetic division or otherwise) we obtain the factored
form $(m - 1)(m + 2)(m + 3)$. Thus the roots of the
auxiliary equation are $m = 1, -2, -3$. The complementary
function is $y_c = c_1e^x + c_2e^{-2x} + c_3e^{-3x}$. The NH term is a
linear combination of the UC functions x^2 and 1.

Step 1: Form the UC set of each of these two UC functions: $S_1 = \{x^2, x, 1\}$, $S_2 = \{1\}$.

Step 2: Set S_2 is completely included in S_1, so S_2 is omitted, leaving just S_1.

Step 3: An examination of the complementary function shows that none of the functions in S_1 is a solution of the corresponding homogeneous D.E. Hence S_1 does not need revision.

Step 4: Thus the original set S_1 remains. We form a linear combination of its three members.

Step 5: Thus we take $y_p = Ax^2 + Bx + C$. Then $y_p{}' = 2Ax + B$, $y_p{}'' = 2A$, $y_p{}''' = 0$. We substitute into the D.E., obtaining

$$0 + 4(2A) + (2Ax + B) - 6(Ax^2 + Bx + C) = -18x^2 + 1$$

or

$$-6Ax^2 + (2A - 6B)x + (8A + B - 6C) = -18x^2 + 1.$$

We equate coefficients of like terms on both sides of this to obtain $-6A = -18$, $2A - 6B = 0$, $8A + B - 6C = 1$. From these, we find $A = 3$, $B = 1$, $C = 4$. Thus we obtain the particular solution

$$y_p = 3x^2 + x + 4.$$

The G.S. of the D.E. is

$$y = c_1e^x + c_2e^{-2x} + c_3e^{-3x} + 3x^2 + x + 4.$$

17. The corresponding homogeneous D.E. is $y''' + y'' + 3y' - 5y = 0$. The auxiliary equation is $m^3 + m^2 + 3m - 5 = 0$. By inspection note that $m = 1$ is a root. From this (by synthetic division or otherwise) we obtain the factored form $(m - 1)(m^2 + 2m + 5) = 0$. From this, the roots of the auxiliary equation are 1 and $-1 \pm 2i$. The complementary function is $y_c = c_1e^x + e^{-x}(c_2\sin 2x + c_3\cos 2x)$. The NH term is a linear combination of the UC functions given by $\sin 2x$, x^2, x, and 1.

Step 1: Form the UC set of each of these four UC functions: $S_1 = \{\sin 2x, \cos 2x\}$, $S_2 = \{x^2, x, 1\}$,

   $S_3 = \{x, 1\}$, $S_4 = \{1\}$.

Step 2: Sets S_3 and S_4 are completely included in S_2; so S_3 and S_4 are omitted, leaving the two sets S_1 and S_2.

Step 3: An examination of the complementary function shows that none of the functions in S_1 or S_2 is a solution of the corresponding homogeneous D.E. Hence neither S_1 nor S_2 needs revision.

Step 4: Thus the original sets S_1 and S_2 remain. We form a linear combination of their five members.

Step 5: Thus we take

$$y_p = A \sin 2x + B \cos 2x + Cx^2 + Dx + E$$

as a particular solution. Then

$$y_p{}' = 2A \cos 2x - 2B \sin 2x + 2Cx + D,$$

$$y_p{}'' = -4A \sin 2x - 4B \cos 2x + 2C,$$

$$y_p{}''' = -8A \cos 2x + 8B \sin 2x.$$

We substitute in the D.E., obtaining

$$(-8A \cos 2x + 8B \sin 2x) + (-4A \sin 2x - 4B \cos 2x + C)$$

$$+ 3(2A \cos 2x - 2B \sin 2x + 2Cx + D)$$

$$- 5(A \sin 2x + B \cos 2x + Cx^2 + Dx + E)$$

$$= 5 \sin 2x + 10x^2 + 3x + 7 \qquad \text{or}$$

$$(-9A + 2B) \sin 2x + (-2A - 9B) \cos 2x$$

$$- 5Cx^2 + (6C - 5D)x + (2C + 3D - 5E)$$

$$= 5 \sin 2x + 10x^2 + 3x + 7.$$

We equate coefficients of like terms on both sides of this to obtain $-9A + 2B = 5$, $-2A - 9B = 0$, $-5C = 10$, $6C - 5D = 3$, $2C + 3D - 5E = 7$. From these, we find $A = -\frac{9}{17}$, $B = \frac{2}{17}$, $C = -2$, $D = -3$, $E = -4$. Thus we obtain the particular integral

$$y_p = -\frac{9 \sin 2x}{17} + \frac{2 \cos 2x}{17} - 2x^2 - \frac{9x}{5} - \frac{82}{25}.$$

The G.S. of the D.E. is

$$y = c_1 e^x + e^{-x}(c_2 \sin 2x + c_3 \cos 2x)$$

$$- \frac{9 \sin 2x}{17} + \frac{2 \cos 2x}{17} - 2x^2 - 3x - 4.$$

19. The corresponding homogeneous D.E. is $y'' + y' - 6y = 0$. The auxiliary equation is $m^2 + m - 6 = 0$, with roots 2, -3. The complementary function is $y_c = c_1 e^{2x} + c_2 e^{-3x}$.

The NH term is a linear combination of the UC functions e^{2x}, e^{3x}, x and 1.

Step 1: Form the UC set of each of these four functions:
$$S_1 = \{e^{2x}\}, \ S_2 = \{e^{3x}\}, \ S_3 = \{x,1\}, \ S_4 = \{1\}.$$

Step 2: Set S_4 is completely included in S_3; so S_4 is omitted, leaving the three sets S_1, S_2, and S_3.

Step 3: Observe that the only member e^{2x} of $S_1 = \{e^{2x}\}$ is included in the complementary function and so is a solution of the corresponding homogeneous D.E. Thus we multiply the member e^{2x} of S_1 by x to obtain the revised set
$$S_1{}' = \{xe^{2x}\},$$

whose only member is not a solution of the homogeneous D.E.

Step 4: We now have the three sets $S_1{}' = \{xe^{2x}\}$, $S_2 = \{e^{3x}\}$, $S_3 = \{x,1\}$. We form a linear combination of their four elements.

Step 5: Thus we take

$$y_p = Axe^{2x} + Be^{3x} + Cx + D$$

as a particular solution. Then

$$y_p' = 2Axe^{2x} + Ae^{2x} + 3Be^{3x} + C,$$

$$y_p'' = 4Axe^{2x} + 4Ae^{2x} + 9Be^{3x},$$

We substitute in the D.E., obtaining

$$(4Axe^{2x} + 4Ae^{2x} + 9Be^{3x})$$
$$+ (2Axe^{2x} + Ae^{2x} + 3Be^{3x} + C)$$
$$- 6(Axe^{2x} + Be^{3x} + Cx + D)$$
$$= 10e^{2x} - 18e^{3x} - 6x - 11 \qquad \text{or}$$
$$5Ae^{2x} + 6Be^{3x} - 6Cx + (C - 6D)$$
$$= 10e^{2x} - 18e^{3x} - 6x - 11.$$

We equate coefficients of like terms on both sides of this to obtain $5A = 10$, $6B = -18$, $-6C = -6$, $C - 6D = -11$. From this, we find $A = 2$, $B = -3$, $C = 1$, $D = 2$. We thus obtain the particular integral

$$y_p = 2xe^{2x} - 3e^{3x} + x + 2.$$

The G.S. of the D.E. is

$$y = c_1 e^{2x} + c_2 e^{-3x} + 2xe^{2x} - 3e^{3x} + x + 2.$$

22.   The corresponding homogeneous D.E. is $y'' - 4y' + 5y = 0$.
The auxiliary equation is $m^2 - 4m + 5 = 0$, with roots
$2 \pm i$.   The complementary function is
$y_c = e^{2x}(c_1 \sin x + c_2 \cos x)$.   The NH term is a constant
multiple of the UC function given by $e^{2x} \cos x$.

Step 1: Form the UC set of $e^{2x} \cos x$.   It is
$$S_1 = \{e^{2x} \cos x, \; e^{2x} \sin x\}.$$

Step 2: This step does not apply.

Step 3: Both members $e^{2x} \cos x$ and $e^{2x} \sin x$ of S_1 are
included in the complementary function and so are
solutions of the corresponding homogeneous D.E.   Thus we
multiply each member of set S_1 by x to obtain the revised
set $S_1' = \{xe^{2x} \cos x, \; xe^{2x} \sin x\}$, which has no members
which are solutions of the homogeneous D.E.

Step 4: We now have the set $S_1' = \{xe^{2x} \cos x, \; xe^{2x} \sin x\}$,
and we form a linear combination of its two members.

Step 5   Thus we take

$$y_p = Axe^{2x} \sin x + Bxe^{2x} \cos x$$

as a particular solution.   Then

$$y_p' = (2A - B)xe^{2x} \sin x + (A + 2B)xe^{2x} \cos x$$
$$+ Ae^{2x} \sin x + Be^{2x} \cos x,$$
$$y_p'' = (3A - 4B)xe^{2x} \sin x + (4A + 3B)xe^{2x} \cos x$$
$$+ (4A - 2B)e^{2x} \sin x + (2A + 4B)e^{2x} \cos x.$$

We substitute in the D.E., obtaining

$$[(3A - 4B)xe^{2x}\sin x + (4A + 3B)xe^{2x}\cos x$$
$$+ (4A - 2B)e^{2x}\sin x + (2A + 4B)e^{2x}\cos x]$$
$$- 4[(2A - B)xe^{2x}\sin x + (A + 2B)xe^{2x}\cos x$$
$$+ Ae^{2x}\sin x + Be^{2x}\cos x]$$
$$+ 5[Axe^{2x}\sin x + Bxe^{2x}\cos x] = 6e^{2x}\cos x \quad \text{or}$$
$$(3A - 4B - 8A + 4B + 5A)xe^{2x}\sin x$$
$$+ (4A + 3B - 4A - 8B + 5B)xe^{2x}\cos x$$
$$+ (4A - 2B - 4A)e^{2x}\sin x + (2A + 4B - 4B)e^{2x}\cos x$$
$$= 6e^{2x}\cos x \qquad \text{or simply}$$

$$-2Be^{2x}\sin x + 2Ae^{2x}\cos x = 6e^{2x}\cos x.$$

We equate coefficients of like terms on both sides of this to obtain

$$-2B = 0, \quad 2A = 6.$$

From this, we find $A = 3$, $B = 0$. We thus obtain the particular solution

$$y_p = 3xe^{2x}\sin x.$$

The G.S. of the D.E. is

$$y = e^{2x}(c_1\sin x + c_2\cos x) + 3xe^{2x}\sin x.$$

24. The corresponding homogeneous D.E. is

$y''' - 2y'' - y' + 2y = 0$. The auxiliary equation is
$m^3 - 2m^2 - m + 2 = 0$, that is $(m^2 - 1)(m - 2) = 0$. From
this, the roots are 1, -1, 2. The complementary function
is $y_c = c_1 e^x + c_2 e^{-x} + c_3 e^{2x}$. The NH term is a linear
combination of the UC functions given by e^{2x} and e^{3x}.

Step 1: Form the UC set of each of these two functions:
$$S_1 = \{e^{2x}\}, \ S_2 = \{e^{3x}\}.$$

Step 2: Neither set is identical with nor included in the
other, so both are retained.

Step 3: Observe that the only member e^{2x} of $S_1 = \{e^{2x}\}$ is
included in the complementary function, and so is a
solution of the corresponding homogeneous D.E. Thus we
multiply the member e^{2x} of S_1 by x to obtain the revised
set $S_1' = \{xe^{2x}\}$ whose only member is not a solution of
the homogeneous D.E.

Step 4: We now have the two sets $S_1' = \{xe^{2x}\}$, $S_2 = \{e^{3x}\}$.
We form a linear combination of their two elements.

Step 5: Thus we take $y_p = Axe^{2x} + Be^{3x}$ as a particular
solution. Then $y_p' = 2Axe^{2x} + Ae^{2x} + 3Be^{3x}$.
$$y_p'' = 4Axe^{2x} + 4Ae^{2x} + 9Be^{3x},$$
$$y_p''' = 8Axe^{2x} + 12Ae^{2x} + 27Be^{3x}.$$

We substitute in the D.E., obtaining

$$(8Axe^{2x} + 12Ae^{2x} + 27e^{3x}) - 2(4Axe^{2x} + 4Ae^{2x} + 9Be^{3x})$$

$$- (2Axe^{2x} + Ae^{2x} + 3Be^{3x}) + 2(Axe^{2x} + Be^{3x})$$

$$= 9e^{2x} - 8e^{3x} \qquad \text{or}$$

$$3Ae^{2x} + 8Be^{3x} = 9e^{2x} - 8e^{3x}.$$

We equate coefficients of like terms on both sides of this to obtain $3A = 9$, $8B = -8$, and hence $A = 3$, $B = -1$. Thus we obtain the particular integral $y_p = 3xe^{2x} - e^{3x}$. The G.S. of the D.E. is

$$y = c_1 e^x + c_2 e^{-x} + c_3 e^{2x} + 3xe^{2x} - e^{3x}.$$

25. The corresponding homogeneous D.E. is $y''' + y' = 0$. The auxiliary equation is $m^3 + m = 0$, with roots $m = 0, \pm i$. The complementary function is $y_c = c_1 + c_2 \sin x + c_3 \cos x$.

The NH term is a linear combination of the UC functions given by x^2 and $\sin x$.

Step 1: Form the UC set of each of these two UC functions: $S_1 = \{x^2, x, 1\}$, $S_2 = \{\sin x, \cos x\}$.

Step 2: Neither set is identical with nor included in the other, so both are retained.

Step 3: Observe that the member 1 of S_1 is included in the complementary function (in the $c_1 = c_1 \cdot 1$ term) and so is a solution of the corresponding homogeneous D.E. Thus we multiply each member of set S_1 by x to obtain the revised set $S_1' = \{x^3, x^2, x\}$, which has no members which are members of S_2 are included in the complementary function. Thus we multiply each member of S_2 by x to obtain the revised set $S_2' = \{x \sin x, \ x \cos x\}$, which has no members which are solutions of the homogeneous D.E.

Step 4: We now have the two sets $S_1' = \{x^3, x^2, x\}$, $S_2' = \{x \sin x, \ x \cos x\}$. We form a linear combination of their five elements.

Step 5: Thus we take $y_p = Ax^3 + Bx^2 + Cx + Dx \sin x + Ex \cos x$. Then

$$y_p' = 3Ax^2 + 2Bx + C + Dx \cos x - Ex \sin x$$
$$+ D \sin x + E \cos x,$$

$$y_p'' = 6Ax + 2B - Dx \sin x - Ex \cos x$$
$$+ 2D \cos x - 2E \sin x,$$

$$y_p''' = 6A - Dx \cos x + Ex \sin x - 3D \sin x$$
$$- 3E \cos x.$$

We substitute in the D.E., obtaining

$$(6A - Dx \cos x + Ex \sin x - 3D \sin x - 3E \cos x)$$
$$+ (3Ax^2 + 2Bx + C + Dx \cos x - Ex \sin x + D \sin x$$
$$+ E \cos x) = 2x^2 + 4 \sin x \qquad\qquad \text{or}$$

$$3Ax^2 + 2Bx + (6A + C) - 2D \sin x - 2E \cos x$$
$$= 2x^2 + 4 \sin x.$$

We equate coefficients of like terms on both sides of this
to obtain $3A = 2$, $2B = 0$, $6A + C = 0$, $-2D = 4$, $-2E = 0$,
and hence $A = \frac{2}{3}$, $B = 0$, $C = -4$, $D = -2$, $E = 0$. Thus we
obtain the particular integral $y_p = \left(\frac{2}{3}\right)x^3 - 4x - 2x \sin x$.
The G.S. of the D.E. is

$$y = c_1 + c_2 \sin x + c_3 \cos x + \left(\frac{2}{3}\right)x^3 - 4x - 2x \sin x.$$

26. The corresponding homogeneous D.E. is $y^{iv} - y''' + 2y'' = 0$.
The auxiliary equation is $m^4 - 3m^3 + 2m^2 = 0$, or
$m^2(m - 1)(m - 2) = 0$, with roots 0, 0 (real double root),
1, 2. The complementary function is
$y_c = c_1 + c_2 x + c_3 e^x + c_4 e^{2x}$. The NH term is a linear
combination of the UC functions given by e^{-x}, e^{2x}, and x.

Step 1: Form the UC set of each of these three UC
functions: $S_1 = \{e^{-x}\}$, $S_2 = \{e^{2x}\}$, $S_3 = \{x,1\}$.

Step 2: No set is identical with nor included in any
other, so each is retained.

Step 3: Observe that the member e^{2x} of S_2 is included in
the complementary function and so is a solution of the
corresponding homogeneous D.E. Thus we multiply each
member of S_2 by x to obtain the revised set $S_2' = \{xe^{2x}\}$,
whose member is not a solution of the homogeneous D.E.
Next observe that both members x and 1 of S_3 are included
in the complementary function. Thus we multiply each
member of S_3 by x^2 to obtain the revised set

$S_3' = \{x^3, x^2\}$, whose members are not solutions of the homogeneous D.E. (note that multiplication by x, instead of x^2, is not sufficient here).

Step 4: We now have the three sets $S_1' = \{e^{-x}\}$, $S_2' = \{xe^{2x}\}$, $S_3' = \{x^3, x^2\}$. We form a linear combination of their four elements.

Step 5: Thus we take $y_p = Ae^{-x} + Bxe^{2x} + Cx^3 + Dx^2$ as a particular solution. Then

$$y_p' = -Ae^{-x} + 2Bxe^{2x} + Be^{2x} + 3Cx^2 + 2Dx,$$

$$y_p'' = Ae^{-x} + 4Bxe^{2x} + 4Be^{2x} + 6Cx + 2D,$$

$$y_p''' = -Ae^{-x} + 8Bxe^{2x} + 12Be^{2x} + 6C,$$

$$y_p^{iv} = Ae^{-x} + 16Bxe^{2x} + 32Be^{2x}.$$

We substitute into the D.E., obtaining

$$(Ae^{-x} + 16Bxe^{2x} + 32Be^{2x}) - 3(-Ae^{-x} + 8Bxe^{2x}$$
$$+ 12Be^{2x} + 6C) + 2(Ae^{-x} + 4Bxe^{2x} + 4Be^{2x}$$
$$+ 6Cx + 2D) = 3e^{-x} + 6e^{2x} - 6x \text{ or}$$
$$6Ae^{-x} + 4Be^{2x} + 12Cx + (-18C + 4D)$$
$$= 3e^{-x} + 6e^{2x} - 6x.$$

We equate coefficients of like terms on both sides of this to obtain $6A = 3$, $4B = 6$, $12C = -6$, $-18C + 4D = 0$, and hence $A = \frac{1}{2}$, $B = \frac{3}{2}$, $C = -\frac{1}{2}$, $D = -\frac{9}{4}$. Thus we obtain the particular integral $y_p = \frac{1}{2}e^{-x} + \frac{3}{2}xe^{2x} - \frac{1}{2}x^3 - \frac{9}{4}x^2$.

The G.S. of the D.E. is $y = c_1 + c_2x + c_3e^x + c_4e^{2x}$

$+ \frac{1}{2}e^{-x} + \frac{3}{2}xe^{2x} - \frac{1}{2}x^3 - \frac{9}{4}x^2$.

27. The corresponding homogeneous D.E. is
$y''' - 6y'' + 11y' - 6y = 0$. The auxiliary equation is
$m^3 - 6m^2 + 11m - 6 = 0$. By inspection we find that $m = 1$
is a root. Then by synthetic division, we have
$(m - 1)(m^2 - 5m + 6) = 0$ or $(m - 1)(m - 2)(m - 3) = 0$.
Thus the roots are $m = 1, 2, 3$. The complementary
function is $y_c = c_1e^x + c_2e^{2x} + c_3e^{3x}$. The NH member is a
linear combination of the UC functions xe^x, e^{2x}, and e^{4x}.

Step 1: Form the UC set of each of these three UC
functions: $S_1 = \{xe^x, e^x\}$, $S_2 = \{e^{2x}\}$, $S_3 = \{e^{4x}\}$.

Step 2: No set is identical with nor included in another,
so each is retained.

Step 3: Observe that member e^x of S_1 is included in the
complementary function and so is a solution of the
corresponding homogeneous equation. Thus we multiply each
member of S_1 by x to obtain the revised set
$S_1' = \{x^2e^x, xe^x\}$, whose members are not solutions of the
homogeneous D.E. Similarly, member e^{2x} of S_2 is included
in y_c, so we multiply each member of S_2 by x to obtain the
revised set $S_2' = \{xe^{2x}\}$.

Step 4: We now have the two revised sets $S_1' = \{x^2e^x, xe^x\}$ and $S_2' = \{xe^{2x}\}$ and also the original set $S_3 = \{e^{4x}\}$. We form a linear combination $Ax^2e^x + Bxe^x + Cxe^{2x} + De^{4x}$ of their four elements.

Step 5: Thus we take $y_p = Ax^2e^x + Bxe^x + Cxe^{2x} + De^{4x}$ as a particular solution. Then

$$y_p' = Ax^2e^x + (2A + B)xe^x + Be^x + 2Cxe^{2x}$$
$$+ Ce^{2x} + 4De^{4x},$$

$$y_p'' = Ax^2e^x + (4A + B)xe^x + (2A + 2B)e^x$$
$$+ 4Cxe^{2x} + 4Ce^{2x} + 16De^{4x},$$

$$y_p''' = Ax^2e^x + (6A + B)xe^x + (6A + 3B)e^x$$
$$+ 8Cxe^{2x} + 12Ce^{2x} + 64De^{4x}.$$

We substitute in the D.E., obtaining

$$[Ax^2e^x + (6A + B)xe^x + (6A + 3B)e^x + 8Cxe^{2x}$$
$$+ 12Ce^{2x} + 64De^{4x}] - 6[Ax^2e^x + (4A + B)xe^x$$
$$+ (2A + 2B)e^x + 4Cxe^{2x} + 4Ce^{2x} + 16De^{4x}]$$
$$+ 11[Ax^2e^x + (2A + B)xe^x + Be^x + 2Cxe^{2x}$$
$$+ Ce^{2x} + 4De^{4x}] - 6[Ax^2e^x + Bxe^x$$
$$+ Cxe^{2x} + De^{4x}] = xe^x - 4e^{2x} + 6e^{4x} \qquad \text{or}$$

$$4Axe^x + (-6A + 2B)e^x - Ce^{2x} + 6De^{4x}$$
$$= xe^x - 4e^{2x} + 6e^{4x}.$$

We equate coefficients of like terms on both sides of this to obtain $4A = 1$, $-6A + 2B = 0$, $-C = -4$, $6D = 6$.   From these equations, we find $A = \frac{1}{4}$, $B = \frac{3}{4}$, $C = 4$, $D = 1$.   Thus we obtain the particular integral

$$y_p = \frac{1}{4} x^2 e^x + \frac{3}{4} xe^x + 4xe^{2x} + e^{4x}.$$

The G.S. of the D.E. is

$$y = c_1 e^x + c_2 e^{2x} + c_3 e^{3x} + \frac{1}{4} x^2 e^x$$

$$+ \frac{3}{4} xe^x + 4xe^{2x} + e^{4x}.$$

28.   The corresponding homogeneous D.E. is $y''' - 4y'' + 5y' - 2y = 0$.   The auxiliary equation is $m^3 - 4m^2 + 5m - 2 = 0$.   By inspection we find that $m = 1$ is a root.   Then by synthetic division, we have $(m - 1)(m^2 - 3m + 2) = 0$ or $(m - 1)^2(m - 2) = 0$.   Thus the roots are $m = 1$, 1 (real double root), and 2.   The complementary function is $y_c = (c_1 + c_2 x)e^x + c_3 e^{2x}$.   The NH member is a linear combination of the UC functions $x^2 e^x$ and e^x.

Step 1: Form the UC set of each of these two UC functions: $S_1 = \{x^2 e^x, xe^x, e^x\}$, $S_2 = \{e^x\}$.

Step 2: We have $S_2 \subset S_1$, so we discard S_2, leaving only S_1.

Step 3: Observe that the members e^x and xe^x of S_1 are both included in the complementary function and so are solutions of the corresponding homogeneous equation. Thus we multiply each member of S_1 by x^2 to obtain the revised set $S_1' = \{x^4e^x, x^3e^x, x^2e^x\}$, whose members are not solutions of the homogeneous D.E.

Step 4: We now have the set $S_1' = \{x^4e^x, x^3e^x, x^2e^x\}$, which was revised in Step 3. We form the linear combination $Ax^4e^x + Bx^3e^x + Cx^2e^x$ of the three elements of S_1'.

Step 5: Thus we take $y_p = Ax^4e^x + Bx^3e^x + Cx^2e^x$ as a particular integral. Then

$$y_p' = Ax^4e^x + (4A + B)x^3e^x + (3B + C)x^2e^x + 2Cxe^x,$$

$$y_p'' = Ax^4e^x + (8A + B)x^3e^x + (12A + 6B + C)x^2e^x$$
$$+ (6B + 4C)xe^x + 2Ce^x,$$

$$y_p''' = Ax^4e^x + (12A + B)x^3e^x + (36A + 9B + C)x^2e^x$$
$$+ (24A + 18B + 6C)xe^x + (6B + 6C)e^x.$$

We substitute into the D.E., obtaining

$$[Ax^4e^x + (12A + B)x^3e^x + (36A + 9B + C)x^2e^x$$
$$+ (24A + 18B + 6C)xe^x + (6B + 6C)e^x] - 4[Ax^4e^x$$
$$+ (8A + B)x^3e^x + (12A + 6B + C)x^2e^x + (6B + 4C)xe^x$$
$$+ 2Ce^x] + 5[Ax^4e^x + (4A + B)x^3e^x + (3B + C)x^2e^x$$
$$+ 2Cxe^x] - 2[Ax^4e^x + Bx^3e^x + Cx^2e^x]$$
$$= 3x^2e^x - 7e^x$$

or

$$-12Ax^2e^x + (24A - 6B)xe^x + (6B - 2C)e^x$$
$$= 3x^2e^x - 7e^x.$$

We equate coefficients of like terms on both sides of this to obtain $-12A = 3$, $24A - 6B = 0$, $6B - 2C = -7$. From these equations, we find $A = -1/4$, $B = -1$, $C = 1/2$. Thus we obtain the particular integral

$$y_p = -\frac{1}{4}x^4e^x - x^3e^x + \frac{1}{2}x^2e^x.$$

The G.S. of the D.E. is

$$y = (c_1 + c_2x)\,e^x + c_3e^{2x} - \frac{1}{4}x^4e^x - x^3e^x + \frac{1}{2}x^2e^x.$$

31. The corresponding homogeneous D.E. is $y'' + y = 0$. The auxiliary equation is $m^2 + 1 = 0$, with roots $\pm i$. The complementary function is $y_c = c_1\sin x + c_2\cos x$. The NH term is the UC function $x\sin x$.

Step 1: Form the UC set of this UC function. It is $S_1 = \{x\sin x,\ x\cos x,\ \sin x,\ \cos x\}$.

Step 2: This step does not apply.

Step 3: Observe that the members $\sin x$ and $\cos x$ of S_1 are included in the complementary function and so are solutions of the corresponding homogeneous D.E. Thus we multiply each member of S_1 by x to obtain the revised set $S_1' = \{x^2\sin x,\ x^2\cos x,\ x\sin x,\ x\cos x\}$ whose members are not solutions of the homogeneous D.E.

Step 4: We now have the set S_1'. We form a linear combination of its four elements $x^2\sin x$, $x^2\cos x$, $x\sin x$, and $x\cos x$.

Step 5: Thus we take

$$y_p = Ax^2\sin x + Bx^2\cos x + Cx\sin x + Dx\cos x$$

as a particular solution. Then

$$y_p' = Ax^2\cos x - Bx^2\sin x + (2A - D)x\sin x$$
$$+ (2B + C)x\cos x + C\sin x + D\cos x \qquad \text{and}$$
$$y_p'' = -Ax^2\sin x - Bx^2\cos x + (-4B - C)x\sin x$$
$$+ (4A - D)x\cos x + (2A - 2D)\sin x$$
$$+ (2B + 2C)\cos x.$$

We substitute into the D.E., obtaining

$$-4Bx\sin x + 4Ax\cos x + (2A - 2D)\sin x$$
$$+ (2B + 2C)\cos x = x\sin x.$$

We equate coefficients of like terms on both sides of this to obtain $-4B = 1$, $4A = 0$, $2A - 2D = 0$, $2B + C = 0$. From this, $A = 0$, $B = -\frac{1}{4}$, $C = \frac{1}{4}$, $D = 0$. Thus we obtain the particular integral $y_p = -\frac{1}{4}x^2\cos x + \frac{1}{4}x\sin x$. The G.S. of the D.E. is

$$y = c_1\sin x + c_2\cos x - \frac{1}{4}x^2\cos x + \frac{1}{4}x\sin x .$$

33. The corresponding homogeneous D.E. is $y^{iv} + 2y''' - 3y'' = .$
The auxiliary equation is $m^4 + 2m^3 - 3m^2 = 0$ or
$m^2(m - 1)(m + 3) = 0$, with roots 0, 0 (real double root),
1, -3. The complementary function is
$y_c = c_1 + c_2x + c_3e^x + c_4e^{-3x}$. The NH term is a linear
combination of the UC functions given by x^2, xe^x, e^{3x}, and
1.

Step 1: Form the UC set of each of these four UC
functions: $S_1 = \{x^2, x, 1\}$, $S_2 = \{xe^x, e^x\}$,
$S_3 = \{e^{3x}\}$, $S_4 = \{1\}$.

Step 2: Set S_4 is completely included in S_1, so S_4 is
omitted, leaving the three sets S_1, S_2, S_3.

Step 3: Observe that members x and 1 of S_1 are included in
the complementary function and so are solutions of the
corresponding homogeneous D.E. Thus we multiply each
member of S_1 by x^2 to obtain the revised set $S_1' =$
$\{x^4, x^3, x^2\}$, whose members are not solutions of the
homogeneous D.E. (Note that multiplication by x, instead
of x^2, is not sufficient here.) Next observe that the
member e^x of S_2 is included in the complementary function.
Thus we multiply each member of S_2 by x to obtain the
revised set $S_2' = \{x^2e^x, xe^x\}$, whose members are not
solutions of the homogeneous D.E.

Step 4: We now have the three sets $S_1' = \{x^4, x^3, x^2\}$,
$S_2' = \{x^2e^x, xe^x\}$, $S_3 = \{e^{3x}\}$. We form the linear

combination $x^4 + Bx^3 + Cx^2 + Dx^2e^x + Exe^x + Fe^{3x}$ of their six elements.

<u>**Step 5**</u>: Thus we take

$$y_p = Ax^4 + Bx^3 + Cx^2 + Dx^2e^x + Exe^x + Fe^{3x}.$$

Then

$$y_p' = 4Ax^3 + 3Bx^2 + 2Cx + Dx^2e^x$$
$$+ (2D + E)xe^x + Ee^x + 3Fe^{3x},$$
$$y_p'' = 12Ax^2 + 6Bx + 2C + Dx^2e^x$$
$$+ (4D + E)xe^x + (2D + 2E)e^x + 9Fe^{3x},$$
$$y_p''' = 24Ax + 6B + Dx^2e^x + (6D + E)xe^x$$
$$+ (6D + 3E)e^x + 27Fe^{3x},$$
$$y_p^{iv} = 24A + Dx^2e^x + (8D + E)xe^x$$
$$+ (12D + 4E)e^x + 81Fe^{3x}.$$

We substitute into the D.E., obtaining

$$24A + Dx^2e^x + (8D + E)xe^x + (12D + 4E)e^x$$
$$+ 81Fe^{3x} + 2[24Ax + 6B + Dx^2e^x + (6D + E)xe^x$$
$$+ (6D + 3E)e^x + 27Fe^{3x}] - 3[12Ax^2 + 6Bx + 2C$$
$$+ Dx^2e^x + (4D + E)xe^x + (2D + 2E)e^x + 9Fe^{3x}]$$
$$= 18x^2 + 16xe^x + 4e^{3x} - 9$$

or

$$-36Ax^2 + (48A - 18B)x + (24A + 12B - 6C)$$
$$+ 8Dxe^x + (18D + 4E)e^x + 108Fe^{3x}$$
$$= 18x^2 + 16xe^x + 4e^{3x} - 9.$$

We equate coefficients of like terms on both sides of this to obtain $-36A = 18$, $48A - 18B = 0$, $24A + 12B - 6C = -9$, $8D = 16$, $18D + 4E = 0$, $108F = 4$. From this, we find $A = -1/2$, $B = -4/3$, $C = -19/6$, $D = 2$, $E = -9$, $F = 1/27$. Thus we obtain the particular integral

$$y_p = -\frac{1}{2} x^4 - \frac{4}{3} x^3 - \frac{19}{6} x^2 + 2x^2 e^x$$

$$- 9xe^x + \frac{1}{27} e^{3x}.$$

The G.S. of the D.E. is

$$y = c_1 + c_2 x + c_3 e^x + c_4 e^{-3x} - \frac{1}{2} x^4 - \frac{4}{3} x^3$$

$$- \frac{19}{6} x^2 + 2x^2 e^x - 9xe^x + \frac{1}{27} e^{3x}.$$

34. The corresponding homogeneous D.E. is $y^{iv} - 5y''' + 7y'' - 5y' + 6y = 0$. The auxiliary equation is $m^4 - 5m^3 + 7m^2 - 5m + 6 = 0$. By inspection we find that $m = 2$ is a root. Then by synthetic division, we have $(m - 2)(m^3 - 3m^2 + m - 3) = 0$ or $(m - 2)(m - 3)(m^2 + 1) = 0$. Thus the roots are $m = 2, 3, \pm i$. The complementary function is $y_c = c_1 e^{2x} + c_2 e^{3x} + c_3 \sin x + c_4 \cos x$. The NH term is a linear combination of the UC functions given by $\sin x$ and $\sin 2x$.

__Step 1__: Form the UC set of each of these two UC functions: $S_1 = \{\sin x, \cos x\}$, $S_2 = \{\sin 2x, \cos 2x\}$.

__Step 2__: Neither set is identical with nor included in the other, so both are retained.

Step 3: Observe that both members of S_1 are included in the complementary function and so are solutions of the corresponding homogeneous D.E. Thus we multiply each member of S_1 by x to obtain the revised set

$S_1' = \{x \sin x, \ x \cos x\}$, whose members are not solutions of the homogeneous D.E.

Step 4: We now have the two sets $S_1' = \{x \sin x, \ x \cos x\}$, $S_2 = \{\sin 2x, \ \cos 2x\}$. We form a linear combination of their four elements.

Step 5: Thus we take $y_p = Ax \sin x + Bx \cos x + C \sin 2x + D \cos 2x$ as a particular solution. Then $y_p' =$
$Ax \cos x - Bx \sin x + A \sin x + B \cos x + 2C \cos 2x - 2D \sin 2x$, $y_p'' = -Ax \sin x - Bx \cos x + 2A \cos x - 2B \sin x - 4C \sin 2x - 4D \cos 2x$,
$y_p''' = -Ax \cos x + Bx \sin x - 3A \sin x - 3B \cos x - 8C \cos 2x + 8D \sin 2x$, $y_p^{iv} = Ax \sin x + Bx \cos x - 4A \cos x + 4B \sin x + 16C \sin 2x + 16D \cos 2x.$
We 9ubstitute into the D.E., obtaining $(Ax \sin x + Bx \cos x - 4A \cos x + 4B \sin x + 16C \sin 2x + 16D \cos 2x) - 5(-Ax \cos x + Bx \sin x - 3A \sin x - 3B \cos x - 8C \cos 2x + 8D \sin 2x) + 7(-Ax \sin x - Bx \cos x + 2A \cos x - 2B \sin x - 4C \sin 2x - 4D \cos 2x) - 5(Ax \cos x - Bx \sin x + A \sin x + B \cos x + 2C \cos 2x - 2D \sin 2x) + 6(Ax \sin x + Bx \cos x + C \sin 2x + D \cos 2x) = 5 \sin x - 12 \sin 2x$
or $(10A - 10B)\sin x + (10A + 10B)\cos x + (-6C - 30D)\sin 2x + (30C - 6D)\cos 2x$
$= 5 \sin x - 12 \sin 2x$. We equate coefficients of like terms

on both sides of this to obtain

$10A - 10B = 5$, $10A + 10B = 0$, $-6C - 30D = -12$,

$30C - 6D = 0$. The equations in A and B are equivalent to

$2A - 2B = 1$, $A + B = 0$, from which $A = \frac{1}{4}$, $B = -\frac{1}{4}$. The

equations in C and D are equivalent to $C + 5D = 2$, $5C - D$

$= 0$, from which $C = \frac{1}{13}$, $D = \frac{5}{13}$. Thus we obtain the

particular integral $y_p = \frac{1}{4} x \sin x - \frac{1}{4} x \cos x + \frac{1}{13} \sin 2x +$

$\frac{5}{13} \cos 2x$. The G.S. of the D.E. is

$$y = c_1 e^{2x} + c_2 e^{3x} + c_3 \sin x + c_4 \cos x$$

$$+ \frac{1}{4} x \sin x - \frac{1}{4} x \cos x + \frac{1}{13} \sin 2x.$$

$$+ \frac{5}{13} \cos 2x.$$

37. The auxiliary equation of the corresponding homogeneous
D.E. is $m^2 - 8m + 15 = 0$, with roots 3, 5. The
complementary function is $y_c = c_1 e^{3x} + c_2 e^{5x}$. The UC set
of UC function xe^{2x} is $S_1 = \{xe^{2x}, e^{2x}\}$. This does not
need revision, so we take $y_p = Axe^{2x} + Be^{2x}$ as a
particular integral. Then $y_p' = 2Axe^{2x} + (A + 2B)e^{2x}$ and
$y_p'' = 4Axe^{2x} + (4A + 4B)e^{2x}$. We substitute in the D.E.
obtaining

$$4Axe^{2x} + (4A + 4B)e^{2x} - 8[2Axe^{2x} + (A + 2B)e^{2x}]$$
$$+ 15(Axe^{2x} + Be^{2x} = 9xe^{2x} \qquad\qquad \text{or}$$
$$3Axe^{2x} + (-4A + 3B)e^{2x} = 9xe^{2x}.$$

We equate coefficients of like terms on both sides of this to obtain $3A = 9$, $-4A + 3B = 0$. From these, we find $A = 3$, $B = 4$. Thus we obtain the particular integral $y_p = 3xe^{2x} + 4e^{2x}$. The G.S. of the D.E. is

$$y = c_1 e^{2x} + c_2 e^{5x} + 3xe^{2x} + 4e^{2x}.$$

We apply the I.C. $y(0) = 5$ to this. We find $c_1 + c_2 + 4 = 5$ or $c_1 + c_2 = 1$. (*) We next differentiate the G.S. to obtain $y' = 3c_1 e^{3x} + 5c_2 e^{5x} + 6xe^{2x} + 11e^{2x}$. We apply the I.C. $y'(0) = 10$ to this. We find $3c_1 + 5c_2 + 11 = 10$ or $3c_1 + 5c_2 = -1$. (**) From (*) and (**) we find $c_1 = 3$, $c_2 = -2$. Thus we obtain the particular solution

$$y = 3e^{3x} - 2e^{5x} + 3xe^{2x} + 4e^{2x}.$$

38. The corresponding homogeneous equation is $y'' + 7y' + 10y = 0$; and the auxiliary equation is $m^2 + 7m + 10 = 0$, with roots $m = -2, -5$. Thus the complementary function is $y_c = c_1 e^{-2x} + c_2 e^{-5x}$. The UC set of UC function $4xe^{-3x}$ is $\{xe^{-3x}, e^{-3x}\}$. We assume a particular integral $y_p = Axe^{-3x} + Be^{-3x}$. Then

$$y_p' = -3Axe^{-3x} + (A - 3B)e^{-3x},$$

$$y_p'' = 9Axe^{-3x} + (-6A + 9B)e^{-3x}.$$ Substituting in the given D.E., we obtain

$$[9Axe^{-3x} + (-6A + 9B)e^{-3x}] + 7[-3Axe^{-3x}$$
$$+ (A - 3B)e^{-3x}] + 10[Axe^{-3x} + Be^{-3x}] = 4xe^{-3x}.$$

Simplifying, we find

$$-2Axe^{-3x} - (A + 2B)e^{-3x} = 4xe^{-3x}.$$

From this, we have the equations $-2A = 4$, $A - 2B = 0$, with the solution $A = -2$, $B = -1$. Thus we find the particular integral $y_p = -2xe^{-3x} - e^{-3x}$ and the general solution

$$y = c_1e^{-2x} + c_2e^{-5x} - 2xe^{-3x} - e^{-3x}.$$

We now apply the initial condition $y(0) = 0$ to this. We have

$$c_1 + c_2 = 1.$$

Differentiating the general solution, we find

$$y' = -2c_1e^{-2x} - 5c_2e^{-5x} + 6xe^{-3x} + e^{-3x}.$$

Applying the initial condition $y'(0) = -1$ to this, we find

$$2c_1 + 5c_2 = 2.$$

The solution of the two equations in c_1 and c_2 is $c_1 = 1$, $c_2 = 0$. Thus the solution is

$$y = e^{-2x} - 2xe^{-3x} - e^{-3x}.$$

39. The auxiliary equation of the corresponding homogeneous
     equation is $m^2 + 8m + 16 = 0$ or $(m + 4)^2 = 0$, with double
     real root -4. Thus the complementary function is
     $y_c = (c_1 + c_2 x)e^{-4x}$. The UC set of UC function $8e^{-2x}$ is
     $\{e^{-2x}\}$. We assume the particular integral $y_p = Ae^{-2x}$.
     Then $y_p{}' = -2Ae^{-2x}$, $y_p{}'' = 4Ae^{-2x}$. Substituting into the
     given D.E., we find

     $$4Ae^{-2x} - 16Ae^{-2x} + 16Ae^{-2x} = 8e^{-2x}$$

     or $4Ae^{-2x} = 8e^{-2x}$. From this $4A = 8$, and so $A = 2$. Thus
     we obtain the particular integral $y_p = 2e^{-2x}$ and the
     general solution $y = (c_1 + c_2 x)e^{-4x} + 2e^{-2x}$.

     We apply the initial condition $y(0) = 2$ to this,
     obtaining $c_1 = 0$. We differentiate the general solution
     to obtain $y' = (-4c_1 + c_2 - 4c_2 xe^{-2x}) e^{-4x} - 4e^{-2x}$. From
     this, $(-4c_1 + c_2) - 4 = 0$. Since $c_1 = 0$, this gives
     $c_2 = 4$. Thus we obtain

     $$y = 4xe^{-4x} + 2e^{-2x}.$$

40. The auxiliary equation of the corresponding homogeneous
     D.E. is $m^2 + 6m + 9 = 0$, with double real root -3. The
     complementary function is $y = (c_1 + c_2 x)e^{-3x}$. The UC set
     of the UC function $27e^{-6x}$ is $\{e^{-6x}\}$. We assume the

particular integral $y_p = Ae^{-6x}$. Then $y_p' = -6Ae^{-6x}$, $y_p'' = 36Ae^{-6x}$. Substituting in the given D.E., we obtain

$$36Ae^{-6x} - 36Ae^{-6x} + 9Ae^{-6x} = 27e^{-6x}$$

or $9A = 27e^{-6x}$. From this, $9A = 27$, and so $A = 3$. Thus we obtain the particular integral $y_p = 3e^{-6x}$ and the general solution $y = (c_1 + c_2x)e^{-3x} + 3e^{-6x}$. Now apply the I.C. $y(0) = -2$ to the general solution. We obtain $c_1 + 3 = -2$, so $c_1 = -5$. Differentiate the general solution to obtain $y' = (-3c_1 + c_2 - 3c_2x)e^{-3x} - 18e^{-6x}$. Apply the I.C. $y'(0) = 0$ to this. We obtain $-3c_1 + c_2 - 18 = 0$. Since $c_1 = -5$, we find $c_2 = 3$. Thus we obtain the solution $y = (3x - 5)e^{-3x} + 3e^{-6x}$.

42. The auxiliary equation of the corresponding homogeneous D.E. is $m^2 - 10m + 29 = 0$, with roots $5 \pm 2i$. The complementary function is $y_c = e^{5x}(c_1\sin 2x + c_2\cos 2x)$. The UC set of the UC function e^{5x} is $S_1 = \{e^{5x}\}$. This does not need revision, so we take $y_p = Ae^{5x}$ as a particular integral. Then $y_p' = 5Ae^{5x}$ and $y_p'' = 25Ae^{5x}$. We substitute in the D.E., obtaining

$$25Ae^{5x} - 10(5Ae^{5x}) + 29Ae^{5x} = 8e^{5x}$$

or $4Ae^{5x} = 8e^{5x}$. We at once see that $A = 2$ and hence obtain the particular integral $y_p = 2e^{5x}$. The G.S. of the D.E. is

$$y = e^{5x}(c_1 \sin 2x + c_2 \cos 2x) + 2e^{5x}.$$

We apply the I.C. $y(0) = 0$ to this. We find $c_2 + 2 = 0$, so $c_2 = -2$. We next differentiate the G.S. to obtain

$$y' = e^{5x}[(5c_1 - 2c_2)\sin 2x + (2c_1 + 5c_2)\cos 2x] + 10e^{5x}.$$

We apply the I.C. $y'(0) = 8$ to this. We find $2c_1 + 5c_2 + 10 = 8$ or $2c_1 + 5c_2 = -2$. Since $c_2 = -2$, this gives $c_1 = 4$. Thus we obtain the particular solution $y = 2e^{5x}(2 \sin 2x - \cos 2x + 1)$.

44.   The auxiliary equation of the corresponding homogeneous D.E. is $m^2 - m - 6 = 0$ or $(m - 3)(m + 2) = 0$, with roots $3, -2$. The complementary function is $y_c = c_1 e^{3x} + c_2 e^{-2x}$. The UC sets of the UC functions in the right member of the D.E. are $S_1 = \{e^{2x}\}$, $S_2 = \{e^{3x}\}$. Neither is completely contained in the other so both are retained. The member e^{3x} of S_2 is contained in the complementary function and so is a solution of the corresponding homogeneous D.E. Thus we multiply the member of S_2 by x to obtain the revised set $S_2' = \{xe^{3x}\}$, whose member is not a solution of the homogeneous D.E.

Now we have the two sets $S_1 = \{e^{2x}\}$, $S_2' = \{xe^{3x}\}$. We take a linear combination of their two members as a particular integral. That is, we take $y_p = Ae^{2x} + Bxe^{3x}$. Then $y_p' = 2Ae^{2x} + Be^{3x} + 3Bxe^{3x}$ and $y_p'' = 4Ae^{2x} + 6Be^{3x} + 9Bxe^{3x}$. We substitute into the D.E., obtaining

$$4Ae^{2x} + 6Be^{3x} + 9Bxe^{3x} - (2Ae^{2x} + Be^{3x} + 3Bxe^{3x})$$
$$- 6(Ae^{2x} + Bxe^{3x}) = 8e^{2x} - 5e^{3x} \qquad \text{or}$$
$$-4Ae^{2x} + 5Be^{3x} = 8e^{2x} - 5e^{3x}.$$

We equate coefficients of like terms on both sides of this to obtain $-4A = 8$, $5B = -5$. From these, we find $A = -2$, $B = -1$. Thus we obtain the particular integral $y_p = -2e^{2x} - xe^{3x}$. The G.S. of the D.E. is

$$y = c_1 e^{3x} + c_2 e^{-2x} - 2e^{2x} - xe^{3x}.$$

We apply the I.C. $y(0) = 3$ to this. We find $c_1 + c_2 - 2 = 3$ or $c_1 + c_2 = 5$ (*). We next differentiate the G.S. to obtain $y' = 3c_1 e^{3x}$ $2c_2 e^{-2x}$ $- 4e^{2x} - 3xe^{3x} - e^{3x}$. We apply the I.C. $y'(0) = 5$ to this. We find

$$3c_1 - 2c_2 - 4 - 1 = 5$$

or

$$3c_1 - 2c_2 = 10 \qquad (**).$$

From the two equations (*) and (**) in the unknowns c_1 and
c_2, we find that $c_1 = 4$, $c_2 = 1$. Thus we obtain the
particular solution $y = 4e^{3x} + e^{-2x} - 2e^{2x} - xe^{3x}$.

45.   The auxiliary equation of the corresponding homogeneous
D.E. is $m^2 - 2m + 1 = 0$ or $(m - 1)^2 = 0$, with roots 1, 1
(double root). The complementary function is
$y_c = c_1e^x + c_2xe^x$. The UC sets of the UC function in the
right member of the D.E. are $S_1 = \{xe^{2x}, e^{2x}\}$, $S_2 = \{e^x\}$.
Neither is completely contained in the other, so both are
retained. The member e^x of S_2 is included in the
complementary function and so is a solution of the
corresponding homogeneous D.E. Thus we multiply the
member of S_2 by x^2 to obtain the revised set $S_2' = \{x^2e^x\}$,
whose member is not a solution of the homogeneous D.E.
(note that multiplication by x, instead of x^2, is not
sufficient here).

   Now we have the two sets $S_1 = \{xe^{2x}, e^{2x}\}$,
$S_2' = \{x^2e^x\}$. We take a linear combination of their three
members as a particular integral. That is, we take
$y_p = Axe^{2x} + Be^{2x} + Cx^2e^x$. Then $y_p' = 2Axe^{2x} +$
$(A + 2B)e^{2x} + Cx^2e^x + 2Cxe^x$ and $y_p'' = 4Axe^{2x} +$
$(4A + 4B)e^{2x} + Cx^2e^x + 4Cxe^x + 2Ce^x$. We substitute into
the D.E., obtaining

$$4Axe^{2x} + (4A + 4B)e^{2x} + Cx^2e^x + 4Cxe^x + 2Ce^x$$
$$- 2[2Axe^{2x} + (A + 2B)e^{2x} + Cx^2e^x + 2Cxe^x]$$
$$+ Axe^{2x} + Be^{2x} + Cx^2e^x = 2xe^{2x} + 6e^x \quad \text{or}$$
$$Axe^{2x} + (2A + B)e^{2x} + 2Ce^x = 2xe^{2x} + 6e^x.$$

We equate coefficients of like terms on both sides of this to obtain $A = 2$, $2A + B = 0$, $2C = 6$. From these, we find $A = 2$, $B = -4$, $C = 3$. Thus we obtain the particular integral $y_p = 2xe^{2x} - 4e^{2x} + 3x^2e^x$. The G.S. of the D.E. is $y = c_1e^x + c_2xe^x + 2xe^{2x} - 4e^{2x} + 3x^2e^x$.

We apply the I.C. $y(0) = 1$ to this. We find $c_1 - 4 = 1$ or $c_1 = 5$. We next differentiate the G.S. to obtain $y' = c_1e^x + c_2xe^x + c_2e^x + 4xe^{2x} + 2e^{2x} - 8e^{2x}$ $+ 3x^2e^x + 6xe^x$. We apply the I.C. $y'(0) = 0$ to this. We find $c_1 + c_2 + 2 - 8 = 0$ or $c_1 + c_2 = 6$. Since $c_1 = 5$, this gives $c_2 = 1$. Thus we obtain the particular solution

$$y = 5e^x + xe^x + 2xe^{2x} - 4e^{2x} + 3x^2e^x \quad \text{or}$$
$$y = (x + 5)e^x + 3x^2e^x + 2xe^{2x} - 4e^{2x}.$$

47. The auxiliary equation of the corresponding homogeneous D.E. is $m^2 + 1 = 0$, with roots $\pm i$. The complementary function is $y_c = c_1\sin x + c_2\cos x$. The UC sets of the UC functions in the right member of the D.E. are $S_1 = \{x^2, x, 1\}$, $S_2 = \{\sin x, \cos x\}$. Neither is completely contained in the other, so both are retained. The members $\sin x$ and $\cos x$ of S_2 are included in the complementary

function and so are solutions of the corresponding homogeneous D.E. Thus we multiply each member of S_2 by x to obtain the revised set $S_2' = \{x \sin x, \ x \cos x\}$, whose members are not solutions of the homogeneous D.E.

Now we have the two sets $S_1 = \{x^2, x, 1\}$, $S_2' = \{x \sin x, \ x \cos x\}$. We take a linear combination of their five members as a particular integral. That is, we take $y_p = Ax^2 + Bx + C + Dx \sin x + Ex \cos x$. then

$$y_p' = 2Ax + B + Dx \cos x + D \sin x - Ex \sin x + E \cos x \text{ and}$$

$$y_p'' = 2A - Dx \sin x + 2D \cos x - Ex \cos x - 2E \sin x. \quad \text{We}$$

substitute into the D.E., obtaining

$$2A - Dx \sin x + 2D \cos x - Ex \cos x$$

$$- 2E \sin x + Ax^2 + Bx + C + Dx \sin x$$

$$+ Ex \cos x = 3x^2 - 4 \sin x \qquad \qquad \text{or}$$

$$Ax^2 + Bx + (2A + C) + 2D \cos x - 2E \sin x$$
$$= 3x^2 - 4 \sin x.$$

We equate coefficients of like terms on both sides of this to obtain $A = 3$, $B = 0$, $2A + C = 0$, $2D = 0$, $-2E = -4$. From these we find $A = 3$, $B = 0$, $C = -6$, $D = 0$, $E = 2$. Thus we obtain the particular integral $y_p = 3x^2 - 6 + 2x \cos x$. The G.S. of the D.E. is $y = c_1 \sin x + c_2 \cos x + 3x^2 - 6 + 2x \cos x$.

We apply the I.C. $y(0) = 0$ to this. We find $c_2 - 6 = 0$ or $c_2 = 6$. We next differentiate the G.S. to obtain $y' = c_1 \cos x - c_2 \sin x + 6x - 2x \sin x + 2 \cos x$. We apply the I.C. $y'(0) = 1$ to this. We find $c_1 + 2 = 1$ or $c_1 = -1$. Thus we obtain the particular solution
$y = 6 \cos x - \sin x + 3x^2 - 6 + 2x \cos x$.

48. The auxiliary equation of the corresponding homogeneous D.E. is $m^2 + 4 = 0$, with pure imaginary conjugate complex roots $\pm 2i$. Thus the complementary function is $y_c = c_1 \sin 2x + c_2 \cos 2x$. The UC set of the UC function $8 \sin 2x$ is $S = \{\sin 2x, \cos 2x\}$. Since each member of this set is a solution of the corresponding homogeneous equation, we multiply each by x and replace S by $S' = \{x \sin 2x, x \cos 2x\}$. We then form the particular integral $y_p = Ax \sin 2x + Bx \cos 2x$. Then

$y_p' = -2Bx \sin 2x + 2Ax \cos 2x + A \sin 2x + B \cos 2x$ and

$y_p'' = -4Ax \sin 2x - 4Bx \cos 2x - 4B \sin 2x + 4A \cos 2x$.

Substituting in the given D.E., we obtain

$$(-4Ax \sin 2x - 4Bx \cos 2x - 4B \sin 2x$$
$$+ 4A \cos 2x) + (4Ax \sin 2x + 4Bx \cos 2x)$$
$$= 8 \sin 2x \qquad\qquad \text{or}$$
$$-4B \sin 2x + 4A \cos 2x = 8 \sin 2x.$$

From this, $4A = 0$, $-4B = 8$, and hence $A = 0$, $B = -2$. Thus we obtain the particular integral $y_p = -2x \cos 2x$ and the general solution $y = c_1 \sin 2x + c_2 \cos 2x - 2x \cos 2x$.

We apply the I.C. $y(0) = 6$ to this, obtaining $c_2 = 6$.

Differentiating the general solution, we obtain

$y' = 2c_1 \cos 2x - 2c_2 \sin 2x + 4x \sin 2x - 2 \cos 2x$. Applying

the I.C. $y'(0) = 8$ to this, we find $2c_1 - 2 = 8$, from

which $c_1 = 5$. Thus we obtain the solution

$y = 5 \sin 2x + 6 \cos 2x - 2x \cos 2x$.

49. The auxiliary equation of the corresponding homogeneous
D.E. is $m^3 - 4m^2 + m + 6 = 0$ or $(m + 1)(m - 3)(m - 2) = 0$,
with roots -1, 2, 3. The complementary function is
$y_c = c_1 e^{-x} + c_2 e^{2x} + c_3 e^{3x}$. The UC sets of the UC

functions in the right member of the D.E. are
$S_1 = \{xe^x, e^x\}$, $S_2 = \{e^x\}$, and $S_3 = \{\sin x, \cos x\}$. Since

$S_1 \supset S_2$, we omit S_2, retaining S_1 and S_3. Neither of

these need revision, so we take a linear combination of
their four members as a particular integral. That is, we
take $y_p = Axe^x + Be^x + C \sin x + D \cos x$. Then

$y_p' = Axe^x + (A + B)e^x + C \cos x - D \sin x$,

$y_p'' = Axe^x + (2A + B)e^x - C \sin x - D \cos x$,

$y_p''' = Axe^x + (3A + B)e^x - C \cos x + D \sin x$. We substitute

into the D.E., obtaining $Axe^x + (3A + B)e^x - C \cos x +$

$D \sin x - 4[Axe^x + (2A + B)e^x - C \sin x - D \cos x] +$

$[Axe^x + (A + B)e^x + C \cos x - D \sin x] + 6[Axe^x + Be^x$

$+ C \sin x + D \cos x] = 3xe^x + 2e^x - \sin x$ or $4Axe^x$

$+ (-4A + 4B)e^x + 10C \sin x + 10D \cos x = 3xe^x + 2e^x - \sin x$.
We equate coefficients of like terms on both sides of this
to obtain $4A = 3$, $-4A + 4B = 2$, $10C = -1$, $10D = 0$. From

these, we find A = 3/4, B = 5/4, C = -1/10, D = 0. Thus we obtain the particular integral,

$$y_p = \frac{3}{4} x e^x + \frac{5}{4} e^x - \frac{1}{10} \sin x.$$

The G.S. of the D.E. is

$$y = c_1 e^{-x} + c_2 e^{2x} + c_3 e^{3x} + \frac{3}{4} x e^x$$

$$+ \frac{5}{4} e^x - \frac{1}{10} \sin x.$$

We apply the I.C. y(0) = 33/40 to this. We find $c_1 + c_2 + c_3 + 5/4 = 33/40$ or

$$c_1 + c_2 + c_3 = - 17/40. \qquad (*)$$

We next differentiate the G.S. to obtain

$$y' = -c_1 e^{-x} + 2c_2 e^{2x} + 3c_3 e^{3x} + \frac{3}{4} x e^x$$

$$+ 2e^x - \frac{1}{10} \cos x.$$

We apply the I.C. y'(0) = 0 to this. We find

$$-c_1 + 2c_2 + 3c_3 + 2 - \frac{1}{10} = 0$$

or

$$-c_1 + 2c_2 + 3c_3 = - \frac{19}{10}. \qquad (**)$$

We differentiate once more, obtaining

$$y'' = c_1 e^{-x} + 4c_2 e^{2x} + 9c_3 e^{3x} + \frac{3}{4} x e^x$$

$$+ \frac{11}{4} e^x + \frac{1}{10} \sin x.$$

We apply the I.C. $y''(0) = 0$ to this. We find

$$c_1 + 4c_2 + 9c_3 + \frac{11}{4} = 0$$

or

$$c_1 + 4c_2 + 9c_3 = -\frac{11}{4}. \qquad\qquad (***)$$

The three equations (*), (**), (***) determine c_1, c_2, c_3. Adding (*) and (**), we have $3c_2 + 4c_3 = -93/40$. Adding (**) and (***), we have $6c_2 + 12c_3 = -93/20$. Solving these two resulting equations in c_2 and c_3, we find $c_2 = -31/40$, $c_3 = 0$. Then (*) gives $c_1 = -c_2 - c_3 - 17/40$ $= 7/20$. Thus we obtain the desired particular solution

$$y = \frac{7}{20} e^{-x} - \frac{31}{40} e^{2x} + \frac{3}{4} x e^x + \frac{5}{4} e^x - \frac{1}{10} \sin x.$$

50. The auxiliary equation of the corresponding homogeneous D.E. is $m^3 - 6m^2 + 9m - 4 = 0$. Observe by inspection that 1 is a root. Then, by synthetic division, we obtain

$$
\begin{array}{r|rrrr}
1 & 1 & -6 & 9 & -4 \\
  &   & 1 & -5 & 4 \\
\hline
  & 1 & -5 & 4 & \boxed{0}
\end{array} \; .
$$

Thus we find the factorization $(m - 1)(m^2 - 5m + 4) = 0$ or $(m - 1)^2(m - 4) = 0$. Thus the roots are 1, 1 (double real root) and the simple real root 4. The complementary function is $y_c = (c_1 + c_2 x)e^x + c_3 e^{4x}$. The UC sets of the UC functions in the right member of the D.E. are

$S_1 = \{x^2, x, 1\}$, $S_2 = \{1\}$, $S_3 = \{e^{2x}\}$. Since $S_1 \supset S_2$, we omit S_2, retaining S_1 and S_3. Neither of these need revision, so we take a linear combination of their four members as a particular integral. That is, we take $y_p = Ax^2 + Bx + C + De^{2x}$. Then $y_p' = 2Ax + B + 2De^{2x}$, $y_p'' = 2A + 4De^{2x}$, $y_p''' = 8De^{2x}$. We substitute into the D.E, obtaining

$$8De^{2x} - 6(2A + 4De^{2x}) + 9(2Ax + B + 2De^{2x})$$
$$- 4(Ax^2 + Bx + C + De^{2x}) = 8x^2 + 3 - 6e^{2x}.$$

or

$$-4Ax^2 + (18A - 4B)x + (-12A + 9B - 4C) - 2De^{2x}$$
$$= 8x^2 + 3 - 6e^{2x}.$$

We equate coefficients of like terms on both sides of this to obtain $-4A = 8$, $18A - 4B = 0$, $-12A + 9B - 4C = 3$, $-2D = -6$. From these, we find $A = -2$, $B = -9$, $C = -15$, $D = 3$. Thus we obtain the particular integral $y_p = -2x^2 - 9x - 15 + 3e^{2x}$. The G.S. of the D.E. is

$$y = (c_1 + c_2 x)e^x + c_3 e^{4x} - 2x^2 - 9x - 15 + 3e^{2x}.$$

We apply the I.C. $y(0) = 1$ to this. We find

$$c_1 + c_3 - 15 + 3 = 1$$

or

$$c_1 + c_3 = 13. \qquad\qquad (*)$$

We next differentiate the G.S. to obtain

$$y' = (c_1 + c_2 + c_2 x)e^x + 4c_3 e^{4x} - 4x - 9 + 6e^{2x}.$$

We apply the I.C. $y'(0) = 7$ to this. We find

$$c_1 + c_2 + 4c_3 - 9 + 6 = 7$$

or

$$c_1 + c_2 + 4c_3 = 10. \qquad\qquad (**)$$

We differentiate once more, obtaining

$$y'' = (c_1 + 2c_2 + c_2 x)e^x + 16c_3 e^{4x} - 4 + 12e^{2x}.$$

We apply the I.C. $y''(0) = 10$ to this. We find

$$c_1 + 2c_2 + 16c_3 - 4 + 12 = 10$$

or

$$c_1 + 2c_2 + 16c_3 = 2. \qquad\qquad (***)$$

Solving the three equations $(*)$, $(**)$, $(***)$ for c_1, c_2, c_3, we find $c_1 = 122/9$, $c_2 = -4/3$, $c_3 = -5/9$.

Thus we obtain the desired particular solution

$$y = \left(\frac{122}{9} - \frac{4}{3}x\right)e^x - \frac{5}{9}e^{4x} - 2x^2 - 9x - 15 + 3e^{2x}.$$

53. The auxiliary equation of the corresponding homogeneous D.E. is $m^2 + 4m + 5 = 0$ with conjugate complex roots $-2 \pm i$. The complementary function is $y_c = e^{-2x}$ $(c_1 \sin x + c_2 \cos x)$. The NH member can be written as the linear combination $e^{-2x} + e^{-2x}\cos x$ of the two UC functions e^{-2x} and $e^{-2x}\cos x$. The UC sets of these functions are, respectively, $S_1 = \{e^{-2x}\}$ and $S_2 = \{e^{-2x}\sin x,\ e^{-2x}\cos x\}$. Neither is contained in the other, so each is retained.

Both members of S_2 are contained in the complementary function and so are solutions of the corresponding homogeneous D.E. Thus we multiply each member of S_2 by the lowest integral power of x so that the resulting revised set will contain no members that are solutions of the corresponding homogeneous D.E. It turns out that the first power of x, x itself, will accomplish this. So we multiply each member of S_2 by x to obtain the revised set $S_2' = \{xe^{-2x}\sin x,\ xe^{-2x}\cos x\}$, whose members are not solutions of the homogeneous D.E.

We now have sets $S_1 = \{e^{-2x}\}$ and $S_2' = \{xe^{-2x}\sin x, xe^{-2x}\cos x\}$. We form a linear combination of their members

$$y_p = Ae^{-2x} + Bxe^{-2x}\sin x + Cxe^{-2x}\cos x.$$

54. The auxiliary equation of the corresponding homogeneous D.E. is $m^2 - 6m + 9 = 0$, with roots 3, 3 (double root). The complementary function is $y_c = (c_1 + c_2 x)e^{3x}$. The NH member is a linear combination of the three UC functions $x^4 e^x$, $x^3 e^{2x}$, $x^2 e^{3x}$. The UC sets of these functions are

$$S_1 = \{x^4 e^x, x^3 e^x, x^2 e^x, xe^x, e^x\},$$

$$S_2 = \{x^3 e^{2x}, x^2 e^{2x}, xe^{2x}, e^{2x}\}, \quad \text{and}$$

$$S_3 = \{x^2 e^{3x}, xe^{3x}, e^{3x}\}, \quad \text{respectively.}$$

None is completely contained in any other, so each is retained. The members xe^{3x} and e^{3x} of S_3 are included in the complementary function and so are solutions of the corresponding homogeneous D.E. Thus we multiply each member of S_3 by x^2 to obtain the revised set $S_3' = \{x^4 e^{3x}, x^3 e^{3x}, x^2 e^{3x}\}$, whose members are not solutions of the homogeneous D.E. (Note that multiplication by x, instead of x^2, is not sufficient here.) We now have the three sets S_1, S_2, and S_3'. We form a linear combination of their twelve members. Thus:

$$\begin{aligned}
y_p = {}& Ax^4 e^x + Bx^3 e^x + Cx^2 e^x + Dxe^x + Ee^x \\
& + Fx^3 e^{2x} + Gx^2 e^{2x} + Hxe^{2x} + Ie^{2x} \\
& + Jx^4 e^{3x} + Kx^3 e^{3x} + Lx^2 e^{3x}.
\end{aligned}$$

55. The auxiliary equation of the corresponding homogeneous
     D.E. is $m^2 + 6m + 13 = 0$, with roots $-3 \pm 2i$.   The
     complementary function is $y_c = e^{-3x}(c_1 \sin 2x + c_2 \cos 2x)$.
     The NH member is a linear combination of the two UC
     functions $xe^{-3x}\sin 2x$ and $x^2 e^{-2x}\sin 3x$.   The UC sets of
     these functions are

$$S_1 = \{xe^{-3x}\sin 2x,\ xe^{-3x}\cos 2x,\ e^{-3x}\sin 2x,\ e^{-3x}\cos 2x\}$$

and

$$S_2 = \{x^2 e^{-2x}\sin 3x,\ x^2 e^{-2x}\cos 3x,\ xe^{-2x}\sin 3x,$$
$$xe^{-2x}\cos 3x,\ e^{-2x}\sin 3x,\ e^{-2x}\cos 3x\},$$

respectively.   Neither is completely contained in the
other, so each is retained.

The members $e^{-3x}\sin 2x$ and $e^{-3x}\cos 2x$ of S_1 are
included in the complementary function and so are
solutions of the corresponding homogeneous D.E.   Thus we
multiply each member of S_1 by x to obtain the revised set.

$$S_1{}' = \{x^2 e^{-3x}\sin 2x,\ x^2 e^{-3x}\cos 2x,\ xe^{-3x}\sin 2x,$$
$$xe^{-3x}\cos 2x\},$$

whose members are not solutions of the homogeneous D.E.

We now have the two sets $S_1{}'$ and S_2.   We form a linear
combination of their ten members.   Thus:

$$y_p = Ax^2e^{-3x}\sin 2x + Bx^2e^{-3x}\cos 2x + Cxe^{-3x}\sin 2x$$

$$+ Dxe^{-3x}\cos 2x + Ex^2e^{-2x}\sin 3x + Fx^2e^{-2x}\cos 3x$$

$$+ Gxe^{-2x}\sin 3x + Hxe^{-2x}\cos 3x + Ie^{-2x}\sin 3x$$

$$+ Je^{-2x}\cos 3x.$$

57. The auxiliary equation of the corresponding homogeneous
     D.E. is $m^3 - 6m^2 + 12m - 8 = 0$ or $(m - 2)^3 = 0$, with roots
     2, 2, 2 (three-fold root). The complementary function is
     $y_c = (c_1 + c_2x + c_3x^2)e^{2x}$. The NH member is a linear
     combination of the two UC functions xe^{2x} and x^2e^{3x}. The
     UC sets of these functions are, respectively,
     $S_1 = \{xe^{2x}, e^{2x}\}$ and $S_2 = \{x^2e^{3x}, xe^{3x}, e^{3x}\}$. Neither is
     contained in the other, so each is retained.

     Both members of S_1 are contained in the complementary
     function and so are solutions of the corresponding
     homogeneous D.E. Thus we multiply each member of S_1 by
     the lowest integral power of x so that the resulting
     revised set will contain no members that are solutions of
     the corresponding homogeneous D.E. By trial, we see that
     multiplying by x or x^2 will *not* accomplish this. Rather,
     we must multiply each member of S_1 by x^3, obtaining
     $S_1' = \{x^4e^{2x}, x^3e^{2x}\}$, whose members are not solutions of
     the homogeneous D.E.

     We now have sets $S_1' = \{x^4e^{2x}, x^3e^{2x}\}$ and
     $S_2 = \{x^2e^{3x}, xe^{3x}, e^{3x}\}$. We form a linear combination of

their five members $y_p = Ax^4e^{2x} + Bx^3e^{2x} + Cx^2e^{3x} + Dxe^{3x} + Ee^{3x}$.

60. The auxiliary equation of the corresponding homogeneous D.E. is $m^6 + 2m^5 + 5m^4 = 0$, with roots 0, 0, 0, 0 (four-fold root) and $-1 \pm 2i$. The complementary function is $y_c = c_1 + c_2x + c_3x^2 + c_4x^3 + e^{-x}(c_5\sin 2x + c_6\cos 2x)$. The NH member is a linear combination of the three UC functions x^3, x^2e^{-x}, and $e^{-x}\sin 2x$. The UC sets of these functions are $S_1 = \{x^3, x^2, x, 1\}$, $S_2 = \{x^2e^{-x}, xe^{-x}, e^{-x}\}$, and $S_3 = \{e^{-x}\sin 2x, e^{-x}\cos 2x\}$. None is completely contained in any other, so all three are retained.

All four members of S_1 are contained in the complementary function and so are solutions of the corresponding homogeneous D.E. Thus we multiply each member of S_1 by the lowest positive integral power of x so that the resulting revised set will contain no members that are solutions of the corresponding homogeneous D.E. By actual trial, we see that multiplying by x, x^2, or x^3 will *not* accomplish this. Rather, we must multiply each member of S_1 by x^4, obtaining

$$S_1' = \{x^7, x^6, x^5, x^4\}$$

whose members are not solutions of the homogeneous D.E.

Also, both members of S_3 are included in the complementary function and so are solutions of the corresponding homogeneous D.E. Thus we multiply each member of S_3 by x to obtain the revised set

$$S_3' = \{xe^{-x}\sin 2x, \ xe^{-x}\cos 2x\}.$$

whose members are not solutions of the homogeneous D.E. We now have the three sets S_1', S_2, and S_3'. We form a linear combination of their nine members. Thus:

$$y_p = Ax^7 + Bx^6 + Cx^5 + Dx^4 + Ex^2e^{-x} + Fxe^{-x}$$
$$+ \ Ge^{-x} + Hxe^{-x}\sin 2x + Ixe^{-x}\cos 2x.$$

63. The auxiliary equation of the corresponding homogeneous D.E. is $m^4 + 3m^2 - 4 = 0$ or $(m^2 + 4)(m^2 - 1) = 0$, with roots $1, -1, \pm 2i$. The complementary function is $y_c = c_1e^x + c_2e^{-x} + c_3\sin 2x + c_4\cos 2x$. The NH member is a linear combination of $\cos^2 x$ and $\cosh x$. Since $\cos^2 x = \dfrac{1 + \cos 2x}{2}$ and $\cosh x = \dfrac{e^x + e^{-x}}{2}$, we see that the NH member is in fact the linear combination

$$\frac{1}{2} + \frac{1}{2}\cos 2x - \frac{1}{2}e^x - \frac{1}{2}e^{-x}$$

of the four UC functions $1, \cos 2x, e^x$, and e^{-x}. The UC sets of these are $S_1 = \{1\}$, $S_2 = \{\cos 2x, \sin 2x\}$, $S_3 = \{e^x\}$, and $S_4 = \{e^{-x}\}$, respectively.

Both members cos 2x and sin 2x of S_2 are contained in the complementary function and so are solutions of the corresponding homogeneous D.E. Thus we multiply each member of S_2 by x to obtain the revised set $S_2' = \{x \cos 2x, \ x \sin 2x\}$, whose members are not solutions of the homogeneous D.E. Thus we replace S_2 by the revised set S_2'. The situation is exactly the same for each of S_3 and S_4, so we replace S_3 by the revised set $S_3' = \{xe^x\}$ and S_4 by the revised set $S_4' = \{xe^{-x}\}$. Thus we have the sets $S_1 = \{1\}$, $S_2' = \{x \sin 2x, \ x \cos 2x\}$, $S_3' = \{xe^x\}$, and $S_4' = \{xe^{-x}\}$. We form a linear combination of their five members. Thus

$$y_p = A + Bx \sin 2x + Cx \cos 2x + Dxe^x + Exe^{-x}.$$

Alternatively, we regard the NH member of the D.E. as a linear combination of the UC function $\cos^2 x = (\cos x)(\cos x)$ and the UC combination $\cosh x = \dfrac{e^x + e^{-x}}{2}$. The UC sets of these are respectively $S_1 = \{\sin^2 x, \ \cos^2 x, \ \sin x \cos x\}$ and $S_2 = \{\sinh x, \ \cosh x\}$. One finds by direct substitution that the member $\sin x \cos x$ of S_1 is a solution of the corresponding homogeneous D.E. Thus we multiply each member of S_1 by x to obtain the revised set $S_1' = \{x \sin^2 x, \ x \cos^2 x, \ x \sin x \cos x\}$, whose members are not solutions of the homogeneous D.E. So we replace S_1 by the revised set S_1'. In like manner, one finds $\cosh x$ is a solution of the homogeneous D.E.; and so we replace S_2 by the revised set $S_2' = \{x \sinh x, \ x \cosh x\}$. So we now have the two revised sets S_1' and S_2'. We form a linear combination of their five members. Thus:

$$y_p = Ax \sin^2 x + Bx \cos^2 x + Cx \sin x \cos x$$
$$+ \ Dx \sinh x + Ex \cosh x.$$

64. The auxiliary equation of the corresponding homogeneous D.E. is $m^4 + 10m^2 + 9 = 0$ or $(m^2 + 9)(m^2 + 1) = 0$ with roots $\pm i, \pm 3i$. The complementary function is $y_c = c_1 \sin x + c_2 \cos x + c_3 \sin 3x + c_4 \cos 3x$. The NH member is the UC function $\sin x \sin 2x$. The UC set of this function is $S_1 = \{\sin x \sin 2x, \ \sin x \cos 2x, \ \cos x \sin 2x,$

$\cos x \cos 2x\}$. One finds by direct substitution that the member $\sin x \sin 2x$ of S_1 is a solution of the corresponding homogeneous D.E. Thus we multiply each member of S_1 by x to obtain the revised set

$S_1' = \{x \sin x \sin 2x, \ x \sin x \cos 2x, \ x \cos x \sin 2x,$

$x \cos x \cos 2x\}$, whose members are not solutions of the homogeneous D.E. So we replace S_1 by the revised set S_1'. We form a linear combination of its four members. Thus:

$$y_p = Ax \sin x \sin 2x + Bx \sin x \cos 2x$$
$$+ \ Cx \cos x \sin 2x + Dx \cos x \cos 2x. \qquad (*)$$

Alternatively, using the trigonometric identify $\sin u \sin v = \frac{1}{2}[\cos(u - v) - \cos(u + v)]$ with $u = 2x$, v = x, we see that the NH member is equal to

$\frac{1}{2}[\cos x - \cos 3x]$ and thus is a linear combination of the

simple UC functions $\cos x$ and $\cos 3x$. The UC sets of these

two functions are $S_2 = \{\cos x,\ \sin x\}$ and $S_3 = \{\cos 3x,$

$\sin 3x\}$, respectively.

Both members of S_2 are contained in the complementary

function and so are solutions of the corresponding

homogeneous D.E. Thus we multiply each member of S_2 by x

to obtain the revised set $S_2' = \{x \cos x,\ x \sin x\}$ whose

members are not solutions of the homogeneous D.E. Thus we

replace S_2 by the revised set S_2'. The situation is

exactly the same for the set S_3, so we replace S_3 by the

revised set $S_3' = \{x \cos 3x,\ x \sin 3x\}$. We now have the two

revised sets S_2' and S_3'. We form a linear combination of

their four members. Thus:

$$y_p = Ex \cos x + Fx \sin x + Gx \cos 3x$$

$$+ Hx \sin 3x \qquad\qquad (**)$$

The student should convince himself that the y_p given by

(*) can be expressed in the form given by (**), and vice

versa.

Section 4.4, Page 169.

3. The complementary function is defined by

$y_c(x) = c_1 \sin x + c_2 \cos x$. We assume a particular integral

of the form

$$y_p(x) = v_1(x)\sin x + v_2(x)\cos x. \qquad (1)$$

Then $y_p'(x) = v_1(x)\cos x - v_2(x)\sin x + v_1'(x)\sin x$
$+ v_2'(x)\cos x.$ We impose the condition

$$v_1'(x)\sin x + v_2'(x)\cos x = 0, \qquad (2)$$

leaving $y_p' = v_1(x)\cos x - v_2(x)\sin x.$ Then from this

$$y_p''(x) = -v_1(x)\sin x - v_2(x)\cos x + v_1'(x)\cos x$$
$$- v_2'(x)\sin x. \qquad (3)$$

Substituting (1) and (3) into the given D.E., we obtain

$$v_1'(x)\cos x - v_2'(x)\sin x = \sec x. \qquad (4)$$

We now have conditions (2) and (4) from which to determine
$v_1'(x)$ and $v_2'(x)$:

$$\begin{cases} \sin x \; v_1'(x) + \cos x \; v_2'(x) = 0, \\ \\ \cos x \; v_1'(x) - \sin x \; v_2'(x) = \sec x . \end{cases}$$

Solving, we find

$$v_1'(x) = \frac{\begin{vmatrix} 0 & \cos x \\ \sec x & -\sin x \end{vmatrix}}{\begin{vmatrix} \sin x & \cos x \\ \cos x & -\sin x \end{vmatrix}} = \frac{-\sec x \cos x}{-1} = 1.$$

$$v_2' = \frac{\begin{vmatrix} \sin x & 0 \\ \cos x & \sec x \end{vmatrix}}{\begin{vmatrix} \sin x & \cos x \\ \cos x & -\sec x \end{vmatrix}} = \frac{\sin x \sec x}{-1} = -\tan x.$$

Integrating, we find $v_1(x) = x$, $v_2(x) = \ln|\cos x|$.

Substituting into $y_p(x) = v_1(x)\sin x + v_2(x)\cos x$, we find

$y_p(x) = x\sin x + \cos x\,[\ln|\cos x\,|]$.   The G.S. of the D.E.

is

$$y = c_1 \sin x + c_2 \cos x + x \sin x + \cos x\,[\ln|\cos x\,|].$$

6. The complementary function is defined by

$y_c(x) = c_1 \sin x + c_2 \cos x$.   We assume a particular

integral of the form

$$y_p(x) = v_1(x)\sin x + v_2(x)\cos x. \qquad (1)$$

Then $y_p'(x) = v_1(x)\cos x - v_2(x)\sin x + v_1'(x)\sin x$

$+ v_2'(x)\cos x$.   We impose the condition

$$v_1'(x)\sin x + v_2'(x)\cos x = 0. \qquad (2)$$

leaving $y_p'(x) = v_1(x)\cos x - v_2(x)\sin x$. Then from this

$$y_p''(x) = -v_1(x)\sin x - v_2(x)\cos x + v_1'(x)\cos x$$
$$- v_2'(x)\sin x. \qquad (3)$$

Substituting (1) and (3) into the given D.E., we obtain

$$v_1'(x)\cos x - v_2'(x)\sin x = \tan x \sec x. \qquad (4)$$

We now have conditions (2) and (4) from which to determine $v_1'(x)$ and $v_2'(x)$:

$$\begin{cases} \sin x \; v_1'(x) + \cos x \; v_2'(x) = 0 \\ \cos x \; v_1'(x) - \sin x \; v_2'(x) = \tan x \sec x. \end{cases}$$

Solving, we find

$$v_1'(x) = \frac{\begin{vmatrix} 0 & \cos x \\ \tan x \sec x & -\sin x \end{vmatrix}}{\begin{vmatrix} \sin x & \cos x \\ \cos x & -\sin x \end{vmatrix}} = \frac{-\cos x \tan x \sec x}{-1}$$

$$= \tan x.$$

$$v_2'(x) = \frac{\begin{vmatrix} \sin x & 0 \\ \cos x & \tan x \sec x \end{vmatrix}}{\begin{vmatrix} \sin x & \cos x \\ \cos x & -\sin x \end{vmatrix}} = \frac{\sin x \tan x \sec x}{-1}$$

$$= -\tan^2 x = 1 - \sec^2 x.$$

Integrating, we find $v_1(x) = -\ln|\cos x|$, $v_2(x) =$
$x - \tan x$.   Substituting into $y_p(x) = v_1(x)\sin x$
$+ v_2(x)\cos x$, we find $y_p(x) = -\sin x \ln|\cos x|$
$+ (x - \tan x)\cos x = -\sin x \ln|\cos x| + x\cos x - \sin x$.
The G.S. of the D.E. is

$$y = c_1\sin x + c_2\cos x - \sin x \ln|\cos x| + x\cos x - \sin x,$$

which may be more simply written as

$$y = c_0\sin x + c_2\cos x - \sin x \ln|\cos x| + x\cos x,$$

where $c_0 = c_1 - 1$.

7. The auxiliary equation $m^2 + 4m + 5 = 0$ has the conjugate
complex roots $m = \dfrac{-4 \pm \sqrt{16 - 20}}{2} = -2 \pm i$, so the
complementary function is $y_c(x) = c_1 e^{-2x}\sin x$
$+ c_2 e^{-2x}\cos x$.   We assume a particular integral of the
form

$$y_p(x) = v_1(x)e^{2x}\sin x + v_2(x)e^{-2x}\cos x. \qquad (1)$$

Then $y_p'(x) = v_1(x)e^{-2x}\cos x - 2v_1(x)e^{-2x}\sin x$
$+ v_1'(x)e^{-2x}\sin x - v_2(x)e^{-2x}\sin x - 2v_2(x)e^{-2x}\cos x$
$+ v_2'(x)e^{-2x}\cos x$.   We impose the condition

$$v_1'(x)e^{-2x}\sin x + v_2'(x)e^{-2x}\cos x = 0, \qquad (2)$$

leaving

$$y_p'(x) = v_1(x)e^{-2x}\cos x - 2v_1(x)e^{-2x}\sin x$$
$$- v_2(x)e^{-2x}\sin x - 2v_2(x)e^{-2x}\cos x . \qquad (3)$$

Then, differentiating this and collecting like terms, we find

$$y_p''(x) = 3v_1(x)e^{-2x}\sin x - 4v_1(x)e^{-2x}\cos x$$
$$+ v_1'(x)[e^{-2x}\cos x - 2e^{-2x}\sin x]$$
$$+ 3v_2(x)e^{-2x}\cos x + 4v_2(x)e^{-2x}\sin x$$
$$- v_2'(x)[e^{-2x}\sin x + 2e^{-2x}\cos x]. \qquad (4)$$

Then substituting (1), (3), and (4) into the given D.E. and collecting like terms, we obtain

$$v_1'(x)[e^{-2x}\cos x - 2e^{-2x}\sin x] - v_2'(x)[e^{-2x}\sin x$$
$$+ 2e^{-2x}\cos x] = e^{-2x}\sec x. \qquad (5)$$

We now have conditions (2) and (5) from which to determine $v_1'(x)$ and $v_2'(x)$:

$$e^{-2x}\sin x \ v_1'(x) + e^{-2x}\cos x \ v_2'(x) = 0,$$

$$e^{-2x}(\cos x - 2\sin x) \ v_1'(x) + e^{-2x}(-\sin x$$
$$- 2\cos x) \ v_2'(x) = e^{-2x}\sec x.$$

Solving, we find:

$$v_1'(x) = \frac{\begin{vmatrix} 0 & e^{-2x}\cos x \\ e^{-2x}\sec x & e^{-2x}(-\sin x - 2\cos x) \end{vmatrix}}{\begin{vmatrix} e^{-2x}\sin x & e^{-2x}\cos x \\ e^{-2x}(\cos x - 2\sin x) & e^{-2x}(-\sin x - 2\cos x) \end{vmatrix}}$$

$$= \frac{-e^{-4x}\cos x \sec x}{e^{-4x}(-1)} = 1.$$

$$v_2'(x) = \frac{\begin{vmatrix} e^{-2x}\sin x & 0 \\ e^{-2x}(\cos x - 2\sin x) & e^{-2x}\sec x \end{vmatrix}}{\begin{vmatrix} e^{-2x}\sin x & e^{-2x}\cos x \\ e^{-2x}(\cos x - 2\sin x) & e^{-2x}(-\sin x - 2\cos x) \end{vmatrix}}$$

$$= \frac{e^{-4x}\sec x \sin x}{e^{-4x}(-1)} = -\tan x.$$

Integrating, we find $v_1(x) = x$, $v_2(x) = \ln|\cos x|$.
Substituting into $y_p(x) = v_1(x)e^{-2x}\sin x + v_2(x)e^{-2x}\cos x$,
we find $y_p(x) = xe^{-2x}\sin x + [\ln|\cos x|] e^{-2x}\cos x$.   The
G.S. of the D.E. is

$$y = e^{-2x}(c_1\sin x + c_2\cos x) + xe^{-2x}\sin x$$
$$+ [\ln|\cos x|] e^{-2x}\cos x.$$

10. The auxiliary equation $m^2 - 2m + 1 = 0$ has roots -1, -1,
so the complementary function is $y_c(x) = (c_1 + c_2x)e^x$,

which we rewrite slightly as $y_c(x) = c_1 e^x + c_2 x e^x$. We assume a particular integral of the form

$$y_p(x) = v_1(x)e^x + v_2(x)xe^x. \tag{1}$$

Then $y'_p(x) = v_1(x)e^x + v_2(x)[(x + 1)e^x] + v_1'(x)e^x + v_2'(x)xe^x$. We impose the condition

$$v_1'(x)e^x + v_2'(x)xe^x = 0 \tag{2}$$

leaving

$$y_p'(x) = v_1(x)e^x + v_2(x)[(x + 1)e^x]. \tag{3}$$

Then

$$y_p''(x) = v_1(x)e^x + v_2(x)[(x + 2)e^x] + v_1'(x)e^x + v_2'(x)[(x + 1)e^x]. \tag{4}$$

Then substituting (1), (3), and (4) into the given D.E., we obtain

$$v_1(x)e^x + v_2(x)[(x + 2)e^x] + v_1'(x)e^x$$
$$+ v_2'(x)[(x + 1)e^x] - 2\{v_1(x)e^x$$
$$+ v_2(x)[(x + 1)e^x]\} + v_1(x)e^x + v_2(x)xe^x$$
$$= xe^x \ln x$$

or

$$v_1{}'(x)e^x + v_2{}'(x)[(x + 1)e^x] = xe^x \ln x. \qquad (5)$$

We now have conditions(2) and (5) from which to determine $v_1{}'(x)$ and $v_2{}'(x)$:

$$\begin{cases} v_1{}'(x)e^x + v_2{}'(x)xe^x = 0, \\ v_1{}'(x)e^x + v_2{}'(x)[(x + 1)e^x] = xe^x \ln x. \end{cases}$$

Dividing out $e^x \neq 0$, these simplify to

$$\begin{cases} v_1{}'(x) + xv_2{}'(x) = 0, \\ v_1{}'(x) + (x + 1)v_2{}'(x) = x \ln x. \end{cases}$$

Subtracting the first from the second, we find

$$v_2{}'(x) = x \ln x;$$

and then the first equation gives

$$v_1{}'(x) = -x^2 \ln x.$$

Integrating (using either tables or integration by parts), we find

$$v_1(x) = \frac{1}{9} x^3 - \frac{1}{3} x^3 \ln x$$

$$v_2(x) = \frac{1}{2} x^2 \ln x - \frac{1}{4} x^2$$

Substituting into $y_p(x) = v_1(x)e^x + v_2(x)xe^x$, we find

$$y_p = \left(\frac{1}{9}x^3 - \frac{1}{3}x^3\ln x\right)e^x + \left(\frac{1}{2}x^2\ln x - \frac{1}{4}x^2\right)xe^x \quad \text{or}$$

$$y_p = -\frac{5}{36}x^3 e^x + \frac{1}{6}x^3 e^x \ln x. \quad \text{The G.S. of the D.E. is}$$

$$y = c_1 e^x + c_2 x e^x - \frac{5}{36}x^3 e^x + \frac{1}{6}x^3 e^x \ln x.$$

12. The complementary function is defined by
$y_c(x) = c_1 \sin x + c_2 \cos x$. We assume a particular
integral of the form

$$y_p(x) = v_1(x)\sin x + v_2(x)\cos x. \quad (1)$$

Then $y_p'(x) = v_1(x)\cos x - v_2(x)\sin x + v_1'(x)\sin x + v_2'(x)\cos x$. We impose the condition

$$v_1'(x)\sin x + v_2'(x)\cos x = 0, \quad (2)$$

leaving $y_p'(x) = v_1(x)\cos x - v_2(x)\sin x$. Then from this,

$$y_p''(x) = -v_1(x)\sin x - v_2(x)\cos x$$
$$+ v_1'(x)\cos x - v_2'(x)\sin x. \quad (3)$$

Substituting (1) and (3) into the given D.E., we obtain

$$v_1'(x)\cos x - v_2'(x)\sin x = \tan^3 x. \quad (4)$$

We now have conditions (2) and (4) from which to determine
$v_1'(x)$ and $v_2'(x)$:

$$\begin{cases} \sin x \; v_1{}'(x) + \cos x \; v_2{}'(x) = 0, \\ \cos x \; v_1{}'(x) - \sin x \; v_2{}'(x) = \tan^3 x. \end{cases}$$

Solving, we find

$$v_1{}'(x) = \frac{\begin{vmatrix} 0 & \cos x \\ \tan^3 x & -\sin x \end{vmatrix}}{\begin{vmatrix} \sin x & \cos x \\ \cos x & -\sin x \end{vmatrix}} = \frac{-\tan^3 x \cos x}{-1}$$

$$= \frac{(1 - \cos^2 x)\sin x}{\cos^2 x} = [(\cos x)^{-2} - 1]\sin x,$$

$$v_2{}'(x) = \frac{\begin{vmatrix} \sin x & 0 \\ \cos x & \tan^3 x \end{vmatrix}}{\begin{vmatrix} \sin x & \cos x \\ \cos x & -\sin x \end{vmatrix}} = \frac{\tan^3 x \sin x}{-1} = -\frac{\sin^4 x}{\cos^3 x}$$

$$= -\frac{(1 - \cos^2 x)^2}{\cos^3 x} = -\frac{1}{\cos^3 x} + \frac{2}{\cos x} - \cos x$$

$$= -\sec^3 x + 2 \sec x - \cos x.$$

Integrating (we recommend using tables), we find

$$v_1(x) = (\cos x)^{-1} + \cos x,$$

$$v_2(x) = -\frac{1}{2} \tan x \sec x$$

$$+ \frac{3}{2} \ln|\sec x + \tan x| - \sin x.$$

Substituting into $y_p(x) = v_1(x)\sin x + v_2(x)\cos x$, we find

$$y_p(x) = [(\cos x)^{-1} + \cos x]\sin x$$
$$+ \left[\frac{1}{2}\tan x \sec x + \frac{3}{2}\ln|\sec x + \tan x|\right.$$
$$\left. - \sin x\right]\cos x = \frac{\tan x}{2} + \frac{3}{2}\cos x \ln|\sec x + \tan x|.$$

The G.S. of the D.E. is

$$y = c_1\sin x + c_2\cos x + \frac{\tan x}{2} + \frac{3}{2}\cos x \ln|\sec x + \tan x|.$$

13.   The auxiliary equation $m^2 + 3m + 2 = 0$ has roots -1, -2, so the complementary function is $y_c(x) = c_1 e^{-x} + c_2 e^{-2x}$. We assume a particular integral of the form

$$y_p(x) = v_1(x)e^{-x} + v_2(x)e^{-2x} \qquad (1)$$

Then $y_p'(x) = -v_1(x)e^{-x} - 2v_2(x)e^{-2x} + v_1'(x)e^{-x} + v_2'(x)e^{-2x}$. We impose the condition

$$v_1'(x)e^{-x} + v_2'(x)e^{-2x} = 0 \qquad (2)$$

leaving

$$y_p'(x) = -v_1(x)e^{-x} - 2v_2(x)e^{-2x}. \qquad (3)$$

From this,

$$y_p''(x) = v_1(x)e^{-x} + 4v_2(x)e^{-2x} - v_1'(x)e^{-x}$$

$$- 2v_2'(x)e^{-2x} \qquad (4)$$

Substituting (1), (3), and (4) into the given D.E., we obtain

$$v_1(x)e^{-x} + 4v_2(x)e^{-2x} - v_1'(x)e^{-x} - 2v_2'(x)e^{-2x}$$

$$+ 3[-v_1(x)e^{-x} - 2v_2(x)e^{-2x}] + 2[v_1(x)e^{-x}$$

$$+ v_2(x)e^{-2x}] = \frac{1}{1 + e^x}$$

or

$$-v_1'(x)e^{-x} - 2v_2'(x)e^{-2x} = \frac{1}{1 + e^x} \qquad (5)$$

We now have conditions (2) and (5) from which to obtain $v_1'(x)$ and $v_2'(x)$:

$$\begin{cases} e^{-x}v_1'(x) + e^{-2x}v_2'(x) = 0, \\[2mm] -e^{-x}v_1'(x) - 2e^{-2x}v_2'(x) = \dfrac{1}{1 + e^x}. \end{cases}$$

Solving, we find

$$v_1'(x) = \frac{\begin{vmatrix} 0 & e^{-2x} \\[2mm] \dfrac{1}{1 + e^x} & -2e^{-2x} \end{vmatrix}}{\begin{vmatrix} e^{-x} & e^{-2x} \\[2mm] -e^{-x} & -2e^{-2x} \end{vmatrix}} = \frac{e^x}{1 + e^x},$$

$$v_2'(x) = \frac{\begin{vmatrix} e^{-x} & 0 \\[2mm] -e^{-x} & \dfrac{1}{1 + e^x} \end{vmatrix}}{\begin{vmatrix} e^{-x} & e^{-2x} \\[2mm] -e^{-x} & -2e^{-2x} \end{vmatrix}} = -\frac{e^{2x}}{1 + e^x}$$

Integrating, we find

$$v_1(x) = \ln(1 + e^x)$$

$$v_2(x) = \int \frac{-e^{2x}}{1 + e^x}\, dx = \int \left(\frac{e^x}{1 + e^x} - e^x \right) dx$$

$$= \ln(1 + e^x) - e^x.$$

Substituting into $y_p(x) = v_1(x)e^{-x} + v_2(x)e^{-2x}$, we find
$y_p(x) = [\ln(1 + e^x)]e^{-x} + [\ln(1 + e^x) - e^x]e^{-2x}$. The G.S.
of the D.E. is

$$y = c_1 e^{-x} + c_2 e^{-2x} + [\ln(1 + e^x)]e^{-x}$$
$$+ [\ln(1 + e^x) - e^x]e^{-2x}$$

or

$$y = c_3 e^{-x} + c_2 e^{-2x} + (e^{-x} + e^{-2x})[\ln(1 + e^x)],$$

where $c_3 = c_1 - 1$.

15.   The complementary function is defined by
$y_c(x) = c_1 \sin x + c_2 \cos x$.   We assume a particular
integral of the form

$$y_p(x) = v_1(x)\sin x + v_2(x) \cos x. \qquad (1)$$

Then $y_p'(x) = v_1(x)\cos x - v_2(x) \sin x + v_1'(x)\sin x$
$+ v_2'(x)\cos x$.   We impose the condition

$$v_1'(x)\sin x + v_2'(x) \cos x = 0, \qquad (2)$$

leaving $y_p'(x) = v_1(x)\cos x - v_2(x) \sin x$.   Then from this

$$y_p''(x) = -v_1(x)\sin x - v_2(x) \cos x$$
$$+ v_1'(x)\cos x - v_2'(x) \sin x. \qquad (3)$$

Substituting (1) and (3) into the given D.E., we obtain

$$v_1'(x)\cos x - v_2'(x)\sin x = \frac{1}{1 + \sin x}. \qquad (4)$$

We now have conditions (2) and (4) from which to determine
$v_1'(x)$ and $v_2'(x)$:

$$\begin{cases} \sin x \; v_1'(x) + \cos x \; v_2'(x) = 0, \\ \cos x \; v_1'(x) - \sin x \; v_2'(x) = \dfrac{1}{1 + \sin x}. \end{cases}$$

Solving, we find

$$v_1'(x) = \frac{\begin{vmatrix} 0 & \cos x \\ \dfrac{1}{1 + \sin x} & -\sin x \end{vmatrix}}{\begin{vmatrix} \sin x & \cos x \\ \cos x & -\sin x \end{vmatrix}} = \frac{\cos x}{1 + \sin x},$$

$$v_2'(x) = \frac{\begin{vmatrix} \sin x & 0 \\ \cos x & \dfrac{1}{1 + \sin x} \end{vmatrix}}{\begin{vmatrix} \sin x & \cos x \\ \cos x & -\sin x \end{vmatrix}} = -\frac{\sin x}{1 + \sin x}.$$

Integrating, we find

$$v_1(x) = \ln(1 + \sin x),$$

$$v_2(x) = -\int \frac{\sin x \, dx}{1 + \sin x} = \int \left(-1 + \frac{1}{1 + \sin x} \right) dx$$

$$= -x - \frac{\cos x}{1 + \sin x}.$$

Substituting into $y_p(x) = v_1(x) \sin x + v_2(x) \cos x$, we find

$$y_p(x) = \sin x \, [\ln(1 + \sin x)] - x \cos x - \frac{\cos^2 x}{1 + \sin x}.$$

The G.S. of the D.E. is

$$y = c_1 \sin x + c_2 \cos x + \sin x \left[\ln(1 + \sin x)\right]$$

$$- x \cos x - \frac{\cos^2 x}{1 + \sin x}.$$

16. The auxiliary equation $m^2 - 2m + 1 = 0$ has roots 1, 1
    (double root), so the complementary function is
    $y_c(x) = (c_1 + c_2 x)e^x$, which we rewrite slightly as
    $y_c(x) = c_1 e^x + c_2 x e^x$. We assume a particular integral of
    the form

$$y_p(x) = v_1(x)e^x + v_2(x)x e^x. \tag{1}$$

Then

$$y_p'(x) = v_1(x)e^x + v_2(x)\left[(x + 1)e^x\right]$$

$$+ v_1'(x)e^x + v_2'(x)x e^x.$$

We impose the condition

$$v_1'(x)e^x + v_2'(x)x e^x = 0 \tag{2}$$

leaving

$$y_p'(x) = v_1(x)e^x + v_2(x)\left[(x + 1)e^x\right]. \tag{3}$$

Then

$$y_p''(x) = v_1(x)e^x + v_2(x)\left[(x + 2)e^x\right] + v_1'(x)e^x$$

$$+ v_2'(x)\left[(x + 1)e^x\right]. \tag{4}$$

Then substituting (1), (3), and (4) into the given D.E.,
we obtain

$$v_1(x)e^x + v_2(x)[(x + 2)e^x] + v_1'(x)e^x$$

$$+ v_2'(x)[(x + 1)e^x] - 2\{v_1(x)e^x$$

$$+ v_2(x)[(x + 1)e^x]\} + v_1(x)e^x + v_2(x)xe^x$$

$$= e^x\sin^{-1}x$$

or

$$v_1'(x)e^x + v_2'(x)[(x + 1)e^x] = e^x\sin^{-1}x. \qquad (5)$$

We now have conditions (2) and (5) from which to determine
$v_1'(x)$ and $v_2'(x)$:

$$\begin{cases} e^x v_1'(x) + xe^x v_2'(x) = 0, \\ e^x v_1'(x) + (x + 1)e^x v_2'(x) = e\sin^{-1}x. \end{cases}$$

Dividing out $e^x \neq 0$, these simplify to

$$\begin{cases} v_1'(x) + x v_2'(x) = 0, \\ v_1'(x) + (x + 1)v_2'(x) = \sin^{-1}x. \end{cases}$$

Subtracting the first from the second, we find

$$v_2'(x) = \sin^{-1}x;$$

and then the first equation gives

$$v_1'(x) = -xv_2'(x) = -x \sin^{-1}x.$$

Integrating (using tables), we find

$$v_1(x) = -\left(\frac{x^2}{2}\right)\sin^{-1}x + \left(\frac{1}{4}\right)\sin^{-1}x - \left(\frac{x}{4}\right)\sqrt{1 - x^2},$$

$$v_2(x) = x \sin^{-1}x + \sqrt{1 - x^2}.$$

Substituting into $y_p(x) = v_1(x)e^x + v_2(x)xe^x$, we find

$$y_p(x) = -\left(\frac{x^2 e^x}{2}\right)\sin^{-1}x + \left(\frac{1}{4}\right)xe^x\sin^{-1}x$$

$$- \left(\frac{1}{4}\right)xe^x\sqrt{1 - x^2} + x^2 e^x\sin^{-1}x$$

$$+ xe^x\sqrt{1 - x^2}$$

$$= \frac{e^x\sin^{-1}x}{4} + \frac{x^2 e^x\sin^{-1}x}{2} + \frac{3xe^x\sqrt{1 - x^2}}{4}.$$

The G.S. of the D.E. is

$$y = (c_1 + c_2 x)e^x + \frac{e^x\sin^{-1}x}{4} + \frac{x^2 e^x\sin^{-1}x}{2}$$

$$+ \frac{3xe^x\sqrt{1 - x^2}}{4}.$$

17. The auxiliary equation $m^2 + 3m + 2 = 0$ has the roots -1, -2, so the complementary function is $y_c(x) = c_1 e^{-x} + c_2 e^{-2x}$. We assume a particular integral of the form

$$y_p(x) = v_1(x)e^{-x} + v_2(x)e^{-2x}. \qquad (1)$$

Then

$$y_p'(x) = -v_1(x)e^{-x} - 2v_2(x)e^{-2x} + v_1'(x)e^{-x} + v_2'(x)e^{-2x}.$$

We impose the condition

$$v_1'(x)e^{-x} + v_2'(x)e^{-2x} = 0 \qquad (2)$$

leaving

$$y_p'(x) = -v_1(x)e^{-x} - 2v_2(x)e^{-2x}. \qquad (3)$$

Then from this,

$$y_p''(x) = v_1(x)e^{-x} + 4v_2(x)e^{-2x} - v_1'(x)e^{-x}$$

$$- 2v_2'(x)e^{-2x}. \qquad (4)$$

Substituting (1), (3), and (4) into the given D.E., we obtain

$$v_1(x)e^{-x} + 4v_2(x)e^{-2x} - v_1'(x)e^{-x} - 2v_2'(x)e^{-2x}$$

$$+ 3[-v_1(x)e^{-x} - 2v_2(x)e^{-2x}] + 2[v_1(x)e^{-x}$$

$$+ v_2(x)e^{-2x}] = \frac{e^{-x}}{x}$$

or

$$-v_1'(x)e^{-x} - 2v_2'(x)e^{-2x} = \frac{e^{-x}}{x}. \qquad (5)$$

We now have conditions (2) and (5) from which to determine $v_1'(x)$ and $v_2'(x)$:

$$\begin{cases} e^{-x} v_1'(x) + e^{-2x} v_2'(x) = 0 \\ -e^{-x} v_1'(x) - 2e^{-2x} v_2'(x) = \dfrac{e^{-x}}{x}. \end{cases}$$

Solving, we find

$$v_1'(x) = \frac{\begin{vmatrix} 0 & e^{-2x} \\ \dfrac{e^{-x}}{x} & -2e^{-2x} \end{vmatrix}}{\begin{vmatrix} e^{-x} & e^{-2x} \\ -e^{-x} & -2e^{-2x} \end{vmatrix}} = \frac{1}{x},$$

$$v_2'(x) = \frac{\begin{vmatrix} e^{-x} & 0 \\ -e^{-x} & \dfrac{e^{-x}}{x} \end{vmatrix}}{\begin{vmatrix} e^{-x} & e^{-2x} \\ -e^{-x} & -2e^{-2x} \end{vmatrix}} = -\frac{e^{x}}{x}.$$

Integrating, we find $v_1(x) = \ln|x|$. We also have $v_2(x) = -\int \left(\dfrac{e^x}{x}\right) dx$. Since $\int \left(\dfrac{e^x}{x}\right) dx$ cannot be expressed in closed form in terms of a finite number of elementary functions, we simply leave it as indicated. Substituting into $y_p(x) = v_1(x)e^{-x} + v_2(x)e^{-2x}$, we find

$y_p(x) = e^{-x}\ln|x| - e^{-2x}\int \left(\dfrac{e^x}{x}\right) dx$. The G.S. of the D.E. is

$$y = c_1 e^{-x} + c_2 e^{-2x} + e^{-x}\ln|x| - e^{-2x}\int \left(\frac{e^x}{x}\right) dx.$$

18. The auxiliary equation $m^2 - 2m + 1 = 0$ has the roots 1, 1
     (double root), so the complementary function is
     $y_c(x) = (c_1 + c_2 x)e^x$, which we rewrite slightly as
     $y_c(x) = c_1 e^x + c_2 x e^x$. We assume a particular integral of
     the form

$$y_p(x) = v_1(x)e^x + v_2(x)xe^x. \tag{1}$$

Then

$$y_p{}'(x) = v_1(x)e^x + v_2(x)[(x + 1)e^x] + v_1{}'(x)e^x$$
$$+ v_2{}'(x)xe^x.$$

We impose the condition

$$v_1{}'(x)e^x + v_2{}'(x)xe^x = 0, \tag{2}$$

leaving

$$y_p{}'(x) = v_1(x)e^x + v_2(x)[(x + 1)e^x]. \tag{3}$$

Then

$$y_p{}''(x) = v_1(x)e^x + v_2(x)[(x + 2)e^x] + v_1{}'(x)e^x$$
$$+ v_2{}'(x)[(x + 1)e^x]. \tag{4}$$

Then substituting (1), (3), and (4) into the given D.E.,
we obtain

$$v_1(x)e^x + v_2(x)[(x + 2)e^x] + v_1'(x)e^x$$
$$+ v_2'(x)[(x + 1)e^x] - 2\{v_1(x)e^x$$
$$+ v_2(x)[(x + 1)e^x]\}$$
$$+ v_1(x)e^x + v_2(x)xe^x = x \ln x$$

or

$$v_1'(x)e^x + v_2'(x)[(x + 1)e^x] = x \ln x. \qquad (5)$$

We now have conditions (2) and (5) from which to determine $v_1'(x)$ and $v_2'(x)$:

$$\begin{cases} e^x v_1'(x) + xe^x v_2'(x) = 0, \\ e^x v_1'(x) + (x + 1)e^x v_2'(x) = x \ln x. \end{cases}$$

Dividing out $e^x \neq 0$, these simplify to

$$v_1'(x) + xv_2'(x) = 0,$$
$$v_1'(x) + (x + 1)v_2'(x) = e^{-x}x \ln x.$$

Subtracting the first from the second, we find

$$v_2'(x) = e^{-x}x \ln x;$$

and then the first equation gives

$$v_1'(x) = -e^{-x}x^2\ln x.$$

Thus we have $v_1(x) = -\int \dfrac{x^2 \ln x}{e^x}\, dx$ and $v_2(x) = \int \dfrac{x \ln x}{e^x}\, dx$;

and since neither of these integrals can be expressed in closed form in terms of a finite number of elementary functions, we simply leave them as indicated. Substituting into

$$y_p(x) = v_1(x)e^x + v_2(x)xe^x,$$

we find

$$y_p(x) = -e^x \int \dfrac{x^2 \ln x}{e^x}\, dx + xe^x \int \dfrac{x \ln x}{e^x}\, dx.$$

The G.S. of the D.E. is

$$y = (c_1 + c_2 x)e^x - e^x \int \dfrac{x^2 \ln x}{e^x}\, dx + xe^x \int \dfrac{x \ln x}{e^x}\, dx.$$

21. Since $x + 1$ and x^2 are linearly independent solutions of the corresponding homogeneous equation, the complementary function is defined by $y_c(x) = c_1(x + 1) + c_2 x^2$. We assume a particular integral of the form

$$y_p(x) = v_1(x)(x + 1) + v_2(x)x^2. \qquad (1)$$

Then

$$y_p'(x) = v_1(x) + 2v_2(x)x + v_1'(x)(x + 1) + v_2'(x)x^2.$$

We impose the condition

$$v_1'(x)(x + 1) + v_2'(x)x^2 = 0, \qquad (2)$$

leaving

$$y_p'(x) = v_1(x) + 2v_2(x)x. \qquad (3)$$

Then from this,

$$y_p''(x) = v_1'(x) + 2v_2(x) + 2v_2'(x)x. \qquad (4)$$

Substituting (1), (3), and (4) into the given D.E., we obtain

$$(x^2 + 2x)[v_1'(x) + 2v_2(x) + 2v_2'(x)x]$$
$$- 2(x + 1)[v_1(x) + 2v_2(x)x] + 2[v_1(x)(x + 1)$$
$$+ v_2(x)x^2] = (x + 2)^2,$$

or

$$(x^2 + 2x)[v_1'(x) + 2v_2'(x)x] = (x + 2)^2,$$

or finally,

$$v_1'(x) + 2v_2'(x)x = \frac{x + 2}{x}. \qquad (5)$$

We now have conditions (2) and (5) from which to determine $v_1'(x)$ and $v_2'(x)$:

$$\begin{cases} (x + 1)v_1'(x) + x^2 v_2'(x) = 0, \\ \\ v_1'(x) + 2x\, v_2'(x) = \dfrac{x + 2}{x}. \end{cases}$$

Solving, we find

$$v_1'(x) = \frac{\begin{vmatrix} 0 & x^2 \\ \dfrac{x + 2}{x} & 2x \end{vmatrix}}{\begin{vmatrix} x + 1 & x^2 \\ 1 & 2x \end{vmatrix}} = \frac{-x(x + 2)}{x(x + 2)} = -1,$$

$$v_2'(x) = \frac{\begin{vmatrix} x + 1 & 0 \\ 1 & \dfrac{x + 2}{x} \end{vmatrix}}{\begin{vmatrix} x + 1 & x^2 \\ 1 & 2x \end{vmatrix}} = \frac{\left[\dfrac{(x + 1)(x + 2)}{x}\right]}{x(x + 2)} = \frac{1}{x} + \frac{1}{x^2}.$$

Integrating we find $v_1(x) = -x$, $v_2(x) = \ln|x| - x^{-1}$.

Substituting into $y_p(x) = v_1(x)(x + 1) + v_2(x)x^2$, we have

$y_p(x) = -x(x + 1) + (\ln|x| - x^{-1})x^2 = -x^2 - 2x + x^2\ln|x|$.

The G.S. of the D.E. is

$$y = c_1(x + 1) + c_2x^2 - x^2 - 2x + x^2\ln|x|.$$

22. Since x and xe^x are linearly independent solutions of the corresponding homogeneous equation, the complementary function is defined by $y_c(x) = c_1x + c_2xe^x$. We assume a particular integral of the form

$$y_p(x) = v_1(x)x + v_2(x)xe^x. \qquad (1)$$

Then

$$y_p'(x) = v_1(x) + v_2(x)[(x + 1)e^x] + v_1'(x)x + v_2'(x)xe^x.$$

We impose the condition

$$v_1'(x)x + v_2'(x)xe^x = 0, \qquad (2)$$

leaving

$$y_p'(x) = v_1(x) + v_2(x)[(x + 1)e^x]. \qquad (3)$$

Then from this,

$$y_p''(x) = v_1'(x) + v_2(x)[(x + 2)e^x]$$
$$+ v_2'(x)[(x + 1)e^x]. \qquad (4)$$

Substituting (1), (3), and (4) into the given D.E., we obtain

$$x^2\{v_1'(x) + v_2(x)[(x + 2)e^x] + v_2'(x)[(x + 1)e^x]\}$$
$$- x(x + 2)\{v_1(x) + v_2(x)[(x + 1)e^x]\}$$
$$+ (x + 2)[v_1(x)x + v_2(x)xe^x] = x^3$$

or

$$x^2\{v_1'(x) + v_2'(x)[(x + 1)e^x]\} = x^3$$

or finally,

$$v_1'(x) + v_2'(x)[(x + 1)e^x] = x. \qquad (5)$$

We now have conditions (2) and (5) from which to obtain $v_1'(x)$ and $v_2'(x)$:

$$\begin{cases} v_1'(x)x + v_2'(x)xe^x = 0, \\ v_1'(x) + v_2'(x)[(x + 1)e^x] = x. \end{cases}$$

Solving, we find

$$v_1'(x) = \frac{\begin{vmatrix} 0 & xe^x \\ x & (x + 1)e^x \end{vmatrix}}{\begin{vmatrix} x & xe^x \\ 1 & (x + 1)e^x \end{vmatrix}} = -1,$$

$$v_2'(x) = \frac{\begin{vmatrix} x & 0 \\ 1 & x \end{vmatrix}}{\begin{vmatrix} x & xe^x \\ 1 & (x + 1)e^x \end{vmatrix}} = e^{-x}.$$

Integrating, we find $v_1(x) = -x$, $v_2(x) = -e^{-x}$.

Substituting into $y_p(x) = v_1(x)x + v_2(x)xe^x$, we have $y_p(x) = -x^2 - x$. The G.S. of the D.E. is

$$y = c_1x + c_2xe^x - x^2 - x.$$

24.   Since x and $(x + 1)^{-1}$ are linearly independent solutions
of the corresponding homogeneous equation, the
complementary function is defined by
$y_c(x) = c_1 x + c_2(x + 1)^{-1}$.   We assume a particular
integral of the form

$$y_p(x) = v_1(x)x + v_2(x)(x + 1)^{-1}. \qquad (1)$$

Then

$$y_p'(x) = v_1(x) - v_2(x)(x + 1)^{-2} + v_1'(x)x$$
$$+ v_2'(x)(x + 1)^{-1}.$$

We impose the condition

$$v_1'(x)x + v_2'(x)(x + 1)^{-1} = 0, \qquad (2)$$

leaving

$$y_p'(x) = v_1(x) - v_2(x)(x + 1)^{-2}. \qquad (3)$$

Then from this,

$$y_p''(x) = v_1'(x) + 2v_2(x)(x + 1)^{-3}$$
$$- v_2'(x)(x + 1)^{-2}. \qquad (4)$$

Substituting (1), (3), and (4) into the given D.E., we
obtain

$$(2x + 1)(x + 1)[v_1'(x) + 2v_2(x)(x + 1)^{-3}$$

$$- v_2'(x)(x + 1)^{-2}] + 2x[v_1(x) - v_2(x)(x + 1)^{-2}]$$

$$- 2[v_1(x)x + v_2(x)(x + 1)^{-1}] = (2x + 1)^2,$$

or

$$(2x + 1)(x + 1)[v_1'(x) - v_2'(x)(x + 1)^{-2}] = (2x + 1)^2,$$

or finally,

$$v_1'(x) - v_2'(x)(x + 1)^{-2} = (2x + 1)(x + 1)^{-1}. \quad (5)$$

We now have conditions (2) and (5) from which to determine $v_1'(x)$ and $v_2'(x)$:

$$\begin{cases} x v_1'(x) + (x + 1)^{-1} v_2'(x) = 0, \\ v_1'(x) - (x + 1)^{-2} v_2'(x) = (2x + 1)(x + 1)^{-1}. \end{cases}$$

Solving, we find

$$v_1'(x) = \frac{\begin{vmatrix} 0 & (x + 1)^{-1} \\ (2x + 1)(x + 1)^{-1} & -(x + 1)^{-2} \end{vmatrix}}{\begin{vmatrix} x & (x + 1)^{-1} \\ 1 & -(x + 1)^{-2} \end{vmatrix}}$$

$$= \frac{-(2x + 1)(x + 1)^{-2}}{-x(x + 1)^{-2} - (x + 1)^{-1}} = 1,$$

$$v_2'(x) = \frac{\begin{vmatrix} x & 0 \\ 1 & (2x + 1)(x + 1)^{-1} \end{vmatrix}}{\begin{vmatrix} x & (x + 1)^{-1} \\ 1 & -(x + 1)^{-2} \end{vmatrix}}$$

$$= \frac{x(2x + 1)(x + 1)^{-1}}{-x(x + 1)^{-2} - (x + 1)^{-1}} = -x(x + 1).$$

Integrating, we find $v_1(x) = x$, $v_2(x) = -\dfrac{x^3}{3} - \dfrac{x^2}{2}$.

Substituting into

$$y_p(x) = v_1(x)x + v_2(x)(x + 1)^{-1},$$

we have

$$y_p(x) = x^2 - \frac{(2x^3 + 3x^2)(x + 1)^{-1}}{6}.$$

The G.S. of the D.E. is

$$y = c_1 x + c_2(x + 1)^{-1} + x^2 - \frac{(2x^3 + 3x^2)(x + 1)^{-1}}{6}.$$

25. Since $\sin x$ and $x \sin x$ are linearly independent solutions
of the corresponding homogeneous equation, the
complementary function is defined by $y_c(x) = c_1 \sin x$
+ $c_2 x \sin x$. We assume a particular integral of the form

$$y_p(x) = v_1(x)\sin x + v_2(x)x \sin x. \qquad (1)$$

Then

$$y_p'(x) = v_1(x)\cos x + v_2(x)[x \cos x + \sin x]$$
$$+ v_1'(x)\sin x + v_2'(x)x \sin x.$$

We impose the condition

$$v_1'(x)\sin x + v_2'(x)x \sin x = 0, \qquad (2)$$

leaving

$$y_p'(x) = v_1(x)\cos x + v_2(x)[x \cos x + \sin x]. \qquad (3)$$

Then from this,

$$y_p''(x) = -v_1(x)\sin x + v_1'(x)\cos x$$
$$+ v_2(x)[-x \sin x + 2 \cos x]$$
$$+ v_2'(x)[x \cos x + \sin x]. \qquad (4)$$

Substituting (1), (3), and (4) into the given D.E., we obtain

$$\sin^2 x\{-v_1(x)\sin x + v_1'(x)\cos x$$
$$+ v_2(x)[-x \sin x + 2 \cos x]$$
$$+ v_2'(x)[x \cos x + \sin x]\}$$
$$- 2 \sin x \cos x\{v_1(x)\cos x$$

$$+ v_2(x) [x \cos x + \sin x]\}$$
$$+ (\cos^2 x + 1) [v_1(x) \sin x$$
$$+ v_2(x) x \sin x] = \sin^3 x,$$

or

$$\sin^2 x \{ v_1'(x) \cos x + v_2'(x) [x \cos x + \sin x] \} = \sin^3 x,$$

or finally

$$v_1'(x) \cos x + v_2'(x) [x \cos x + \sin x] = \sin x . \quad (5)$$

We now have conditions (2) and (5) from which to obtain $v_1'(x)$ and $v_2'(x)$:

$$\begin{cases} v_1'(x) \sin x + v_2'(x) x \sin x = 0, \\ v_1'(x) \cos x + v_2'(x) [x \cos x + \sin x] = \sin x. \end{cases}$$

Solving, we find

$$v_1'(x) = \frac{\begin{vmatrix} 0 & x \sin x \\ \sin x & x \cos x + \sin x \end{vmatrix}}{\begin{vmatrix} \sin x & x \sin x \\ \cos x & x \cos x + \sin x \end{vmatrix}} = -x,$$

$$v_2'(x) = \frac{\begin{vmatrix} \sin x & 0 \\ \cos x & \sin x \end{vmatrix}}{\begin{vmatrix} \sin x & x \sin x \\ \cos x & x \cos x + \sin x \end{vmatrix}} = 1.$$

Integrating, we find $v_1(x) = -\frac{x^2}{2}$, $v_2(x) = x$.

Substituting into $y_p(x) = v_1(x)\sin x + v_2(x)x\sin x$, we have

$$y_p(x) = -\frac{x^2\sin x}{2} + x^2\sin x$$

or

$$y_p(x) = \frac{x^2\sin x}{2}.$$

The G.S. of the D.E. is

$$y = c_1\sin x + c_2 x\sin x + \frac{x^2\sin x}{2}.$$

Section 4.5, Page 176

1. Let $x = e^t$; then $t = \ln x$; and (as on page 172 of the text)

$$x\frac{dy}{dx} = \frac{dy}{dt}, \qquad x^2\frac{d^2y}{dx^2} = \frac{d^2y}{dt^2} - \frac{dy}{dt}.$$

The D.E. transforms into

$$\frac{d^2y}{dt^2} - \frac{dy}{dt} - 3\frac{dy}{dt} + 3y = 0$$

or $\frac{d^2y}{dt^2} - 4\frac{dy}{dt} + 3y = 0$. The auxiliary equation of this is $m^2 - 4m + 3 = 0$, and it has the roots $m = 1, 3$. Thus the general solution of the D.E. in y and t is $y = c_1 e^t + c_2 e^{3t}$. We return to the original independent variable x and replace e^t by x and e^{3t} by x^3. Doing this, we find that the general solution of the given D.E. is $y = c_1 x + c_2 x^3$.

3. Let $x = e^t$; then $t = \ln x$, $x \frac{dy}{dx} = \frac{dy}{dt}$, $x^2 \frac{d^2 y}{dx^2} = \frac{d^2 y}{dt^2} - \frac{dy}{dt}$.

The D.E. transforms into $4\left(\frac{d^2 y}{dt^2} - \frac{dy}{dt}\right) - 4 \frac{dy}{dt} + 3y = 0$ or

$4 \frac{d^2 y}{dt^2} - 8 \frac{dy}{dt} + 3y = 0$. The auxiliary equation of this is

$4m^2 - 8m + 3 = 0$, and it has the roots $m = \frac{1}{2}, \frac{3}{2}$. Thus the

general solution of the D.E. in y and t is
$y = c_1 e^{t/2} + c_2 e^{3t/2}$. We return to the original

independent variable x and replace $e^{t/2}$ by $x^{1/2}$ and $e^{3t/2}$
by $x^{3/2}$. Doing this, we find that the general solution of
the given D.E. is $y = c_1 x^{1/2} + c_2 x^{3/2}$.

6. Let $x = e^t$; then $t = \ln x$, $x \frac{dy}{dx} = \frac{dy}{dt}$, $x^2 \frac{d^2 y}{dx^2} = \frac{d^2 y}{dt^2} - \frac{dy}{dt}$.

The D.E. transforms into $\frac{d^2 y}{dt^2} - \frac{dy}{dt} - 3 \frac{dy}{dt} + 13y = 0$ or

$\frac{d^2 y}{dt^2} - 4 \frac{dy}{dt} + 13y = 0$. The auxiliary equation of this is

$m^2 - 4m + 13 = 0$, and it has the conjugate complex roots
$2 \pm 3i$. The general solution of the D.E. in y and t is
$y = e^{2t}(c_1 \sin 3t + c_2 \cos 3t)$. We return to the original

independent variable x and replace e^{2t} by x^2 and t by
$\ln x$. Doing this, we find that the general solution of
the given D.E. is $y = x^2[c_1 \sin(3 \ln x) + c_2 \cos(3 \ln x)]$, or

$y = x^2[c_1 \sin(\ln x^3) + c_2 \cos(\ln x^3)]$.

8. Let $x = e^t$; then $t = \ln x$, $x\dfrac{dy}{dx} = \dfrac{dy}{dt}$, $x^2\dfrac{d^2y}{dx^2} = \dfrac{d^2y}{dt^2} - \dfrac{dy}{dt}$.

The D.E. transforms into $\dfrac{d^2y}{dt^2} - \dfrac{dy}{dt} + \dfrac{dy}{dt} + 9y = 0$ or

$\dfrac{d^2y}{dt^2} + 9y = 0$.   The auxiliary equation of this is

$m^2 + 9 = 0$, and it has roots $\pm 3i$.   Thus the general

solution of the D.E. in y and t is y =

$c_1\sin 3t + c_2\cos 3t$.   We return to the original

independent variable x and replace t by $\ln x$.   Doing this,

we find that the general solution of the given D.E. is y =

$c_1\sin(3\ln x) + c_2\cos(3\ln x)$ or $y = c_1\sin(\ln x^3)$

$+ c_2\cos(\ln x^3)$.

9. Let $x = e^t$; then $t = \ln x$, $x\dfrac{dy}{dx} = \dfrac{dy}{dt}$, $x^2\dfrac{d^2y}{dx^2} = \dfrac{d^2y}{dt^2} - \dfrac{dy}{dt}$.

The D.E. transforms into $9\left(\dfrac{d^2y}{dt^2} - \dfrac{dy}{dt}\right) + 3\dfrac{dy}{dt} + y = 0$ or

$9\dfrac{d^2y}{dt^2} - 6\dfrac{dy}{dt} + y = 0$.   The auxiliary equation of this

is $9m^2 - 6m + 1 = 0$, and it has the real double root $\dfrac{1}{3}$.

The general solution of the D.E. in y and t is
$y = (c_1 + c_2 t)e^{t/3}$.   We return to the original independent

variable x and replace $e^{t/3}$ by $x^{1/3}$ and t by $\ln x$.   Doing

this, we find that the general solution of the given D.E.

is $y = (c_1 + c_2\ln x)\, x^{1/3}$.

11. Let $x = e^t$; then $t = \ln x$, $x\frac{dy}{dx} = \frac{dy}{dt}$, $x^2\frac{d^2y}{dx^2} = \frac{d^2y}{dt^2} - \frac{dy}{dt}$,

and (as on page 174 of the text) $x^3\frac{d^3y}{dx^3} = \frac{d^3y}{dt^3} - 3\frac{d^2y}{dt^2}$

$+ 2\frac{dy}{dt}$. The D.E. transforms into $\frac{d^3y}{dt^3} - 3\frac{d^2y}{dt^2} + 2\frac{dy}{dt}$

$- 3\left(\frac{d^2y}{dt^2} - \frac{dy}{dt}\right) + 6\frac{dy}{dt} - 6y = 0$ or $\frac{d^3y}{dt^3} - 6\frac{d^2y}{dt^2} + 11\frac{dy}{dt}$

$- 6y = 0$. The auxiliary equation of this is
$m^3 - 6m^2 + 11m - 6 = 0$, and its roots are 1, 2, 3. The
general solution of the D.E. in y and t is
$y = c_1e^t + c_2e^{2t} + c_3e^{3t}$. We return to the original

independent variable x and replace e^t by x, e^{2t} by x^2, and
e^{3t} by x^3. Doing this, we find that the general solution
of the given D.E. is $y = c_1x + c_2x^2 + c_3x^3$.

14. Let $x = e^t$; then $t = \ln x$, $x\frac{dy}{dx} = \frac{dy}{dt}$, and

$x^2\frac{d^2y}{dx^2} = \frac{d^2y}{dt^2} - \frac{dy}{dt}$. To transform $x^4\frac{d^4y}{dx^4}$, we apply the

four-step procedure outlined in the Remarks on page 175 of
the text. With n = 4, we determine
$r(r - 1)(r - 2)(r - 3)$. We expand this to obtain

$r^4 - 6r^3 + 11r^2 - 6r$. Replacing r^k by $\frac{d^ky}{dt^k}$ for k = 1, 2,

3, 4, we have $\frac{d^4y}{dt^4} - 6\frac{d^3y}{dt^3} + 11\frac{d^2y}{dt^2} - 6\frac{dy}{dt}$. This is

$x^4\frac{d^4y}{dx^4}$. That is, $x^4\frac{d^4y}{dx^4} = \frac{d^4y}{dt^4} - 6\frac{d^3y}{dt^3} + 11\frac{d^2y}{dt^2} - 6\frac{dy}{dt}$.

The D.E. transforms into $\dfrac{d^4y}{dt^4} - 6\dfrac{d^3y}{dt^3} + 11\dfrac{d^2y}{dt^2} - 6\dfrac{dy}{dt}$

$- 4\left[\dfrac{d^2y}{dt^2} - \dfrac{dy}{dt}\right] + 8\dfrac{dy}{dt} - 8y = 0$ or $\dfrac{d^4y}{dt^4} - 6\dfrac{d^3y}{dt^3} + 7\dfrac{d^2y}{dt^2}$

$+ 6\dfrac{dy}{dt} - 8y = 0$. The auxiliary equation of this is

$m^4 - 6m^3 + 7m^2 + 6m - 8 = 0$, and its roots are 1, 2, 4,

and -1. The G.S. of the D.E. in y and t is

$y = c_1 e^t + c_2 e^{2t} + c_3 e^{4t} + c_4 e^{-t}$. We return to the

original variable x and replace e^t by x, e^{2t} by x^2, e^{4t} by

x^4, and e^{-t} by x^{-1}. Doing this, we find that the G.S. of

the given D.E. is $y = c_1 x + c_2 x^2 + c_3 x^4 + c_4 x^{-1}$.

15. Let $x = e^t$; then $t = \ln x$, $x\dfrac{dy}{dx} = \dfrac{dy}{dt}$, $x^2\dfrac{d^2y}{dx^2} = \dfrac{d^2y}{dt^2} - \dfrac{dy}{dt}$.

The D.E. transforms into

$$\dfrac{d^2y}{dt^2} - 5\dfrac{dy}{dt} + 6y = 4e^t - 6 \qquad (1)$$

(note that x in the right member has transformed into e^t;

see Remark 1, page 174 of text). The auxiliary equation

of the corresponding homogeneous D.E. is $m^2 - 5m + 6 = 0$,

with roots $m = 2, 3$. Thus the complementary function of

(1) is $y_c = c_1 e^{2t} + c_2 e^{3t}$. We find a particular integral

by the method of undetermined coefficients. We assume

$y_p = Ae^t + B$. Then $y_p' = Ae^t$, $y_p'' = Ae^t$, and substituting

into D.E. (1), we quickly have $2Ae^t + 6B = 4e^t - 6$. From

this, $2A = 4$, $6B = -6$, and hence $A = 2$, $B = -1$. Thus the

particular integral of (1) is $y_p = 2e^t - 1$, and its

general solution is $y = c_1e^{2t} + c_2e^{3t} + 2e^t - 1$.

Returning to the original independent variable x by replacing e^t by x, etc., we find the general solution of the given D.E.: $y = c_1x^2 + c_2x^3 + 2x - 1$.

16. Let $x = e^t$; then $t = \ln x$, $x\frac{dy}{dx} = \frac{dy}{dt}$, $x^2\frac{d^2y}{dx^2} = \frac{d^2y}{dt^2} - \frac{dy}{dt}$.

The D.E. transforms into

$$\frac{d^2y}{dt^2} - 6\frac{dy}{dt} + 8y = 2e^{3t}. \tag{1}$$

The auxiliary equation of the corresponding homogeneous D.E. is $m^2 - 6m + 8 = 0$, with roots $m = 2, 4$. Thus the complementary function of (1) is $y_c = c_1e^{2t} + c_2e^{4t}$. We find a particular integral by the method of undetermined coefficients. We assume $y_p = Ae^{3t}$. Then $y_p' = 3Ae^{3t}$, $y_p'' = 9Ae^{3t}$, and substituting into D.E. (1), we quickly have $-Ae^{3t} = 2e^{3t}$. From this, $-A = 2$ and so $A = -2$. Thus the particular integral of (1) is $y_p = -2e^{3t}$, and its general solution is $y = c_1e^{2t} + c_2e^{4t} - 2e^{3t}$. Returning to the original independent variable x by replacing e^t by x, etc., we find the general solution of the given D.E.: $y = c_1x^2 + c_2x^4 - 2x^3$.

18. Let $x = e^t$; then $t = \ln x$, $x\frac{dy}{dx} = \frac{dy}{dt}$, and $x^2\frac{d^2y}{dx^2} = \frac{d^2y}{dt^2} - \frac{dy}{dt}$. The D.E. transforms into

$$\frac{d^2y}{dt^2} + 4y = 2t\, e^t. \tag{1}$$

The auxiliary equation of the corresponding homogeneous
D.E. is $m^2 + 4 = 0$ with roots $m = \pm 2i$.   Thus the
complementary function of (1) is $y_c = c_1 \sin 2t + c_2 \cos 2t$.

We find a particular integral by the method of
undetermined coefficients. We assume $y_p = Ate^t + Be^t$.

Then $y_p' = Ate^t + (A + B)e^t$, $y_p'' = Ate^t + (2A + B)e^t$, and

substituting into the D.E. (1), we have
$5Ate^t + (2A + 5B)e^t = 2te^t$.   From this $5A = 2$,

$2A + 5B = 0$, hence $A = \frac{2}{5}$, $B = \frac{-4}{25}$.   Thus the particular

integral of (1) is $y_p = \frac{2}{5} te^t - \frac{4}{25} e^t$, and its general

solution is $y = c_1 \sin 2t + c_2 \cos 2t + \frac{2}{5} te^t - \frac{4}{25} e^t$.

Returning to the original independent variable by
replacing e^t by x, t by $\ln x$, we find the general solution
of the given D.E.:

$$y = c_1 \sin(2 \ln x) + c_2 \cos(2 \ln x)$$

$$+ \frac{2}{5} x \ln x - \frac{4}{25} x$$

or

$$y = c_1 \sin(\ln x^2) + c_2 \cos(\ln x^2) + \frac{x \ln x^2}{5} - \frac{4x}{25}.$$

20.   Let $x = e^t$; then $t = \ln x$, $x \frac{dy}{dx} = \frac{dy}{dt}$, and,

$x^2 \frac{d^2y}{dx^2} = \frac{d^2y}{dt^2} - \frac{dy}{dt}$.   The D.E. transforms into

$$\frac{d^2y}{dt^2} - \frac{dy}{dt} - 3\frac{dy}{dt} + 5y = 5e^{2t} \qquad \text{or}$$

$$\frac{d^2y}{dt^2} - 4\frac{dy}{dt} + 5y = 5e^{2t}. \qquad (1)$$

The homogeneous equation corresponding to this has auxiliary equation $m^2 - 4m + 5 = 0$, with roots $2 \pm i$. The complementary function of (1) is $y_c = e^{2t}(c_1 \sin t + c_2 \cos t)$. We find a particular integral of (1) by the method of undetermined coefficients. We assume $y_p = Ae^{2t}$ and readily find $A = 5$. Thus the G.S. of (1) is $y = e^{2t}(c_1 \sin t + c_2 \cos t) + 5e^{2t}$. We return to the original independent variable x by replacing e^{2t} by x^2 and t by $\ln x$. Thus the G.S. of the given D.E. is $y = x^2[c_1 \sin(\ln x) + c_2 \cos(\ln x)] + 5x^2$.

21. Let $x = e^t$; then $t = \ln x$, $x\frac{dy}{dx} = \frac{dy}{dt}$, $x^2\frac{d^2y}{dx^2} = \frac{d^2y}{dt^2} - \frac{dy}{dt}$,

and (as on page 174 of the text) $x^3\frac{d^3y}{dx^3} = \frac{d^3y}{dt^3} - 3\frac{d^2y}{dt^2}$

$+ 2\frac{dy}{dt}$. The D.E. transforms into $\frac{d^3y}{dt^3} - 3\frac{d^2y}{dt^2} + 2\frac{dy}{dt}$

$$- 8\frac{d^2y}{dt^2} + 8\frac{dy}{dt} + 28\frac{dy}{dt} - 40y = -9e^{-t} \qquad \text{or}$$

$$\frac{d^3y}{dt^3} - 11\frac{d^2y}{dt^2} + 38\frac{dy}{dt} - 40y = -9e^{-t}. \qquad (1)$$

The auxiliary equation of the homogeneous D.E. corresponding to this is $m^3 - 11m^2 + 38m - 40 = 0$, with roots 2, 4, 5. The complementary function of (1) is $y_c = c_1 e^{2t} + c_2 e^{4t} + c_3 e^{5t}$. We find a particular integral of (1) by the method of undetermined coefficients. We assume $y_p = Ae^{-t}$. Differentiating and substituting into (1), we find $-A - 11A - 38A - 40A = -9$ or $-90A = -9$; and so $A = \frac{1}{10}$. Thus the G.S. of (1) is

$y = c_1 e^{2t} + c_2 e^{4t} + c_3 e^{5t} + \frac{1}{10} e^{-t}$. Returning to the original independent variable x by replacing e^{2t} by x^2, etc., we have the G.S. of the given D.E., that is,

$$y = c_1 x^2 + c_2 x^4 + c_3 x^5 + \frac{1}{10} x^{-1}.$$

23. Let $x = e^t$; then $t = \ln x$, $x \frac{dy}{dx} = \frac{dy}{dt}$, $x^2 \frac{d^2 y}{dx^2} = \frac{d^2 y}{dt^2} - \frac{dy}{dt}$.

The D.E. transforms into $\frac{d^2 y}{dt^2} - 3 \frac{dy}{dt} - 10y = 0$. The auxiliary equation of this is $m^2 - 3m - 10 = 0$, with roots $m = 5, -2$. The general solution of the D.E. in y and t is $y = c_1 e^{5t} + c_2 e^{-2t}$. Returning to the original independent variable x, we replace e^t by x, etc., and obtain the general solution of the given D.E. in the form

$$y = c_1 x^5 + c_2 x^{-2}. \qquad (1)$$

Differentiating (1), we find

$$y' = 5c_1 x^4 - 2c_2 x^{-3}. \qquad (2)$$

Now apply the I.C.'s. Applying $y(1) = 5$ to (1), we have $c_1 + c_2 = 5$; and applying $y'(1) = 4$ to (2), we have $5c_1 - 2c_2 = 4$. Solving these two equations in c_1 and c_2, we find $c_1 = 2$, $c_2 = 3$. Substituting these values back into (1), we find the particular solution of the stated I.V.P.: $y = 2x^5 + 3x^{-2}$.

26. Let $x = e^t$; then $t = \ln x$, $x \frac{dy}{dx} = \frac{dy}{dt}$, and $x^2 \frac{d^2y}{dx^2} = \frac{d^2y}{dt^2} - \frac{dy}{dt}$. The D.E. transforms into

$$\frac{d^2y}{dt^2} - \frac{dy}{dt} - 2y = 4e^t - 8. \tag{1}$$

The auxiliary equation of the corresponding homogeneous D.E. is $m^2 - m - 2 = 0$ with roots $m = -1, 2$. Thus the complementary function of (1) is $y_c = c_1e^{-t} + c_2e^{2t}$. We find a particular integral by the method of undetermined coefficients. We assume $y_p = Ae^t + B$. Then $y_p{}' = Ae^t$, $y_p{}'' = Ae^t$, and substituting into the D.E. (1), we have $-2Ae^t - 2B = 4e^t - 8$. From this, $A = -2$, $B = 4$. Thus the particular integral of (1) is $y_p = -2e^t + 4$, and its general solution is $y = c_1e^{-t} + c_2e^{2t} - 2e^t + 4$.

Returning to the original variable by replacing e^t by x, etc., we find the general solution of the given D.E.

$$y = c_1x^{-1} + c_2x^2 - 2x + 4. \tag{2}$$

Differentiating (2), we find:

$$y' = -c_1 x^{-2} + 2c_2 x - 2. \tag{3}$$

Now apply the I.C.'s. Applying $y(1) = 4$ to (2), we have $c_1 + c_2 = 2$; and applying $y'(1) = -1$ to (3), we have $-c_1 + 2c_2 = 1$. Solving these two equations in c_1 and c_2, we find $c_1 = 1$, $c_2 = 1$. Substituting these values back into (2), we find the particular solution of the stated I.V.P.: $y = x^{-1} + x^2 - 2x + 4$.

27. Let $x = e^t$; then $t = \ln x$, $x \dfrac{dy}{dx} = \dfrac{dy}{dt}$, and $x^2 \dfrac{d^2 y}{dx^2} = \dfrac{d^2 y}{dt^2} - \dfrac{dy}{dt}$. The D.E. transforms into

$$\frac{d^2 y}{dt^2} - \frac{dy}{dt} - 4\frac{dy}{dt} + 4y = 4e^{2t} - 6e^{3t} \qquad \text{or}$$

$$\frac{d^2 y}{dt^2} - 5\frac{dy}{dt} + 4y = 4e^{2t} - 6e^{3t}. \tag{1}$$

The auxiliary equation of the homogeneous D.E. corresponding to (1) is $m^2 - 5m + 4 = 0$ with roots 1, 4. The complementary function of (1) is $y_c = c_1 e^t + c_2 e^{4t}$.

We find a particular integral of (1) by the method of undetermined coefficients. We assume $y_p = Ae^{2t} + Be^{3t}$.

Differentiating, substituting into (1), and simplifying, we obtain $-2Ae^{2t} - 2Be^{3t} = 4e^{2t} - 6e^{3t}$. Thus $A = -2$, $B = 3$. The G.S. of (1) is $y = c_1 e^t + c_2 e^{4t} - 2e^{2t} + 3e^{3t}$.

Returning to the original independent variable x, we

replace e^t by x, etc., and obtain the G.S. of the given
D.E.,

$$y = c_1 x + c_2 x^4 - 2x^2 + 3x^3. \qquad (2)$$

Applying the I.C. $y(2) = 4$ to this, we obtain $2c_1 + 16c_2 =$
-12. Differentiating (2), we have $y' =$
$c_1 + 4c_2 x^3 - 4x + 9x^2$. Applying the I.C. $y'(2) = -1$ to
this, we obtain $c_1 + 32c_2 = -29$. The system of

equations $\begin{cases} c_1 + 8c_2 = -6 \\ c_1 + 32c_2 = -29 \end{cases}$ has the solution $c_1 = 5/3$,

$c_2 = -23/24$. Thus we obtain the solution

$$y = \frac{5}{3} x - 2x^2 + 3x^3 - \frac{23}{24} x^4.$$

30. Let $x = e^t$; then $t = \ln x$ and $x^2 \frac{d^2y}{dx^2} = \frac{d^2y}{dt^2} - \frac{dy}{dt}$. The D.E.

transforms into

$$\frac{d^2y}{dt^2} - \frac{dy}{dt} - 6y = t. \qquad (1)$$

The auxiliary equation of the corresponding homogeneous
equation is $m^2 - m - 6 = 0$, with roots 3, -2. The
complementary function of (1) is $y_c = c_1 e^{3t} + c_2 e^{-2t}$. We
find a particular integral of (1) by the method of
undetermined coefficients. We assume $y_p = At + B$.
Differentiating and substituting into (1), we find
$-A - 6At - 6B = t$. From this, $-6A = 1$, $-A - 6B = 0$.

Thus $A = -\frac{1}{6}$, $B = \frac{1}{36}$. The G.S. of (1) is $y = c_1 e^{3t}$
$+ c_2 e^{-2t} - \frac{1}{6} t + \frac{1}{36}$. Returning to the original
independent variable x, we replace e^t by x, etc., and
obtain the G.S. of the given D.E.

$$y = c_1 x^3 + c_2 x^{-2} - \frac{1}{6} \ln x + \frac{1}{36}. \qquad (2)$$

We apply the I.C. $y(1) = 1/6$ to (2), obtaining
$c_1 + c_2 = \frac{5}{36}$. Differentiating (2), we have
$y' = 3c_1 x^2 - 2c_2 x^{-3} - \frac{1}{6x}$. Applying the I.C. $y'(1) = -1/6$
to this, we obtain $3c_1 - 2c_2 = 0$. From the pair
$c_1 + c_2 = 5/36$, $3c_1 - 2c_2 = 0$, we find $c_1 = 1/18$,
$c_2 = 1/12$. Thus we obtain the solution

$$y = \frac{1}{18} x^3 + \frac{1}{12} x^{-2} - \frac{1}{6} \ln x + \frac{1}{36} \qquad \text{or}$$

$$y = \frac{1}{6} \left(\frac{x^3}{3} + \frac{x^{-2}}{2} - \ln x + \frac{1}{6} \right).$$

31. Although this is not a Cauchy-Euler equation, it is
similar to one. Let $x + 2 = e^t$. Then $t = \ln(x + 2)$,
$(x + 2) \frac{dy}{dx} = \frac{dy}{dt}$, and $(x + 2)^2 \frac{d^2y}{dx^2} = \frac{d^2y}{dt^2} - \frac{dy}{dt}$. The D.E.
transforms into $\frac{d^2y}{dt^2} - \frac{dy}{dt} - \frac{dy}{dt} - 3y = 0$ or
$\frac{d^2y}{dt^2} - 2\frac{dy}{dt} - 3y = 0$. The corresponding auxiliary equation
is $m^2 - 2m - 3 = 0$, with roots 3, -1. The G.S. of the
D.E. in y and t is $y = c_1 e^{3t} + c_2 e^{-t}$. Returning to the

original independent variable x, we replace e^{3t} by $(x + 2)^3$ and e^{-t} by $(x + 2)^{-1}$. Thus the G.S. of the given D.E. is $y = c_1(x + 2)^3 + c_2(x + 2)^{-1}$.

Chapter 5

1. This is an example of free, undamped motion; and equation
    (5.8) applies. Since the 12 lb. weight stretches the
    spring 1.5 in. $= \frac{1}{8}$ ft., Hooke's Law $F = ks$ gives $12 = \left(\frac{1}{8}\right)$,
    so $k = 96$ lb./ft. Also, $m = \frac{w}{g} = \frac{12}{32} = \frac{3}{8}$ (slug). Thus by
    (5.8) we have the D.E.

    $$\frac{3}{8} x'' + 96x = 0 \quad \text{or} \quad x'' + 256x = 0. \quad (1)$$

    Since the weight was released from rest from a position 2
    in. $= \frac{1}{6}$ ft. below its equilibrium position, we also have
    the I.C. $x(0) = \frac{1}{6}$, $x'(0) = 0$. The auxiliary equation of
    the D.E. (1) is $r^2 + 256 = 0$ with roots $r = \pm 16i$. The
    G.S. of the D.E. is

    $$x = c_1 \sin 16t + c_2 \cos 16t. \quad (2)$$

    Differentiating this, we obtain

    $$x' = 16c_1 \cos 16t - 16c_2 \sin 16t. \quad (3)$$

    Applying the first I.C. to (2), we find $c_2 = \frac{1}{6}$; and
    applying the second to (3) gives $c_1 = 0$. Thus the
    solution of the D.E. satisfying the stated I.C.'s is

$x = \left(\frac{1}{6}\right) \cos 16t$. The amplitude is $\frac{1}{6}$ (ft.); the period $\frac{2\pi}{16} =$ $\frac{\pi}{8}$ (sec); and the frequency is $\frac{1}{(\pi/8)} = \frac{8}{\pi}$ oscillations/sec.

2. This is an example of free, undamped motion; and equation (5.8) applies. Since the 16 lb. weight stretches the spring 6 in. $= \frac{1}{2}$ foot, Hooke's Law $F = ks$ gives $16 = k\left(\frac{1}{2}\right)$, so $k = 32$ lb./ft. Also, $m = \frac{w}{g} = \frac{16}{32} = \frac{1}{2}$ (slug). Thus by (5.8), we have the D.E.

$$\frac{1}{2} x'' + 32x = 0 \quad \text{or} \quad x'' + 64x = 0. \qquad (1)$$

The auxiliary equation corresponding to (1) is $r^2 + 64 = 0$ with roots $r = \pm 8i$. The G.S. of the D.E. (1) is

$$x = c_1 \sin 8t + c_2 \cos 8t. \qquad (2)$$

Differentiating this, we obtain

$$x' = 8c_1 \cos 8t - 8c_2 \sin 8t. \qquad (3)$$

The three cases (a), (b), (c) lead to different I.C.'s.

In (a), the weight is released from a position 4 in. $= \frac{1}{3}$ ft. *below* its equilibrium position, so we have the first I.C. $x(0) = \frac{1}{3}$. Since it is released with initial velocity of 2 ft./sec., directed *downward*, we have the second I.C. $x'(0) = 2$. Applying the first I.C. to (2), we find $c_2 = \frac{1}{3}$; and applying the second to (3), we have $8c_1 = 2$, so

$c_1 = \frac{1}{4}$. Thus we obtain the particular solution

$x = \left(\frac{1}{4}\right) \sin 8t + \left(\frac{1}{3}\right) \cos 8t$.

In (b), we again have the first I.C. $x(0) = \frac{1}{3}$. But here the weight is released with initial velocity of 2 ft./sec., directed *upward*, so the second I.C. is $x'(0) = -2$. We again find $c_2 = \frac{1}{3}$; but applying the second I.C. to (3) gives $c_1 = -\frac{1}{4}$. Thus we obtain the particular solution $x = -\left(\frac{1}{4}\right) \sin 8t + \left(\frac{1}{3}\right) \cos 8t$.

In (c), the weight is released from a position 4 in. = $\frac{1}{3}$ ft. *above* its equilibrium position, so we have the first I.C. $x(0) = -\frac{1}{3}$. Since it is released with initial velocity of 2 ft./sec., directed *downward*, we have the second I.C. $x'(0) = 2$, just as in part (a). Applying the first I.C. to (2), we find $c_2 = -\frac{1}{3}$, and just as in part (a), the second I.C. gives $c_1 = \frac{1}{4}$. Thus we obtain the particular solution

$$x = \left(\frac{1}{4}\right) \sin 8t - \left(\frac{1}{3}\right) \cos 8t.$$

5. Equation (5.8) applies. Since the 4 lb. weight stretches the spring 6 in. = $\frac{1}{2}$ ft., Hooke's law $F = ks$ gives $4 = k\left(\frac{1}{2}\right)$, so $k = 8$ lb./ft. Also, $m = \frac{w}{g} = \frac{4}{32} = \frac{1}{8}$ (slug). By (5.8) we have the D.E.

$$\frac{1}{8} x'' + 8x = 0 \quad \text{or} \quad x'' + 64x = 0. \qquad (1)$$

Since the weight was released from its equilibrium position with an initial velocity of 2 ft./sec., directed downward, we have the I.C.'s $x(0) = 0$, $x'(0) = 2$. The auxiliary equation of the D.E. (1) is $r^2 + 64 = 0$ with roots $r = \pm 8\,i$. The G.S. of the D.E. is

$$x = c_1 \sin 8t + c_2 \cos 8t. \qquad (2)$$

Differentiating this, we obtain

$$x' = 8c_1 \cos 8t - 8c_2 \sin 8t. \qquad (3)$$

Applying the first I.C. $x(0) = 0$ to (2), we find $c_2 = 0$; and applying the second $x'(0) = 2$ to (3), we get $8c_1 = 2$, so $c_1 = \frac{1}{4}$.

Thus the solution of the D.E. satisfying the given I.C.'s is

$$x = \left(\frac{1}{4}\right) \sin 8t, \qquad (4)$$

and its derivative is

$$x' = 2 \cos 8t. \qquad (5)$$

These are the displacement and velocity, respectively, and hence provide the answer to part (a). From the solution

(4), we see that the amplitude is $\frac{1}{4}$ (ft.), the period is $\frac{2\pi}{8} = \frac{\pi}{4}$ (sec.), and the frequency is $\frac{1}{(\pi/4)} = \frac{4}{\pi}$ oscillations/sec. These are the answers to (b).

To answer (c), we seek times t at which x = 1.5 in. = $\frac{1}{8}$ ft. and x$'$ > 0. Thus we first let x = $\frac{1}{8}$ in solution (4) and find $\sin 8t = \frac{1}{2}$. From this, $8t = \frac{\pi}{6} + 2n\pi$ or $8t = \frac{5\pi}{6} + 2n\pi$, and hence $t = \frac{\pi}{48} + \frac{n\pi}{4}$ or $t = \frac{5\pi}{48} + \frac{n\pi}{4}$, where n = 0, 1, 2,.... We must choose these t for which x$'$ > 0. From (5), we see that $x'\left(\frac{\pi}{48} + \frac{n\pi}{4}\right) = 2\cos\left(\frac{\pi}{6} + 2n\pi\right) > 0$, but $x'\left(\frac{5\pi}{48} + \frac{n\pi}{4}\right) = 2\cos\left(\frac{5\pi}{6} + 2n\pi\right) < 0$. Thus the answer to part (c) is $t = \frac{\pi}{48} + \frac{n\pi}{4}$. (n = 0, 1, 2, ...). To answer (d), we seek times t at which x = 1.5 in. = $\frac{1}{8}$ ft. and x$'$ < 0. From our work in (c), we see these are given by $t = \frac{5\pi}{48} + \frac{n\pi}{4}$, (n = 0, 1, 2, ...).

8.   Equation (5.8) applies, with m = $\frac{w}{g} = \frac{8}{32} = \frac{1}{4}$ (slug) and the spring constant k > 0 to be determined. Thus (5.8) becomes

$$\frac{1}{4} x'' + kx = 0 \quad \text{or} \quad x'' + 4kx = 0. \qquad (1)$$

Since the weight is released from a position A ft. *below* its equilibrium position with an initial velocity of 3 ft./sec., directed *downward*, we have the I.C.'s $x(0) = A > 0$, $x'(0) = 3$. The auxiliary equation of D.E. (1) is $r^2 + 4k = 0$ with roots $r = \pm 2\sqrt{k}i$. The G.S. of the D.E. is

$$x = c_1 \sin 2\sqrt{k}t + c_2 \cos 2\sqrt{k}t. \qquad (2)$$

Differentiating we obtain

$$x' = 2\sqrt{k}c_1 \cos 2\sqrt{k}t - 2\sqrt{k}c_2 \sin 2\sqrt{k}t. \qquad (3)$$

Applying the first I.C. $x(0) = A$ to (2), we find $c_2 = A$; and applying the second $x'(0) = 3$ to (3), we have $2\sqrt{k} \ c_1 = 3$, so $c_1 = \dfrac{3}{2\sqrt{k}}$. Thus the solution of the D.E. satisfying the given I.C.'s is $x = \left(\dfrac{3}{2\sqrt{k}}\right) \sin 2\sqrt{k} \ t + A \cos 2\sqrt{k} \ t$. We express this in the form (5.18) of the text. Multiplying and dividing by

$$c = \sqrt{\left(\frac{3}{2\sqrt{k}}\right)^2 + A^2} = \frac{\sqrt{9 + 4kA^2}}{2\sqrt{k}},$$

we have $x = c\left[\left(\dfrac{3}{2\sqrt{k} \ c}\right) \sin 2\sqrt{k} \ t + \left(\dfrac{A}{C}\right) \cos 2\sqrt{k} \ t\right]$. Then letting $\dfrac{A}{C} = \cos \phi$, $\dfrac{3}{2\sqrt{k} \ c} = -\sin \phi$, we have $x =$ $c \cos(2\sqrt{k} \ t + \phi)$. From this, the period is $\dfrac{2\pi}{2\sqrt{k}} = \dfrac{\pi}{\sqrt{k}}$; but the period is *given* to be $\dfrac{\pi}{2}$. Thus $\sqrt{k} = 2$, and hence $k =$

4.   The amplitude is $c = \sqrt{\left(\dfrac{3}{2\sqrt{k}}\right)^2 + A^2} = \sqrt{\left(\dfrac{3}{4}\right)^2 + A^2}$; but

the amplitude is *given* to be $\sqrt{\frac{10}{2}} = \sqrt{5}$. Hence $\left(\frac{3}{4}\right)^2 + A^2 = 5$, from which $A = \frac{\sqrt{71}}{4}$.

9. There are two different D.E.'s of form (5.8) here, one involving the 8 lb. weight and the other involving the other weight. Concerning the 8 lb. weight, $m = \frac{w}{g} = \frac{8}{32} = \frac{1}{4}$ (slug), and the corresponding D.E. of form (5.8) is

$$\frac{1}{4} x'' + kx = 0 \quad \text{or} \quad x'' + 4kx = 0,$$

where $k > 0$ is the spring constant. The auxiliary equation is $r^2 + 4k = 0$ with roots $r = \pm 2\sqrt{k}\, i$. The G.S. of the D.E. is $x = c_1 \sin 2\sqrt{k}\, t + c_2 \cos 2\sqrt{k}\, t$. From this, the period of the motion is $\frac{2\pi}{2\sqrt{k}}$. But the period is given as 4. Thus $\frac{2\pi}{2\sqrt{k}} = 4$, from which $k = \frac{\pi^2}{16}$.

Now let w be the other weight. For this, $m = \frac{w}{g} = \frac{w}{32}$ (slugs), and the corresponding D.E. of form (5.8) is

$$\frac{w}{32} x'' + \frac{\pi^2}{16} x = 0 \quad \text{or} \quad x'' + \frac{2\pi^2}{w} x = 0.$$

The auxiliary equation is $r^2 + \frac{2\pi^2}{w} = 0$ with roots $r = \pm \sqrt{\frac{2}{w}}\, \pi\, i$. The G.S. of the D.E. is

$$x = c_1 \sin \sqrt{\frac{2}{w}}\, \pi\, t + c_2 \cos \sqrt{\frac{2}{w}}\, \pi\, t.$$

From this the period of the motion is $\dfrac{2\pi}{\sqrt{\dfrac{2}{w}}}$ $\pi = \sqrt{2w}$.   But the

period of this motion is given to be 6.   Thus $\sqrt{2w} = 6$, from which we find $w = 18$ (lb.).

Section 5.3, Page 208

1. (a) This is a free damped motion, and Equation (5.27) applies.   Since the 8 lb. weight stretches the spring 0.4 ft., Hooke's Law $F = ks$ gives $8 = k(0.4)$, so $k = 20$ lb./ft.   Also, $m = \dfrac{w}{g} = \dfrac{8}{32} = \dfrac{1}{4}$ (slug), and $a = 2$.

Thus equation (5.27) becomes

$$\frac{1}{4} x'' + 2 x' + 20x = 0. \tag{1}$$

Since the weight is then pulled down 6 in. $= \dfrac{1}{2}$ ft. below its equilibrium position and released from rest at $t = 0$, we have the I.C.'s

$$x(0) = \frac{1}{2}, \; x'(0) = 0. \tag{2}$$

(b) The D.E. (1) may be written as $x'' + 8 x' + 80x = 0$. The auxiliary equation is $r^2 + 8r + 80 = 0$, with roots $r = -4 \pm 8i$.   The G.S. of the D.E. is
$$x = e^{-4t}(c_1 \sin 8t + c_2 \cos 8t). \tag{3}$$

Differentiating this, we find

$$x' = e^{-4t}[(-4c_1 - 8c_2) \sin 8t$$
$$+ (8c_1 - 4c_2) \cos 8t]. \tag{4}$$

Applying the first I.C. (2) to (3), we find $c_2 = \frac{1}{2}$,

and applying the second to (4), we have $8c_1 - 4c_2 = 0$,

from which $c_1 = \frac{c_2}{2} = \frac{1}{4}$. Thus the solution is

$$x = e^{-4t}\left[\left(\frac{1}{4}\right) \sin 8t + \left(\frac{1}{2}\right) \cos 8t\right] \qquad (5)$$

(c) We first multiply and divide (5) by $c = \sqrt{\left(\frac{1}{4}\right)^2 + \left(\frac{1}{2}\right)^2} = $

$\frac{\sqrt{5}}{4}$, obtaining $x = \left(\frac{\sqrt{5}}{4}\right) e^{-4t} \left[\left(\frac{1}{\sqrt{5}}\right) \sin 8t + \left(\frac{2}{\sqrt{5}}\right) \cos 8t\right]$.

We can now write this as $x = \left(\frac{\sqrt{5}}{4}\right) e^{-4t} \cos(8t - \phi)$,

where ϕ is such that $\cos\phi = \frac{2}{\sqrt{5}}$, $\sin\phi = \frac{1}{\sqrt{5}}$, and hence

$\phi \approx 0.46$ (rad.).

(d) The period is $\frac{2\pi}{8} = \frac{\pi}{4}$ (sec.).

3. This is a free damped motion, and Equation (5.27) applies.
    Since the 8 lb. weight stretches the spring 6 in. $= \frac{1}{2}$ ft.,
    Hooke's Law $F = ks$ gives $8 = k\left(\frac{1}{2}\right)$, so $k = 16$ lb./ft.
    Also, $m = \frac{w}{g} = \frac{8}{32} = \frac{1}{4}$ (slug), and $a = 4$. Thus equation
    (5.27) becomes

$$\frac{1}{4} x'' + 4 x' + 16x = 0.$$

The I.C.'s are $x(0) = \frac{3}{4}$, $x'(0) = 0$. The D.E. may be written in the form

$$x'' + 16\,x' + 64x = 0.$$

The auxiliary equation is $r^2 + 16r + 64 = 0$ with roots -8, -8 (double root). The G.S. of the D.E. is $x = (c_1 + c_2 t)e^{-8t}$. Differentiating this, we find $\frac{dx}{dt} = (-8c_1 - 8c_2 t + c_2)e^{-8t}$. Applying the first I.C. to the G.S., we find $c_1 = \frac{3}{4}$; and applying the second to its derivative, we have $-8c_1 + c_2 = 0$, from which $c_2 = 6$.

Thus we find the solution $x = \left(\frac{3}{4} + 6t\right) e^{-8t}$.

6. This is a free damped motion, and Equation (5.27) applies. The mks system of units is used. Since a force of 4 newtons stretches the spring 5 cm $= 0.05$ meters, Hooke's Law $F = ks$ gives $4 = k(0.05)$, so $k = 80$ newtons/meter. Also, we are given that $m = 2$ and $a = 16$. Thus equation (5.27) becomes

$$2\,x'' + 16\,x' + 80x = 0$$

The I.C.'s are $x(0) = 0.02$ (in meters), $x'(0) = 0.04$ (in meters). Dividing the D.E. by 2, we have the auxiliary equation $r^2 + 8r + 40 = 0$, with roots $r = -4 \pm 2\sqrt{6}\ i$. The G.S. of the D.E. is

$$x = e^{-4t}(c_1 \sin 2\sqrt{6}\ t + c_2 \cos 2\sqrt{6}\ t). \qquad (1)$$

Differentiating this, we find

$$x' = e^{-4t}[(-4c_1 - 2\sqrt{6} \ c_2) \sin 2\sqrt{6} \ t$$
$$+ (2\sqrt{6} \ c_1 - 4c_2) \cos 2\sqrt{6} \ t]. \qquad (2)$$

Applying the first I.C. $x(0) = 0.02$ to (1), we find $c_2 = 2/100$; and applying the second, $x'(0) = 0.04$, to (2), we find $2\sqrt{6} \ c_1 - 4 \ c_2 = 4/100$, from which $c_1 = \sqrt{6}/100$. Thus the solution is

$$x = \frac{e^{-4t}}{100} \ [\sqrt{6} \sin 2\sqrt{6} \ t + 2 \cos 2\sqrt{6} \ t].$$

7. This is a free damped motion, and Equation (5.27) applies. Since a force of 20 lb. would stretch the spring 6 in. $= \frac{1}{2}$ ft., Hooke's Law $F = ks$ gives $20 = k\left(\frac{1}{2}\right)$, so $k = 40$ lb./ft. The weight is 4 lb., so $m = \frac{w}{g} = \frac{4}{32} = \frac{1}{8}$ (slug); and $a = 2$. Thus equation (5.27) becomes

$$\frac{1}{8} \ x'' + 2 \ x' + 40x = 0.$$

Since the weight is released from rest from a position 8 inches $= \frac{2}{3}$ ft. below its equilibrium position, we have the I.C.'s $x(0) = \frac{2}{3}$, $x'(0) = 0$. The D.E. may be written $x'' + 16 \ x' + 320x = 0$. The auxiliary equation is $r^2 + 16r + 320 = 0$, with roots $-8 \pm 16i$. The G.S. of the D.E. is

$$x = e^{-8t}(c_1 \sin 16t + c_2 \cos 16t).$$

Differentiating we obtain

$$x' = e^{-8t}[(-8c_1 - 16c_2) \sin 16t + (16c_1 - 8c_2) \cos 16t].$$

Applying the first I.C. to the G.S., we find $c_2 = \frac{2}{3}$; and applying the second to the derived equation, we have $16c_1 - 8c_2 = 0$, from which $c_1 = \frac{c_2}{2} = \frac{1}{3}$. Thus we have the displacement

$$x = \left(\frac{e^{-8t}}{3}\right)(\sin 16t + 2 \cos 16t).$$

To put this in the form (5.32), we multiply and divide by

$$c = \sqrt{(1)^2 + (2)^2} = \sqrt{5}, \text{ obtaining } x = \left(\frac{\sqrt{5}e^{-8t}}{3}\right)$$

$$\left[\left(\frac{1}{\sqrt{5}}\right) \sin 16t + \left(\frac{2}{\sqrt{5}}\right) \cos 16t\right]. \text{ We can now write this as}$$

$$x = \left(\frac{\sqrt{5}e^{-8t}}{3}\right) \cos(16t - \phi). \qquad (1)$$

where $\cos \phi = \frac{2}{\sqrt{5}}$, $\sin \phi = \frac{1}{\sqrt{5}}$, and hence $\phi \approx 0.46$ (rad.).

We have thus answered part (a).

To answer part (b), we see from (1) that the period is $\frac{2\pi}{16} = \frac{\pi}{8}$ (sec.). In the notation of page 202 of the text, the logarithmic decrement is $\dfrac{2\pi b}{\sqrt{\lambda^2 - b^2}}$.   Here $b = \dfrac{a}{2m} =$

$\dfrac{2}{(2)\left(\frac{1}{8}\right)} = 8$ and $\lambda^2 = \dfrac{k}{m} = \dfrac{40}{\left(\frac{1}{8}\right)} = 320$.   Thus we find the

logarithmic decrement is $\dfrac{(2\pi)8}{\sqrt{256}} = \pi$.

To answer part (c), we let $x = 0$ in (1) and solve for
t. We have $\cos(16t - \phi) = 0$, so $16t - \phi = \dfrac{\pi}{2}$, from which

$t = \dfrac{\phi + \frac{\pi}{2}}{16}$.   With $\phi = 0.46$, this gives $t = 0.127$ (sec.).

8. This is a free damped motion, and Equation (5.27) applies.
Since the 24 lb. weight stretches the spring 1 ft.,
Hooke's Law $F = ks$ gives $24 = k(1)$, so $k = 24$ lb./ft.
Also, $m = \dfrac{w}{g} = \dfrac{24}{32} = \dfrac{3}{4}$ (slug), and $a = 6$. Thus equation
(5.27) becomes

$$\frac{3}{4} x'' + 6 x' + 24x = 0.$$

or

$$x'' + 8 x' + 32x = 0.$$

The I.C.'s are $x(0) = 1$, $x'(0) = 0$. The auxiliary
equation is $m^2 + 8m + 32 = 0$, with roots $m = -4 \pm 4i$. The
G.S. of the D.E. is

$$x = e^{-4t}(c_1 \sin 4t + c_2 \cos 4t). \qquad (1)$$

Differentiating this, we find

$$x' = e^{-4t}[(-4c_1 - 4c_2) \sin 4t$$
$$+ (4c_1 - 4c_2) \cos 4t]. \qquad (2)$$

Applying the first I.C. $x(0) = 1$, to (1), we find $c_2 = 1$; and applying the second, $x'(0) = 0$, to (2), we have $4c_1 - 4c_2 = 0$, from which $c_1 = 1$. Thus the solution is

$$x = e^{-4t}(\sin 4t + \cos 4t). \qquad (3)$$

This is the first answer to part (a). Top determine the alternate form (5.32), we multiply and divide (3) by $c = \sqrt{(1)^2 + (1)^2} = \sqrt{2}$, obtaining $x = \sqrt{2} \, e^{-4t} \left(\dfrac{1}{\sqrt{2}} \sin 4t + \dfrac{1}{\sqrt{2}} \cos 4t \right)$. We can express this as

$$x = \sqrt{2} \, e^{-4t} \cos(4t - \pi/4). \qquad (4)$$

This is the displacement in the desired alternate form.

(b) The quasi period is $2\pi/4 = \pi/2$, and the time-varying amplitude is $\sqrt{2} \, e^{-4t}$

(c) The weight first attains a relative maximum displacement above its equilibrium position at the first time $t > 0$ at which the derivative x' of (4) equals zero. Differentiating (4) and equating it to zero, we obtain

$$-4\sqrt{2} \, e^{-4t} [\sin(4t - \pi/4) + \cos(4t - \pi/4)] = 0. \qquad (5)$$

This holds when $\sin(4t - \pi/4) + \cos(4t - \pi/4) = 0$ and hence when $\tan(4t - \pi/4) = -1$. Thus (5) is satisfied at $4t - \pi/4 = -\pi/4 + k\pi$, that is, $t = k\pi/4$, where k is an integer. The first $t > 0$ for which this is valid is for $k = 1$, and this gives $t = \pi/4$. This is the time desired. The displacement at this time is found by letting $t = \pi/4$ in (4). Doing so, we find $x = \sqrt{2} e^{-\pi} \cos 3\pi/4 \approx -0.0432$. The minus sign simply indicates that the mass is above the equilibrium position. Thus the first maximum displacement above it is approximately 0.04 feet.

11. Equation (5.27) applies, with $m = \dfrac{w}{g} = \dfrac{10}{32} = \dfrac{5}{16}$ (slug), $a > 0$, and $k = 20$ lb./ft. Thus we have the D.E.

$$\frac{5}{16} x'' + a x' + 20x = 0.$$

Writing this as $5 x'' + 16ax' + 320x = 0$, the auxiliary equation is $5r^2 + 16ar + 320 = 0$. The roots are given by

$$r = \frac{-16a \pm \sqrt{256a^2 - 6400}}{10} \tag{1}$$

In part (a), we seek the smallest value of a for which damping is nonoscillatory. This is the value of a for which damping is critical. It occurs when the two roots given by (1) are real and equal. Thus we set $256a^2 - 6400 = 0$ and find $a^2 = \dfrac{6400}{256}$ and hence $a = \dfrac{80}{16} = 5$.

In part (b), we let $a = 5$ in (1) and obtain the roots $r = -8, -8$ (double root). Then the displacement is given by $x = (c_1 + c_2 t)e^{-8t}$. Differentiating this, we have $x' = (-8c_1 - 8c_2 t + c_2)e^{-8t}$. The I.C.'s are $x(0) = \frac{1}{2}$, $x'(0) = 1$. Applying them to the preceding expressions for x and x', we find $c_1 = \frac{1}{2}$ and $-8c_1 + c_2 = 1$, from which $c_2 = 5$. Thus we obtain the displacement $x = \left(\frac{1}{2} + 5t\right)e^{-8t}$.

For part (c), we note that the extrema of the displacement are found by setting $x' = (1 - 40t)e^{-8t}$ equal to zero and solving for t. We at once have $t = \frac{1}{40}$. Then $x\left(\frac{1}{40}\right) = \left(\frac{5}{8}\right)e^{-1/5} \approx 0.51(\text{sec.})$. For $t > \frac{1}{40}$, $1 - 40t < 0$ and $x' < 0$. Also using L'Hospital's Rule, $\lim\limits_{t \to \infty} x' = 0$. Thus the weight approaches its position

monotonically.

12. This is a free damped motion, and Equation (5.27) applies. Since the 64 lb. weight stretches the spring 4/3 foot, Hooke's Law $F = ks$ gives $64 = k(4/3)$, so $k = 48$ lb./ft. Also, $m = \frac{w}{g} = \frac{64}{32} = 2$ (slugs). The damping constant is $a > 0$. Thus equation (5.27) becomes

$$2x'' + ax' + 48x = 0. \qquad (1)$$

The I.C.'s are $x(0) = 2$, $x'(0) = 0$. The auxiliary equation is $2m^2 + am + 48 = 0$, with roots

$$m = \frac{-a \pm \sqrt{a^2 - 384}}{4}. \qquad (2)$$

(a) We must determine a so that the resulting motion is critically damped. This occurs when the auxiliary

equation has a double root. This occurs if and only
if $a^2 - 384 = 0$ and hence if $a = \sqrt{384} = 8\sqrt{6}$ (recall
$a > 0$). In this case the auxiliary equation has the
double root $m = -2\sqrt{6}$, and the G.S. of the D.E. (1) is

$$x = (c_1 + c_2 t)e^{-2\sqrt{6}\,t}. \qquad (3)$$

Differentiating this, we obtain

$$x' = [(c_2 - 2\sqrt{6}\,c_1) - 2\sqrt{6}\,c_2\,t]e^{-2\sqrt{6}\,t}. \qquad (4)$$

Applying the I.C. $x(0) = 2$ to (3), we find $c_1 = 2$; and
applying the I.C. $x'(0) = 0$ to (4), we have
$c_2 - 2\sqrt{6}\,c_1 = 0$, from which $c_2 = 4\sqrt{6}$. Thus the
displacement is given by

$$x = (2 + 4\sqrt{6}\,t)e^{-2\sqrt{6}\,t},$$

which is positive for all $t > 0$. The derivative of
this is

$$x' = -48\,t\,e^{-2\sqrt{6}\,t}.$$

Since $x' < 0$ for all $t > 0$, x decreases monotonically
for all $t > 0$.

(b) The motion is underdamped if $0 < a < 8\sqrt{6}$. Then $a^2 -$
384 < 0, the roots (2) are the conjugate complex

numbers $\left(-a \pm \sqrt{384 - a^2}\, i\right)/4$, and the G.S. of D.E.
(1) is of the form

$$x = e^{-at/4}\left[c_1 \sin \frac{\sqrt{384 - a^2}}{4}\, t\right.$$

$$\left. + c_2 \cos \frac{\sqrt{384 - a^2}}{4}\, t\right]. \qquad (5)$$

From the discussion on pages 200-201 of the text, we see that the quasi period is given by $\dfrac{2\pi}{\left(\sqrt{\dfrac{384 - a^2}{4}}\right)}$.

Equating this to the given value $\pi/2$ in the present case, we obtain $\sqrt{384 - a^2} = 16$, from which $a = 8\sqrt{2}$. This is the desired value of a in part (b).

With this value of a, the solution (5) becomes

$$x = e^{-2\sqrt{2}t}(c_1 \sin 4t + c_2 \cos 4t), \qquad (6)$$

with derivative

$$x' = e^{-2\sqrt{2}t}[(-2\sqrt{2}\, c_1 - 4\, c_2) \sin 4t$$

$$+ (4\, c_1 - 2\sqrt{2}\, c_2) \cos 4t].$$

Applying the I.C.'s $x(0) = 2$, $x'(0) = 0$ to these gives $c_2 = 2$, $4\, c_1 - 2\sqrt{2}\, c_2 = 0$, from which $c_1 = \sqrt{2}$. Thus the solution (6) becomes

$$x = e^{-2\sqrt{2}t}(\sqrt{2}\,\sin 4t + 2\cos 4t)$$

$$= \sqrt{6}\, e^{-2\sqrt{2}t}\left(\frac{\sqrt{2}}{\sqrt{6}}\sin 4t + \frac{2}{\sqrt{6}}\cos 4t\right)$$

$$= \sqrt{6}\, e^{-2\sqrt{2}t}\cos(4t - \phi),$$

where $\cos \phi = 2/\sqrt{6}$, $\sin \phi = \sqrt{2}/\sqrt{6}$. From this we see that the time-varying amplitude is $\sqrt{6}e^{-2\sqrt{2}t}$.

13. Here $m = \dfrac{w}{g} = \dfrac{32}{32} = 1$ (slug). If there were no resistance, the D.E. would be m $x'' + kx = 0$, that is, $x'' + kx = 0$, where $k > 0$. The auxiliary equation of this is $r^2 + k = 0$, with roots $\pm\sqrt{k}i$. The G.S. of this D.E. is then $x = c_1 \sin \sqrt{k}t + c_2 \cos \sqrt{k}t$. The period of this undamped motion is $\dfrac{2\pi}{\sqrt{k}}$, and hence the natural frequency is $\dfrac{\sqrt{k}}{2\pi}$. But this is given to be $\dfrac{4}{\pi}$. Hence we have $\dfrac{\sqrt{k}}{2\pi} = \dfrac{4}{\pi}$, from which $\sqrt{k} = 8$ and $k = 64$. This answers (a).

To answer (b), we take the resistance into account and have the D.E. $mx'' + ax' + kx = 0$, that is, $x'' + ax' + 64x = 0$. The auxiliary equation of this is $r^2 + ar + 64 = 0$ with roots $r = \dfrac{-a \pm \sqrt{a^2 - 256}}{2}$, where $a^2 < 256$. The G.S. of this D.E. is

$$x = e^{-at/2}\left(c_1 \sin \sqrt{256 - a^2}\,\frac{t}{2} + c_2 \cos \sqrt{256 - a^2}\,\frac{t}{2}\right).$$

The period of the trigonometric factor of this is

$$\frac{2\pi}{\left[\dfrac{\sqrt{256 - a^2}}{2}\right]} = \frac{4\pi}{\sqrt{256 - a^2}}.$$ Hence the frequency of this

motion is $\dfrac{\sqrt{256 - a^2}}{4\pi}$. But the frequency of this damped

motion is given as half the natural frequency $\dfrac{4}{\pi}$ and so is

$\dfrac{2}{\pi}$. Thus $\dfrac{\sqrt{256 - a^2}}{4\pi} = \dfrac{2}{\pi}$. Thus $256 - a^2 = 64$, from which

$a = 8\sqrt{3}$.

Section 5.4, Page 217

1. The D.E. is of the form (5.58) of the text, with
    $m = \dfrac{w}{g} = \dfrac{6}{32} = \dfrac{3}{16}$ (slugs), $a = 0$ (since damping is
    negligible), $k = 27$ lb./ft., and $F(t) = 12 \cos 20t$.
    Thus we have

    $$\frac{3}{16} x'' + 27 x = 12 \cos 20t$$

    or

    $$x'' + 144. x = 64 \cos 20t. \qquad (1)$$

    The I.C.'s are $x(0) = 0$, $x'(0) = 0$. The auxiliary
    equation of the homogeneous D.E. corresponding to (1) is
    $r^2 + 144 = 0$, with pure imaginary conjugate complex roots
    $r = \pm 12i$. Thus the complementary function of (1) is
    $x = c_1 \sin 12t + c_2 \cos 12t$. We use undetermined

    coefficients to find a particular integral of (1).
    We let $x_p = A \sin 20t + B \cos 20t$. Then $x'_p = 20A \cos 20t -$
    $20B \sin 20t$, $x''_p = -400A \sin 20t - 400B \cos 20t$. Substituting
    into (1), we obtain $-256A \sin 20t - 256B \cos 20t = 64 \cos 20t$.

From this, we find A = 0, B = -1/4; and hence we have the particular integral $x_p = -\frac{1}{4}\cos 20t$ of (1). Thus the G.S. of (1) is

$$x = c_1 \sin 12t + c_2 \cos 12t - \frac{1}{4}\cos 20t. \qquad (2)$$

Differentiating, we find

$$x' = 12\ c_1 \cos 12t - 12\ c_2 \sin 12t + 5 \sin 20t. \qquad (3)$$

Applying the I.C. x(0) = 0 to (2), we have $c_2 - (1/4) = 0$, so $c_2 = 1/4$. Applying the I.C. x'(0) = 0 to (3), $12\ c_1 = 0$, so $c_1 = 0$. Thus we have

$$x = \frac{\cos 12t - \cos 20t}{4}$$

3.   The D.E. is of the form (5.58) of the text, with $m = \frac{w}{g} = \frac{10}{32} = \frac{5}{16}$ (slug), a = 5, k = 20, and F(t) = 10 cos 8t. Thus we have

$$\frac{5}{16}\ x'' + 5\ x' + 20x = 10 \cos 8t$$

or

$$x'' + 16\ x' + 64x = 32 \cos 8t. \qquad (1)$$

The I.C.'s are x(0) = 0, x'(0) = 0. The auxiliary equation of the homogeneous D.E. corresponding to (1) is $r^2 + 16r + 64 = 0$ or $(r + 8)^2 = 0$, with roots -8, -8 (double root). Thus the complementary function of (1) is

$x_c = (c_1 + c_2 t)e^{-8t}$. We use undetermined coefficients to find a particular integral. We let $x_p = A \cos 8t +$ $B \sin 8t$. Then $x_p' = -8A \sin 8t + 8B \cos 8t$ and $x_p'' = -64A \cos 8t - 64B \sin 8t$. Substituting into (1) and simplifying, we find $128B \cos 8t - 128A \sin 8t = 32 \cos 8t$. Hence $-128A = 0$, $128B = 32$, so $A = 0$, $B = \frac{1}{4}$. Thus we have the particular integral $x_p = \left(\frac{1}{4}\right) \sin 8t$, and the G.S. of D.E. (1) is $x = (c_1 + c_2 t)e^{-8t} + \left(\frac{1}{4}\right) \sin 8t$. Differentiating this, we obtain $x' = (-8c_1 - 8c_2 t + c_2)e^{-8t} + 2 \cos 8t$. Applying the stated I.C. to these, we have $c_1 = 0$, $-8c_1 + c_2 + 2 = 0$, and hence $c_2 = -2$. Thus we obtain the solution $x = -2te^{-8t} + \left(\frac{1}{4}\right) \sin 8t$.

4. The D.E. is of the form (5.58) of the text, with $m = \frac{w}{g} = \frac{4}{32} = \frac{1}{8}$ (slug), $a = 2$, and $F(t) = 13 \sin 4t$. Also, by Hooke's Law $F = ks$, we have $4 = k\left(\frac{1}{4}\right)$, so $k = 16$ lb./ft. Thus we have
$$\frac{1}{8} x'' + 2x' + 16x = 13 \sin 4t,$$
or
$$x'' + 16x' + 128x = 104 \sin 4t. \qquad (1)$$

The I.C.'s are $x(0) = \frac{1}{2}$ and $x'(0) = 0$. The auxiliary equation of the homogeneous D.E. corresponding to (1) is $r^2 + 16r + 128 = 0$ with roots $r = -8 \pm 8i$. Thus the complementary function of (1) is

$$x = e^{-8t}(c_1 \sin 8t + c_2 \cos 8t).$$

We use undetermined coefficients to find a particular integral. We let $x_p = A \sin 4t + B \cos 4t$. Then

$x_p' = 4A \cos 4t - 4B \sin 4t$, $x_p'' = -16A \sin 4t - 16B \cos 4t$.

Substituting into (1) and simplifying, we find (112A − 64B) sin 4t + (64A + 112B) cos 4t = 104 sin 4t. Thus we have the equations 112A − 64B = 104 and 64A + 112B = 0. These reduce to 14A − 8B = 13 and 8A + 14B = 0, from which we find $A = \frac{7}{10}$, $B = -\frac{2}{5}$. Thus we have the particular integral $x_p = \left(\frac{7}{10}\right) \sin 4t - \left(\frac{2}{5}\right) \cos 4t$ and the G.S. of the D.E. (1) is

$$x = e^{-8t}(c_1 \sin 8t + c_2 \cos 8t) + \left(\frac{7}{10}\right) \sin 4t - \left(\frac{2}{5}\right) \cos 4t.$$

Differentiating this, we obtain

$$x' = e^{-8t}[(-8c_1 - 8c_2) \sin 8t + (8c_1 - 8c_2) \cos 8t]$$

$$+ \left(\frac{14}{5}\right) \cos 4t + \left(\frac{8}{5}\right) \sin 4t.$$

Applying the stated I.C. to these, we have $c_2 - \frac{2}{5} = \frac{1}{2}$ and $8c_1 - 8c_2 + \frac{14}{5} = 0$. Hence $c_2 = \frac{9}{10}$ and $c_1 = \frac{11}{20}$. Thus we obtain the solution

$$x = e^{-8t}\left[\left(\frac{11}{20}\right) \sin 8t + \left(\frac{9}{10}\right) \cos 8t\right]$$

$$+ \left(\frac{7}{10}\right) \sin 4t - \left(\frac{2}{5}\right) \cos 4t.$$

This is the answer to part (a). Concerning part (b), we
note that the steady-state term is $\left(\frac{7}{10}\right) \sin 4t - \left(\frac{2}{5}\right) \cos 4t$.
The amplitude of this is given by $c = \sqrt{\left(\frac{7}{10}\right)^2 + \left(\frac{2}{5}\right)^2} = \frac{\sqrt{65}}{10}$.

6. The D.E. is of the form (5.58) of the text, with
    $m = \frac{w}{g} = \frac{32}{32} = 1$ (slug), $a = 2$, $k = 10$ lb./ft., and
    $F(t) = \sin t + \frac{1}{4} \sin 2t + \frac{1}{9} \sin 3t$. Thus we have

$$x'' + 2x' + 10x = \sin t + \frac{1}{4} \sin 2t + \frac{1}{9} \sin 3t . \qquad (1)$$

The I.C.'s are $x(0) = 0$, $x'(0) = 0$. The auxiliary
equation of the homogeneous D.E. corresponding to (1) is
$m^2 + 2m + 10 = 0$, with conjugate complex roots $-1 \pm 3i$.
Thus the complementary function of (1) is

$$x_c = e^{-t}(c_1 \sin 3t + c_2 \cos 3t). \qquad (2)$$

The problem tells us to find x_p using Theorem 4.10 of
Chapter 4. We first find the particular integral $x_{p,k}$ of
$$x'' + 2x' + 10x = \sin kt, \qquad (k \text{ constant}), \qquad (3)$$

by the method of undetermined coefficients. We let $x_{p,k} =$
$A \sin kt + B \cos kt$. Then $x'_{p,k} = kA \cos kt - kB \sin kt$,
$x''_{p,k} = -k^2 A \sin kt - k^2 B \cos kt$. Substituting into (3) and
simplifying, we obtain $[(10 - k^2)A - 2kB] \sin kt +$
$[2kA + (10 - k^2)B] \cos kt = \sin kt$. From this,

$$A = \frac{10 - k^2}{(10 - k^2)^2 + 4k^2}, \qquad B = \frac{-2k}{(10 - k^2)^2 + 4k^2}.$$

Thus we find

$$x_{p,k} = \frac{10 - k^2}{(10 - k^2)^2 + 4k^2} \sin kt$$

$$- \frac{2k}{(10 - k^2)^2 + 4k^2} \cos kt \qquad (4)$$

as the particular integral of (3). Letting $k = 1$ in (4), we have the particular integral

$$x_{p,1} = \frac{9}{85} \sin t - \frac{2}{85} \cos t \qquad (5)$$

of the D.E. $x'' + 2x' + 10x = \sin t$. Letting $k = 2$ in (4), we have the particular integral

$$x_{p,2} = \frac{3}{26} \sin 2t - \frac{1}{13} \cos 2t \qquad (6)$$

of the D.E. $x'' + 2x' + 10x = \sin 2t$. Letting $k = 3$ in (4), we have the particular integral

$$x_{p,3} = \frac{1}{37} \sin 3t - \frac{6}{37} \cos 3t \qquad (7)$$

of the D.E. $x'' + 2x' + 10x = \sin 3t$.

We are now ready to apply Theorem 4.10 of Chapter 4, and we essentially do so twice. We first use the theorem with (5) and (6) to obtain the particular integral

$$x_p = \frac{9}{85} \sin t - \frac{2}{85} \cos t$$

$$+ \frac{1}{4} \left[\frac{3}{26} \sin 2t - \frac{1}{13} \cos 2t \right] \qquad (8)$$

of the D.E. $x'' + 2x' + 10x = \sin t + \frac{1}{4} \sin 2t$. We then use the theorem with (7) and (8) to obtain the particular integral

$$x_p = \left[\frac{9}{85} \sin t - \frac{2}{85} \cos t + \frac{1}{4} \left(\frac{3}{26} \sin 2t - \frac{1}{13} \cos 2t \right) \right]$$

$$+ \frac{1}{9} \left[\frac{1}{37} \sin 3t - \frac{6}{37} \cos 3t \right]$$

of the D.E.

$$x'' + 2x' + 10x = \sin t + \frac{1}{4} \sin 2t + \frac{1}{9} \sin 3t. \qquad (1)$$

Thus the G.S. of (1) is

$$x = e^{-t}(c_1 \sin 3t + c_2 \cos 3t)$$

$$+ \frac{9}{85} \sin t - \frac{2}{85} \cos t + \frac{3}{104} \sin 2t - \frac{1}{52} \cos 2t$$

$$+ \frac{1}{333} \sin 3t - \frac{6}{333} \cos 3t. \qquad (9)$$

Differentiating this, we obtain

$$x' = e^{-t}[(-c_1 - 3c_2) \sin 3t + (3c_1 - c_2) \cos 3t]$$

$$+ \frac{9}{85} \cos t + \frac{2}{85} \sin t + \frac{3}{52} \cos 2t + \frac{1}{26} \sin 2t$$

$$+ \frac{1}{111} \cos 3t + \frac{6}{111} \sin 3t. \qquad (10)$$

Applying the I.C. $x(0) = 0$ to (9), we obtain

$$c_2 - \frac{2}{85} - \frac{1}{52} - \frac{6}{333} = 0, \qquad \text{so} \qquad c_2 = \frac{29,819}{490,620}.$$

Applying the I.C. $x'(0) = 0$ to (10), we obtain

$3c_1 - c_2 + \frac{9}{85} + \frac{3}{52} + \frac{1}{111} = 0$, from which $c_1 = -\frac{54,854}{1,471,860}.$

Substituting these values of c_1 and c_2 into (9) gives the

desired solution.

7.   There are two problems here, one for $0 \le t \le \tau$, and the

other for $t > \tau$. We first consider that for which

$0 \le t \le \tau$. Here the D.E. is of the form (5.58), with

$m = \frac{w}{g} = \frac{32}{32} = 1$ (slug), $a = 4$, $k = 20$, and $F(t) = 40 \cos 2t$.

Thus we have

$$x'' + 4x' + 20x = 40 \cos 2t. \qquad (1)$$

The I.C.'s are $x(0) = 0$, $x'(0) = 0$. The auxiliary

equation of the homogeneous D.E. corresponding to (1)

is $r^2 + 4r + 20 = 0$, with roots $-2 \pm 4i$. Thus the

complementary function of (1) is $x = e^{-2t}(c_1 \sin 4t +$

$c_2 \cos 4t)$. We use undetermined coefficients to find a

particular integral. We let $x_p = A \sin 2t + B \cos 2t$.

Then $x'_p = 2A \cos 2t - 2B \sin 2t$, $x''_p = -4A \sin 2t - 4B \cos 2t$.

Substituting into (1) and simplifying, we find

$(16A - 8B)\sin 2t + (8A + 16B)\cos 2t = 40 \cos 2t$. Thus we

have $16A - 8B = 0$ and $8A + 16B = 40$, from which we find

$A = 1$, $B = 2$. Thus we have the particular integral of

(1), $x_p = \sin 2t + 2 \cos 2t$; and the G.S. of D.E. (1) is

$$x = e^{-2t}(c_1 \sin 4t + c_2 \cos 4t) + \sin 2t + 2 \cos 2t.$$

Differentiating this, we obtain

$$x' = e^{-2t}[(-2c_1 - 4c_2) \sin 4t + (4c_1 - 2c_2) \cos 4t]$$
$$+ 2 \cos 2t - 4 \sin 2t.$$

Applying the stated I.C. to these, we have $c_2 + 2 = 0$ and $4c_1 - 2c_2 + 2 = 0$, from which we find $c_1 = -\frac{3}{2}$, $c_2 = -2$. Thus we obtain the solution

$$x = e^{-2t}\left[\left(-\frac{3}{2}\right) \sin 4t - 2 \cos 4t\right]$$
$$+ \sin 2t + 2 \cos 2t \qquad (2)$$

valid for $0 \leq t \leq \pi$.

Now we consider the problem for which $t > \pi$. The D.E. is again of the form (5.58), where m = 1, a = 4, k = 20, but here $F(t) = 0$. Thus we have

$$x'' + 4x' + 20x = 0. \qquad (3)$$

Assuming the displacement is continuous at $t = \pi$, the solution of (3) must take the value given by (2) at $t = \pi$. That is, we must impose the I.C.

$$x(\pi) = -2e^{-2\pi} + 2 \qquad (4)$$

on the solution of (3). Similarly, assuming the velocity is continuous at $t = \pi$, the derivative of the solution of

(3) must take the value given by the derivative of (2) at t = π. The derivative of (2) is x′ = e^{-2t}[11 sin 4t - 2 cos 4t] + 2 cos 2t - 4 sin 2t. From this, we see that we must therefore impose the I.C.

$$x'(\pi) = -2e^{-2\pi} + 2 \qquad (5)$$

on the solution of (3). The auxiliary equation of (3) is $r^2 + 4r + 20 = 0$ with roots -2 ± 4i. Thus the G.S. of (3) is x = e^{-2t}(k_1 sin 4t + k_2 cos 4t), and its derivative is

$$x' = e^{-2t}[(-2k_1 - 4k_2) \sin 4t + (4k_1 - 2k_2) \cos 4t].$$

Applying the I.C. (4) to this G.S., we have $k_2 e^{-2\pi}$ = $-2e^{-2\pi}$ + 2, from which $k_2 = 2(e^{2\pi} - 1)$. Applying the I.C. (5) to the derivative of this G.S., we have $(4k_1 - 2k_2)e^{-2\pi} = -2e^{-2\pi} + 2$, from which $k_1 = \left(\frac{3}{2}\right)(e^{2\pi} - 1)$. Thus we obtain the solution

$$x = (e^{2\pi} - 1)e^{-2t}\left[\left(\frac{3}{2}\right)\sin 4t + 2\cos 4t\right],$$

valid for t ≥ π.

Section 5.5, Page 224

1. (a) Since the 12 lb. weight stretches the spring 6 in. = $\frac{1}{2}$

ft., Hooke's Law, F = ks, gives 12 = $k\left(\frac{1}{2}\right)$, so k = 24 lb./ft. Then since m = $\frac{w}{g}$ = $\frac{12}{32}$ = $\frac{3}{8}$ (slugs), a = 3, and F(t) = 2 cos ωt, The D.E. is

$$\frac{3}{8} x'' + 3x' + 24x = 2 \cos \omega t. \qquad (1)$$

The I.C.'s are $x(0) = 0$, $x'(0) = 0$. The resonance frequency is given by formula (5.69) of the text. It is

$$\frac{1}{2\pi} \sqrt{\frac{24}{\frac{3}{8}} - \frac{9}{2\left(\frac{3}{8}\right)^2}} = \frac{2\sqrt{2}}{\pi}.$$

This is $\frac{\omega_1}{2\pi}$, where ω_1 is the value of ω for which the forcing function is in resonance with the system. Thus $\frac{\omega_1}{2\pi} = \frac{2\sqrt{2}}{\pi}$, and so $\omega_1 = 4\sqrt{2}$.

We now let $\omega = \omega_1 = 4\sqrt{2}$ in (1), obtaining

$$\frac{3}{8} x'' + 3x' + 24x = 2 \cos 4\sqrt{2}t$$

or

$$x'' + 8x' + 64x = \frac{16}{3} \cos 4\sqrt{2}t. \qquad (2)$$

The auxiliary equation of the corresponding homogeneous D.E. is $r^2 + 8r + 64 = 0$, with roots $r = -4 \pm 4\sqrt{3}i$. Thus the complementary function of (2) is

$$x_c = e^{-4t}(c_1 \sin 4\sqrt{3}t + c_2 \cos 4\sqrt{3}t).$$

We use undetermined coefficients to find a particular integral. We let

$$x_p = A \sin 4\sqrt{2}t + B \cos 4\sqrt{2}t.$$

Differentiating twice and substituting into (1), we
find

$$\begin{cases} 32A - 32\sqrt{2}B = 0, \\ 32\sqrt{2}A + 32B = \dfrac{16}{3}; \end{cases}$$

and from these $A = \dfrac{\sqrt{2}}{18}$, $B = \dfrac{1}{18}$.   Thus we obtain the
general solution of (2) in the form

$$x = e^{-4t}(c_1 \sin 4\sqrt{3}t + c_2 \cos 4\sqrt{3}t)$$

$$+ \frac{\sqrt{2}}{18} \sin 4\sqrt{2}t + \frac{1}{18} \cos 4\sqrt{2}t.$$

Differentiating, we find

$$x' = e^{-4t}[(-4c_1 - 4\sqrt{3}c_2) \sin 4\sqrt{3}t$$

$$+ (4\sqrt{3}c_1 - 4c_2) \cos 4\sqrt{3}t]$$

$$+ \frac{4}{9} \cos 4\sqrt{2}t - \frac{2\sqrt{2}}{9} \sin 4\sqrt{2}t.$$

Applying the I.C.'s to these expressions for x and x',
we obtain $c_2 + \dfrac{1}{18} = 0$, $4\sqrt{3}c_1 - 4c_2 + \dfrac{4}{9} = 0$.   From

these, we find $c_1 = \dfrac{-\sqrt{3}}{18}$, $c_2 = \dfrac{-1}{18}$.   Thus we obtain the
solution of (2) in the form

$$x = e^{-4t} \frac{-\sqrt{3}\sin 4\sqrt{3}t - \cos 4\sqrt{3}t}{18}$$

$$+ \frac{\sqrt{2}\sin 4\sqrt{2}t + \cos 4\sqrt{2}t}{18}.$$

(b) In this part m, k, and F(t) are as in (1), but a = 0, so we have the D.E.

$$\frac{3}{8} x'' + 24x = 2\cos \omega t. \qquad (3)$$

The auxiliary equation of the corresponding homogeneous equation is $\left(\frac{3}{8}\right) r^2 + 24 = 0$ with roots r = ±8i. Thus the complementary function is

$$x_c = c_1 \sin 8t + c_2 \cos 8t. \qquad (4)$$

Undamped resonance occurs when the frequency $\frac{\omega}{2\pi}$ of the impressed force equals the natural frequency $\frac{4}{\pi}$ and hence when $\omega = 8$. Letting $\omega = 8$ in (3), we have the D.E.

$$\frac{3}{8} x'' + 24x = 2\cos 8t. \qquad (5)$$

The complementary function is given by (4). We find a particular integral using undetermined coefficients. We modify the UC set {sin 8t, cos 8t} of cos 8t by multiplying each member by t, obtaining {t sin 8t, t cos 8t}. Thus we assume $x_p = At \sin 8t + Bt \cos 8t$. Differentiating twice and substituting into (5), we find $A = \frac{1}{3}$, B = 0. Thus we obtain the general solution of (5) in the form

$$x = c_1 \sin 8t + c_2 \cos 8t + t \sin \frac{8t}{3}.$$

Differentiating this, we find

$$x' = 8c_1 \cos 8t - 8c_2 \sin 8t + 8t \cos \frac{8t}{3} + \sin \frac{8t}{3}.$$

The I.C.'s are the same as in part (a). Applying them to these expressions for x and x', we readily find $c_1 = 0$, $c_2 = 0$. Thus we obtain the solution of (5) in the form $x = t \sin \frac{8t}{3}$.

2. The 20 lb. weight stretches the spring 6 inches; so by Hooke's Law, F = ks, we have $20 = k\left(\frac{1}{2}\right)$, from which k = 40 lb./ft. Also, $m = \frac{w}{g} = \frac{20}{32} = \frac{5}{8}$ (slugs). Thus the D.E. is

$$\frac{5}{8} x'' + ax' + 40x = \cos \omega t$$

or

$$x'' + \frac{8}{5} ax' + 64x = \frac{8}{5} \cos \omega t . \qquad (1)$$

The auxiliary equation of the corresponding homogeneous D.E. is $r^2 + (8a/5)r + 64 = 0$, with roots

$r = -\frac{4}{5} a \pm \sqrt{\frac{16}{25} a^2 - 64}$. For resonance the system must be underdamped, so $\frac{16}{25} a^2 - 64 < 0$. Then the complementary function can be written

$$x_c = ce^{-(4a/5)t} \cos\left(\sqrt{64 - \frac{16a^2}{25}}\, t + \phi\right). \qquad (2)$$

The D.E. (1) is of the form (5.51) on page 212 of the text, and the complementary function (2) is of the form (5.52), where $2b = \frac{8}{5} a$, $\lambda^2 = 64$.

The resonance frequency is given as 0.5 (cycles/sec.). From the text, pages 219-220, this is $\omega_1/2\pi$, where

$\omega_1 = \sqrt{\lambda^2 - 2b^2}$. With the values of 2b and λ^2 just noted in this problem, as have $\lambda^2 - 2b^2 = 64 - \frac{32}{25} a^2$. Thus we obtain

$$\frac{\sqrt{64 - \frac{32}{25} a^2}}{2\pi} = 0.5.$$

Thus $\sqrt{64 - \frac{32}{25} a^2} = \pi$, and from this $a^2 = \frac{25}{32} (64 - \pi^2)$.

Thus recalling that $a > 0$, we find $a = 5\sqrt{2 - \frac{\pi^2}{32}}$.

Section 5.6, Page 232

1. Let i denote the current in amperes at time t. The total electromotive force is 40 V. Using the voltage drop laws 1 and 2 of the text (page 225) we find the following voltage drops:

   1. across the resistor: $E_R = Ri = 10i$.

   2. across the inductor: $E_L = Li' = 0.2i'$.

Applying Kirchhoff's Law, we have the D.E.

$$0.2i' + 10i = 40. \qquad (1)$$

Since the initial current is 0, the I.C. is

$$i(0) = 0.$$

The D.E. (1) is a first order linear D.E. In standard form it is

$$i' + 50 i = 200; \qquad (2)$$

an I.F. is $e^{\int 50t \, d} = e^{50t}$. Multiplying (2) through by this, we obtain

$$e^{50t} i' + 50e^{50t} i = 200e^{50t}$$

or

$$[e^{50t} i]' = 200e^{50t}.$$

Integrating and simplifying, we find

$$i = 4 + ce^{-50t}.$$

Applying the I.C. i = 0 at t = 0 to this, we find c = -4. Thus we obtain the solution

$$i = 4(1 - e^{-50t}).$$

3. Let i denote the current and let q denote the charge on the capacitor at time t. The electromotive force is 100 V. Using the voltage drop laws 1 and 3, we find the following voltage drops:

    1. across the resistor: $E_R = Ri = 10i$.

    2. across the capacitor: $E_C = \dfrac{1}{c} q = (10)^4 \dfrac{q}{2}$.

Applying Kirchhoff's Law, we have the equation

$$10i + (10)^4 \frac{q}{2} = 100.$$

Since $i = q'$, this reduces to

$$10q' + (10)^4 \frac{q}{2} = 100. \qquad (1)$$

Since the charge is initially zero, we have the I.C.

$$q(0) = 0.$$

The D.E. (1) is a first order linear D.E. In standard form it is

$$q' + 500q = 10, \qquad (2)$$

and an I.F. is $e^{\int 500\,dt} = e^{500t}$. Multiplying (2) through by this, we obtain

$$e^{500t}q' + 500e^{500t}q = 10e^{500t}$$

or

$$[e^{500t}q]' = 10e^{500t}.$$

Integrating and simplifying, we find

$$q = \frac{1}{50} + ce^{-500t}.$$

Applying the I.C. $q = 0$ at $t = 0$ to this, we find $c = \frac{-1}{50}$. Thus we obtain the solution

$$q = \frac{1 - e^{-500t}}{50}.$$

This is the charge. To find the current, we return to $10i + (10)^4 \frac{q}{2} = 100$, substitute the expression for q just found, and solve for i. We find $i = -500q + 10$ and hence $i = 10e^{-500t}$.

6. Let i denote the current in amperes at time t. The total electromotive force is $200e^{-100t}$. Using the voltage drop laws 1, 2, and 3, we find the following voltage drops:

   1. across the resistor: $E_R = Ri = 80i$.

   2. across the inductor: $E_L = Li' = 0.2i'$

   3. across the capacitor: $E_c = \frac{1}{c} q = 10^6 \frac{q}{5}$.

Applying Kirchhoff's Law, we have the D.E.

$$0.2i' + 80i + 10^6 \frac{q}{5} = 200e^{-100t}.$$

Since $i = q'$, this reduces to

$$0.2q'' + 80q' + 10^6 \frac{q}{5} = 200e^{-100t}. \qquad (1)$$

Since the charge q is initially zero, we have the first I.C.

$$q(0) = 0.$$

Since the current i is initially zero and $i = q'$, we have the second I.C.

$$q'(0) = 0.$$

The homogeneous D.E. corresponding to D.E. (1) has the auxiliary equation

$$\frac{1}{5} r^2 + 80r + 200,000 = 0,$$

and the roots of this are $-200 \pm 979.8i$. Thus the complementary function of D.E. (1) is

$$q_c = e^{-200t}(c_1 \sin 979.8t + c_2 \cos 979.8t).$$

We use undetermined coefficients to find a particular integral. We write

$$q_p = Ae^{-100t}.$$

Differentiating twice and substituting into (1), we find $A = \frac{1}{970} \approx 0.0010$. Thus the general solution of D.E. (1) is

$$q = e^{-200t}(c_1 \sin 979.8t + c_2 \cos 979.8t) + \frac{e^{-100t}}{970}.$$

Differentiating this, we obtain

$$q' = e^{-200t}[(-200c_1 - 979.8c_2) \sin 979.8t$$
$$+ (979.8c_1 - 200c_2) \cos 979.8t] - \frac{10e^{-100t}}{97}. \quad (2)$$

Applying the I.C.'s to these expressions for q and q', we have

$$c_2 + \frac{1}{970} = 0, \quad 979.8c_1 - 200c_2 - \frac{10}{97} = 0,$$

from these we find that $c_1 = \frac{-10}{(97)(979.8)} \approx -0.0001$ and $c_2 = \frac{-1}{970} \approx -0.0010$. Thus we obtain

$$q = e^{-200t}(-0.0001 \sin 979.8t - 0.0010 \cos 979.8t)$$
$$+ 0.0010 \, e^{-100t}.$$

Since $i = q'$, using formula (2) for q' and the values of c_1 and c_2 determined above, we find

$$i = e^{-200t}(1.0311 \sin 979.8t + 0.1031 \cos 979.8t)$$
$$- 0.1031e^{-100t}.$$

Chapter 6

2. We assume $y = \sum\limits_{n=0}^{\infty} c_n x^n$. Then $y' = \sum\limits_{n=1}^{\infty} n c_n x^{n-1}$,

$y'' = \sum\limits_{n=2}^{\infty} n(n-1) c_n x^{n-2}$. Substituting into the D.E., we

obtain

$$\sum\limits_{n=2}^{\infty} n(n-1) c_n x^{n-2} + 8x \sum\limits_{n=1}^{\infty} n c_n x^{n-1} - 4 \sum\limits_{n=0}^{\infty} c_n x^n = 0$$

or

$$\sum\limits_{n=2}^{\infty} n(n-1) c_n x^{n-2} + 8 \sum\limits_{n=1}^{\infty} n c_n x^n - 4 \sum\limits_{n=0}^{\infty} c_n x^n = 0.$$

We rewrite the first summation so that x has the exponent n. Doing so, we have:

$$\sum\limits_{n=0}^{\infty} (n+2)(n+1) c_{n+2} x^n + 8 \sum\limits_{n=1}^{\infty} n c_n x^n - 4 \sum\limits_{n=0}^{\infty} c_n x^n = 0.$$

The common range of these three summations is from 1 to ∞. We write out the individual terms in each that do *not* belong to this common range. Thus we have

$$(2c_2 - 4c_0) + \sum\limits_{n=1}^{\infty} [(n+2)(n+1) c_{n+2} + 4(2n-1) c_n] x^n = 0.$$

301

Equating to zero the coefficient of each power of x, we obtain

$$2c_2 - 4c_0 = 0, \tag{1}$$

$$(n + 2)(n + 1)c_{n+2} + 4(2n - 1)c_n = 0, \ n \geq 1. \tag{2}$$

From (1) we have $c_2 = 2c_0$. From (2), we obtain

$$c_{n+2} = - \frac{4(2n - 1)c_n}{(n + 1)(n + 2)}, \quad n \geq 1.$$

Using this, we find $c_3 = - \dfrac{2c_1}{3}$, $c_4 = -c_2 = -2c_0$,

$c_5 = -c_3 = \dfrac{2c_1}{3}$. Substituting these values into the assumed solution, we have

$$y = c_0 + c_1 x + 2c_0 x^2 - \frac{2c_1}{3} x^3 - 2c_0 x^4 + \frac{2c_1}{3} x^5 + \cdots$$

or

$$y = c_0(1 + 2x^2 - 2x^4 + \cdots) + c_1\left(x - \frac{2}{3} x^3 + \frac{2}{3} x^5 + \cdots\right).$$

4. We assume $y = \displaystyle\sum_{n=0}^{\infty} c_n x^n$. Then $y' = \displaystyle\sum_{n=1}^{\infty} nc_n x^{n-1}$,

$y'' = \displaystyle\sum_{n=2}^{\infty} n(n - 1)c_n x^{n-2}$. Substituting into the D.E., we obtain

$$\sum_{n=2}^{\infty} n(n - 1)c_n x^{n-2} + \sum_{n=1}^{\infty} nc_n x^{n-1} + 3x^2 \sum_{n=0}^{\infty} c_n x^n = 0$$

or

$$\sum_{n=2}^{\infty} n(n-1)c_n x^{n-2} + \sum_{n=1}^{\infty} nc_n x^{n-1} + 3\sum_{n=0}^{\infty} c_n x^{n+2} = 0.$$

We rewrite the summations so that x has the same exponent in each. We choose n – 2 as this common exponent, leave the first summation alone, and rewrite the second and third accordingly. We have

$$\sum_{n=2}^{\infty} n(n-1)c_n x^{n-2} + \sum_{n=2}^{\infty} (n-1)c_{n-1} x^{n-2}$$

$$+ 3\sum_{n=4}^{\infty} c_{n-4} x^{n-2} = 0.$$

The common range of these three summations is from 4 to ∞. We write out the individual terms in each that do *not* belong to this common range. Thus we have

$$(2c_2 + c_1) + (6c_3 + 2c_2)x$$

$$+ \sum_{n=4}^{\infty} [n(n-1)c_n + (n-1)c_{n-1} + 3c_{n-4}]x^{n-2} = 0.$$

Equating to zero the coefficient of each power of x, we obtain

$$2c_2 + c_1 = 0, \quad 6c_3 + 2c_2 = 0, \tag{1}$$

$$n(n-1)c_n + (n-1)c_{n-1} + 3c_{n-4} = 0, \quad n \geq 4. \tag{2}$$

From (1), we have $c_2 = -\dfrac{c_1}{2}$, $c_3 = -\dfrac{c_2}{3} = \dfrac{c_1}{6}$. From (2), we obtain

$$c_n = -\frac{(n-1)c_{n-1} + 3c_{n-4}}{n(n-1)}, \quad n \geq 4.$$

Using this, we find $c_4 = -\dfrac{c_3 + c_0}{4} = -\dfrac{c_0}{4} - \dfrac{c_1}{24}$,

$$c_5 = -\frac{4c_4 + 3c_1}{20} = -\frac{1}{5}\left(-\frac{c_0}{4} - \frac{c_1}{24}\right) - \frac{3}{20}c_1 = \frac{c_0}{20} - \frac{17c_1}{120}.$$

Substituting these values into the assumed solution, we have

$$y = c_0 + c_1 x - \frac{c_1}{2}x^2 + \frac{c_1}{6}x^3 - \left(\frac{c_0}{4} + \frac{c_1}{24}\right)x^4$$
$$+ \left(\frac{c_0}{20} - \frac{17c_1}{120}\right)x^5 + \cdots$$

or

$$y = c_0\left(1 - \frac{1}{4}x^4 + \frac{1}{20}x^5 + \cdots\right)$$
$$+ c_1\left(x - \frac{1}{2}x^2 + \frac{1}{6}x^3 - \frac{1}{24}x^4 - \frac{17}{120}x^5 + \cdots\right).$$

5. We assume $y = \displaystyle\sum_{n=0}^{\infty} c_n x^n$. Then $y' = \displaystyle\sum_{n=1}^{\infty} nc_n x^{n-1}$,

$y'' = \displaystyle\sum_{n=2}^{\infty} n(n-1)c_n x^{n-2}$. Substituting into the D.E., we

obtain

$$\sum_{n=2}^{\infty} n(n-1)c_n x^{n-2} + x\sum_{n=1}^{\infty} nc_n x^{n-1}$$
$$+ 2x^2\sum_{n=0}^{\infty} c_n x^n + \sum_{n=0}^{\infty} c_n x^n = 0$$

or

$$\sum_{n=2}^{\infty} n(n-1)c_n x^{n-2} + \sum_{n=1}^{\infty} nc_n x^n + \sum_{n=0}^{\infty} 2c_n x^{n+2} + \sum_{n=0}^{\infty} c_n x^n = 0.$$

We rewrite the first and third summations so that x has the exponent n in each.   Thus we have

$$\sum_{n=0}^{\infty} (n+2)(n+1)c_{n+2} x^n + \sum_{n=1}^{\infty} nc_n x^n$$

$$+ \sum_{n=2}^{\infty} 2c_{n-2} x^n + \sum_{n=0}^{\infty} c_n x^n = 0.$$

The common range of these four summations is from 2 to ∞. We write out the individual terms in each that do *not* belong to this range.   Thus we have:

$$(c_0 + 2c_2) + (2c_1 + 6c_3)x$$

$$+ \sum_{n=2}^{\infty} [(n+2)(n+1)c_{n+2} + (n+1)c_n + 2c_{n-2}]x^n = 0.$$

Equating to zero the coefficient of each power of x, we obtain

$$c_0 + 2c_2 = 0, \quad 2c_1 + 6c_3 = 0, \tag{1}$$

$$(n+2)(n+1)c_{n+2} + (n+1)c_n$$

$$+ 2c_{n-2} = 0, \quad n \geq 2. \tag{2}$$

From (1), we find $c_2 = \dfrac{-c_0}{2}$, $c_3 = \dfrac{-c_3}{3} = \dfrac{-c_1}{3}$.   From (2), we find

$$c_{n+2} = - \frac{(n + 1)c_n + 2c_{n-2}}{(n + 1)(n + 2)}, \quad n \geq 2.$$

Using this, we find $c_4 = - \dfrac{3c_2 + 2c_0}{12} = \dfrac{-c_0}{24}$;

$c_5 = - \dfrac{4c_3 + 2c_1}{20} = \dfrac{-c_1}{30}.$ Substituting these values into

the assumed solution, we have

$$y = c_0 + c_1 x - \left(\frac{c_0}{2}\right)x^2 - \left(\frac{c_1}{3}\right)x^3 - \left(\frac{c_0}{24}\right)x^4 - \left(\frac{c_1}{30}\right)x^5 + \cdots$$

or

$$y = c_0\left(1 - \frac{x^2}{2} - \frac{x^4}{24} + \cdots\right) + c_1\left(x - \frac{x^3}{3} - \frac{x^5}{30} + \cdots\right).$$

9. We assume $y = \displaystyle\sum_{n=0}^{\infty} c_n x^n.$ Then $y' = \displaystyle\sum_{n=1}^{\infty} n c_n x^{n-1},$

$y'' = \displaystyle\sum_{n=2}^{\infty} n(n - 1)c_n x^{n-2}.$ Substituting into the D.E., we

obtain

$$\sum_{n=2}^{\infty} n(n - 1)c_n x^{n-2} - x^3 \sum_{n=1}^{\infty} n c_n x^{n-1}$$

$$- 2\sum_{n=1}^{\infty} n c_n x^{n-1} - 6x^2 \sum_{n=0}^{\infty} c_n x^n = 0$$

or

$$\sum_{n=2}^{\infty} n(n-1)c_n x^{n-2} - \sum_{n=1}^{\infty} nc_n x^{n+2} - 2\sum_{n=1}^{\infty} nc_n x^{n-1}$$

$$- 6\sum_{n=0}^{\infty} c_n x^{n+2} = 0.$$

We write the first and third summations so that x has the exponent n + 2 in each, as it already has in the second and fourth summations. We have

$$\sum_{n=-2}^{\infty} (n+4)(n+3)c_{n+4} x^{n+2} - \sum_{n=1}^{\infty} nc_n x^{n+2}$$

$$- 2\sum_{n=-2}^{\infty} (n+3)c_{n+3} x^{n+2} - 6\sum_{n=0}^{\infty} c_n x^{n+2} = 0.$$

The common range of these four summations is from 1 to ∞. We write out the individual terms in each that do *not* belong to this common range. We have

$$(2c_2 - 2c_1) + (6c_3 - 4c_2)x + (12c_4 - 6c_3 - 6c_0)x^2$$

$$+ \sum_{n=1}^{\infty} [(n+4)(n+3)c_{n+4} - 2(n+3)c_{n+3}$$

$$- (n+6)c_n]x^{n+2} = 0.$$

Equating to zero the coefficient of each power of x, we obtain

$$2c_2 - 2c_1 = 0, \quad 6c_3 - 4c_2 = 0,$$
$$12c_4 - 6c_3 - 6c_0 = 0, \tag{1}$$

$$(n + 4)(n + 3)c_{n+4} - 2(n + 3)c_{n+3}$$
$$- (n + 6)c_n = 0, \quad n \geq 1. \tag{2}$$

From (1), we find $c_2 = c_1$, $c_3 = \dfrac{2c_2}{3} = \dfrac{2c_1}{3}$,

$$c_4 = \frac{c_0 + c_3}{2} = \frac{c_0}{2} + \frac{c_1}{3}. \quad \text{From (2), we obtain}$$

$$c_{n+4} = \frac{(n + 6)c_n + 2(n + 3)c_{n+3}}{(n + 3)(n + 4)}, \quad n \geq 1.$$

Using this, we find $c_5 = \dfrac{7c_1 + 8c_4}{20} = \dfrac{c_0}{5} + \dfrac{29c_1}{60}$.

Substituting these values into the assumed solution, we have

$$y = c_0 + c_1 x + c_1 x^2 + \frac{2c_1}{3} x^3 + \left(\frac{c_0}{2} + \frac{c_1}{3}\right) x^4$$
$$+ \left(\frac{c_0}{5} + \frac{29c_1}{60}\right) x^5 + \cdots$$

or

$$y = c_0 \left(1 + \frac{x^4}{2} + \frac{x^5}{5} + \cdots\right)$$
$$+ c_1 \left(x + x^2 + \frac{2x^3}{3} + \frac{x^4}{3} + \frac{29x^5}{60} + \cdots\right).$$

10. We assume $y = \displaystyle\sum_{n=0}^{\infty} c_n x^n$. Then $y' = \displaystyle\sum_{n=1}^{\infty} nc_n x^{n-1}$,

$y'' = \displaystyle\sum_{n=2}^{\infty} n(n - 1)c_n x^{n-2}$. Substituting into the D.E., we

obtain

$$\sum_{n=2}^{\infty} n(n - 1)c_n x^{n-2} - x^2 \sum_{n=1}^{\infty} nc_n x^{n-1}$$

$$- x\sum_{n=1}^{\infty} nc_n x^{n-1} + \sum_{n=0}^{\infty} c_n x^n = 0$$

or

$$\sum_{n=2}^{\infty} n(n - 1)c_n x^{n-2} - \sum_{n=1}^{\infty} nc_n x^{n+1} - \sum_{n=1}^{\infty} nc_n x^n + \sum_{n=0}^{\infty} c_n x^n = 0.$$

We rewrite the summations so that x has the common exponent n in each. We have

$$\sum_{n=0}^{\infty} (n + 2)(n + 1)c_{n+2} x^n - \sum_{n=2}^{\infty} (n - 1)c_{n-1} x^n$$

$$- \sum_{n=1}^{\infty} nc_n x^n + \sum_{n=0}^{\infty} c_n x^n = 0.$$

The common range of these four summations is from 2 to ∞. We write out the individual terms in each that do *not* belong to this common range. Thus we have

$$(2c_2 + c_0) + (6c_3 - c_1 + c_1)x$$

$$+ \sum_{n=2}^{\infty} [(n + 2)(n + 1)c_{n+2} - (n - 1)c_{n-1}$$

$$- (n - 1)c_n]x^n = 0.$$

Equating to zero the coefficients of each power of x, we obtain

$$2c_2 + c_0 = 0, \; 6c_3 = 0, \tag{1}$$

$$(n + 1)(n + 2)c_{n+2} - (n - 1)c_{n-1}$$
$$- (n - 1)c_n = 0, \; n \geq 2. \tag{2}$$

From (1), we have $c_2 = -\dfrac{c_0}{2}$ and $c_3 = 0$. From (2), we obtain

$$c_{n+2} = \frac{(n - 1)(c_{n-1} + c_n)}{(n + 1)(n + 2)}, \quad n \geq 2.$$

Using this, we find $c_4 = \dfrac{c_1 + c_2}{12} = -\dfrac{c_0}{24} + \dfrac{c_1}{12}$,

$$c_5 = \frac{c_2 + c_3}{10} = -\frac{c_0}{20}, \; c_6 = \frac{c_3 + c_4}{10} = -\frac{c_0}{240} + \frac{c_1}{120}.$$

Substituting these values into the assumed solution, we have

$$y = c_0 + c_1 x - \frac{c_0}{2}x^2 + \left(-\frac{c_0}{24} + \frac{c_1}{12}\right)x^4$$
$$- \frac{c_0}{20}x^5 + \left(-\frac{c_0}{240} + \frac{c_1}{120}\right)x^6 + \cdots$$

or

$$y = c_0\left(1 - \frac{1}{2}x^2 - \frac{1}{24}x^4 - \frac{1}{20}x^5 - \frac{1}{240}x^6 \cdots\right)$$
$$+ c_1\left(x + \frac{1}{12}x^4 + \frac{1}{120}x^6 + \cdots\right).$$

11.   We assume $y = \displaystyle\sum_{n=0}^{\infty} c_n x^n$. Then $y' = \displaystyle\sum_{n=1}^{\infty} n c_n x^{n-1}$,

$$y'' = \sum_{n=2}^{\infty} n(n - 1)c_n x^{n-2}.$$ Substituting into the D.E., we

obtain

$$x^2 \sum_{n=2}^{\infty} n(n-1)c_n x^{n-2} + \sum_{n=2}^{\infty} n(n-1)c_n x^{n-2}$$

$$+ x \sum_{n=1}^{\infty} nc_n x^{n-1} + x \sum_{n=0}^{\infty} c_n x^n = 0$$

or

$$\sum_{n=2}^{\infty} n(n-1)c_n x^n + \sum_{n=2}^{\infty} n(n-1)c_n x^{n-2} + \sum_{n=1}^{\infty} nc_n x^n$$

$$+ \sum_{n=0}^{\infty} c_n x^{n+1} = 0.$$

We rewrite the second and fourth summations so that x has the exponent n in each. We have:

$$\sum_{n=2}^{\infty} n(n-1)c_n x^n + \sum_{n=0}^{\infty} (n+2)(n+1)c_{n+2} x^n$$

$$+ \sum_{n=1}^{\infty} nc_n x^n + \sum_{n=1}^{\infty} c_{n-1} x^n = 0.$$

The common range of these four summations is from 2 to ∞. We write out the individual terms in each that do *not* belong to this range. Thus we have:

$$2c_2 + (c_0 + c_1 + 6c_3)x$$

$$+ \sum_{n=2}^{\infty} \{(n+2)(n+1)c_{n+2} + [n(n-1) + n]c_n$$

$$+ c_{n-1}\}x^n = 0.$$

Equating to zero the coefficients of each power of x, we
obtain

$$2c_2 = 0, \quad c_0 + c_1 + 6c_3 = 0, \tag{1}$$

$$(n + 2)(n + 1)c_{n+2} + [n(n - 1) + n]c_n$$
$$+ c_{n-1} = 0, \quad n \geq 2 \tag{2}$$

From (1), we find $c_2 = 0$, $c_3 = -\dfrac{c_0 + c_1}{6}$. From (2), we
find

$$c_{n+2} = -\frac{n^2 c_n + c_{n-1}}{(n + 1)(n + 2)}, \quad n \geq 2.$$

Using this, we find $c_4 = -\dfrac{4c_2 + c_1}{12} = -\dfrac{c_1}{12}$;

$$c_5 = -\frac{9c_3 + c_2}{20} = \left(\frac{3}{40}\right)(c_0 + c_1).$$ Substituting these

values into the assumed solution, we have

$$y = c_0 + c_1 x - \frac{(c_0 + c_1)x^3}{6} - \frac{c_1 x^4}{12} + \frac{3(c_0 + c_1)x^5}{40} + \cdots$$

or

$$y = c_0\left(1 - \frac{x^3}{6} + \frac{3x^5}{40} + \cdots\right) + c_1\left(x - \frac{x^3}{6} - \frac{x^4}{12} + \frac{3x^5}{40} + \cdots\right).$$

13. We assume $y = \displaystyle\sum_{n=0}^{\infty} c_n x^n$. Then $y' = \displaystyle\sum_{n=1}^{\infty} n c_n x^{n-1}$,

$y'' = \displaystyle\sum_{n=2}^{\infty} n(n - 1)c_n x^{n-2}$. Substituting into the D.E., we

obtain

$$x^3 \sum_{n=2}^{\infty} n(n-1)c_n x^{n-2} - \sum_{n=2}^{\infty} n(n-1)c_n x^{n-2}$$

$$+ x^2 \sum_{n=1}^{\infty} nc_n x^{n-1} + x \sum_{n=0}^{\infty} c_n x^n = 0$$

or

$$\sum_{n=2}^{\infty} n(n-1)c_n x^{n+1} - \sum_{n=2}^{\infty} n(n-1)c_n x^{n-2}$$

$$+ \sum_{n=1}^{\infty} nc_n x^{n+1} + \sum_{n=0}^{\infty} c_n x^{n+1} = 0.$$

We rewrite the second summation so that x has the exponent n + 1 in it (since this is the exponent of x in all the other summations). Thus we have:

$$\sum_{n=2}^{\infty} n(n-1)c_n x^{n+1} - \sum_{n=-1}^{\infty} (n+3)(n+2)c_{n+3} x^{n+1}$$

$$+ \sum_{n=1}^{\infty} nc_n x^{n+1} + \sum_{n=0}^{\infty} c_n x^{n+1} = 0.$$

The common range of these four summations is from 2 to ∞. We write out the individual terms in each that do *not* belong to this range. Thus we have:

$$-2c_2 + (c_0 - 6c_3)x + (2c_1 - 12c_4)x^2$$

$$+ \sum_{n=2}^{\infty} \{-(n + 3)(n + 2)c_{n+3}$$

$$+ [n(n - 1) + n + 1]c_n\}x^n = 0.$$

Equating to zero the coefficients of each power of x, we obtain

$$-2c_2 = 0, \quad c_0 - 6c_3 = 0, \quad 2c_1 - 12c_4 = 0, \tag{1}$$

$$-(n + 3)(n + 2)c_{n+3} + [n(n - 1)$$
$$+ n + 1]c_n = 0, \quad n \geq 2. \tag{2}$$

From (1), we find $c_2 = 0$, $c_3 = \dfrac{c_0}{6}$, $c_4 = \dfrac{c_1}{6}$. From (2), we find

$$c_{n+3} = \frac{(n^2 + 1)c_n'}{(n + 2)(n + 3)}, \quad n \geq 2.$$

Using this, we find $c_5 = \dfrac{5c_2}{20} = 0$; $c_6 = \dfrac{10c_3}{30} = \dfrac{c_3}{3} = \dfrac{c_0}{18}$;

and $c_7 = \dfrac{17c_4}{42} = \dfrac{17c_1}{252}$. Substituting these values into the assumed solution, we have

$$y = c_0 + c_1 x + \frac{c_0 x^3}{6} + \frac{c_1 x^4}{6} + \frac{c_0 x^6}{18} + \frac{17c_1 x^7}{252} + \cdots$$

or

$$y = c_0 \left(1 + \frac{x^3}{6} + \frac{x^6}{18} + \cdots\right) + c_1 \left(x + \frac{x^4}{6} + \frac{17x^7}{252} + \cdots\right).$$

15.   We assume $y = \sum\limits_{n=0}^{\infty} c_n x^n$. Then $y' = \sum\limits_{n=1}^{\infty} nc_n x^{n-1}$,

$y'' = \sum\limits_{n=2}^{\infty} n(n-1)c_n x^{n-2}$. Substituting into the D.E., we obtain

$$\sum_{n=2}^{\infty} n(n-1)c_n x^{n-2} - x\sum_{n=1}^{\infty} nc_n x^{n-1} - \sum_{n=0}^{\infty} c_n x^n = 0$$

or

$$\sum_{n=2}^{\infty} n(n-1)c_n x^{n-2} + \sum_{n=1}^{\infty} (-nc_n)x^n + \sum_{n=0}^{\infty} (-c_n)x^n = 0.$$

We rewrite the first summation so that x has an exponent n. Thus we have:

$$\sum_{n=0}^{\infty} (n+1)(n+2)c_{n+2} x^n + \sum_{n=1}^{\infty} (-nc_n)x^n + \sum_{n=0}^{\infty} (-c_n)x^n = 0.$$

The common range of these three summations is from 1 to ∞.
We write out the terms from the 1st and 3rd that do *not*
belong in this range. Thus we have:

$(2c_2 - c_0)$

$$+ \sum_{n=1}^{\infty} [(n+1)(n+2)c_{n+2} + (-n-1)c_n]x^n = 0.$$

Equating the coefficients of each power of x to zero we obtain

$$2c_2 - c_0 = 0, \tag{1}$$

$$(n + 1)(n + 2)c_{n+2} + (-n - 1)c_n = 0, \quad n \geq 1. \tag{2}$$

From (1), we find $c_2 = \frac{1}{2} c_0$. From (2), we find

$$c_{n+2} = \frac{1}{n + 2} c_n, \quad n \geq 1.$$

Using this, we find $c_3 = \frac{1}{3} c_1$; $c_4 = \frac{1}{4} c_2 = \frac{1}{8} c_0$;

$c_5 = \frac{1}{5} c_3 = \frac{1}{15} c_1$; $c_6 = \frac{1}{6} c_4 = \frac{1}{48} c_0$. Thus the G.S. of

the D.E. is

$$y = c_0 \left(1 + \frac{1}{2} x^2 + \frac{1}{8} x^4 + \frac{1}{48} x^6 + \cdots \right)$$
$$+ c_1 \left(x + \frac{1}{3} x^3 + \frac{1}{15} x^5 + \cdots \right).$$

We now apply the I.C. $y(0) = 1$ to this, obtaining $c_0 = 1$. Differentiating we find

$$y' = x + \frac{1}{2} x^3 + \frac{1}{8} x^5 + \cdots + c_1 \left(1 + x^2 + \frac{1}{3} x^4 + \cdots \right).$$

Applying the I.C. $y'(0) = 0$ to this, we have $c_1 = 0$. Thus we have the solution

$$y = 1 + \frac{1}{2} x^2 + \frac{1}{8} x^4 + \frac{1}{48} x^6 + \cdots.$$

Writing this as

$$y = 1 + \frac{1}{2^1 1!} x^2 + \frac{1}{2^2 2!} x^4 + \frac{1}{2^3 3!} x^6 + \cdots$$

and observing that for $n \geq 1$,

$$c_{2n} = \frac{1}{2 \cdot 4 \cdot 6 \cdots (2n)} = \frac{1}{2^n n!},$$

we express the solution simply as

$$y = \sum_{n=0}^{\infty} \frac{x^{2n}}{2^n n!}.$$

18. We assume $y = \sum_{n=0}^{\infty} c_n x^n$. Then $y' = \sum_{n=1}^{\infty} n c_n x^{n-1}$,

$y'' = \sum_{n=2}^{\infty} n(n-1) c_n x^{n-2}$. Substituting into the D.E., we obtain

$$x^2 \sum_{n=2}^{\infty} n(n-1) c_n x^{n-2} + \sum_{n=2}^{\infty} n(n-1) c_n x^{n-2}$$

$$+ x \sum_{n=1}^{\infty} n c_n x^{n-1} + 2x \sum_{n=0}^{\infty} c_n x^n = 0$$

or

$$\sum_{n=2}^{\infty} n(n-1) c_n x^n + \sum_{n=2}^{\infty} n(n-1) c_n x^{n-2}$$

$$+ \sum_{n=1}^{\infty} n c_n x^n + \sum_{n=0}^{\infty} 2 c_n x^{n+1} = 0.$$

We rewrite the second and fourth summations so that x has the exponent n in each. Thus we have:

$$\sum_{n=2}^{\infty} n(n - 1)c_n x^n + \sum_{n=0}^{\infty} (n + 2)(n + 1)c_{n+2} x^n$$

$$+ \sum_{n=1}^{\infty} nc_n x^n + \sum_{n=1}^{\infty} 2c_{n-1} x^n = 0.$$

The common range of these four summations is from 2 to ∞. We write out the individual terms in each that do *not* belong to this range. Thus we have:

$$2c_2 + (6c_3 + c_1 + 2c_0)x + \sum_{n=2}^{\infty} \{[n(n - 1) + n]c_n$$

$$+ (n + 2)(n + 1)c_{n+2} + 2c_{n-1}\}x^n = 0.$$

Equating to zero the coefficients of each power of x, we obtain

$$2c_2 = 0, \ 2c_0 + c_1 + 6c_3 = 0, \tag{1}$$

$$[n(n - 1) + n]c_n + (n + 2)(n + 1)c_{n+2}$$
$$+ 2c_{n-1} = 0, \ n \geq 2. \tag{2}$$

From (1), we find $c_2 = 0$, $c_3 = -\dfrac{2c_0 + c_1}{6}$. From (2), we find

$$c_{n+2} = -\frac{n^2 c_n + 2c_{n-1}}{(n + 1)(n + 2)}, \quad n \geq 2.$$

Using this, we find $c_4 = -\dfrac{4c_2 + 2c_1}{12} = -\dfrac{c_1}{6}$,

$c_5 = -\dfrac{9c_3 + 2c_2}{20} = \dfrac{3(2c_0 + c_1)}{40}$. Substituting these values

into the assumed solution, we have

$$y = c_0 + c_1 x - \frac{(2c_0 + c_1)x^3}{6} - \frac{c_1 x^4}{6} + \frac{3(2c_0 + c_1)x^5}{40} + \cdots$$

or

$$y = c_0 \left(1 - \frac{x^3}{3} + \frac{3x^5}{20} + \cdots \right)$$

$$+ c_1 \left(x - \frac{x^3}{6} - \frac{x^4}{6} + \frac{3x^5}{40} + \cdots \right).$$

We now apply the I.C. $y(0) = 2$ to this, obtaining $c_0 = 2$. Differentiating we find

$$y' = c_0 \left(-x^2 + \frac{3x^4}{4} + \cdots \right)$$

$$+ c_1 \left(1 - \frac{x^2}{2} - \frac{2x^3}{3} + \frac{3x^4}{8} + \cdots \right).$$

Applying the I.C. $y'(0) = 3$ to this, we have $c_1 = 3$. Thus we have the solution

$$y = 2 \left(1 - \frac{x^3}{3} + \frac{3x^5}{20} + \cdots \right)$$

$$+ 3 \left(x - \frac{x^3}{6} - \frac{x^4}{6} + \frac{3x^5}{40} + \cdots \right)$$

or

$$y = 2 + 3x - \frac{7x^3}{6} - \frac{x^4}{2} + \frac{21x^5}{40} + \cdots.$$

20. We assume $y = \displaystyle\sum_{n=0}^{\infty} c_n x^n$. Then $y' = \displaystyle\sum_{n=1}^{\infty} n c_n x^{n-1}$,

$y'' = \displaystyle\sum_{n=2}^{\infty} n(n-1) c_n x^{n-2}$. Substituting into the D.E., we

obtain

$$x^2 \sum_{n=2}^{\infty} n(n-1) c_n x^{n-2} - \sum_{n=2}^{\infty} n(n-1) c_n x^{n-2}$$

$$+ 4x \sum_{n=1}^{\infty} n c_n x^{n-1} + 2 \sum_{n=0}^{\infty} c_n x^n = 0$$

or

$$\sum_{n=2}^{\infty} n(n-1) c_n x^n - \sum_{n=2}^{\infty} n(n-1) c_n x^{n-2}$$

$$+ 4 \sum_{n=1}^{\infty} n c_n x^n + 2 \sum_{n=0}^{\infty} c_n x^n = 0.$$

We rewrite the second summation so that x has the exponent
n, as in the other three summations. We have:

$$\sum_{n=2}^{\infty} n(n-1) c_n x^n - \sum_{n=0}^{\infty} (n+2)(n+1) c_{n+2} x^n$$

$$+ 4 \sum_{n=1}^{\infty} n c_n x^n + 2 \sum_{n=0}^{\infty} c_n x^n = 0.$$

The common range of these summations is from 2 to ∞.
We write out the individual terms that do *not* belong to
this common range. We have:

$$(-2c_2 + 2c_0) + (-6c_3 + 6c_1)x$$

$$+ \sum_{n=2}^{\infty} \{[n(n-1) + 4n + 2]c_n$$

$$- (n+2)(n+1)c_{n+2}\}x^n = 0.$$

Noting that $n(n-1) + 4n + 2 = (n+1)(n+2)$, we equate to zero the coefficients of each power of x, to obtain

$$-2c_2 + 2c_0 = 0, \quad -6c_3 + 6c_1 = 0, \tag{1}$$

$$(n+1)(n+2)c_n$$
$$- (n+1)(n+2)c_{n+2} = 0, \quad n \geq 2. \tag{2}$$

From (1), we have $c_2 = c_0$ and $c_3 = c_1$. From (2), we obtain $c_{n+2} = c_n$, $n \geq 2$. From this we observe that $c_{2n} = c_0$ and $c_{2n+1} = c_1$ for all $n \geq 1$. Thus we obtain the G.S.

$$y = c_0 \left(1 + x^2 + x^4 + \cdots \right) + c_1 \left(x + x^3 + x^5 + \cdots \right)$$

$$= c_0 \sum_{n=0}^{\infty} x^{2n} + c_1 \sum_{n=0}^{\infty} x^{2n+1}.$$

We now apply the I.C. $y(0) = 1$ to this, obtaining $c_0 = 1$. Then differentiating we have

$$y' = (2x + 4x^3 + \cdots) + c_1(1 + 3x^2 + 5x^4 + \cdots).$$

Applying the I.C. $y'(0) = -1$ to this, we find $c_1 = -1$. Thus we obtain the desired solution

$$y = 1 - x + x^2 - x^3 + x^4 - x^5 + \cdots = \sum_{n=0}^{\infty} (-1)^n x^n.$$

22. We assume $y = \sum_{n=0}^{\infty} c_n (x - 1)^n$. Then $y' = \sum_{n=1}^{\infty} n c_n (x - 1)^{n-1}$,

$y'' = \sum_{n=2}^{\infty} n(n - 1) c_n (x - 1)^{n-2}$. Substituting into the D.E.,

we obtain

$$x^2 \sum_{n=2}^{\infty} n(n - 1) c_n (x - 1)^{n-2} + 3x \sum_{n=1}^{\infty} n c_n (x - 1)^{n-1}$$

$$- \sum_{n=0}^{\infty} c_n (x - 1)^n = 0.$$

Since the summations involve powers of $x - 1$, we must express the respective "coefficients" x^2 and $3x$ of the first two summations in powers of $x - 1$. Thus we have

$$[(x - 1)^2 + 2(x - 1) + 1] \sum_{n=2}^{\infty} n(n - 1) c_n (x - 1)^{n-2}$$

$$+ [3(x - 1) + 3] \sum_{n=1}^{\infty} n c_n (x - 1)^{n-1}$$

$$- \sum_{n=0}^{\infty} c_n (x - 1)^n = 0.$$

or

$$\sum_{n=2}^{\infty} n(n-1)c_n(x-1)^n + \sum_{n=2}^{\infty} 2n(n-1)c_n(x-1)^{n-1}$$

$$+ \sum_{n=2}^{\infty} n(n-1)c_n(x-1)^{n-2} + \sum_{n=1}^{\infty} 3nc_n(x-1)^n$$

$$+ \sum_{n=1}^{\infty} 3nc_n(x-1)^{n-1} - \sum_{n=0}^{\infty} c_n(x-1)^n = 0.$$

We rewrite the second, third, and fifth summations so
that x − 1 has the exponent n in each. Thus we have:

$$\sum_{n=2}^{\infty} n(n-1)c_n(x-1)^n + \sum_{n=1}^{\infty} 2(n+1)nc_{n+1}(x-1)^n$$

$$+ \sum_{n=0}^{\infty} (n+2)(n+1)c_{n+2}(x-1)^n + \sum_{n=1}^{\infty} 3nc_n(x-1)^n$$

$$+ \sum_{n=0}^{\infty} 3(n+1)c_{n+1}(x-1)^n - \sum_{n=0}^{\infty} c_n(x-1)^n = 0.$$

The common range of these six summations is from 2 to ∞.
We write out the individual terms in each that do *not*
belong to this range. Thus we have:

$$(-c_0 + 3c_1 + 2c_2) + (2c_1 + 10c_2 + 6c_3)(x-1)$$

$$+ \sum_{n=2}^{\infty} \{[n(n-1) + 3n - 1]c_n$$

$$+ [2(n+1)n + 3(n+1)]c_{n+1}$$

$$+ (n+2)(n+1)c_{n+2}\} (x-1)^n = 0.$$

Equating to zero the coefficients of each power of x, we obtain

$$-c_0 + 3c_1 + 2c_2 = 0, \quad 2c_1 + 10c_2 + 6c_3 = 0, \quad (1)$$

$$(n^2 + 2n - 1)c_n + (2n + 3)(n + 1)c_{n+1}$$
$$+ (n + 2)(n + 1)c_{n+2} = 0, \quad n \geq 2. \quad (2)$$

From (1), we find $c_2 = \dfrac{c_0 - 3c_1}{2}$, $c_3 = -\dfrac{c_1 + 5c_2}{3}$

$$= \dfrac{-5c_0 + 13c_1}{6}.$$ From (2), we find

$$c_{n+2} = -\dfrac{(n^2 + 2n - 1)c_n + (2n + 3)(n + 1)c_{n+1}}{(n + 1)(n + 2)}, \quad n \geq 2.$$

Using this, we find $c_4 = -\dfrac{7c_2 + 21c_3}{12} = \dfrac{7(2c_0 - 5c_1)}{12}.$

Substituting these values into the assumed solution, we have

$$y = c_0 + c_1(x - 1) + \dfrac{(c_0 - 3c_1)(x - 1)^2}{2}$$

$$+ \dfrac{(-5c_0 + 13c_1)(x - 1)^3}{6}$$

$$+ \dfrac{7(2c_0 - 5c_1)(x - 1)^4}{12} + \cdots$$

or

$$y = c_0\left[1 + \dfrac{(x - 1)^2}{2} - \dfrac{5(x - 1)^3}{6} + \dfrac{7(x - 1)^4}{6} + \cdots\right]$$

$$+ c_1\left[(x - 1) - \dfrac{3(x - 1)^2}{2} + \dfrac{13(x - 1)^3}{6}\right.$$

$$\left. - \dfrac{35(x - 1)^4}{12} + \cdots\right].$$

23. Since the initial value is 1, we assume that

$$y = \sum_{n=0}^{\infty} c_n(x - 1)^n. \quad \text{Then } y' = \sum_{n=1}^{\infty} nc_n(x - 1)^{n-1}, \text{ and}$$

$$y'' = \sum_{n=2}^{\infty} (n - 1)(n)c_n(x - 1)^{n-2}. \quad \text{In order to obtain}$$

multiples of $(x - 1)$ we rewrite the original D.E. as
follows:

$$(x - 1)y'' + 1y'' + y' + 2y = 0$$

Substituting into this D.E. we obtain:

$$(x - 1) \sum_{n=2}^{\infty} (n - 1)(n)c_n(x - 1)^{n-2}$$

$$+ \sum_{n=2}^{\infty} (n - 1)(n)c_n(x - 1)^{n-2}$$

$$+ \sum_{n=1}^{\infty} nc_n(x - 1)^{n-1} + 2\sum_{n=0}^{\infty} c_n(x - 1)^n = 0$$

or

$$\sum_{n=2}^{\infty} (n - 1)(n)c_n(x - 1)^{n-1}$$

$$+ \sum_{n=2}^{\infty} (n - 1)(n)c_n(x - 1)^{n-2}$$

$$+ \sum_{n=1}^{\infty} nc_n(x - 1)^{n-1} + \sum_{n=0}^{\infty} 2c_n(x - 1)^n = 0.$$

We rewrite the first three summations so that $(x - 1)$ has an exponent of n in each one.   Thus we have:

$$\sum_{n=1}^{\infty} (n)(n + 1)c_{n+1}(x - 1)^n$$

$$+ \sum_{n=0}^{\infty} (n + 1)(n + 2)c_{n+2}(x - 1)^n$$

$$+ \sum_{n=0}^{\infty} (n + 1)c_{n+1}(x - 1)^n + \sum_{n=0}^{\infty} 2c_n(x - 1)^n = 0.$$

The common range of these summations is from 1 to ∞. We write out the terms that are not in this range separately.   Thus we have:

$$(2c_2 + c_1 + 2c_0) + \sum_{n=1}^{\infty} [(n + 1)(n + 2)c_{n+2}$$

$$+ (n + 1)(n + 1)c_{n+1} + 2c_n](x - 1)^n = 0.$$

Equating the coefficients of each power of $(x - 1)$ to 0 we obtain

$$2c_2 + c_1 + 2c_0 = 0, \tag{1}$$

$$(n + 1)(n + 2)c_{n+2} + (n + 1)^2c_{n+1} + 2c_n = 0. \tag{2}$$

From (1), we find $c_2 = - c_0 - \dfrac{c_1}{2}$.   From (2), we find

$$c_{n+2} = - \frac{2c_n + (n + 1)^2 c_{n+1}}{(n + 1)(n + 2)}, \qquad n \geq 1.$$

Using this, $c_3 = -\dfrac{c_1 + 2c_2}{3} = \dfrac{2c_0}{3}$, $c_4 = -\dfrac{c_2}{6} - \dfrac{3c_3}{4}$

$= -\dfrac{c_0}{3} + \dfrac{c_1}{12}$, $c_5 = -\dfrac{c_3}{10} - \dfrac{4c_4}{5} = \dfrac{c_0}{5} - \dfrac{c_1}{15}$. Thus the G.S. of

the D.E. is

$$y = c_0\left[1 - (x - 1)^2 + \frac{2}{3}(x - 1)^3 - \frac{1}{3}(x - 1)^4\right.$$

$$\left. + \frac{1}{5}(x - 1)^5 + \cdots\right] + c_1\left[(x - 1) - \frac{1}{2}(x - 1)^2\right.$$

$$\left. + \frac{1}{12}(x - 1)^4 - \frac{1}{15}(x - 1)^5 + \cdots\right].$$

Applying the I.C. $y(1) = 2$ to this, we find $c_0 = 2$. Differentiating, we find

$$y' = [-2(x - 1) + 2(x - 1)^2 - \cdots]$$
$$+ c_1[1 - (x - 1) + \cdots].$$

Applying the I.C. $y'(1) = 4$ to this, we find $c_1 = 4$.   Thus we obtain the solution

$$y = 2\left[1 - (x - 1)^2 + \frac{2}{3}(x - 1)^3 - \frac{1}{3}(x - 1)^4\right.$$

$$\left. + \frac{1}{5}(x - 1)^5 + \cdots\right] + 4\left[(x - 1) - \frac{1}{2}(x - 1)^2\right.$$

$$\left. + \frac{1}{12}(x - 1)^4 - \frac{1}{15}(x - 1)^5 + \cdots\right]$$

or

$$y = 2 + 4(x - 1) - 4(x - 1)^2 + \frac{4}{3}(x - 1)^3$$

$$- \frac{1}{3}(x - 1)^4 + \frac{2}{15}(x - 1)^5 + \cdots.$$

Section 6.2, Page 269

1. We write the D.E. in the normalized form (6.3) of the
    text. This is:

$$y" + \frac{x + 2}{x^2 - 3x} y' + \frac{1}{x^2 - 3x} y = 0.$$

Here $P_1(x) = \frac{x + 2}{x(x - 3)}$ and $P_2(x) = \frac{1}{x(x - 3)}$. We see from
this that the singular points are x = 0 and x = 3.

We consider x = 0, and the functions formed by the
products

$$xP_1(x) = \frac{x + 2}{x - 3} \quad \text{and} \quad x^2 P_2(x) = \frac{x}{x - 3}.$$

Both of the product functions are defined at x = 0, so
x = 0 is a *regular singular point.*

We now consider x = 3, and form the product functions

$$(x - 3)P_1(x) = \frac{x + 2}{x} \quad \text{and} \quad (x - 3)^2 P_2(x) = \frac{x - 3}{x}.$$

Both of these product functions are defined at x = 3, so
x = 3 is a *regular singular point.*

2. We write the D.E. in the normalized form (6.3) of the
    text. This is:

$$y" + \frac{x - 2}{x(x + 1)} y' + \frac{4}{x^2(x + 1)} y = 0.$$

Here $P_1(x) = \frac{x - 2}{x(x + 1)}$ and $P_2(x) = \frac{4}{x^2(x + 1)}$. We see from
this that the singular points are x = 0 and x = -1.

We consider x = 0 and form the functions defined   by
the products

$$xP_1(x) = \frac{x - 2}{x + 1} \quad \text{and} \quad x^2 P_2(x) = \frac{4}{x + 1}.$$

Both of the product functions thus defined are analytic at
$x = 0$; so $x = 0$ is a *regular singular point*.

We now consider $x = -1$ and form the functions defined
by the products

$$(x + 1)P_1(x) = \frac{x - 2}{x} \quad \text{and} \quad (x + 1)^2 P_2(x) = \frac{4(x + 1)}{x^2}.$$

Both of these product functions thus defined are analytic
at $x = -1$; so $x = -1$ is a *regular singular point*.

4. We write the D.E. in the normalized form (6.3) of the
    text. This is:

$$y'' + \frac{1}{x(x + 3)(x - 2)} y' + \frac{1}{x^3(x + 3)} y = 0.$$

Here $P_1(x) = \frac{1}{x(x + 3)(x - 2)}$ and $P_2(x) = \frac{1}{x^3(x + 3)}$.   From

this, we see that the singular points are $x = 0$, $x = -3$
and $x = 2$.

We consider $x = 0$, and form the product functions

$$xP_1(x) = \frac{1}{(x + 3)(x - 2)} \quad \text{and} \quad x^2 P_2(x) = \frac{1}{x(x + 3)}.$$

Although $xP_1(x)$ is defined at $x = 0$, $x^2 P_2(x)$ is *not*

defined at $x = 0$. Therefore $x = 0$ is an *irregular*

singular point.

Consider $x = -3$, and form the product functions:

$$(x + 3)P_1(x) = \frac{1}{x(x - 2)} \quad \text{and} \quad (x + 3)^2 P_2(x) = \frac{x + 3}{x^3}.$$

Both are defined at $x = -3$, so $x = -3$ is a *regular singular point.*

Consider $x = 2$, and form the product functions:

$$(x - 2)P_1(x) = \frac{1}{x(x + 3)} \quad \text{and} \quad (x - 2)^2 P_2(x) = \frac{(x - 2)^2}{x^3(x + 3)}.$$

Both are defined at $x = 2$, so $x = 2$ is a *regular singular point.*

5.   We assume $y = \displaystyle\sum_{n=0}^{\infty} c_n x^{n+r}$ where $c_0 \neq 0$. Then

$$y' = \sum_{n=0}^{\infty} (n + r)c_n x^{n+r-1} \quad \text{and}$$

$$y'' = \sum_{n=0}^{\infty} (n + r)(n + r - 1)c_n x^{n+r-2}.$$ Substituting these

into the D.E., we obtain

$$2x^2 \sum_{n=0}^{\infty} (n + r)(n + r - 1)c_n x^{n+r-2}$$

$$+ x \sum_{n=0}^{\infty} (n + r)c_n x^{n+r-1}$$

$$+ (x^2 - 1) \sum_{n=0}^{\infty} c_n x^{n+r} = 0.$$

Simplifying, we obtain

$$\sum_{n=0}^{\infty} [(n + r)(n + r - 1)c_n + (n + r)c_n - c_n]x^{n+r}$$

$$+ \sum_{n=2}^{\infty} c_{n-2}x^{n+r} = 0$$

or

$$[2r(r - 1) + r - 1]c_0 x^r + [2(r + 1)(r)$$

$$+ (r + 1) - 1]c_1 x^{r+1}$$

$$+ \sum_{n=2}^{\infty} [(2n + 2r + 1)(n + r - 1)c_n$$

$$+ c_n - 2]x^{n+r} = 0. \tag{1}$$

Equating the coefficient of the lowest power of x to zero yields the *indicial equation* $2r(r - 1) + r - 1 = 0$ or $2r^2 - r - 1 = 0$ with roots $r_1 = 1$ and $r_2 = -1/2$. Since the difference between these roots is not zero or a positive integer, Conclusion 1 of Theorem 6.3 tells us that the D.E. has two linearly independent solutions of the assumed form, one corresponding to each of the roots r_1 and r_2. Equating the coefficients of the higher powers of x in (1) to zero we obtain

$$[2(r + 1)r + r]c_1 = 0 \tag{2}$$

and

$$(2n + 2r + 1)(n + r - 1)c_n + c_{n-2} = 0, \; n \geq 2. \tag{3}$$

Letting $r = r_1 = 1$ in (2) we obtain $5c_1 = 0$ so $c_1 = 0$.

Letting $r = r_1 = 1$ in (3) we obtain

$$(2n + 3)(n)c_n + c_{n-2} = 0, \quad n \geq 2$$

or

$$c_n = -\frac{c_{n-2}}{(2n + 3)n}, \quad n \geq 2.$$

From this $c_2 = \frac{-1}{14} c_0$, $c_3 = \frac{-1}{27} c_1 = 0$, $c_4 = \frac{-1}{44} c_2 = \frac{1}{616} c_0$

$c_5 = \frac{-1}{65} c_3 = 0$. Using these values, we obtain the

solution corresponding to the larger root $r_1 = 1$:

$$y_1(x) = c_0 x \left(1 - \frac{1}{14} x^2 + \frac{1}{616} x^4 + \cdots \right).$$

Letting $r = r_2 = -1/2$ in (2), we obtain $-c_1 = 0$ so $c_1 = 0$.

Letting $r = r_2 = -1/2$ in (3) we obtain

$$(n - 3/2)(2n)c_n + c_{n-2} = 0$$

or

$$c_n = -\frac{c_{n-2}}{(2n - 3)n}, \quad n \geq 2.$$

From this, $c_2 = -\frac{1}{2} c_0$, $c_3 = -\frac{1}{9} c_1 = 0$,

$c_4 = -\frac{1}{20} c_2 = \frac{1}{40} c_0$, $c_5 = 0$. Using these values, we

obtain the solution corresponding to the smaller root

$r_2 = -1/2$

$$y_2(x) = c_0 x^{-1/2} \left(1 - \frac{x^2}{2} + \frac{x^4}{40} + \cdots \right).$$

The G.S. of the D.E. is $y = C_1 y_1(x) + C_2 y_2(x)$.

7.   We assume $y = \displaystyle\sum_{n=0}^{\infty} c_n x^{n+r}$ where $c_0 \neq 0$.   Then

$y' = \displaystyle\sum_{n=0}^{\infty} (n + r)c_n x^{n+r-1}$ and

$y'' = \displaystyle\sum_{n=0}^{\infty} (n + r)(n + r - 1)c_n x^{n+r-2}$.   Substituting these

into the D.E., we obtain

$$\sum_{n=0}^{\infty} (n + r)(n + r - 1)c_n x^{n+r}$$

$$- x \sum_{n=0}^{\infty} (n + r)c_n x^{n+r}$$

$$+ \sum_{n=0}^{\infty} c_n x^{n+r+2} + \sum_{n=0}^{\infty} \frac{8}{9} c_n x^{n+r} = 0.$$

Simplifying as in the solutions of Section 6.1, we write
this as

$$\sum_{n=0}^{\infty} \left[(n + r)(n + r - 1) - (n + r) + \frac{8}{9} \right] c_n x^{n+r}$$

$$+ \sum_{n=2}^{\infty} c_{n-2} x^{n+r} = 0$$

or

$$\left[r(r-1) - r + \frac{8}{9}\right]c_0 x^r + \left[(r+1)r - (r+1)\right.$$

$$\left. + \frac{8}{9}\right]c_1 x^{r+1} + \sum_{n=2}^{\infty}\left\{\left[(n+r)(n+r-1)\right.\right.$$

$$\left.\left. - (n+r) + \frac{8}{9}\right]c_n + c_{n-2}\right\}x^{n+r} = 0. \qquad (1)$$

Equating to zero the coefficient of the lowest power of x we have the *indicial equation* $r(r-1) - r + \left(\frac{8}{9}\right) = 0$ or $r^2 - 2r + \left(\frac{8}{9}\right) = 0$ with roots $r_1 = \frac{4}{3}$ and $r_2 = \frac{2}{3}$. Since the difference between these roots is *not* zero or a positive integer, Conclusion 1 of Theorem 6.3 tells us that the D.E. has *two* linearly independent solutions of the assumed form, one corresponding to each of the roots r_1 and r_2.

Equating to zero the coefficients of the higher powers of x in (1), we have

$$\left[(r+1)r - (r+1) + \left(\frac{8}{9}\right)\right]c_1 = 0 \qquad (2)$$

and

$$\left[(n+r)(n+r-1) - (n+r) + \frac{8}{9}\right]c_n$$

$$+ c_{n-2} = 0, \; n \geq 2. \qquad (3)$$

Letting $r = r_1 = \frac{4}{3}$ in (2) we obtain $\left(\frac{5}{3}\right)c_1 = 0$ so $c_1 = 0$.

Letting $r = r_1 = \frac{4}{3}$ in (3) we obtain $n\left(n + \frac{2}{3}\right)c_n + c_{n-2} = 0,$

or

$$c_n = -\frac{3c_{n-2}}{n(3n+2)}, \quad n \geq 2.$$

From this $c_2 = -\dfrac{3c_0}{16}$, $c_3 = -\dfrac{c_1}{11} = 0$, $c_4 = -\dfrac{3c_2}{56} = \dfrac{9c_0}{896}$.

Using these values, we obtain the solution corresponding to the larger root $r_1 = \dfrac{4}{3}$:

$$y_1(x) = c_0 x^{4/3}\left(1 - \frac{3x^2}{16} + \frac{9x^4}{896} - \cdots\right).$$

Letting $r = r_2 = \dfrac{2}{3}$ in (2), we obtain $\dfrac{c_1}{3} = 0$; so $c_1 = 0$.

Letting $r = r_2 = \dfrac{2}{3}$ in (3) we obtain $n\left(n - \dfrac{2}{3}\right)c_n + c_{n-2} = 0$

or

$$c_n = -\frac{3c_{n-2}}{n(3n - 2)}, \qquad n \geq 2.$$

From this, $c_2 = -\dfrac{3c_0}{8}$, $c_3 = -\dfrac{c_1}{7} = 0$, $c_4 = -\dfrac{3c_2}{40} = \dfrac{9c_0}{320}$.

Using these values, we obtain the solution corresponding to the smaller root $r_2 = \dfrac{2}{3}$:

$$y_2(x) = c_0 x^{2/3}\left(1 - \frac{3x^2}{8} + \frac{9x^4}{320} - \cdots\right).$$

The G.S. of the D.E. is $y = C_1 y_1(x) + C_2 y_2(x)$.

9. We assume $y = \displaystyle\sum_{n=0}^{\infty} c_n x^{n+r}$ where $c_0 \neq 0$. Then

$$y' = \sum_{n=0}^{\infty} (n + r)c_n x^{n+r-1} \text{ and}$$

$$y'' = \sum_{n=0}^{\infty} (n + r)(n + r - 1)c_n x^{n+r-2}.$$ Substituting these

into the D.E., we obtain

$$\sum_{n=0}^{\infty} (n + r)(n + r - 1)c_n x^{n+r}$$

$$+ \sum_{n=0}^{\infty} (n + r)c_n x^{n+r} + \sum_{n=0}^{\infty} c_n x^{n+r+2} - \frac{1}{9} \sum_{n=0}^{\infty} c_n x^{n+r} = 0.$$

Simplifying, as in the solutions of Section 6.1, we write
this as

$$\sum_{n=0}^{\infty} \left[(n + r)(n + r - 1) + (n + r) - \frac{1}{9} \right] c_n x^{n+r}$$

$$+ \sum_{n=2}^{\infty} c_{n-2} x^{n+r} = 0$$

or

$$\left[r(r - 1) + r - \frac{1}{9} \right] c_0 x^r + \left[(r + 1)r + (r + 1) - \frac{1}{9} \right] c_1 x^{r+1}$$

$$+ \sum_{n=2}^{\infty} \left\{ \left[(n + r)(n + r - 1) + (n + r) - \frac{1}{9} \right] c_n \right.$$

$$\left. + c_{n-2} \right\} x^{n+r} = 0. \tag{1}$$

Equating to zero the coefficient of the lowest power of x
we have the *indicial equation* $r^2 - \frac{1}{9} = 0$ with roots

$r_1 = \frac{1}{3}$, $r_2 = -\frac{1}{3}$. Since the difference between these

roots is *not* zero or a positive integer, Conclusion 1 of
Theorem 6.3 tells us that the D.E. has *two* linearly

independent solutions of the assumed form, one corresponding to each of the roots r_1 and r_2. Equating to zero the coefficients of the higher powers of x in (1) we have

$$\left[(r + 1)r + (r + 1) - \frac{1}{9}\right]c_1 = 0 \tag{2}$$

and

$$\left[(n + r)(n + r - 1) + (n + r) - \frac{1}{9}\right]c_n$$
$$+ c_{n-2} = 0, \quad n \geq 2. \tag{3}$$

Letting $r = r_1 = \frac{1}{3}$ in (2) we obtain $\frac{5}{3}c_1 = 0$; so $c_1 = 0$. Letting $r = r_1 = \frac{1}{3}$ in (3) we obtain

$$n\left(n + \frac{2}{3}\right)c_n + c_{n-2} = 0$$

or

$$c_n = -\frac{3c_{n-2}}{n(3n + 2)}, \quad n \geq 2.$$

From this, $c_2 = -\frac{3c_0}{16}$, $c_3 = -\frac{c_1}{11} = 0$, $c_4 = -\frac{3c_2}{56} = \frac{9c_0}{896}$, $c_{ODD} = 0$ for $n \geq 1$. Using these values, we obtain the solution corresponding to the larger root $r_1 = \frac{1}{3}$:

$$y_1(x) = c_0 x^{1/3}\left[1 - \frac{3x^2}{16} + \frac{9x^4}{896} - \cdots\right].$$

Letting $r = r_2 = -\frac{1}{3}$ in (2), we obtain $\frac{1}{3}c_1 = 0$; so $c_1 = 0$. Letting $r = r_2 = -\frac{1}{3}$ in (3) we obtain

$$n\left(n - \frac{2}{3}\right)c_n + c_{n-2} = 0$$

or

$$c_n = - \frac{3c_{n-2}}{n(3n - 2)}, \quad n \geq 2.$$

From this, $c_2 = - \frac{3c_0}{8}$, $c_3 = - \frac{c_1}{7}$, $c_4 = - \frac{3c_2}{40} = \frac{9c_0}{320}$,

$c_{ODD} = 0$ for $n \geq 1$. Using these values, we obtain the

solution corresponding to the smaller root $r_2 = - \frac{1}{3}$:

$$y_2(x) = c_0 x^{-1/3} \left[1 - \frac{3x^2}{8} + \frac{9x^4}{320} - \cdots \right].$$

The G.S. of the D.E. is

$$y = C_1 y_1(x) + C_2 y_2(x).$$

12. We assume $y = \sum_{n=0}^{\infty} c_n x^{n+r}$ where $c_0 \neq 0$. Then

$$y' = \sum_{n=0}^{\infty} c_n (n + r) x^{n+r-1} \text{ and}$$

$$y'' = \sum_{n=0}^{\infty} c_n (n + r)(n + r - 1) x^{n+r-2}. \text{ Substituting these}$$

into the D.E., we obtain

$$2 \sum_{n=0}^{\infty} c_n (n + r)(n + r - 1)x^{n+r}$$

$$+ 5 \sum_{n=0}^{\infty} c_n (n + r)x^{n+r}$$

$$+ 2 \sum_{n=0}^{\infty} c_n x^{n+r+1} - 2 \sum_{n=0}^{\infty} c_n x^{n+r} = 0.$$

Simplifying, as in the solutions of Section 6.1, we write this as

$$\sum_{n=0}^{\infty} [2(n + r)(n + r - 1) + 5(n + r) - 2]c_n x^{n+r}$$

$$+ 2 \sum_{n=1}^{\infty} c_{n-1} x^{n+r} = 0$$

or

$$[2r(r - 1) + 5r - 2]c_0 x^r$$

$$+ \sum_{n=1}^{\infty} \{[2(n + r)(n + r - 1) + 5(n + r) - 2]c_n$$

$$+ 2c_{n-1}\}x^{n+r} = 0. \tag{1}$$

Equating to zero the coefficient of the lowest power of x, we have the *indicial equation* $2r^2 + 3r - 2 = 0$, that is $(2r - 1)(r + 2) = 0$, with roots $r_1 = \frac{1}{2}$ and $r_2 = -2$. Since the difference between these roots is *not* zero or a positive integer, Conclusion 1 of Theorem 6.3 tells us that the D.E. has *two* linearly independent solutions of the assumed form, one corresponding to each of the roots r_1 and r_2. Equating to zero the coefficients of the higher powers of x in (1), we have

$$[2(n + r)(n + r - 1) + 5(n + r) - 2]c_n$$
$$+ 2c_{n-1} = 0, \ n \geq 1. \tag{2}$$

Letting $r = r_1 = \frac{1}{2}$ in (2) we obtain $(2n^2 + 5n)c_n$ $+ 2c_{n-1} = 0$ or

$$c_n = -\frac{2c_{n-1}}{n(2n + 5)}, \quad n \geq 1.$$

From this $c_1 = -\frac{2c_0}{7}$, $c_2 = -\frac{2c_1}{2(9)} = -\frac{c_1}{9} = \frac{2c_0}{63}$,

$c_3 = -\frac{2c_2}{33} = -\frac{4c_0}{2079}$. Using these values, we obtain the solution corresponding to the larger root $r_1 = \frac{1}{2}$:

$$y_1(x) = c_0 x^{1/2}\left[1 - \frac{2x}{7} + \frac{2x^2}{63} - \frac{4x^3}{2079} + \cdots\right].$$

Letting $r = r_2 = -2$ in (2), we obtain $(2n^2 - 5n)c_n$ $+ 2c_{n-1} = 0$ so $c_1 = 0$ or

$$c_n = -\frac{2c_{n-1}}{n(2n - 5)}, \quad n \geq 1.$$

From this, $c_1 = -\frac{2c_0}{(-3)} = \frac{2c_0}{3}$, $c_2 = -\frac{2c_1}{2(-1)} = c_1 = \frac{2c_0}{3}$,

$c_3 = -\frac{2c_2}{3} = -\frac{4c_0}{9}$. Using these values, we obtain the solution corresponding to the smaller root $r_2 = -2$:

$$y_2(x) = c_0 x^{-2}\left[1 + \frac{2}{3}x + \frac{2}{3}x^2 - \frac{4}{9}x^3 + \cdots\right].$$

The G.S. of the D.E. is

$$y = C_1 y_1(x) + C_2 y_2(x).$$

14. We assume $y = \displaystyle\sum_{n=0}^{\infty} c_n x^{n+r}$ where $c_0 \neq 0$. Then

$$y' = \sum_{n=0}^{\infty}(n + r)c_n x^{n+r-1} \text{ and}$$

$$y'' = \sum_{n=0}^{\infty}(n + r)(n + r - 1)c_n x^{n+r-2}.$$ Substituting these

into the D.E., we obtain

$$\sum_{n=0}^{\infty}(n + r)(n + r - 1)c_n x^{n+r+1}$$

$$+ 2\sum_{n=0}^{\infty}(n + r)(n + r - 1)c_n x^{n+r}$$

$$+ \sum_{n=0}^{\infty}(n + r)c_n x^{n+r+1} + \sum_{n=0}^{\infty}(n + r)c_n x^{n+r}$$

$$- 10\sum_{n=0}^{\infty} c_n x^{n+r} = 0.$$

Simplifying, as in the solutions of Section 6.1, we write this as

$$\sum_{n=0}^{\infty} [2(n + r)(n + r - 1) + (n + r) - 10]c_n x^{n+r}$$

$$+ \sum_{n=1}^{\infty} [(n + r - 1)(n + r - 2) +$$

$$(n + r - 1)]c_{n-1} x^{n+r} = 0$$

or

$$[2r(r - 1) + r - 10]c_0 x^r$$

$$+ \sum_{n=1}^{\infty} \{[2(n + r)(n + r - 1) + (n + r) - 10]c_n$$

$$+ (n + r - 1)^2 c_{n-1}\}x^{n+r} = 0. \tag{1}$$

Equating to zero the coefficient of the lowest power of x, we have the *indicial equation* $2r^2 - r - 10 = 0$ or $(2r - 5)(r + 2) = 0$, with roots $r_1 = 5/2$ and $r_2 = -2$.

Since the difference between these roots is *not* zero or a positive integer, Conclusion 1 of Theorem 6.3 tells us that the D.E. has *two* linearly independent solutions of the assumed form, one corresponding to each of the roots r_1 and r_2. Equating to zero the coefficients of the higher powers of x in (1), we have

$$[2(n + r)(n + r - 1) + (n + r) - 10]c_n$$

$$+ (n + r - 1)^2 c_{n-1} = 0, \quad n > 1. \tag{2}$$

Letting $r = r_1 = 5/2$ in (2) we obtain $n(2n + 9)c_n$

$$+ \left(n + \frac{3}{2}\right)^2 c_{n-1} = 0 \text{ or}$$

$$c_n = - \frac{(2n + 3)^2 c_{n-1}}{4n(2n + 9)}, \quad n \geq 1.$$

From this $c_1 = - \frac{25c_0}{44}$, $c_2 = - \frac{49c_1}{104} = \frac{1225c_0}{4576}$.   Using these

values, we obtain the solution corresponding to the larger

root $r_1 = 5/2$:

$$y = c_0 x^{5/2}\left[1 - \frac{25x}{44} + \frac{1225x^2}{4576} - \cdots\right].$$

Letting $r = r_2 = -2$ in (2), we obtain $n(2n - 9)c_n$

$+ (n - 3)^2 c_{n-1} = 0$ or

$$c_n = - \frac{(n - 3)^2 c_{n-1}}{n(2n - 9)}, \quad n \geq 1.$$

From this, $c_1 = \frac{4c_0}{7}$, $c_2 = \frac{c_1}{10} = \frac{2c_0}{35}$, $c_3 = \frac{0c_2}{9} = 0$; and

hence $c_n = 0$ for all $n \geq 3$.   The solution corresponding to

the smaller root $r_2 = -2$ is the finite sum

$$y(x) = c_0 x^{-2}\left(1 + \frac{4x}{7} + \frac{2x^2}{35}\right).$$

The G.S. of the D.E. is

$$y = C_1 y_1(x) + C_2 y_2(x).$$

15.   We assume $y = \sum_{n=0}^{\infty} c_n x^{n+r}$, where $c_0 \neq 0$.   Then

$$y' = \sum_{n=0}^{\infty} (n + r)c_n x^{n+r-1},$$

$y'' = \sum_{n=0}^{\infty} (n + r)(n + r - 1)c_n x^{n+r-2}$.   Substituting these

into the D.E., we obtain

$$\sum_{n=0}^{\infty} (n + r)(n + r - 1)c_n x^{n+r-1} + \sum_{n=0}^{\infty} 2(n + r)c_n x^{n+r-1}$$

$$+ \sum_{n=0}^{\infty} c_n x^{n+r+1} = 0,$$

Simplifying, as in the solutions of Section 6.1, we write this as

$$\sum_{n=0}^{\infty} [(n + r)(n + r - 1) + 2(n + r)]c_n x^{n+r-1}$$

$$+ \sum_{n=2}^{\infty} c_{n-2} x^{n+r-1} = 0$$

or

$$[r(r - 1) + 2r]c_0 x^{r-1} + [(r + 1)r$$

$$+ 2(r + 1)]c_1 x^r + \sum_{n=2}^{\infty} \{[(n + r)(n + r - 1)$$

$$+ 2(n + r)]c_n + c_{n-2}\}x^{n+r-1} = 0. \tag{1}$$

Equating to zero the coefficient of the lowest power of x, we have the indicial equation $r^2 - r + 2r = 0$ or $r^2 + r = 0$, with roots $r_1 = 0$, $r_2 = -1$. Note that the difference between these roots is a positive integer. By Conclusion 2 of Theorem 6.3, the D.E. has a solution of

the assumed form corresponding to the larger root $r_1 = 0$.
Equating to zero the coefficients of the higher powers of
x in (1), we have

$$[(r + 1)r + 2(r + 1)]c_1 = 0, \qquad\qquad (2)$$

$$[(n + r)(n + r - 1) + 2(n + r)]c_n + c_{n-2} = 0,$$
$$n \geq 2. \qquad\qquad (3)$$

Letting $r = r_1 = 0$ in (2), we obtain $2c_1 = 0$; so
$c_1 = 0$. Letting $r = r_1 = 0$ in (3), we obtain

$$n(n + 1)c_n + c_{n-2} = 0 \qquad \text{or}$$

$$c_n = - \frac{c_{n-2}}{n(n + 1)}, \qquad n \geq 2.$$

From this, $c_2 = - \dfrac{c_0}{3!}, \; c_3 = - \dfrac{c_1}{4!} = 0, \; c_4 = - \dfrac{c_2}{(4)(5)}$
$= \dfrac{c_0}{5!}, \; \cdots$. Using these values, we obtain the solution
corresponding to the larger root $r_1 = 0$.

$$y_1(x) = c_0\left(1 - \frac{x^2}{3!} + \frac{x^4}{5!} - \cdots\right)$$

$$= c_0 x^{-1}\left(x - \frac{x^3}{3!} + \frac{x^5}{5!} - \cdots\right)$$

or

$$y_1(x) = c_0 x^{-1}\sin x. \qquad\qquad (4)$$

Letting $r = r_2 = -1$ in (2), we obtain $0c_1 = 0$; so c_1
is arbitrary. Letting $r = r_2 = -1$ in (3), we obtain

$$n(n - 1)c_n + c_{n-2} = 0 \quad \text{or}$$

$$c_n = -\frac{c_{n-2}}{n(n-1)}, \quad n \geq 2.$$

From this, $c_2 = -\frac{c_0}{2!}$, $c_3 = -\frac{c_1}{3!}$, $c_4 = -\frac{c_2}{(4)(3)} = \frac{c_0}{4!}$,

$c_5 = -\frac{c_3}{(5)(4)} = \frac{c_1}{5!}$, ... Using these values, we obtain

the solution corresponding to the smaller root $r_2 = -1$:

$$y_2(x) = c_0 x^{-1}\left(1 - \frac{x^2}{2!} + \frac{x^4}{4!} - \cdots\right)$$

$$+ c_1 x^{-1}\left(x - \frac{x^3}{3!} + \frac{x^5}{5!} - \cdots\right)$$

or

$$y_2(x) = c_0 x^{-1}\cos x + c_1 x^{-1}\sin x. \tag{5}$$

The situation here is analogous to that of Example 6.13 of the text (in particular, see text, page 261). In like manner to the case of that example, we see from (4) and (5) that the G.S. of the given D.E. is of the form

$$y = C_1 x^{-1}\left(1 - \frac{x^2}{2!} + \frac{x^4}{4!} - \cdots\right)$$

$$+ C_2 x^{-1}\left(x - \frac{x^3}{3!} + \frac{x^5}{5!} - \cdots\right)$$

or

$$y = x^{-1}(C_1 \cos x + C_2 \sin x).$$

16. We assume $y = \sum\limits_{n=0}^{\infty} c_n x^{n+r}$, where $c_0 \neq 0$. Then

$$y' = \sum_{n=0}^{\infty} (n + r) c_n x^{n+r-1},$$

$$y'' = \sum_{n=0}^{\infty} (n + r)(n + r - 1) c_n x^{n+r-2}.$$ Substituting these

into the D.E., we obtain

$$x^2 \sum_{n=0}^{\infty} (n + r)(n + r - 1) c_n x^{n+r-2}$$

$$+ x \sum_{n=0}^{\infty} (n + r) c_n x^{n+r-1}$$

$$+ (x^2 - 1/4) \sum_{n=0}^{\infty} c_n x^{n+r} = 0.$$

Simplifying, we obtain

$$\sum_{n=0}^{\infty} [(n + r)(n + r - 1) + (n + r) - 1/4] c_n x^{n+r}$$

$$+ \sum_{n=2}^{\infty} c_{n-2} x^{n+r} = 0$$

or

$$[r(r - 1) + r - 1/4]c_0 x^r + [(1 + r)(r)$$

$$+ (1 + r) - 1/4]c_1 x^{1+r} + \sum_{n=2}^{\infty} \{[(n + r)(n + r - 1)$$

$$+ (n + r) - 1/4]c_n + c_{n-2}\}x^{n+r} = 0. \qquad (1)$$

Equating the coefficient of the lowest power of x to zero yields the indicial equation $r(r - 1) + r - 1/4 = 0$ or $r^2 - 1/4 = 0$ with roots $r_1 = 1/2$ and $r_2 = -1/2$. Note that the difference between these roots is a positive integer. By Conclusion 2 of Theorem 6.3, the D.E. has a solution of the assumed form corresponding to the larger root $r_1 = 1/2$. However, as in Example 6.13 of the text, if the smaller root $r_2 = -1/2$ has a solution, it will include that of the larger root. Therefore let us try the smaller root first.

Letting $r = r_2 = -1/2$ in (1), we obtain $[-1/4 + 1/2 - 1/4]c_1 = 0$ so c_1 is arbitrary. Next we obtain $[(n - 1/2)(n - 3/2) + (n - 3/4)]c_n + c_{n-2} = 0$ or $c_n = -\dfrac{1}{n^2 - n} c_{n-2}$; $n \geq 2$. Since $n \geq 2$, we have no undefined terms and we will obtain two linearly independent solutions from this one root.

Using successive values of n we obtain $c_2 = -\dfrac{1}{2} c_0$, $c_3 = -\dfrac{1}{6} c_1$, $c_4 = -\dfrac{1}{12} c_2 = \dfrac{1}{24} c_0$, $c_5 = -\dfrac{1}{20} c_3 = \dfrac{1}{120} c_1$,

$$c_6 = -\frac{1}{30} c_4 = -\frac{1}{720} c_0, \; c_7 = -\frac{1}{42} c_5 = -\frac{1}{5040} c_1. \quad \text{Thus}$$

$$y = c_0 x^{-1/2} \left(1 - \frac{x^2}{2} + \frac{x^4}{24} - \frac{x^6}{720} + \cdots \right)$$

$$+ c_1 x^{-1/2} \left(x - \frac{x^3}{6} + \frac{x^5}{120} - \cdots \right)$$

or

$$y = c_0 x^{-1/2} \left(1 - \frac{x^2}{2} + \frac{x^4}{24} - \cdots \right)$$

$$+ c_1 x^{1/2} \left(1 - \frac{x^2}{6} + \frac{x^5}{120} - \cdots \right).$$

The second part of this solution is the same expression that $r = r_1 = 1/2$ will yield.

17. We assume $y = \displaystyle\sum_{n=0}^{\infty} c_n x^{n+r}$, where $c_0 \neq 0$. Then

$$y' = \sum_{n=0}^{\infty} (n + r) c_n x^{n+r-1},$$

$$y'' = \sum_{n=0}^{\infty} (n + r)(n + r - 1) c_n x^{n+r-2}. \quad \text{Substituting these}$$

into the D.E., we obtain

$$\sum_{n=0}^{\infty} (n + r)(n + r - 1) c_n x^{n+r} + \sum_{n=0}^{\infty} (n + r) c_n x^{n+r+3}$$

$$+ \sum_{n=0}^{\infty} (n + r) c_n x^{n+r} - \sum_{n=0}^{\infty} c_n x^{n+r} = 0.$$

Simplifying, as in the solutions of Section 6.1, we write this as

$$\sum_{n=0}^{\infty} [(n + r)(n + r - 1) + (n + r) - 1]c_n x^{n+r}$$

$$+ \sum_{n=3}^{\infty} (n + r - 3)c_{n-3} x^{n+r} = 0$$

or

$$[r(r - 1) + r - 1]c_0 x^r + [(r + 1)r + r]c_1 x^{r+1}$$

$$+ [(r + 2)(r + 1) + r + 1]c_2 x^{r+2}$$

$$+ \sum_{n=3}^{\infty} [(n + r + 1)(n + r - 1)c_n$$

$$+ (n + r - 3)c_{n-3}]x^{n+r} = 0. \tag{1}$$

Equating to zero the coefficient of the lowest power of x, we have the indicial equation $r(r - 1) + r - 1 = 0$ or $r^2 - 1 = 0$, with roots $r_1 = 1$, $r_2 = -1$. Equating to zero the coefficients of the higher powers of x in (1), we have

$$(r^2 + 2r)c_1 = 0, \tag{2}$$

$$(r^2 + 4r + 3)c_2 = 0, \tag{3}$$

$$(n + r + 1)(n + r - 1)c_n$$
$$+ (n + r - 3)c_{n-3} = 0, \ n \geq 3. \tag{4}$$

Letting $r = r_1 = 1$ in (2), we obtain $3c_1 = 0$, so $c_1 = 0$. Letting $r = r_1 = 1$ in (3), we obtain $8c_2 = 0$, so $c_2 = 0$. Letting $r = r_1 = 1$ in (4), we obtain

$$(n + 2)nc_n + (n - 2)c_{n-3} = 0 \qquad \text{or}$$

$$c_n = -\frac{(n - 2)c_{n-3}}{n(n + 2)}, \qquad n \geq 3.$$

From this, $c_3 = -\frac{c_0}{15}$, $c_4 = -\frac{c_1}{12} = 0$, $c_5 = -\frac{3c_2}{35} = 0$,

$c_6 = -\frac{c_3}{12} = \frac{c_0}{180}$, ... Using these values we obtain the

solution corresponding to the larger root $r_1 = 1$:

$$y_1(x) = c_0 x \left(1 - \frac{x^3}{15} + \frac{x^6}{180} - \cdots \right). \tag{5}$$

Letting $r = r_2 = -1$ in (2), we obtain $-c_1 = 0$, so $c_1 = 0$.
Letting $r = r_2 = -1$ in (3), we obtain $0c_2 = 0$, so c_2 is
arbitrary. Letting $r = r_2 = 1$ in (4), we obtain

$$n(n - 2)c_n + (n - 4)c_{n-3} = 0 \qquad \text{or}$$

$$c_n = -\frac{(n - 4)c_{n-3}}{n(n - 2)}, \qquad n \geq 3.$$

From this, $c_3 = \frac{c_0}{3}$, $c_4 = 0$, $c_5 = \frac{c_2}{15}$, $c_6 = -\frac{c_3}{12} = -\frac{c_0}{36}$,

$c_7 = -\frac{3c_4}{35} = 0$, $c_8 = -\frac{c_5}{12} = \frac{c_2}{180}$, ... Using these values

we obtain the solution corresponding to the smaller root
$r_2 = -1$:

$$y_2(x) = c_0 x^{-1} \left(1 + \frac{x^3}{3} - \frac{x^6}{36} + \cdots \right)$$

$$+ c_2 x^{-1} \left(x^2 - \frac{x^5}{15} + \frac{x^8}{180} - \cdots \right). \tag{6}$$

The situation here is analogous to that of Example 6.13 of the text. In like manner to the case of that example, we see from (5) and (6) that the G.S. of the D.E. is of the form

$$y = C_1 x \left(1 - \frac{x^3}{15} + \frac{x^6}{180} - \cdots \right)$$

$$+ C_2 x^{-1} \left(1 + \frac{x^3}{3} - \frac{x^6}{36} + \cdots \right).$$

22.   We assume $y = \sum_{n=0}^{\infty} c_n x^{n+r}$, where $c_0 \neq 0$. Then

$$y' = \sum_{n=0}^{\infty} (n + r) c_n x^{n+r-1},$$

$$y'' = \sum_{n=0}^{\infty} (n + r)(n + r - 1) c_n x^{n+r-2}. \text{ Substituting into the}$$

D.E., we obtain

$$\sum_{n=0}^{\infty} (n + r)(n + r - 1) c_n x^{n+r} + \sum_{n=0}^{\infty} (n + r) c_n x^{n+r+1}$$

$$+ 5 \sum_{n=0}^{\infty} (n + r) c_n x^{n+r} + 2 \sum_{n=0}^{\infty} c_n x^{n+r+1}$$

$$+ 3 \sum_{n=0}^{\infty} c_n x^{n+r} = 0.$$

Simplifying, as in the solutions of Section 6.1, we obtain

$$\sum_{n=0}^{\infty} (n + r)(n + r - 1)c_n x^{n+r} + \sum_{n=1}^{\infty} (n + r - 1)c_{n-1} x^{n+r}$$

$$+ 5\sum_{n=0}^{\infty} (n + r)c_n x^{n+r} + 2\sum_{n=1}^{\infty} c_{n-1} x^{n+r} + 3\sum_{n=0}^{\infty} c_n x^{n+r} = 0$$

or

$$[r(r - 1) + 5r + 3]c_0 x^r + \sum_{n=1}^{\infty} \{[(n + r)(n + r - 1)$$

$$+ 5(n + r) + 3]c_n + [(n + r - 1)$$

$$+ 2]c_{n-1}\}x^{n+r} = 0. \tag{1}$$

Equating to zero the coefficient of the lowest power of x in (1), we have the indicial equation $r^2 + 4r + 3 = 0$, with roots $r_1 = -1$, $r_2 = -3$. Equating to zero the coefficients of the higher powers of x in (1), we obtain

$$(n + r + 1)(n + r + 3)c_n + (n + r + 1)c_{n-1} = 0$$

or

$$(n + r + 1)[(n + r + 3)c_n + c_{n-1}] = 0, \ n \geq 1. \tag{2}$$

Letting $r - r_1 - -1$ in (2), we obtain $n[(n + 2)c_n$

$+ c_{n-1}] = 0$ and so $c_n = -\dfrac{c_{n-1}}{(n + 2)}$, $n \geq 1$. From this,

$c_1 = -\dfrac{c_0}{3} = -\dfrac{2c_0}{3!}$, $c_2 = -\dfrac{c_1}{4} = \dfrac{2c_0}{4!}$, and in general,

$c_n = \dfrac{(-1)^n 2c_0}{(n + 2)!}$. From this we obtain the solution corresponding to the larger root $r_1 = -1$:

$$y_1(x) = 2c_0 x^{-1} \sum_{n=0}^{\infty} \frac{(-1)^n x^n}{(n+2)!}.$$

Letting $r = r_2 = -3$ in (2), we obtain

$(n - 2)(nc_n + c_{n-1}) = 0$, $n \geq 1$, and so

$$c_n = -\frac{c_{n-1}}{n}, \quad n \geq 1, n \neq 2. \tag{3}$$

From (3) with $n = 1$, we find $c_1 = -c_0$. For $n = 2$, we go
back to $(n - 2)(nc_n + c_{n-1}) = 0$. When $n = 2$, this becomes
$0(2c_2 + c_1) = 0$. Thus c_2 is independent of c_1 and is a
second arbitrary constant. Then from (3) with $n = 3$,

$c_3 = -c_2/3 = -\frac{2c_2}{3!}$; with $n = 4$, $c_4 = -c_3/4 = \frac{2c_2}{4!}$; and in

general $c_n = -c_{n-1}/n = \frac{(-1)^n 2c_2}{n!}$. From this we obtain the

solution corresponding to the smaller root $r_2 = -3$:

$$y_2(x) = c_0 x^{-3}(1 - x) + x^{-3} \sum_{n=2}^{\infty} \frac{(-1)^n 2c_2 x^n}{n!}. \tag{4}$$

The situation here is analogous to that of Example
6.13 of the text. In like manner to the case of that
example, we see from (3) and (4) that the G.S. of the D.E.
is of the form

$$y = C_1 x^{-3}(1 - x) + C_2 x^{-1} \sum_{n=0}^{\infty} \frac{(-1)^n x^n}{(n+2)!}.$$

24. We assume $y = \displaystyle\sum_{n=0}^{\infty} c_n x^{n+r}$, where $c_0 \neq 0$.   Then

$$y' = \sum_{n=0}^{\infty} (n + r) c_n x^{n+r-1},$$

$$y'' = \sum_{n=0}^{\infty} (n + r)(n + r - 1) c_n x^{n+r-2}.$$ Substituting into the

D.E., we obtain

$$\sum_{n=0}^{\infty} (n + r)(n + r - 1) c_n x^{n+r} + 2 \sum_{n=0}^{\infty} (n + r) c_n x^{n+r+2}$$

$$- \sum_{n=0}^{\infty} c_n x^{n+r+2} - \frac{15}{4} \sum_{n=0}^{\infty} c_n x^{n+r} = 0.$$

Simplifying, as in the solutions of Section 6.1, we obtain

$$\sum_{n=0}^{\infty} (n + r)(n + r - 1) c_n x^{n+r} + 2 \sum_{n=2}^{\infty} (n + r - 2) c_{n-2} x^{n+r}$$

$$- \sum_{n=2}^{\infty} c_{n-2} x^{n+r} - \frac{15}{4} \sum_{n=0}^{\infty} c_n x^{n+r} = 0$$

or

$$\left[r(r - 1) - \frac{15}{4} \right] c_0 x^r + \left[(r + 1)r - \frac{15}{4} \right] c_1 x^{r+1}$$

$$+ \sum_{n=2}^{\infty} \left\{ \left[(n + r)(n + r - 1) - \frac{15}{4} \right] c_n \right.$$

$$+ \left. [2(n + r - 2) - 1] c_{n-2} \right\} x^{n+r} = 0. \qquad (1)$$

Equating to zero the coefficient of the lowest power of x in (1), we have the indicial equation $r^2 - r - \frac{15}{4} = 0$, with roots $r_1 = 5/2$, $r_2 = -3/2$. Equating to zero the coefficients of the higher powers of x in (1), we obtain

$$\left(r^2 + r - \frac{15}{4}\right)c_1 = 0, \tag{2}$$

$$\left[(n + r)(n + r - 1) - \frac{15}{4}\right]c_n$$
$$+ [2(n + r - 2) - 1]c_{n-2} = 0, \; n \geq 2. \tag{3}$$

Letting $r = r_1 = \frac{5}{2}$ in (2), we obtain $5c_1 = 0$, so $c_1 = 0$. Letting $r = r_1 = \frac{5}{2}$ in (3), we obtain $n(n + 4)c_n + 2nc_{n-2} = 0$ and hence $c_n = -\frac{2c_{n-2}}{n + 4}$, $n \geq 2$. From this, $c_2 = -\frac{c_0}{3}$, $c_4 = -\frac{c_2}{4} = \frac{c_0}{12}$, and $c_{2n+1} = 0$ for $n \geq 0$. Using these values we obtain the solution corresponding to the larger root $r_1 = \frac{5}{2}$:

$$y_1(x) = c_0 x^{5/2}\left(1 - \frac{x^2}{3} + \frac{x^4}{12} - \cdots\right). \tag{4}$$

Letting $r = r_2 = -\frac{3}{2}$ in (2), we obtain $-3c_1 = 0$, since $c_1 = 0$. Letting $r = r_2 = -\frac{3}{2}$ in (3), we obtain $n(n - 4)c_n + 2(n - 4)c_{n-2} = 0$, $n \geq 2$, and hence

$$c_n = -\frac{2c_{n-2}}{n}, \; n \geq 2, \; n \neq 4. \tag{5}$$

From (5) with $n = 2$, we find $c_2 = -c_0$; and with $n = 3$, $c_3 = -2c_1/3$. For $n = 4$, we return to $n(n - 4)c_n$

+ $2(n - 4)c_{n-2} = 0$. With $n = 4$, this becomes $0c_4 + 0c_2$
= 0. Thus c_4 is independent of c_2 and so is a second
arbitrary constant. Then from (5) with $n = 6$, $c_6 = -c_4/3$;
and with $n = 8$, $c_8 = -c_6/4 = c_4/12$. Also note that
$c_{2n+1} = 0$ for $n \geq 0$. Using these values we obtain the
solution corresponding to the smaller root $r_2 = -\dfrac{3}{2}$:

$$y_2(x) = c_0 x^{-3/2}(1 - x^2)$$

$$+ c_4 x^{-3/2}\left(x^4 - \frac{x^6}{3} + \frac{x^8}{12} - \cdots\right). \tag{6}$$

The situation here is analogous to that of Example
6.13 of the text. In like manner to the case in that
example, we see from (4) and (6) that the G.S. of the D.E.
is of the form

$$y = C_1 x^{-3/2}(1 - x^2) + C_2 x^{5/2}\left(1 - \frac{x^2}{3} + \frac{x^4}{12} - \cdots\right).$$

25. We assume $y = \displaystyle\sum_{n=0}^{\infty} c_n x^{n+r}$, where $c_0 \neq 0$. Then

$$y' = \sum_{n=0}^{\infty} (n + r)c_n x^{n+r-1},$$

$$y'' = \sum_{n=0}^{\infty} (n + r)(n + r - 1)c_n x^{n+r-2}. \quad \text{Substituting these}$$

into the D.E., we obtain

$$\sum_{n=0}^{\infty} (n + r)(n + r - 1)c_n x^{n+r} + \sum_{n=0}^{\infty} (n + r)c_n x^{n+r}$$

$$+ \sum_{n=0}^{\infty} c_n x^{n+r+1} - \sum_{n=0}^{\infty} c_n x^{n+r} = 0.$$

Simplifying, as in the solutions of Section 6.1, we write this as

$$\sum_{n=0}^{\infty} [(n + r)(n + r - 1) + (n + r) - 1]c_n x^{n+r}$$

$$+ \sum_{n=1}^{\infty} c_{n-1} x^{n+r} = 0$$

and hence as

$$[r(r - 1) + r - 1]c_0 + \sum_{n=1}^{\infty} [(n + r + 1)(n + r - 1)c_n$$

$$+ c_{n-1}]x^{n+r} = 0.$$

Equating to zero the coefficient of the lowest power of x, we have the indicial equation $r(r - 1) + r - 1 = 0$ or $(r + 1)(r - 1) = 0$, with roots $r_1 = 1$, $r_2 = -1$.

Equating to zero the coefficients of the higher powers of x, we have

$$(n + r + 1)(n + r - 1)c_n + c_{n-1} = 0, \quad n \geq 1. \qquad (1)$$

Letting $r = r_1 = 1$ in (1), we find $(n + 2)nc_n + c_{n-1} = 0$ and hence

$$c_n = - \frac{c_{n-1}}{n(n+2)}, \quad n \geq 1.$$

From this, $c_1 = - \dfrac{c_0}{(1)(3)} = - \dfrac{2c_0}{1!3!}$, $c_2 = - \dfrac{c_1}{(2)(4)} = \dfrac{2c_0}{2!4!}$,

$c_3 = - \dfrac{c_2}{(3)(5)} = - \dfrac{2c_0}{3!5!}$, and, in general,

$\cdots$, $c_n = (-1)^n \dfrac{2c_0}{n!(n+2)!}$. Using these values, we obtain

the solution corresponding to the larger root $r_1 = 1$ and

given by $y = c_0 x \left[1 - \dfrac{2x}{1!3!} + \dfrac{2x^2}{2!4!} - \dfrac{2x^3}{3!5!} + \cdots \right]$. Choosing

$c_0 = 1$, we obtain the particular solution denoted by $y_1(x)$

and defined by

$$y_1(x) = x \left[1 - \frac{2x}{1!3!} + \frac{2x^2}{2!4!} - \frac{2x^3}{3!5!} + \cdots \right]$$

$$= x \left[1 + 2 \sum_{n=1}^{\infty} \frac{(-1)^n x^n}{n!(n+2)!} \right]. \tag{2}$$

Letting $r = r_2 = -1$ in (1), we find

$$n(n-2)c_n + c_{n-1} = 0 \tag{3}$$

and hence

$$c_n = - \frac{c_{n-1}}{n(n-2)}, \quad n \geq 1, \ n \neq 2. \tag{4}$$

For $n = 1$, (4) gives $c_1 = c_0$. For $n = 2$, (4) does not

apply and we must use (3). For $n = 2$, (3) is $0c_2 + c_1$

$= 0$, and hence $c_1 = 0$. But then, since $c_1 = c_0$, we must

have $c_0 = 0$. However, we assumed $c_0 \neq 0$ at the start.

This contradiction shows there is no solution of the

assumed form with $c_0 \neq 0$ corresponding to the smaller root $r_2 = -1$. Moreover, use of (4) for $n \geq 3$ will only lead us to the solution $y_1(x)$ already obtained and given by (2) (the student should check that this is true).

We now seek a solution that is linearly independent of the solution $y_1(x)$. From Theorem 6.3, we know that it is of the form $\displaystyle\sum_{n=0}^{\infty} c_n {}^* x^{n-1} + C y_1(x) \ln x$, where $c_0{}^* \neq 0$ and $C \neq 0$. We shall use reduction of order to find it. We let

$$y = y_1(x) v.$$

From this we obtain

$$y' = y_1(x) v' + y_1'(x) v \quad \text{and}$$
$$y'' = y_1(x) v'' + 2 y_1'(x) v' + y_1''(x) v.$$

Substituting these into the given D.E., after some simplifications we obtain

$$x y_1(x) v'' + [2 x y_1'(x) + y_1(x)] v' = 0. \tag{5}$$

Letting $w = v'$, this reduces to

$$x y_1(x) \frac{dw}{dx} + [2 x y_1'(x) + y_1(x)] w = 0.$$

From this, we have

$$\frac{dw}{w} = - \left[2 \frac{y_1'(x)}{y_1(x)} + \frac{1}{x} \right] dx,$$

and integrating we obtain the particular solution

$$w = \frac{1}{x[y_1(x)]^2}.$$

Writing out the first three terms of $y_1(x)$ defined by (2), we have

$$y_1(x) = x - \frac{x^2}{3} + \frac{x^3}{24} + \cdots. \tag{6}$$

From this, using basic multiplication and division, we find

$$[y_1(x)]^2 = x^2 - \frac{2x^3}{3} + \frac{7x^4}{36} + \cdots$$

and $\dfrac{1}{[y_1(x)]^2} = \dfrac{1}{x^2} + \dfrac{2}{3x} + \dfrac{1}{4} + \cdots.$ Hence we obtain

$$w = \frac{1}{x[y_1(x)]^2} = \frac{1}{x^3} + \frac{2}{3x^2} + \frac{1}{4x} + \cdots.$$

Integrating we obtain the particular solution of (5) given by

$$v = -\frac{1}{2x^2} - \frac{2}{3x} + \frac{\ln|x|}{4} + \cdots. \tag{7}$$

Multiplying (6) by (7), we obtain the desired linearly independent solution of the given D.E. We thus find

$$y = x^{-1}\left(-\frac{1}{2} - \frac{x}{2} + \frac{29x^2}{144} + \cdots\right) + \frac{y_1(x)\ln|x|}{4}. \tag{8}$$

The general solution is a linear combination of (2) and (8).

28. We assume $y = \sum_{n=0}^{\infty} c_n x^{n+r}$, where $c_0 \neq 0$. Then

$$y' = \sum_{n=0}^{\infty} (n + r) c_n x^{n+r-1},$$

$$y'' = \sum_{n=0}^{\infty} (n + r)(n + r - 1) c_n x^{n+r-2}. \quad \text{Substituting these}$$

into the D.E., we obtain

$$\sum_{n=0}^{\infty} (n + r)(n + r - 1) c_n x^{n+r} + \sum_{n=0}^{\infty} (n + r) c_n x^{n+r+1}$$

$$- \sum_{n=0}^{\infty} \frac{3}{4} c_n x^{n+r} = 0.$$

Simplifying, as in the solutions of Section 6.1, we write this as

$$\sum_{n=0}^{\infty} \left[(n + r)(n + r - 1) - \frac{3}{4} \right] c_n x^{n+r}$$

$$+ \sum_{n=1}^{\infty} (n + r - 1) c_{n-1} x^{n+r} = 0$$

and hence as

$$\left[r(r - 1) - \frac{3}{4} \right] c_0 x^r + \sum_{n=1}^{\infty} \left\{ \left[(n + r)(n + r - 1) - \frac{3}{4} \right] c_n \right.$$

$$\left. + (n + r - 1) c_{n-1} \right\} x^{n+r} = 0.$$

Equating to zero the coefficient of the lowest power of x, we have the indicial equation $r^2 - r - \frac{3}{4} = 0$, with roots, $r_1 = \frac{3}{2}$, $r_2 = -\frac{1}{2}$. Equating to zero the coefficients of the higher powers of x, we have

$$\left[(n + r)(n + r - 1) - \frac{3}{4}\right]c_n$$
$$+ (n + r - 1)c_{n-1} = 0, \quad n \geq 1. \tag{1}$$

Letting $r = r_1 = \frac{3}{2}$ in (1), we obtain $(n^2 + 2n)c_n + \left(n + \frac{1}{2}\right)c_{n-1} = 0$ and hence

$$c_n = -\frac{(2n + 1)c_{n-1}}{2n(n + 2)}, \quad n \geq 1.$$

From this, $c_1 = -\frac{c_0}{2}$, $c_2 = -\frac{5c_1}{16} = \frac{5c_0}{32}$, $c_3 = -\frac{7c_2}{30}$
$= -\frac{7c_0}{192}, \quad \cdots$. Observe that we can also write

$$c_1 = -\frac{3c_0}{2^0 1! 3!}, \quad c_2 = \frac{(3)(5)c_0}{2^1 2! 4!}, \quad c_3 = -\frac{(3)(5)(7)c_0}{2^2 3! 5!}, \quad \cdots,$$

and in general,

$$c_n = (-1)^n \frac{(3)(5)(7) \cdots (2n + 1)c_0}{2^{n-1} n! (n + 2)!}, \quad n > 1.$$

Using these values we obtain the solution corresponding to the larger root $r_1 = \frac{3}{2}$ and given by

$$y = c_0 x^{3/2}\left(1 - \frac{x}{2} + \frac{5x^2}{32} - \frac{7x^3}{192} + \cdots\right).$$ Choosing $c_0 = 1$, we obtain the particular solution denoted by $y_1(x)$ and defined by

$$y_1(x) = x^{3/2}\left(1 - \frac{x}{2} + \frac{5x^2}{32} - \frac{7x^3}{192} + \cdots\right)$$

$$= x^{3/2}\left[1 + \sum_{n=1}^{\infty} \frac{(-1)^n[(3)(5)(7)\cdots(2n+1)]}{2^{n-1}n!(n+2)!}x^n\right] \quad (2)$$

Letting $r = r_2 = -\frac{1}{2}$ in (1), we obtain $(n^2 - 2n)c_n$ $+ \left(n - \frac{3}{2}\right)c_{n-1} = 0$, or

$$n(n-2)c_n + \frac{1}{2}(2n-3)c_{n-1} = 0 \quad (3)$$

and hence

$$c_n = -\frac{(2n-3)c_{n-1}}{2n(n-2)}, \quad n \geq 1, \ n \neq 2. \quad (4)$$

For $n = 1$, (4) gives $c_1 = -\frac{c_0}{2}$. For $n = 2$, (4) does not apply and we must use (3). For $n = 2$, (3) is

$0c_2 + \frac{c_1}{2} = 0$, and hence $c_1 = 0$. But then, since

$c_1 = -\frac{c_0}{2}$, we must have $c_0 = 0$. However, we assumed

$c_0 \neq 0$ at the start. This contradiction shows there is no solution of the assumed form with $c_0 \neq 0$ corresponding to the smaller root $r_2 = -\frac{1}{2}$. Moreover, use of (4) for $n \geq 3$ will only lead us to the solution $y_1(x)$ already obtained and given by (2) (the student should check that this is true).

We now seek a solution that is linearly independent of the solution $y_1(x)$. From Theorem 6.3, we know that it is of the form

$$\sum_{n=0}^{\infty} c_n {}^* x^{n-1/2} + C y_1(x) \ln x,$$

where $c_0{}^* \neq 0$ and $C \neq 0$.

We shall use reduction of order to find it. We let $y = y_1(x)v$. From this we obtain $y' = y_1(x)v' + y_1'(x)v$ and $y'' = y_1(x)v'' + 2y_1'(x)v' + y_1''(x)v$. Substituting these into the given D.E., after some simplifications, we obtain

$$y_1(x)v'' + [2y_1'(x) + y_1(x)]v' = 0. \tag{5}$$

Letting $w = v'$, this reduces to

$$y_1(x)\frac{dw}{dx} + [2y_1'(x) + y_1(x)]w = 0.$$

From this, we have $\dfrac{dw}{w} = -\left[\dfrac{2y_1'(x)}{y_1(x)} + 1\right]dx$, and integrating we obtain the particular solution

$$w = \frac{e^{-x}}{[y_1(x)]^2}.$$

From (2), using basic multiplication and division, we find

$$[y_1(x)]^2 = x^3 - x^4 + \frac{9x^5}{16} - \frac{11x^6}{48} + \cdots$$

and

$$\frac{1}{[y_1(x)]^2} = x^{-3} + x^{-2} + \frac{7x^{-1}}{16} + \frac{5}{48} + \cdots.$$

Thus using $e^{-x} = 1 - x + \dfrac{x^2}{2} - \dfrac{x^3}{6} + \cdots$, we obtain

$$w = \frac{e^{-x}}{[y_1(x)]^2} = x^{-3} - \frac{x^{-1}}{16} + \cdots.$$

Integrating we obtain the particular solution of (5) given by

$$v = -\frac{x^{-2}}{2} - \frac{\ln|x|}{16} + \cdots, \tag{6}$$

where the next nonzero term in this expansion is the term in x^2. Multiplying (2) by (6), we obtain the desired linearly independent solution of the given D.E. We thus find

$$y = x^{-1/2}\left(-\frac{1}{2} + \frac{x}{4} - \frac{5x^2}{64} + \cdots\right) - \frac{y_1(x)\ln|x|}{16}. \tag{7}$$

The general solution is a linear combination of (2) and (7).

31. We assume $y = \displaystyle\sum_{n=0}^{\infty} c_n x^{n+r}$, where $c_0 \neq 0$. Then

$$y' = \sum_{n=0}^{\infty} (n + r)c_n x^{n+r-1},$$

$$y'' = \sum_{n=0}^{\infty} (n + r)(n + r - 1)c_n x^{n+r-2}.$$ Substituting these into the D.E., we obtain

$$\sum_{n=0}^{\infty} (n + r)(n + r - 1)c_n x^{n+r} - \sum_{n=0}^{\infty} (n + r)c_n x^{n+r}$$

$$+ \sum_{n=0}^{\infty} c_n x^{n+r+2} + \sum_{n=0}^{\infty} c_n x^{n+r} = 0.$$

Simplifying, as in the solutions of Section 6.1, we write this as

$$\sum_{n=0}^{\infty} [(n + r)(n + r - 1) - (n + r) + 1]c_n x^{n+r}$$

$$+ \sum_{n=2}^{\infty} c_{n-2} x^{n+r} = 0$$

and hence as

$$[r(r - 1) - r + 1]c_0 x^r + [(r + 1)r$$

$$- (r + 1) + 1]c_1 x^{r+1}$$

$$+ \sum_{n=2}^{\infty} \{[(n + r)(n + r - 1)$$

$$- (n + r) + 1]c_n + c_{n-2}\}x^{n+r} = 0.$$

Equating to zero the coefficient of the lowest power of x, we have the indicial equation $r^2 - 2r + 1 = 0$, with the double root $r = 1$. Equating to zero the coefficients of the higher powers of x, we have

$$[(r + 1)r - (r + 1) + 1]c_1 = 0, \tag{1}$$

$$[(n + r)(n + r - 1) - (n + r) + 1]c_n$$
$$+ c_{n-2} = 0, \ n \geq 2. \tag{2}$$

Letting $r = 1$ in (1), we have $c_1 = 0$. Letting $r = 1$ in (2), we obtain $n^2 c_n + c_{n-2} = 0$ and hence

$$c_n = -\frac{c_{n-2}}{n^2}, \quad n \geq 2.$$

From this, $c_2 = -\frac{c_0}{4}$, $c_3 = -\frac{c_1}{9} = 0$, $c_4 = -\frac{c_2}{16} = \frac{c_0}{64}$, $\cdots$.

Note that all odd coefficients are zero and we can write the general even coefficients as

$$c_{2n} = \frac{(-1)^n}{[(2)(4)(6)\cdots(2n)]^2}, \quad n \geq 1.$$

Using these values we obtain the solution of the assumed form corresponding to the double root $r = 1$ and given by

$$y = c_0 x \left(1 - \frac{x^2}{4} + \frac{x^4}{64} - \frac{x^6}{2304} + \cdots \right).$$

Choosing $c_0 = 1$, we obtain the particular solution denoted by $y_1(x)$ and defined by

$$y_1(x) = x \left(1 - \frac{x^2}{4} + \frac{x^4}{64} - \frac{x^6}{2304} + \cdots \right)$$

$$= x \left[1 + \sum_{n=1}^{\infty} \frac{(-1)^n x^{2n}}{[(2)(4)(6)\cdots(2n)]^2} \right]. \tag{3}$$

Since the indicial equation has the double root $r = 1$, by Conclusion 3 of Theorem 6.3, a linearly independent solution is of the form

$$\sum_{n=0}^{\infty} c_n{}^* x^{n+2} + y_1(x) \ln x.$$

We shall use reduction of order to find it. We let $y = y_1(x)v$. From this we obtain

$$y' = y_1(x)v' + y_1'(x)v \quad \text{and}$$

$$y'' = y_1(x)v'' + 2y_1'(x)v' + y_1''(x)v.$$

Substituting these into the given D.E., after some simplifications, we obtain

$$xy_1(x)v'' + [2xy_1'(x) - y_1(x)]v' = 0. \tag{4}$$

Letting $w = \dfrac{dv}{dx}$, this reduces to

$$xy_1(x)\frac{dw}{dx} + [2xy_1'(x) - y_1(x)]w = 0.$$

From this, we have

$$\frac{dw}{w} = \left[\frac{1}{x} - \frac{2y_1'(x)}{y_1(x)}\right]dx$$

and integrating, we obtain the particular solution

$$w = \frac{x}{[y_1(x)]^2}.$$

From (3), using basic multiplication and division, we find

$$[y_1(x)]^2 = x^2 - \frac{x^4}{2} + \frac{3x^6}{32} - \frac{5x^8}{576} + \cdots \quad \text{and}$$

$$\frac{1}{[y_1(x)]^2} = x^{-2} + \frac{1}{2} + \frac{5x^2}{32} + \frac{23x^4}{576} + \cdots.$$

Thus we obtain

$$w = x^{-1} + \frac{x}{2} + \frac{5x^3}{32} + \frac{23x^5}{576} + \cdots.$$

Integrating we obtain the particular solution of (4) given by

$$v = \ln|x| + \frac{x^2}{4} + \frac{5x^4}{128} + \frac{23x^6}{3456} + \cdots . \tag{5}$$

Multiplying (3) by (5) we obtain the desired linearly independent solution of the given D.E. We thus find

$$y = \frac{x^3}{4} - \frac{3x^5}{128} + \frac{11x^7}{13824} + \cdots + y_1(x)\ln|x| . \tag{6}$$

The general solution is a linear combination of (3) and (6).

Section 6.3, Page 280

2. Differentiating $y = \dfrac{u}{\sqrt{x}}$ twice and simplifying, we obtain

$$y' = \frac{2xu' - u}{2x^{3/2}}, \qquad y'' = \frac{4x^2u'' - 4xu' + 3u}{4x^{5/2}} .$$

Substituting these into $x^2y'' + xy' + (x^2 - p^2)y = 0$, we have

$$\frac{4x^2u'' - 4xu' + 3u}{4\sqrt{x}} + \frac{2xu' - u}{2\sqrt{x}} + \frac{(x^2 - p^2)u}{\sqrt{x}} = 0$$

which reduces to

$$\frac{4x^2u'' + u + 4(x^2 - p^2)u}{4\sqrt{x}} = 0 .$$

Now dividing through by $x^{3/2}$, this quickly reduces to

$$u'' + \left[1 + \left(\frac{1}{4} - p^2 \right) \frac{1}{x^2} \right] u = 0 .$$

3. The Bessel equation of order $p = \frac{1}{2}$ is

$$x^2 y'' + xy' + \left(x^2 - \frac{1}{4}\right)y = 0. \tag{1}$$

To use the result of Exercise 2, we let $y = \dfrac{u}{\sqrt{x}}$ and reduce (1) to

$$u'' + \left[1 + \left(\frac{1}{4} - \frac{1}{4}\right)\frac{1}{x^2}\right]u = 0$$

or simply $u'' + u = 0$. The auxiliary equation of this is $m^2 + 1 = 0$, with roots $m = \pm i$, and the G.S. is $u = c_1 \sin x + c_2 \cos x$. Thus, since $y = u/\sqrt{x}$, the G.S. of (1) is

$$y = \frac{c_1 \sin x + c_2 \cos x}{\sqrt{x}}.$$

4. From the series definition (6.124), we at once have

$$x^P J_p(kx) = \sum_{n=0}^{\infty} \frac{(-1)^n}{n!(n+p)!} \left(\frac{k}{2}\right)^{2n+p} x^{2n+2p}.$$

We may differentiate this series term-by-term to obtain

$$\frac{d}{dx}\left[x^P J_p(kx)\right] = \sum_{n=0}^{\infty} \frac{d}{dx}\left[\frac{(-1)^n}{n!(n+p)!}\left(\frac{k}{2}\right)^{2n+p} x^{2n+2p}\right]$$

$$= \sum_{n=0}^{\infty} \frac{(-1)^n 2(n+p)}{n!(n+p)!} \left(\frac{k}{2}\right)^{2n+p} x^{2n+2p-1}$$

$$= \sum_{n=0}^{\infty} \frac{(-1)^n 2x^p k}{n!(n+p-1)!2} \left(\frac{k}{2}\right)^{2n+p-1} x^{2n+p-1}$$

$$= kx^p \sum_{n=0}^{\infty} \frac{(-1)^n}{n!(n+p-1)!} \left(\frac{k}{2}\right)^{2n+p-1} x^{2n+p-1}$$

$$= kx^p J_{p-1}(kx).$$

The other identity is verified in a similar manner.

Chapter 7

1. We introduce operator notation and write the system in the form

$$\begin{cases} (D - 2)x + (D - 4)y = e^t, \\ \qquad\quad Dx + (D - 1)y = e^{4t}. \end{cases} \tag{1}$$

We apply the operator $(D - 1)$ to the first equation and the operator $(D - 4)$ to the second, obtaining

$$\begin{cases} (D - 1)(D - 2)x + (D - 1)(D - 4)y = (D - 1)e^t, \\ \qquad (D - 4)Dx + (D - 4)(D - 1)y = (D - 4)e^{4t}. \end{cases}$$

Subtracting the second equation from the first, we obtain $[(D - 1)(D - 2) - (D - 4)D]x = (D - 1)e^t - (D - 4)e^{4t}$ or $(D + 2)x = 0$. The G.S. of this D.E. is

$$x = c\,e^{-2t}. \tag{2}$$

Now there are two ways to obtain y, and we give both here. First, we proceed by returning to the system (1), applying the operator D to the first operator and the operator $(D - 2)$ to the second equation, obtaining

$$\begin{cases} D(D - 2)x + D(D - 4)y = De^t, \\ (D - 2)Dx + (D - 2)(D - 1)y = (D - 2)e^{4t}. \end{cases}$$

373

Subtracting the first equation from the second, we obtain
$[(D - 2)(D - 1) - D(D - 4)]y = (D - 2)e^{4t} - De^t$ or
$(D + 2)y = 2e^{4t} = e^t$. Using undetermined coefficients, we
find the G.S. of this D.E. to be

$$y = ke^{-2t} + \frac{e^{4t}}{3} - \frac{e^t}{3}. \tag{3}$$

Now the determinant of the operator "coefficients" of
x and y in (1) is

$$\begin{vmatrix} D - 2 & D - 4 \\ D & D - 1 \end{vmatrix} = D + 2.$$

Since this is of order 1, only one of the two constants c
and k in (2) and (3) can be independent. To determine the
relation which must thus exist between c and k, we
substitute x given by (2) and y given by (3) into system
(1). Substituting into the second equation of (1), we
have

$$-2ce^{-2t} + \left[-2ke^{-2t} + \frac{4e^{4t}}{3} - \frac{e^t}{3} \right]$$

$$- \left[ke^{-2t} + \frac{e^{4t}}{3} - \frac{e^t}{3} \right] = e^{4t}.$$

or $(-2c - 3k)e^{-2t} = 0$. Thus $k = -\frac{2c}{3}$. Hence we obtain
the G.S. of (1) in the form

$$\begin{cases} x = ce^{-2t}, \\ y = -\dfrac{2ce^{-2t}}{3} + \dfrac{e^{4t}}{3} - \dfrac{e^{t}}{3} \end{cases} \tag{4}$$

We now obtain y by the alternative procedure of the text, pages 293-295. We return to system (1) and subtract the second equation from the first, thereby eliminating Dy but not y, and thus obtaining $-2x - 3y = e^{t} - e^{4t}$. From this, $y = -\dfrac{2x}{3} + \dfrac{e^{4t}}{3} - \dfrac{e^{t}}{3}$. Substituting x into this from (2), we at once obtain $y = -\dfrac{2ce^{-2t}}{3} + \dfrac{e^{4t}}{3} - \dfrac{e^{t}}{3}$. Using this and (2), we again have the G.S. (4) of the original system.

2. We introduce operator notation and write the system in the form

$$\begin{cases} (D - 1)x + Dy = -2t \\ (D - 3)x + (D - 1)y = t^{2}. \end{cases} \tag{1}$$

We apply the operator (D - 1) to the first equation and the operator D to the second to obtain

$$\begin{cases} (D - 1)^{2}x + (D - 1)Dy = (D - 1)(-2t), \\ D(D - 3)x + D(D - 1)y = Dt^{2}. \end{cases}$$

Subtracting the second equation from the first, we obtain $[(D - 1)^{2} - D(D - 3)]x = (D-1)(-2t) - Dt^{2}$ or $(D + 1)x = -2$. The G.S. of this D.E. is

$$x = ce^{-t} - 2 \tag{2}$$

There are two ways to obtain y, and we give both.
First, we return to (1) and apply the operator $(D - 3)$ to
the first equation and the operator $(D - 1)$ to the second
equation.   We obtain

$$\begin{cases} (D - 3)(D - 1)x + (D - 3)Dy = (D - 3)(-2t) \\ (D - 1)(D - 3)x + (D - 1)^2 y = (D - 1)t^2. \end{cases}$$

Subtract the first equation from the second, to obtain
$[(D - 1)^2 - (D - 3)D]y = (D - 1)t^2 - (D - 3)(-2t)$, or
$(D + 1)y = -t^2 - 4t + 2.$

The G.S. of this D.E. is

$$y = k e^{-t} - t^2 - 2t + 4. \tag{3}$$

The determinant of the operator "coefficients" of x
and y in (1) is

$$\begin{vmatrix} D - 1 & D \\ D - 3 & D - 1 \end{vmatrix} = D + 1.$$

Since this is of order 1, only one of the two constants c
and k in (2) and (3) can be independent.   To determine the
relation which must thus exist between c and k, we
substitute x given by (2) and y given by (3) into system
(1).   Substituting into the first equation of (1), we have
$-c e^{-t} - c e^{-t} + 2 - k e^{-t} - 2t - 2 = -2t$ or $(2c + k)e^{-t} = 0.$ Thus $k = -2c$.   Hence we obtain the G.S. of (1) in the
form

$$\begin{cases} x = c\,e^{-t} - 2, \\ y = -2c\,e^{-t} - t^2 - 2t + 4. \end{cases} \qquad (4)$$

We now obtain y by the alternative procedure of the text, pages 293-295. We return to the system (1) and subtract the second equation from the first, thereby eliminating Dy but not y, and thus obtaining

$$2x + y = -t^2 - 2t.$$

Substituting x into this from (2) and solving for y, we at once find

$$y = -2c\,e^{-t} - t^2 - 2t + 4.$$

Using this and (2), we again have the G.S. (4) of the original system.

4. We introduce operator notation and write the system in the form

$$\begin{cases} (2D - 1)x + (D + 2)y = 0, \\ (3D - 2)x + (2D + 1)y = 4e^{2t}. \end{cases} \qquad (1)$$

We apply the operator $(2D + 1)$ to the first equation and the operator $(D + 2)$ to the second, obtaining

$$\begin{cases} (2D + 1)(2D - 1)x + (2D + 1)(D + 2)y = (2D + 1)0, \\ (D + 2)(3D - 2)x + (D + 2)(2D + 1)y = (D + 2)e^{2t}. \end{cases}$$

Subtracting the second equation from the first, we obtain
$(D^2 - 4D + 3)x = -(D + 2)e^{2t}$ or $(D^2 - 4D + 3)x = -4e^{2t}$.

The G.S. of this is

$$x = c_1 e^t + c_2 e^{3t} + 4e^{2t}. \qquad (2)$$

We find y using the alternative procedure of the text, pages 293-295. We subtract the second equation of (1) from 2 times the first, thereby eliminating Dy but not y. We have

$$Dx + 3y = -e^{2t}. \qquad (3)$$

Differentiating (2), we find

$$dx = c_1 e^t + 3c_2 e^{3t} + 8e^{2t}.$$

We substitute this into (3) and solve for y. We find

$$y = \frac{1}{3}\left[-c_1 e^t - 3c_2 e^{3t} - 9e^{2t}\right]$$

or

$$y = -\frac{c_1}{3} e^t - c_2 e^{3t} - 3e^{2t}.$$

The G.S. of (1) is this

$$\begin{cases} x = \quad c_1 e^t + c_2 e^{3t} + 4e^{2t}, \\[2mm] y = -\dfrac{c_1}{3} e^t - c_2 e^{3t} - 3e^{2t}. \end{cases}$$

6. We introduce operator notation and write the system in the form

$$\begin{cases} Dx + (D + 2)y = \sin t, \\ (D - 1)x + (D - 1)y = 0 \end{cases} \tag{1}$$

We apply the operator $(D - 1)$ to the first equation and the operator $(D + 2)$ to the second, obtaining

$$\begin{cases} (D - 1)Dx + (D - 1)(D + 2)y = (D - 1)\sin t, \\ (D + 2)(D - 1)x + (D + 2)(D - 1)y = (D + 2)0. \end{cases}$$

Subtracting the first equation from the second, we obtain $[(D + 2)(D - 1) - (D - 1)D]x = -(D - 1)\sin t$ or $(2D - 2)x = -(D - 1)\sin t$ or finally $(D - 1)x = \frac{1}{2}\sin t - \frac{1}{2}\cos t$.
The G.S. of this is

$$x = c\,e^t - \frac{1}{2}\sin t \tag{2}.$$

We find y using the alternative procedure of the text, pages 293 - 295. We subtract the second equation of (1) from the first, thereby eliminating Dy but not y. We have $x + 3y = \sin t$. We substitute x given by (2) into this and solve for y, obtaining $y = -\frac{1}{3}c\,e^t + \frac{1}{2}\sin t$. The G.S. of (1) is thus

$$\begin{cases} x = c\,e^t - \frac{1}{2}\sin t, \\ y = -\frac{1}{3}c\,e^t + \frac{1}{2}\sin t. \end{cases}$$

7.   We introduce operator notation and write the system in the
form

$$\begin{cases} (5D - 5)x + (D - 1)y = 0, \\ (4D - 3)x + \qquad D\,y = t. \end{cases} \tag{1}$$

We apply the operator D to the first equation and the
operator (D - 1) to the second, thereby obtaining

$$\begin{cases} \qquad D(5D - 5)x + D(D - 1)y = D0, \\ (D - 1)(4D - 3)x + (D - 1)Dy = (D - 1)t. \end{cases}$$

We subtract the second equation from the first, obtaining
$(D^2 + 2D - 3)x = -(D - 1)t$, that is, $(D^2 + 2D - 3)x = t -$
1.   The G.S. of this is

$$x = c_1 e^t + c_2 e^{-3t} - \frac{1}{3}t + \frac{1}{9}. \tag{2}$$

We find y using the alternative procedure of the text,
pages 293-295.   We subtract the second equation of (1)
from the first, obtaining

$$Dx - 2x - y = -t,$$

which involves y but does not involve Dy.   From this,

$$y = Dx - 2x + t. \tag{3}$$

Differentiating (2), we have

$$Dx = c_1 e^t - 3c_2 e^{-3t} - \frac{1}{3}. \tag{4}$$

Now substituting x from (2) and Dx from (4) into (3), we find

$$y = c_1 e^t - 3c_2 e^{-3t} - \frac{1}{3} - 2c_1 e^t - 2c_2 e^{-3t} + \frac{2}{3}t - \frac{2}{9} + t$$

or

$$y = -c_1 e^t - 5c_2 e^{-3t} + \frac{5}{3}t - \frac{5}{9}.$$

The G.S. of system (1) is thus

$$\begin{cases} x = c_1 e^t + c_2 e^{-3t} - \frac{1}{3}t + \frac{1}{9}, \\ \\ y = -c_1 e^t - 5c_2 e^{-3t} + \frac{5}{3}t - \frac{5}{9}. \end{cases}$$

9. We introduce operator notation and write the system in the form

$$\begin{cases} (D - 1)x + (D - 6)y = e^{3t}, \\ (D - 2)x + (2D - 6)y = t. \end{cases} \tag{1}$$

We apply the operator $(2D - 6)$ to the first equation and the operator $(D - 6)$ to the second equation, obtaining

$$\begin{cases} (2D - 6)(D - 1)x + (2D - 6)(D - 6)y = (2D - 6)e^{3t}, \\ (D - 6)(D - 2)x + (D - 6)(2D - 6)y = (D - 6)t. \end{cases}$$

Subtracting the second equation from the first, we obtain $[(2D - 6)(D - 1) - (D - 6)(D - 2)]x = (2D - 6)e^{3t} - (D - 6)t$ or $(D^2 - 6)x = 6t - 1$. Using the undetermined coefficients, the G.S. of this D.E. is found to be

$$x = c_1 e^{\sqrt{6}\,t} + c_2 e^{-\sqrt{6}\,t} - t + \frac{1}{6}. \qquad (2)$$

We find y using the alternate procedure of the text, pages 293-295. We multiply the first equation of (1) by 2, obtaining $(2D - 2)x + (2D - 12)y = 2e^{3t}$. We now subtract the second equation of (1) from this, obtaining $Dx - 6y = 2e^{3t} - t$, which involves y but not Dy. From this,

$$y = \frac{Dx - 2e^{3t} + t}{6}. \qquad (3)$$

From (2), $Dx = \sqrt{6}\,c_1 e^{\sqrt{6}\,t} - \sqrt{6}\,c_2 e^{-\sqrt{6}\,t} - 1$.

Substituting this into (3), we get

$$y = \frac{\sqrt{6}\,c_1 e^{\sqrt{6}\,t}}{6} - \frac{\sqrt{6}\,c_2 e^{-\sqrt{6}\,t}}{6} + \frac{t}{6} - \frac{1}{6} - \frac{e^{3t}}{3}. \qquad (4)$$

The pair (2) and (4) together constitute the G.S. of system (1).

11.   We introduce operator notation and write the system in the form

$$\begin{cases} (2D - 1)x + (D - 1)y = e^{-t}, \\ (D + 2)x + (D + 1)y = e^{t}. \end{cases} \qquad (1)$$

We apply the operator $(D + 1)$ to the first equation and the operator $(D - 1)$ to the second, obtaining

$$\begin{cases} (D + 1)(2D - 1)x + (D + 1)(D - 1)y = (D + 1)e^{-t} \\ (D - 1)(D + 2)x + (D - 1)(D + 1)y = (D - 1)e^{t}. \end{cases}$$

Subtracting the second from the first, we obtain $[(D + 1)$
$(2D - 1) - (D - 1)(D + 2)]x = (D + 1)e^{-t} - (D - 1)e^{t}$ or
$(D^2 + 1)x = 0$.   The G.S. of this D.E. is

$$x = c_1 \sin t + c_2 \cos t. \tag{2}$$

We now find y using the alternate procedure of the
text, pages 293–295.   We subtract the second equation of
(1) from the first, thereby eliminating Dy but not y, and
thus obtaining $(D - 3)x - 2y = e^{-t} - e^{t}$.   From this,
$y = \dfrac{Dx - 3x - e^{-t} + e^{t}}{2}$.   From (2), $Dx = c_1 \cos t - c_2 \sin t$.
Substituting into the preceding expression for y, we get

$$y = \frac{c_1 \cos t - c_2 \sin t - 3c_1 \sin t - 3c_2 \cos t + e^{t} - e^{t}}{2} \quad \text{or}$$

$$y = - \frac{[(3c_1 + c_2)\sin t]}{2} + \frac{[c_1 - 3c_2)\cos t]}{2} + \frac{e^{t}}{2} - \frac{e^{-t}}{2}.$$

The G.S. of system (1) is thus:

$$\begin{cases} x = c_1 \sin t + c_2 \cos t, \\ y = - \dfrac{3c_1 + c_2}{2}\sin t + \dfrac{c_1 - 3c_2}{2}\cos t + \dfrac{e^{t}}{2} - \dfrac{e^{-t}}{2}. \end{cases}$$

12.   We introduce operator notation and write the system in the
form

$$\begin{cases} (3D - 1)x + (2D + 1)y = t - 1, \\ (D - 1)x + \qquad\qquad Dy = t + 2. \end{cases} \tag{1}$$

We apply the operator D to the first equation and the operator $(2D + 1)$ to the second, thereby obtaining

$$\begin{cases} D(3D - 1)x + D(2D + 1)y = D(t - 1), \\ (2D + 1)(D - 1)x + (2D + 1)Dy = (2D + 1)(t + 2). \end{cases}$$

We subtract the second equation from the first, obtaining $[D(3D - 1) - (2D + 1)(D - 1)]x = D(t - 1) - (2D + 1)(t + 2)$ or $(D^2 + 1)x = - t - 3$. The auxiliary equation of this D.E. is $m^2 + 1 = 0$ with roots $\pm i$, and the complementary function is $x_c = c_1 \sin t + c_2 \cos t$. A particular integral is $- t - 3$, and so the G.S. is

$$x = c_1 \sin t + c_2 \cos t - t - 3. \qquad (2)$$

We find y using the alternative procedure of the text, pages 293-295. We multiply the second equation of (1) by 2 to obtain $(2D - 2)x + 2Dy = 2t + 4$. Then we subtract this from the first equation of (1) to obtain $Dx + x + y = - t - 5$, which involves y but does not involve Dy. From this,

$$y = - x - Dx - t - 5. \qquad (3)$$

Differentiating (2), we have

$$Dx = c_1 \cos t - c_2 \sin t - 1. \qquad (4)$$

Now substituting x from (2) and Dx from (4) into (3), we find $y = - c_1 \sin t - c_2 \cos t + t + 3 - c_1 \cos t + c_2 \sin t + 1 - t - 5$ or $y = (c_2 - c_1) \sin t - (c_1 + c_2) \cos t - 1$.

The G.S. of system (1) is thus

$$\begin{cases} x = c_1 \sin t + c_2 \cos t - t - 3, \\ y = (c_2 - c_1)\sin t - (c_1 + c_2)\cos t - 1. \end{cases}$$

15. We introduce operator notation and write the system in the form

$$\begin{cases} (2D + 1)x + (D + 5)y = 4t, \\ (D + 2)x + (D + 2)y = 2. \end{cases} \tag{1}$$

We apply the operator $(D + 2)$ to the first equation and the operator $(D + 5)$ to the second, thereby obtaining

$$\begin{cases} (D + 2)(2D + 1)x + (D + 2)(D + 5)y = (D + 2)4t, \\ (D + 5)(D + 2)x + (D + 5)(D + 2)y = (D + 5)2. \end{cases}$$

We subtract the second equation from the first, obtaining $(D^2 - 2D - 8)x = 8t - 6$. The auxiliary equation of this D.E. is $m^2 - 2m - 8 = 0$, that is, $(m - 4)(m + 2) = 0$, with roots $m = 4, -2$; and the complementary function is $x_c = c_1 e^{4t} + c_2 e^{-2t}$. A particular integral is $x_p = -t + 1$, so the G.S. is

$$x = c_1 e^{4t} + c_2 e^{-2t} - t + 1. \tag{2}$$

We find y using the alternative procedure of the text, pages 293-295. We subtract the second equation of (1) from the first to obtain $Dx - x + 3y = 4t - 2$, which involves y but does not involve Dy. From this,

$$y = \frac{1}{3} [x - Dx + 4t - 2]. \qquad (3)$$

Differentiating (2), we have

$$Dx = 4c_1 e^{4t} - 2c_2 e^{-2t} - 1. \qquad (4)$$

Now substituting x from (2) and Dx from (4) into (3), we

find $y = \frac{1}{3} \Bigg[c_1 e^{4t} + c_2 e^{-2t} - t + 1$

$$- 4c_1 e^{4t} + 2c_2 e^{-2t} + 1 + 4t - 2 \Bigg]$$

or $y = -c_1 e^{4t} + c_2 e^{-2t} + t.$ The G.S. of system (1) is

thus

$$\begin{cases} x = c_1 e^{4t} + c_2 e^{-2t} - t + 1, \\ y = -c_1 e^{4t} + c_2 e^{-2t} + t. \end{cases}$$

17. We introduce operator notation and write the system in the
form

$$\begin{cases} (2D + 1)x + (4D - 1)y = 3e^t, \\ (D + 2)x + (D + 2)y = e^t. \end{cases} \qquad (1)$$

We apply the operator $(D + 2)$ to the first equation and
the operator $(4D - 1)$ to the second equation, obtaining

$$\begin{cases} (D + 2)(2D + 1)x + (D + 2)(4D - 1)y = (D + 2)3e^t, \\ (4D - 1)(D + 2)x + (4D - 1)(D + 2)y = (4D - 1)e^t. \end{cases}$$

We subtract the first equation from the second, obtaining
$(2D^2 + 2D - 4)x = -6e^t,$ that is, $(D^2 + D - 2)x = -3e^t.$

The auxiliary equation of this D.E. is $m^2 + m - 2 = 0$, with roots $m = 1, -2$; and the complementary function is $x_c = c_1 e^t + c_2 e^{-2t}$. Using the method of undetermined coefficients, we assume a particular integral of the form $x_p = A t e^t$. We quickly find $A = -1$, so $x_p = -t e^t$. Thus, the G.S. of the D.E. is

$$x = c_1 e^t + c_2 e^{-2t} - t e^t . \qquad (2)$$

We find y using the alternative procedure of the text, pages 293–295. We multiply the second equation of (1) by 4 to obtain

$$(4D + 8)x + (4D + 8)y = 4 e^t .$$

We then subtract the first equation of (1) from this, obtaining $2 Dx + 7x + 9y = e^t$, which involves y but does not involve Dy. From this

$$y = \frac{1}{9} [- 7x - 2 Dx + e^t]. \qquad (3)$$

Differentiating (2), we have

$$Dx = c_1 e^t - 2c_2 e^{-2t} - t e^t - e^t. \qquad (4)$$

Now substituting x from (2) and Dx from (4) into (3), we find $y = \frac{1}{9} \left[- 7c_1 e^t - 7c_2 e^{-2t} + 7t e^t \right.$
$$\left. - 2c_1 e^t + 4c_2 e^{-2t} + 2t e^t + 2e^t + e^t \right]$$

or $y = - c_1 e^t - \dfrac{c_2}{3} e^{-2t} + t e^t + \dfrac{1}{3} e^t$. The G.S. of system

(1) is thus

$$
\begin{cases}
x = c_1 e^t + c_2 e^{-2t} - t\, e^t, \\
y = (1/3 - c_1) e^t - (1/3)\, c_2 e^{-2t} + t\, e^t.
\end{cases}
$$

20. We introduce operator notation and write the system in the form

$$
\begin{cases}
(D - 5)x + \qquad Dy = 2t - 8, \\
\qquad - 8x + (D - 1)y = -t^2.
\end{cases}
\tag{1}
$$

We apply the operator $(D - 1)$ to the first equation and the operator D to the second, obtaining

$$
\begin{cases}
(D + 2)(D - 5)x + (D - 1)Dy = (D - 1)(2t - 8), \\
\qquad - 8Dx + D(D - 1)y = - Dt^2.
\end{cases}
$$

We subtract the second equation from the first, obtaining $(D^2 + 2D + 5)x = 10.$

The auxiliary equation of this D.E. is $m^2 + 2m + 5 = 0$, with roots $m = -1 \pm 2i$; and the complementary function is $x_c = e^{-t}(c_1 \sin 2t + c_2 \cos 2t)$. A particular integral is $x_p = 2$; and the G.S. is

$$
x = e^{-t}(c_1 \sin 2t + c_2 \cos 2t) + 2.
\tag{2}
$$

We find y using the alternative procedure of the text, pages 293-295. We subtract the second equation of (1) from the first to obtain $Dx + 3x + y = t^2 + 2t - 8$, which involves y but does not involve Dy. From this,

$$y = -3x - Dx + t^2 + 2t - 8. \qquad (3)$$

Differentiating (2), we find

$$Dx = e^{-t}\left[(-c_1 - 2c_2)\sin 2t + (2c_1 - c_2)\cos 2t\right]. \qquad (4)$$

Now substituting x given by (2) and Dx given by (4) into (3), we obtain

$$y = -3e^{-t}(c_1\sin 2t + c_2\cos 2t) - 6$$
$$+ e^{-t}\left[(c_1 + 2c_2)\sin 2t + (c_2 - 2c_1)\cos 2t\right]$$
$$+ t^2 + 2t - 8$$

or

$$y = e^{-t}\left[2(c_2 - c_1)\sin 2t - 2(c_1 + c_2)\cos 2t\right]$$
$$+ t^2 + 2t - 14.$$

The G.S. of the system (1) is thus

$$y = e^{-t}(c_1\sin 2t + c_2\cos 2t) + 2,$$
$$y = e^{-t}\left[2(c_2 - c_1)\sin 2t - 2(c_1 + c_2)\cos 2t\right]$$
$$+ t^2 + 2t - 14.$$

21. We introduce operator notation and write the system in the
form

$$\begin{cases} (2D - 1)x + (D - 1)y = 1, \\ (D + 2)x + (D - 1)y = t. \end{cases} \quad (1)$$

For this system we merely have to subtract the second
equation from the first to eliminate both y and Dy. We
have

$$[(2D - 1) - (D + 2)]x = 1 - t \text{ or } (D - 3)x = 1 - t.$$

The G.S. of this is

$$x = c_1 e^{3t} + \frac{t}{3} - \frac{2}{9}. \quad (2)$$

We cannot apply the alternate procedure of pages
293–295 to find y here; for, as noted above, elimination
of Dy from (1) also results in elimination of y itself.
Thus we proceed to eliminate x and Dx from (1). We apply
the operator $(D + 2)$ to the first equation of (1) and the
operator $(2D - 1)$ to the second equation, obtaining

$$\begin{cases} (D + 2)(2D - 1)x + (D + 2)(D - 1)y = (D + 2)1, \\ (2D - 1)(D + 2)x + (2D - 1)(D - 1)y = (2D - 1)t. \end{cases}$$

Subtracting the first from the second, we obtain

$$[(2D - 1)(D - 1) - (D + 2)(D - 1)]y = (2D - 1)t$$
$$- (D + 2)1$$

or $(D^2 - 4D + 3)y = -t$. Using undetermined coefficients, the G.S. of this D.E. is found to be

$$y = k_1 e^t + k_2 e^{3t} - \frac{t}{3} - \frac{4}{9}. \qquad (3)$$

The determinant of the operator "coefficients" of x and y in (1) is

$$\begin{vmatrix} 2D - 1 & D - 1 \\ D + 2 & D - 1 \end{vmatrix} = D^2 - 4D + 3.$$

Since this is of order 2, only two of the three constants c_1, k_1, k_2 in (2) and (3) can be independent. To determine the relations which must thus exist among these three constants, we substitute x given by (2) and y given by (3) into system (1). Substituting into the first equation of (1), we have

$$2\left(3c_1 e^{3t} + \frac{1}{3}\right) + \left(k_1 e^t + 3k_2 e^{3t} - \frac{1}{3}\right)$$

$$- \left(c_1 e^{3t} + \frac{t}{3} - \frac{2}{9}\right) - \left(k_1 e^t + k_2 e^{3t} - \frac{t}{3} - \frac{4}{9}\right) = 1$$

or

$$(5c_1 + 2k_2)e^{3t} + 0k_1 e^t = 0.$$

Thus $k_2 = -\dfrac{5c_1}{2}$ and k_1 is arbitrary. Thus k_1 is the "second" arbitrary constant of the G.S., and we now write it as c_2. Hence we have the G.S. in the form

$$\begin{cases} x = c_1 e^{3t} + \dfrac{t}{3} - \dfrac{2}{9}, \\[3mm] y = c_2 e^t - \dfrac{5c_1 e^{3t}}{2} - \dfrac{t}{3} - \dfrac{4}{9}. \end{cases}$$

22. We introduce operator notation and write the system in the form

$$\begin{cases} D^2 x + Dy = e^{2t}, \\[3mm] (D - 1)x + (D - 1)y = 0. \end{cases} \qquad (1)$$

We apply the operator $(D - 1)$ to the first equation and the operator D to the second equation, obtaining

$$\begin{cases} (D - 1)D^2 x + (D - 1)Dy = (D - 1)e^{2t}, \\[3mm] D(D - 1)x + D(D - 1)y = D0. \end{cases}$$

Subtracting the second equation from the first, we obtain

$$[(D - 1)D^2 - D(D - 1)]x = (D - 1)e^{2t} - 0$$

or

$$(D^3 - 2D^2 + D)x = e^{2t}.$$

The auxiliary equation of this D.E. is $m^3 - 2m^2 + m = 0$ with roots $m = 0, 1, 1$ (double root); and the G.S., found by using undetermined coefficients, is

$$x = c_1 + (c_2 + c_3 t)e^t + \dfrac{e^{2t}}{2}. \qquad (2)$$

We find using the alternate procedure of pages 293–295 of the text. We subtract the second equation of (1) from the first, thereby eliminating Dy but not y, and thus obtaining $D^2x - Dx + x + y = e^{2t}$. From this,

$$y = -x + Dx - D^2x + e^{2t}. \tag{3}$$

From (2), we find $Dx = (c_2 + c_3 + c_3t)e^t + e^{2t}$, $D^2x = (c_2 + 2c_3 + c_3t)e^t + 2e^{2t}$. Substituting x from (2) and these derivatives into (3), we get

$$y = - \left[c_1 + (c_2 + c_3t)e^t + \frac{e^{2t}}{2} \right]$$

$$+ \, [(c_2 + c_3 + c_3t)e^t + e^{2t}]$$

$$- \, [(c_2 + 2c_3 + c_3t)\,e^t + 2e^{2t}] + e^{2t}$$

or

$$y = -c_1 - (c_2 + c_3 + c_3t)e^t - \frac{e^{2t}}{2}. \tag{4}$$

The pair (2) and (4) together constitute the G.S. of system (1).

26. We introduce operator notation and write the system in the form

$$\begin{cases} (D^2 + 1)x + (4D - 4)y = 0, \\[2mm] (D - 1)x + (D + 9)y = e^{2t}. \end{cases} \tag{1}$$

We apply the operator $(D + 9)$ to the first equation and the operator $(4D - 4)$ to the second equation, obtaining

$$\begin{cases} (D + 9)(D^2 + 1)x + (D + 9)(4D - 4)y = (D + 9)0, \\ (4D - 4)(D - 1)x + (4D - 4)(D + 9)y = (4D - 4)e^{2t}. \end{cases}$$

Subtracting the second equation from the first, we obtain

$$[(D + 9)(D^2 + 1) - (4D - 4)(D - 1)]x = 0 - (4D - 4)e^{2t}$$

or

$$(D^3 + 5D^2 + 9D + 5)x = -4e^{2t}.$$

The auxiliary equation of this D.E. is $m^3 + 5m^2 + 9m + 5 = 0$ or $(m + 1)(m^2 + 4m + 5) = 0$, with roots $m = -1$, $m = -2 \pm i$. Using this information and undetermined coefficients, the G.S. of this D.E. is found to be

$$x = c_1 e^{-t} + e^{-2t}(c_2 \sin t + c_3 \cos t) - \frac{4e^{2t}}{51}. \qquad (2)$$

We find y using the alternate procedure of pages 293-295 of the text. We multiply the second equation of (1) by 4, obtaining $(4D - 4)x + (4D + 36)y = 4e^{2t}$. We now subtract this from the first equation of (1), obtaining $D^2x - 4Dx + 5x - 40y = -4e^{2t}$. From this,

$$y = \frac{D^2x - 4Dx + 5x + 4e^{2t}}{40}. \qquad (3)$$

From (2),

$$Dx = -c_1 e^{-t} + e^{-2t}(-2c_2 - c_3)\sin t + e^{-2t}(c_2 - 2c_3)\cos t$$

$$- \frac{8e^{2t}}{51},$$

$$D^2 x = c_1 e^{-t} + e^{-2t}(3c_2 + 4c_3)\sin t + e^{-2t}(-4c_2 + 3c_3)\cos t$$

$$- \frac{16e^{2t}}{51}.$$

Substituting x from (2) and these derivatives into (3), we get

$$y = \frac{\left[c_1 e^{-t} + e^{-2t}(3c_2 + 4c_3)\sin t + e^{-2t}(-4c_2 + 3c_3)\cos t - \frac{16e^{2t}}{51} \right]}{40}$$

$$+ \frac{4\left[-c_1 e^{-t} - e^{-2t}(-2c_2 - c_3)\sin t + e^{-2t}(c_2 - 2c_3)\cos t - \frac{8e^{2t}}{51} \right]}{40}$$

$$+ \frac{5\left[c_1 e^{-t} + e^{-2t}(c_2 \sin t + c_3 \cos t) - \frac{4e^{2t}}{51} \right]}{40} + \frac{4e^{2t}}{40}$$

or

$$y = \frac{(c_1 + 4c_1 + 5c_1)e^{-t}}{40}$$

$$+ \frac{e^{-2t}(3c_2 + 4c_3 + 8c_2 + 4c_3 + 5c_2)(\sin t)}{40}$$

$$+ \frac{e^{-2t}(-4c_2 + 3c_3 - 4c_2 + 8c_3 + 5c_3)(\cos t)}{40}$$

$$+ \frac{(-16 + 32 - 20 + 204)e^{2t}}{(40)(51)}$$

or finally

$$y = \frac{c_1 e^{-t}}{4} + e^{-2t}\left[\left(\frac{2c_2}{5} + \frac{c_3}{5}\right)\sin t\right.$$

$$\left. + \left(-\frac{c_2}{5} + \frac{2c_3}{5}\right)\cos t \right. + \frac{5e^{2t}}{51}. \tag{4}$$

The pair (2) and (4) constitute the G.S. of system (1).

30. By text, page 286, equations (7.7), we let $x_1 = x_1$, $x_2 = x'$, $x_3 = x''$, $x_4 = x'''$. From these and the given fourth order D.E., we have

$$x_1' = x' = x_2, \ x_2' = x'' = x_3, \ x_3' = x''' = x_4,$$

$$x_4' = x^{iv} = -2tx + t^2 x'' + \cos t = -2tx_1 + t^2 x_3 + \cos t.$$

Thus we have the system

$$\begin{cases} x_1' = x_2, \ x_2' = x_3, \ x_3' = x_4, \\ x_4' = -2tx_1 + t^2x_3 + \cos t . \end{cases}$$

Section 7.2, Page 307.

3. We apply Kirchhoff's voltage law (text, Section 5.6) to
    each of the three loops indicated in Figure 7.6. For the
    left hand loop, the voltage drops are as follows:

    1. across the resistor R_1: $20i_1$.

    2. across the inductor L_1: $0.01 \ i'$.

Thus applying the voltage law to this loop, we have the
D.E.

$$0.01 \ i' \ + 20i_1 = 100. \qquad\qquad (1)$$

For the right hand loop, the voltage drops are as follows:

    1. across the resistor R_1: $-20i_1$.

    2. across the resistor R_2: $40i_2$.

    3. across the inductor L_2: $0.02 \ i_2'$.

Thus applying the voltage law to this loop, we have the
D.E.

$$0.02 \ i_2' \ - 20i_1 + 40i_2 = 0. \qquad\qquad (2)$$

For the outside loop, the voltage drops are

    1. across the resistor R_2: $40i_2$.

    2. across the inductor L_1: $0.01 \ i'$.

    3. across the inductor L_2: $0.02 \ i_2'$.

Thus applying the voltage law to this loop, we have the D.E.

$$0.02\, i_2' + 0.01\, i' + 40 i_2 = 100. \qquad (3)$$

The three equations are not all independent. Equation (3) may be obtained by adding equations (1) and (2). Hence we retain only equations (1) and (2).

We now apply Kirchhoff's current law to the upper junction point in the figure. We have $i = i_1 + i_2$. Hence we replace i by $i_1 + i_2$ in (1), retain (2) as is, and obtain the linear system

$$\begin{cases} 0.01\, i_1' + 0.01\, i_2' + 20 i_1 = 100, \\[2mm] 0.02\, i_2' - 20 i_1 + 40 i_2 = 0. \end{cases} \qquad (4)$$

The initially zero currents give the I.C.'s

$$i_1(0) = 0, \quad i_2(0) = 0.$$

We introduce operator notation and write (4) in the form

$$\begin{cases} (0.01D + 20) i_1 + 0.01 D i_2 = 100, \\[2mm] -20 i_1 + (0.02D + 40) i_2 = 0. \end{cases} \qquad (5)$$

We apply the operator $0.02D + 40$ to the first equation of (5) and the operator $0.01D$ to the second and subtract to obtain

$$[(0.01D + 20)(0.02D + 40) + 0.2D]i_1 = (0.02D + 40)100$$

or

$$(.0002D^2 + D + 800)i_1 = 4000$$

or finally

$$(D^2 + 5000D + 4,000,000)i_1 = 20,000,000. \quad (6)$$

The auxiliary equation of the corresponding homogeneous D.E. is $m^2 + 5000m + 4,000,000 = 0$, with roots $m = -1000$, -4000. Using this information and the method of undetermined coefficients, we see that the solution of (6) is

$$i_1 = c_1e^{-1000t} + c_2e^{-4000t} + 5. \quad (7)$$

We use the alternative procedure of page 293 of the text to find i_2. Returning to (5), we multiply the first equation by 2 and then subtract the second equation from this, obtaining

$$.02Di_1 + 60i_1 - 40i_2 = 200.$$

Note that this contains i_2 but not Di_2. From it, we have

$$i_2 = \frac{3i_1}{2} + .0005Di_1 - 5. \quad (8)$$

From (7), we find

$$Di_1 = -1000c_1e^{-1000t} - 4000c_2e^{-4000t}.$$

Now substituting i_1 from (7) and Di_1 just obtained into (8) and simplifying, we obtain

$$i_2 = c_1e^{-1000t} - \frac{c_2e^{-4000t}}{2} + \frac{5}{2}. \qquad (9)$$

Finally, applying the I.C.'s to (7) and (9), we have $c_1 + c_2 + 5 = 0$ and $c_1 - \frac{c_2}{2} + \frac{5}{2} = 0$, from which we find $c_1 = -\frac{10}{3}$, $c_2 = -\frac{5}{3}$. Thus we obtain

$$\begin{cases} i_1 = -\dfrac{10e^{-1000t}}{3} - \dfrac{5e^{-4000t}}{3} + 5, \\[3mm] i_2 = -\dfrac{10e^{-1000t}}{3} + \dfrac{5e^{-4000t}}{6} + \dfrac{5}{2}. \end{cases}$$

6. Let x = the amount of salt in tank X at time t, and let y = the amount of salt in tank Y at time t, each measured in kilograms. Each tank always contains 30 liters of fluid, so the concentration of salt in tank X is $\frac{x}{30}$ (kg/liter) and that in tank Y is $\frac{y}{30}$ (kg/liter) . Salt enters tank X two ways: (a) 1 kg. of salt per liter of brine enters at the rate of 2 liters/min. from outside the system, and (b) salt in the brine pumped from tank Y back to tank X at the rate of 1 liter/min. By (a), salt enters tank X at the

rate of 2 kg./min.; and by (b), it enters at the rate of $\frac{y}{30}$ kg./min. Salt only leaves tank X in the brine flowing from tank X into tank Y at the rate of 4 liters/min. Thus salt leaves tank X at the rate of $\frac{4x}{30}$ kg./min. Hence we obtain the D.E.

$$x' = 2 + \frac{y}{30} - \frac{4x}{30} \tag{1}$$

for the amount of salt in tank X at time t. The D.E. for the amount in tank Y is obtained similarly. It is

$$y' = \frac{4x}{30} - \frac{4y}{30}. \tag{2}$$

Since initially there was 30 kg. of salt in tank X and pure water in tank Y, the I.C.'s are $x(0) = 30$, $y(0) = 0$.

We introduce operator notation and write the D.E.'s (1) and (2) in the forms

$$\begin{cases} \left(D + \frac{2}{15} \right) x - \frac{1}{30} y = 2, \\[2mm] -\frac{2}{15} x + \left(D + \frac{2}{15} \right) y = 0. \end{cases} \tag{3}$$

We apply the operator $\left(D + \frac{2}{15} \right)$ to the first equation (3), multiply the second by $\frac{1}{30}$, and add to obtain

$$\left[\left(D + \frac{2}{15} \right)^2 - \frac{1}{225} \right] x = \frac{4}{15}$$

or

$$\left(D^2 + \frac{4}{15} D + \frac{3}{225}\right)x = \frac{4}{15}. \qquad (4)$$

The auxiliary equation of the corresponding homogeneous D.E. is

$$m^2 + \frac{4m}{15} + \frac{3}{225} = 0$$

with roots $m = -\frac{1}{5}, -\frac{1}{15}$. Using this information and the method of undetermined coefficients, we see that the solution of (4) is

$$x = c_1 e^{-t/5} + c_2 e^{-t/15} + 20. \qquad (5)$$

To find y, we return to (3), multiply the first equation by $\frac{2}{15}$, apply the operator $\left(D + \frac{2}{15}\right)$ to the second, and add. After simplification, we obtain

$$\left(D^2 + \frac{4}{15} D + \frac{3}{225}\right)y = \frac{4}{15}.$$

Comparing this with (4) and its solution (5), we see that its solution is

$$y = c_3 e^{-t/5} + c_4 e^{-t/15} + 20. \qquad (6)$$

Only two of the four constants c_1, c_2, c_3, c_4 in (5) and (6) are independent. To determine the relations which must exist among these, we substitute (5) and (6) into the second equation of (3). We find

$$\frac{(-2c_1 - c_3)e^{-t/5}}{15} + \frac{(-2c_2 + c_4)e^{-t/15}}{15} = 0.$$

Hence we must have $c_3 = -2c_1$, $c_4 = 2c_2$. Thus we have

$$\begin{cases} x = c_1 e^{-t/5} + c_2 e^{-t/15} + 20, \\[2mm] y = -2c_1 e^{-t/5} + 2c_2 e^{-t/15} + 20. \end{cases}$$

Finally we apply the I.C.'s. We have $c_1 + c_2 + 20 = 30$, $-2c_1 + 2c_2 + 20 = 0$, from which we find $c_1 = 10$, $c_2 = 0$. Thus we obtain the solution

$$\begin{cases} x = 10e^{-t/5} + 20, \\[2mm] y = -20e^{-t/5} + 20. \end{cases}$$

Section 7.3, Page 317.

2.   (a) Let $\begin{cases} x = 3e^{7t}, \\[2mm] y = 2e^{7t}. \end{cases}$ and $\begin{cases} x = e^{-t}, \\[2mm] y = -2e^{-t}. \end{cases}$

Consider the first pair: $\dfrac{dx}{dt} = 21e^{7t}$, and $5x + 3y = 5(3e^{7t}) + 3(2e^{7t}) = 15e^{7t} + 6e^{7t} = 21e^{7t}$. So $\dfrac{dx}{dt} = 5x + 3y$. Now, $\dfrac{dy}{dt} = 14e^{7t}$, and $4x + y = 4(3e^{7t}) + 2e^{7t} = 12e^{7t} + 2e^{7t} = 14e^{7t}$. So $\dfrac{dy}{dt} = 4x + y$. Thus the first pair satisfies the linear system.

Consider the second pair: $\frac{dx}{dt} = -e^{-t}$, and $5x + 3y =$

$5(e^{-t}) + 3(-2e^{-t}) = 5e^{-t} - 6e^{-t} = -e^{-t}$. So $\frac{dx}{dt} = 5x +$

$3y$. Now, $\frac{dy}{dt} = 2e^{-t}$, and $4x + y = 4(e^{-t}) + (-2e^{-t}) =$

$4e^{-t} - 2e^{-t} = 2e^{-t}$. So $\frac{dy}{dt} = 4x + y$. Thus the second

pair satisfies the linear system. Hence each pair is

a solution of the system.

(b) Consider $\begin{vmatrix} 3e^{7t} & e^{-t} \\ 2e^{7t} & -2e^{-t} \end{vmatrix} = (3e^{7t})(-2e^{-t}) - (2e^{7t})(e^{-t})$

$$= -6e^{6t} - 2e^{6t} = -8e^{6t} \neq 0.$$

Thus these two solutions are linearly independent for

all $a \leq t \leq b$.

(c) The general solution of the system is

$$\begin{cases} x = 3c_1 e^{7t} + c_2 e^{-t}, \\ y = 2c_1 e^{7t} - 2c_2 e^{-t}. \end{cases}$$

The I.C. $f(0) = 0$ gives $3c_1 + c_2 = 0$; and the I.C.

$g(0) = 8$ gives $2c_1 - 2c_2 = 8$. The unique solution of

this system in c_1 and c_2 is $c_1 = 1$, $c_2 = -3$. Thus

$$\begin{cases} x = 3e^{7t} - 3e^{-t}, \\ y = 2e^{7t} + 6e^{-t}, \end{cases}$$

is the solution of the given linear system that satisfies the two initial conditions.

4. The Wronskian determinant of f_1 and f_2 is

$$W(t) = \begin{vmatrix} f_1(t) & f_2(t) \\ g_1(t) & g_2(t) \end{vmatrix} = f_1(t)g_2(t) - f_2(t)g_1(t).$$

Differentiating, we find

$$W'(t) = f_1(t)g_2'(t) + f_1'(t)g_2(t)$$
$$- f_2(t)g_1'(t) - f_2'(t)g_1(t). \qquad (1)$$

Since $x = f_1(t)$, $y = g_1(t)$ is a solution of system (7.67), we have $f_1'(t) = a_{11}(t)f_1(t) + a_{12}(t)g_1(t)$ and $g_1'(t) = a_{21}(t)f_1(t) + a_{22}(t)g_1(t)$. Since $x = f_2(t)$, $y = g_2(t)$ is a solution of system (7.67), we have $f_2'(t) = a_{11}(t)f_2(t) + a_{12}(t)g_2(t)$ and $g_2'(t) = a_{21}(t)f_2(t) + a_{22}(t)g_2(t)$. Substituting these derivatives into (1), we find

$$W'(t) = f_1(t)[a_{21}(t)f_2(t) + a_{22}(t)g_2(t)]$$
$$+ g_2(t)[a_{11}(t)f_1(t) + a_{12}(t)g_1(t)]$$
$$- f_2(t)[a_{21}(t)f_1(t) + a_{22}(t)g_1(t)]$$
$$- g_1(t)[a_{11}(t)f_2(t) + a_{12}(t)g_2(t)]$$
$$= a_{11}(t)[f_1(t)g_2(t) - f_2(t)g_1(t)]$$
$$+ a_{22}(t)[f_1(t)g_2(t) - f_2(t)g_1(t)]$$
$$= [a_{11}(t) + a_{22}(t)][f_1(t)g_2(t) - f_2(t)g_1(t)].$$

But

$$f_1(t)g_2(t) - f_2(t)g_1(t) = \begin{vmatrix} f_1(t) & f_2(t) \\ g_1(t) & g_2(t) \end{vmatrix} = W(t),$$

so we have proved $W'(t) = [a_{11}(t) + a_{22}(t)] W(t)$, as required.

Section 7.4, Page 328.

1. We assume a solution of the form $x = Ae^{\lambda t}$, $y = Be^{\lambda t}$. Substituting into the given system, we obtain

$$\begin{cases} A\lambda e^{\lambda t} = 5Ae^{\lambda t} - 2Be^{\lambda t}, \\ B\lambda e^{\lambda t} = 4Ae^{\lambda t} - Be^{\lambda t}, \end{cases}$$

which leads at once to the algebraic system

$$\begin{cases} (5 - \lambda)A - 2B = 0, \\ 4A + (-1 - \lambda)B = 0. \end{cases} \tag{1}$$

For nontrivial solutions of this, we must have

$$\begin{vmatrix} 5 - \lambda & -2 \\ 4 & -1 - \lambda \end{vmatrix} = 0.$$

This leads to the characteristic equation $\lambda^2 - 4\lambda + 3 = 0$, whose roots are $\lambda_1 = 1$, $\lambda_2 = 3$.

Setting $\lambda = \lambda_1 = 1$ in (1), we obtain $4A - 2B = 0$

(twice), a nontrivial solution of which is $A = 1$, $B = 2$. With these values of A, B, and λ, we find the nontrivial solution

$$x = e^t, \quad y = 2e^t. \tag{2}$$

Setting $\lambda = \lambda_2 = 3$ in (1), we obtain $2A - 2B = 0$ and $4A - 4B = 0$, a nontrivial solution of which is $A = B = 1$. With these values of A, B, and λ, we find the nontrivial solution

$$x = e^{3t}, \quad y = e^{3t}. \tag{3}$$

The G.S. of the given system is a linear combination of (2) and (3). Thus we find $x = c_1 e^t + c_2 e^{3t}$, $y = 2c_1 e^t + c_2 e^{3t}$.

2. We assume a solution of the form $x = Ae^{\lambda t}$, $y = Be^{\lambda t}$, Substituting into the given system, we obtain

$$\begin{cases} A\lambda e^{\lambda t} = 5Ae^{\lambda t} - Be^{\lambda t}, \\ B\lambda e^{\lambda t} = 3Ae^{\lambda t} + Be^{\lambda t}, \end{cases}$$

which leads at once to the algebraic system

$$\begin{cases} (5 - \lambda)A - B = 0, \\ 3A + (1 - \lambda)B = 0. \end{cases} \tag{1}$$

For nontrivial solutions of this, we must have

$$\begin{vmatrix} 5 - \lambda & -1 \\ 3 & 1 - \lambda \end{vmatrix} = 0.$$

This leads to the characteristic equation $\lambda^2 - 6\lambda + 8 = 0$, whose roots are $\lambda_1 = 4$, $\lambda_2 = 2$.

Setting $\lambda = \lambda_1 = 4$ in (1), we obtain $A - B = 0$, $3A - 3B = 0$, a nontrivial solution of which is $A = B = 1$. With these values of A, B, and λ, we find the nontrivial solution

$$x = e^{4t}, \quad y = e^{4t}. \tag{2}$$

Setting $\lambda = \lambda_2 = 2$ in (1), we obtain $3A - B = 0$ (twice), a nontrivial solution of which is $A = 1$, $B = 3$. With these values of A, B, and λ, we find the nontrivial solution

$$x = e^{2t}, \quad y = 3e^{2t}. \tag{3}$$

The G.S. of the given system is a linear combination of (2) and (3). Thus we find

$$x = c_1 e^{4t} + c_2 e^{2t}, \quad y = c_1 e^{4t} + 3c_2 e^{2t}.$$

5.   We assume a solution of the form $x = Ae^{\lambda t}$, $y = Be^{\lambda t}$. Substituting into the given system, we obtain

$$\begin{cases} A\lambda e^{\lambda t} = 3Ae^{\lambda t} + Be^{\lambda t}, \\ B\lambda e^{\lambda t} = 4Ae^{\lambda t} + 3Be^{\lambda t}, \end{cases}$$

which leads at once to the algebraic system

$$\begin{cases} (3 - \lambda)A + B = 0, \\ 4A + (3 - \lambda)B = 0. \end{cases} \tag{1}$$

For nontrivial solutions of this, we must have

$$\begin{vmatrix} 3 - \lambda & -1 \\ 4 & 3 - \lambda \end{vmatrix} = 0.$$

This leads to the characteristic equation $\lambda^2 - 6\lambda + 5 = 0$, whose roots are $\lambda_1 = 1$, $\lambda_2 = 5$.

Setting $\lambda = \lambda_1 = 1$ in (1), we obtain $2A + B = 0$, $4A + 2B = 0$, a nontrivial solution of which is $A = 1$, $B = -2$. With these values of A, B, and λ, we find the nontrivial solution

$$x = e^t, \ y = -2e^t. \tag{2}$$

Setting $\lambda = \lambda_2 = 5$ in (1), we obtain $-2A + B = 0$, $4A - 2B = 0$, a nontrivial solution of which is $A = 1$, $B = 2$. With these values of A, B, and λ, we find the nontrivial solution

$$x = e^{5t}, \ y = 2e^{5t}. \tag{3}$$

The G.S. of the given system is a linear combination of (2) and (3). Thus we find

$$x = c_1 e^t + c_2 e^{5t}, \quad y = -2c_1 e^t + 2c_2 e^{5t}.$$

6. We assume a solution of the form $x = Ae^{\lambda t}$, $y = Be^{\lambda t}$. Substituting into the given system, we obtain

$$\begin{cases} A\lambda e^{\lambda t} = 6Ae^{\lambda t} - Be^{\lambda t}, \\ B\lambda e^{\lambda t} = 3Ae^{\lambda t} + 2Be^{\lambda t}, \end{cases}$$

which leads at once to the algebraic system

$$\begin{cases} (6 - \lambda)A - B = 0, \\ 3A + (2 - \lambda)B = 0. \end{cases} \qquad (1)$$

For nontrivial solutions of this, we must have

$$\begin{vmatrix} 6 - \lambda & -1 \\ 3 & 2 - \lambda \end{vmatrix} = 0.$$

This leads at once to the characteristic equation $\lambda^2 - 8\lambda + 15 = 0$, whose roots are $\lambda_1 = 5$, $\lambda_2 = 3$.

Setting $\lambda = \lambda_1 = 5$ in (1), we obtain $A - B = 0$, $3A - 3B = 0$, a nontrivial solution of which is $A = 1$, $B = 1$. With these values of A, B, and λ, we find the nontrivial solution

$$x = e^{5t}, \quad y = e^{5t}. \qquad (2)$$

Setting $\lambda = \lambda_2 = 3$ in (1), we obtain $3A - B = 0$
(twice), a nontrivial solution of which is $A = 1$, $B = 3$.
With these values of A, B, and λ, we find the nontrivial
solution

$$x = e^{3t}, \quad y = 3e^{3t} \tag{3}$$

The G.S. of the given system is a linear combination
of (2) and (3). Thus we find

$$x = c_1 e^{5t} + c_2 e^{3t}, \quad y = c_1 e^{5t} + 3c_2 e^{3t}.$$

7. We assume a solution of the form $x = Ae^{\lambda t}$, $y = Be^{\lambda t}$.
Substituting into the given system, we obtain

$$\begin{cases} A\lambda e^{\lambda t} = 3Ae^{\lambda t} - 4Be^{\lambda t}, \\ B\lambda e^{\lambda t} = 2Ae^{\lambda t} - 3Be^{\lambda t}, \end{cases}$$

which leads at once to the algebraic system

$$\begin{cases} (3 - \lambda)A - 4B = 0, \\ 2A + (-3 - \lambda)B = 0. \end{cases} \tag{1}$$

For nontrivial solutions of this, we must have

$$\begin{vmatrix} 3 - \lambda & -4 \\ 2 & -3 - \lambda \end{vmatrix} = 0.$$

This leads to the characteristic equation $\lambda^2 - 1 = 0$,
whose roots are $\lambda_1 = 1$, $\lambda_2 = -1$.

Setting $\lambda = \lambda_1 = 1$ in (1), we obtain $2A - 4B = 0$ (twice), a nontrivial solution of which is $A = 2$, $B = 1$. With these values of A, B, and λ, we find the nontrivial solution

$$x = 2e^t, \ y = e^t. \tag{2}$$

Setting $\lambda = \lambda_2 = -1$ in (1), we obtain $4A - 4B = 0$, $2A - 2B = 0$, a nontrivial solution of which is $A = B = 1$. With these values of A, B, and λ, we find the nontrivial solution

$$x = e^{-t}, \ y = e^{-t}. \tag{3}$$

The G.S. of the given system is a linear combination of (2) and (3). Thus we find

$$x = 2c_1 e^t + c_2 e^{-t}, \ y = c_1 e^t + c_2 e^{-t}.$$

8. We assume a solution of the form $x = Ae^{\lambda t}$, $y = Be^{\lambda t}$. Substituting into the given system, we obtain

$$\begin{cases} A\lambda e^{\lambda t} = 2Ae^{\lambda t} - Be^{\lambda t}, \\ B\lambda e^{\lambda t} = 9Ae^{\lambda t} + 2Be^{\lambda t}, \end{cases}$$

which leads at once to the algebraic system

$$\begin{cases} (2 - \lambda)A - B = 0, \\ 9A + (2 - \lambda)B = 0. \end{cases} \tag{1}$$

For nontrivial solutions of this, we must have

$$\begin{vmatrix} 2 - \lambda & -1 \\ 9 & 2 - \lambda \end{vmatrix} = 0.$$

This leads to the characteristic equation $\lambda^2 - 4\lambda + 13 = 0$, whose roots are $\lambda = 2 \pm 3i$.

Letting $\lambda = 2 + 3i$ in (1), we obtain $-3iA - B = 0$, $9A - 3iB = 0$, a nontrivial solution of which is $A = 1$, $B = -3i$. With these values of A, B, and λ, we find the complex solution

$$\begin{cases} x = e^{(2+3i)t}, \\ y = -3ie^{(2+3i)t}. \end{cases}$$

Using Euler's formula this takes the form

$$\begin{cases} x = e^{2t}(\cos 3t + i \sin 3t), \\ y = e^{2t}(3 \sin 3t - 3i \cos 3t). \end{cases}$$

Both the real and imaginary parts of this are solutions. Thus we obtain the two linearly independent solutions

$$\begin{cases} x = e^{2t}\cos 3t \\ y = 3e^{2t}\sin 3t \end{cases} \quad \text{and} \quad \begin{cases} x = e^{2t}\sin 3t \\ y = -3e^{2t}\cos 3t. \end{cases}$$

The G.S. of the system is thus

$$\begin{cases} x = e^{2t}(c_1 \cos 3t + c_2 \sin 3t), \\ y = e^{2t}(3c_1 \sin 3t - 3c_2 \cos 3t). \end{cases}$$

11. We assume a solution of the form $x = Ae^{\lambda t}$, $y = Be^{\lambda t}$. Substituting into the given system, we obtain

$$\begin{cases} A\lambda e^{\lambda t} = Ae^{\lambda t} - 4Be^{\lambda t}, \\ B\lambda e^{\lambda t} = Ae^{\lambda t} + Be^{\lambda t}, \end{cases}$$

which leads at once to the algebraic system

$$\begin{cases} (1 - \lambda)A - 4B = 0, \\ A + (1 - \lambda)B = 0. \end{cases} \tag{1}$$

For nontrivial solutions of this, we must have

$$\begin{vmatrix} 1 - \lambda & -4 \\ 1 & 1 - \lambda \end{vmatrix} = 0.$$

This leads to the characteristic equation $\lambda^2 - 2\lambda + 5 = 0$, whose roots are $1 \pm 2i$.

Letting $\lambda = 1 + 2i$ in (1), we obtain $-2iA - 4B = 0$, $A - 2iB = 0$, a nontrivial solution of which is $A = 2i$, $B = 1$. With these values of A, B, and λ, we find the complex solution

$$x = 2ie^{(1+2i)t},$$

$$y = e^{(1+2i)t}.$$

Using Euler's formula this takes the form

$$x = 2ie^t(\cos 2t + i \sin 2t),$$

$$y = e^t(\cos 2t + i \sin 2t).$$

Both the real and imaginary parts of this are solutions.
Thus we obtain the two linearly independent solutions

$$\begin{cases} x = -2e^t \sin 2t \\ y = e^t \cos 2t . \end{cases} \quad \text{and} \quad \begin{cases} x = 2e^{2t} \cos 2t \\ y = e^t \sin 2t. \end{cases}.$$

The G.S. of the system is thus

$$\begin{cases} x = 2e^t(-c_1 \sin 2t + c_2 \cos 2t), \\ y = e^t(c_1 \cos 2t + c_2 \sin 2t). \end{cases}$$

15. We assume a solution of the form $x = Ae^{\lambda t}$, $y = Be^{\lambda t}$.
Substituting into the given system, we obtain

$$\begin{cases} A\lambda e^{\lambda t} = 4Ae^{\lambda t} - 2Be^{\lambda t}, \\ B\lambda e^{\lambda t} = 5Ae^{\lambda t} + 2Be^{\lambda t}, \end{cases}$$

which leads at once to the algebraic system

$$\begin{cases} (4 - \lambda)A - 2B = 0, \\ 5A + (2 - \lambda)B = 0. \end{cases} \tag{1}$$

For nontrivial solutions of this, we must have

$$\begin{vmatrix} 4 - \lambda & -2 \\ 5 & 2 - \lambda \end{vmatrix} = 0.$$

This leads to the characteristic equation $\lambda^2 - 6\lambda + 18 = 0$, whose roots are $\lambda = 3 \pm 3i$.

Letting $\lambda = 3 + 3i$ in (1), we obtain $(1 - 3i)A - 2B = 0$, $5A - (1 + 3i)B = 0$, a nontrivial solution of which is $A = 2$, $B = 1 - 3i$. With these values of A, B, and λ, we find the complex solution

$$\begin{cases} x = 2e^{(3+3i)t}, \\ y = (1 - 3i)e^{(3+3i)t}. \end{cases}$$

Using Euler's formula this takes the form

$$\begin{cases} x = 2e^{3t}(\cos 3t + i \sin 3t), \\ y = (1 - 3i)e^{3t}(\cos 3t + i \sin 3t). \end{cases}$$

Both the real and imaginary parts of this are solutions. Thus we obtain the two linearly independent solutions

$$\begin{cases} x = 2e^{3t}\cos 3t \\ y = e^{3t}(\cos 3t + 3 \sin 3t), \end{cases}$$

and

$$\begin{cases} x = 2e^{3t}\sin 3t \\ y = e^{3t}(\sin 3t - 3 \cos 3t). \end{cases}$$

The G.S. of the system is thus

$$\begin{cases} x = 2e^{3t}(c_1 \cos 3t + c_2 \sin 3t), \\ y = e^{3t}[c_1(\cos 3t + 3\sin 3t) + c_2(\sin 3t - 3\cos 3t)]. \end{cases}$$

17.   We assume a solution of the form $x = Ae^{\lambda t}$, $y = Be^{\lambda t}$. Substituting into the given system, we obtain

$$\begin{cases} A\lambda e^{\lambda t} = 3Ae^{\lambda t} - 2Be^{\lambda t}, \\ B\lambda e^{\lambda t} = 2Ae^{\lambda t} + 3Be^{\lambda t}, \end{cases}$$

which leads at once to the algebraic system

$$\begin{cases} (3 - \lambda)A - 2B = 0, \\ 2A + (3 - \lambda)B = 0. \end{cases} \tag{1}$$

For nontrivial solutions, we must have

$$\begin{vmatrix} 3 - \lambda & -2 \\ 2 & 3 - \lambda \end{vmatrix} = 0.$$

This leads to the characteristic equation $\lambda^2 - 6\lambda + 13 = 0$, whose roots are $\lambda = 3 \pm 2i$.

Letting $\lambda = 3 + 2i$ in (1), we obtain $-2iA - 2B = 0$, $2A - 2iB = 0$, a nontrivial solution of which is $A = i$, $B = 1$. With these values of A, B, and λ, we find the complex solution

$$\begin{cases} x = ie^{(3+2i)t}, \\ y = e^{(3+2i)t}. \end{cases}$$

Using Euler's formula this takes the form

$$\begin{cases} x = ie^{3t}(\cos 2t + i \sin 2t), \\ y = e^{3t}(\cos 2t + i \sin 2t). \end{cases}$$

Both the real and imaginary parts of this are solutions. Thus we obtain the two linearly independent solutions

$$\begin{cases} x = -e^{3t} \sin 2t, \\ y = e^{3t} \cos 2t, \end{cases} \quad \text{and} \quad \begin{cases} x = e^{3t} \cos 2t \\ y = e^{3t} \sin 2t. \end{cases}$$

The G.S. of the system is thus

$$\begin{cases} x = e^{3t}(-c_1 \sin 2t + c_2 \cos 2t), \\ y = e^{3t}(c_1 \cos 2t + c_2 \sin 2t). \end{cases}$$

18. We assume a solution of the form $x = Ae^{\lambda t}$, $y = Be^{\lambda t}$. Substituting into the given system, we obtain

$$\begin{cases} A\lambda e^{\lambda t} = 6Ae^{\lambda t} - 5Be^{\lambda t}, \\ B\lambda e^{\lambda t} = Ae^{\lambda t} + 2Be^{\lambda t}, \end{cases}$$

which leads at once to the algebraic system

$$\begin{cases} (6 - \lambda)A - 5B = 0, \\ A + (2 - \lambda)B = 0. \end{cases} \tag{1}$$

For nontrivial solution of this, we must have

$$\begin{vmatrix} 6 - \lambda & -5 \\ 1 & 2 - \lambda \end{vmatrix} = 0.$$

This leads to the characteristic equation $\lambda^2 - 8\lambda + 17 = 0$, whose roots are $\lambda = 4 \pm i$.

Letting $\lambda = 4 + i$ in (1), we obtain $(2 - i)A - 5B = 0$, $A - (2 + i)B = 0$, a nontrivial solution of which is $A = -5$, $B = i - 2$. With these values of A, B, and λ, we find the complex solution

$$\begin{cases} x = -5e^{(4+i)t}, \\ y = (i - 2)e^{(4+i)t}. \end{cases}$$

Using Euler's formula this takes the form

$$\begin{cases} x = -5e^{4t}(\cos t + i \sin t), \\ y = (i - 2)e^{4t}(\cos t + i \sin t). \end{cases}$$

Both the real and imaginary parts of this are solutions. Thus we obtain the two linearly independent solutions

$$\begin{cases} x = -5e^{4t}\cos t, \\ y = e^{4t}(-2 \cos t - \sin t) \end{cases} \quad \text{and} \quad \begin{cases} x = -5e^{4t}\sin t \\ y = e^{4t}(\cos t - 2 \sin t). \end{cases}$$

The G.S. of the system is thus

$$\begin{cases} x = -5e^{4t}(c_1 \cos t + c_2 \sin t), \\ y = e^{4t}[(c_2 - 2c_1)\cos t - (c_1 + 2c_2)\sin t]. \end{cases}$$

19. We assume a solution of the form $x = Ae^{\lambda t}$, $y = Be^{\lambda t}$. Substituting into the given system, we obtain

$$\begin{cases} A\lambda e^{\lambda t} = 3Ae^{\lambda t} - Be^{\lambda t}, \\ B\lambda e^{\lambda t} = 4Ae^{\lambda t} - Be^{\lambda t}, \end{cases}$$

which leads at once to the algebraic system

$$\begin{cases} (3 - \lambda)A - B = 0 \\ 4A + (-1 - \lambda)B = 0. \end{cases} \tag{1}$$

For nontrivial solutions of this, we must have

$$\begin{vmatrix} 3 - \lambda & -1 \\ 4 & -1 - \lambda \end{vmatrix} = 0.$$

This leads to the characteristic equation $\lambda^2 - 2\lambda + 1 = 0$, whose roots are 1, 1 (real and equal).

Setting $\lambda = 1$ in (1), we obtain $2A - B = 0$, $4A - 2B = 0$, a nontrivial solution of which is $A = 1$, $B = 2$. With these values of A, B, and λ, we find the nontrivial solution

$$x = e^t, \quad y = 2e^t. \tag{2}$$

We now seek a second solution of the form $x = (A_1 t + A_2)e^t$, $y = (B_1 t + B_2)e^t$. Substituting these into the given system, we obtain

$$\begin{cases} (A_1 t + A_2 + A_1)e^t = 3(A_1 t + A_2)e^t - (B_1 t + B_2)e^t, \\ (B_1 t + B_2 + B_1)e^t = 4(A_1 t + A_2)e^t - (B_1 t + B_2)e^t. \end{cases}$$

These equations reduce to

$$\begin{cases} (2A_1 - B_1)t + (2A_2 - A_1 - B_2) = 0, \\ (4A_1 - 2B_1)t + (4A_2 - B_1 - 2B_2) = 0. \end{cases}$$

Thus we must have

$$\begin{cases} 2A_1 - B_1 = 0, & 2A_2 - A_1 - B_2 = 0, \\ 4A_1 - 2B_1 = 0, & 4A_2 - B_1 - 2B_2 = 0. \end{cases}$$

A simple nontrivial solution of these is $A_1 = 1$, $B_1 = 2$, $A_2 = 0$, $B_2 = -1$. This leads to the solution

$$x = te^t, \quad y = (2t - 1)e^t. \tag{3}$$

The G.S. of the given system is a linear combination of (2) and (3). Thus we find

$$x = c_1 e^t + c_2 t e^t, \quad y = 2c_1 e^t + c_2(2t - 1)e^t.$$

20. We assume a solution of the form $x = Ae^{\lambda t}$, $y = Be^{\lambda t}$.
Substituting into the given system, we obtain

$$
\begin{cases}
A\lambda e^{\lambda t} = 7Ae^{\lambda t} + 4Be^{\lambda t}, \\
B\lambda e^{\lambda t} = -Ae^{\lambda t} + 3Be^{\lambda t},
\end{cases}
$$

which leads at once to the algebraic system

$$
\begin{cases}
(7 - \lambda)A + 4B = 0, \\
-A + (3 - \lambda)B = 0.
\end{cases}
\tag{1}
$$

For nontrivial solutions of this, we must have

$$
\begin{vmatrix}
7 - \lambda & 4 \\
-1 & 3 - \lambda
\end{vmatrix}
= 0.
$$

This leads to the characteristic equation $\lambda^2 - 10\lambda + 25 = 0$, whose roots are $\lambda = 5, 5$ (real and equal).

Setting $\lambda = 5$ in (1), we obtain $2A + 4B = 0$, $-A - 2B = 0$, a nontrivial solution of which is $A = 2$, $B = -1$. With these values of A, B, and λ, we find the nontrivial solution

$$
x = 2e^{5t}, \quad y = -e^{5t}.
\tag{2}
$$

We now seek a second solution of the form $x = (A_1 t + A_2)e^{5t}$, $y = (B_1 t + B_2)e^{5t}$. Substituting these into the given system, we obtain

$$\begin{cases} (5A_1t + 5A_2 + A_1)e^{5t} = 7(A_1t + A_2)e^{5t} + 4(B_1t + B_2)e^{5t}, \\ (5B_1t + 5B_2 + B_1)e^{5t} = -(A_1t + A_2)e^{5t} + 3(B_1t + B_2)e^{5t}. \end{cases}$$

These equations reduce to

$$\begin{cases} (2A_1 + 4B_1)t + (2A_2 + 4B_2 - A_1) = 0, \\ (A_1 + 2B_1)t + (A_2 + 2B_2 + B_1) = 0. \end{cases}$$

Thus we must have

$$\begin{cases} 2A_1 + 4B_1 = 0, \\ A_1 + 2B_1 = 0, \end{cases} \qquad \begin{cases} 2A_2 + 4B_2 - A_1 = 0, \\ A_2 + 2B_2 + B_1 = 0. \end{cases}$$

A simple nontrivial solution of this is $A_1 = 2$, $B_1 = -1$, $A_2 = 1$, $B_2 = 0$. This leads to the solution

$$x = (2t + 1)e^{5t}, \quad y = -t\, e^{5t}. \tag{3}$$

The G.S. of the given system is a linear combination of (2) and (3). Thus we find the G.S. in the form

$$\begin{cases} x = 2c_1 e^{5t} + c_2(2t + 1)e^{5t}, \\ y = -c_1 e^{5t} - c_2 t\, e^{5t}. \end{cases}$$

22. We assume a solution of the form $x = Ae^{\lambda t}$, $y = Be^{\lambda t}$.
Substituting into the given system, we obtain

$$\begin{cases} A\lambda e^{\lambda t} = Ae^{\lambda t} - 2Be^{\lambda t}, \\ B\lambda e^{\lambda t} = 2Ae^{\lambda t} - 3Be^{\lambda t}, \end{cases}$$

which leads at once to the algebraic system

$$\begin{cases} (1 - \lambda)A - 2B = 0, \\ 2A + (-3 - \lambda)B = 0. \end{cases} \tag{1}$$

For nontrivial solutions of this, we must have

$$\begin{vmatrix} 1 - \lambda & -2 \\ 2 & -3 - \lambda \end{vmatrix} = 0.$$

This leads to the characteristic equation $\lambda^2 + 2\lambda + 1 = 0$,
whose roots are -1, -1 (real and equal).

   Setting $\lambda = -1$ in (1), we obtain $2A - 2B = 0$ (twice),
a nontrivial solution of which is $A = 1$, $B = 1$. With
these values of A, B, and λ, we find the nontrivial
solution

$$x = e^{-t}, \quad y = e^{-t}. \tag{2}$$

   We now seek a second solution of the form $x = (A_1 t + A_2)e^{-t}$, $y = (B_1 t + B_2)e^{-t}$. Substituting these into
the given system, we obtain

$$\begin{cases} (-A_1 t - A_2 + A_1)e^{-t} = (A_1 t + A_2)e^{-t} - 2(B_1 t + B_2)e^{-t}, \\[2mm] (-B_1 t - B_2 + B_1)e^{-t} = 2(A_1 t + A_2)e^{-t} - 3(B_1 t + B_2)e^{-t}. \end{cases}$$

These equations reduce to

$$\begin{cases} (2A_1 - 2B_1)t + (2A_2 - 2B_2 - A_1) = 0, \\[2mm] (2A_1 - 2B_1)t + (2A_2 - 2B_2 - B_1) = 0. \end{cases}$$

Thus we must have

$$\begin{cases} 2A_1 - 2B_1 = 0, \quad 2A_2 - 2B_2 - A_1 = 0, \\[2mm] 2A_1 - 2B_1 = 0, \quad 2A_2 - 2B_2 - B_1 = 0. \end{cases}$$

A simple nontrivial solution of these is $A_1 = 2$, $B_1 = 2$, $A_2 = 1$, $B_2 = 0$. This leads to the solution

$$x = (2t + 1)e^{-t}, \; y = 2te^{-t}. \tag{3}$$

The G.S. of the given system is a linear combination of (2) and (3). Thus we find

$$x = c_1 e^{-t} + c_2(2t + 1)e^{-t}, \; y = c_1 e^{-t} + 2c_2 te^{-t}.$$

23. We assume a solution of the form $x = Ae^{\lambda t}$, $y = Be^{\lambda t}$. Substituting into the given system, we obtain

$$\begin{cases} A\lambda e^{\lambda t} = 3Ae^{\lambda t} - Be^{\lambda t}, \\[2mm] B\lambda e^{\lambda t} = Ae^{\lambda t} + Be^{\lambda t}, \end{cases}$$

which leads at once to the algebraic system

$$\begin{cases} (3 - \lambda)A - B = 0, \\ A + (1 - \lambda)B = 0. \end{cases} \qquad (1)$$

For nontrivial solutions of this, we must have

$$\begin{vmatrix} 3 - \lambda & -1 \\ 1 & 1 - \lambda \end{vmatrix} = 0.$$

This leads to the characteristic equation $\lambda^2 - 4\lambda + 4 = 0$, whose roots are 2, 2 (real and equal).

Setting $\lambda = 2$ in (1), we obtain $A - B = 0$, (twice), a nontrivial solution of which is $A = B = 1$. With these values of A, B, and λ, we find the nontrivial solution

$$x = e^{2t}, \quad y = e^{2t}. \qquad (2)$$

We now seek a second solution of the form $x = (A_1 t + A_2)e^{2t}$, $y = (B_1 t + B_2)e^{2t}$. Substituting these into the given system, we obtain

$$\begin{cases} (2A_1 t + 2A_2 + A_1)e^{2t} = 3(A_1 t + A_2)e^{2t} - (B_1 t + B_2)e^{2t}, \\ (2B_1 t + 2B_2 + B_1)e^{2t} = (A_1 t + A_2)e^{2t} + (B_1 t + B_2)e^{2t}. \end{cases}$$

These equations reduce to

$$\begin{cases} (A_1 - B_1)t + (A_2 - B_2 - A_1) = 0, \\ (A_1 - B_1)t + (A_2 - B_2 - B_1) = 0. \end{cases}$$

Thus we must have

$$\begin{cases} A_1 - B_1 = 0, \\ A_1 - B_1 = 0, \end{cases} \qquad \begin{cases} A_2 - B_2 - A_1 = 0, \\ A_2 - B_2 - B_1 = 0. \end{cases}$$

A simple nontrivial solution of this is $A_1 = B_1 = A_2 = 1$, $B_2 = 0$. This leads to the solution

$$x = (t + 1)e^{2t}, \quad y = te^{2t}. \tag{3}$$

The G.S. of the given system is a linear combination of (2) and (3). Thus we find the G.S. in the form

$$\begin{cases} x = c_1 e^{2t} + c_2(t + 1)e^{2t}, \\ y = c_1 e^{2t} + c_2 te^{2t}. \end{cases}$$

26. We assume a solution of the form $x = Ae^{\lambda t}$, $y = Be^{\lambda t}$. Substituting into the given system, we obtain

$$\begin{cases} A\lambda e^{\lambda t} = 2Ae^{\lambda t} - 4Be^{\lambda t}, \\ B\lambda e^{\lambda t} = Ae^{\lambda t} - 2Be^{\lambda t}, \end{cases}$$

which leads at once to the algebraic system

$$\begin{cases} (2 - \lambda)A - 4B = 0, \\ A + (-2 - \lambda)B = 0. \end{cases} \tag{1}$$

For nontrivial solutions of this, we must have

$$\begin{vmatrix} 2 - \lambda & -4 \\ 1 & -2 - \lambda \end{vmatrix} = 0.$$

This leads to the characteristic equation $\lambda^2 = 0$, whose roots are $\lambda = 0, 0$ (real and equal).

Setting $\lambda = 0$ in (1), we obtain $2A - 4B = 0$, $A - 2B = 0$, a nontrivial solution of which is $A = 2$, $B = 1$. With these values of A, B, and λ, we find the nontrivial solution

$$x = 2e^{0t} = 2, \quad y = e^{0t} = 1. \tag{2}$$

We now seek a second solution of the form $x = (A_1 t + A_2)e^{0t} = A_1 t + A_2$, $y = (B_1 t + B_2)e^{0t} = B_1 t + B_2$. Substituting these into the given system, we obtain

$$\begin{cases} A_1 = 2(A_1 t + A_2) - 4(B_1 t + B_2), \\ B_1 = (A_1 t + A_1) - 2(B_1 t + B_1). \end{cases}$$

These equations reduce to

$$\begin{cases} (2A_1 - 4B_1)t + (2A_2 - 4B_2 - A_1) = 0, \\ (A_1 - 2B_1)t + (A_2 - 2B_2 - B_1) = 0. \end{cases}$$

Thus we must have

$$\begin{cases} 2A_1 - 4B_1 = 0, \\ A_1 - 2B_1 = 0, \end{cases} \qquad \begin{cases} 2A_2 - 4B_2 - A_1 = 0, \\ A_2 - 2B_2 - B_1 = 0. \end{cases}$$

A simple nontrivial solution of this is $A_1 = 2$, $B_1 = 1$, $A_2 = 1$, $B_2 = 0$. This leads to the solution

$$x = 2t + 1, \ y = t. \tag{3}$$

The G.S. of the given system is a linear combination of (2) and (3). Thus we find

$$\begin{cases} x = 2c_1 + c_2(2t + 1), \\ y = c_1 + c_2 t. \end{cases} \tag{4}$$

Here is an alternative way to solve the given system. From the given D.E.'s themselves, we see that $x' = 2y'$. Then integrating, $x = 2y + k_1$, where k_1 is an arbitrary constant. Now substitute $x = 2y + k_1$ into the second given D.E. $y' = x - 2y$. We obtain $y' = 2y + k_1 - 2y = k_1$. Integrating this, we obtain $y = k_1 t + k_2$, where k_2 is a second arbitrary constant. Then since $x = 2y + k_1$, we have $x = 2k_1 t + 2k_2 + k_1$. Thus we have found

$$\begin{cases} x = 2k_1 t + 2k_2 + k_1, \\ y = k_1 t + k_2, \end{cases}$$

which is G.S. (4).

27. We assume a solution of the form $x = Ae^{\lambda t}$, $y = Be^{\lambda t}$.
Substituting into the given system, we obtain

$$\begin{cases} A\lambda e^{\lambda t} = -2Ae^{\lambda t} + 7Be^{\lambda t}, \\ B\lambda e^{\lambda t} = 3Ae^{\lambda t} + 2Be^{\lambda t}, \end{cases}$$

which leads at once to the algebraic system

$$\begin{cases} (-2 - \lambda)A + 7B = 0, \\ 3A + (2 - \lambda)B = 0. \end{cases} \tag{1}$$

For nontrivial solutions of this, we must have

$$\begin{vmatrix} -2 - \lambda & 7 \\ 3 & 2 - \lambda \end{vmatrix} = 0.$$

This leads to the characteristic equation $\lambda^2 = 25$, whose
roots are $\lambda_1 = 5$, $\lambda_2 = -5$.

Setting $\lambda = \lambda_1 = 5$ in (1), we obtain $-7A + 7B = 0$,

$3A - 3B = 0$, a nontrivial solution of which is $A = 1$, $B = 1$. With these values of A, B, and λ, we find the
nontrivial solution

$$x = e^{5t}, \quad y = e^{5t}. \tag{2}$$

Setting $\lambda = \lambda_2 = -5$ in (1), we obtain $3A + 7B = 0$

(twice), a nontrivial solution of which is $A = 7$, $B = -3$.

With these values of A, B, and λ, we find the nontrivial solution

$$x = 7e^{-5t}, \ y = -3e^{-5t}. \tag{3}$$

The G.S. of the given system is a linear combination of (2) and (3). Thus we find

$$x = c_1e^{5t} + 7c_2e^{-5t}, \ y = c_1e^{5t} - 3c_2e^{-5t}. \tag{4}$$

We apply the I.C.'s $x(0) = 9$, $y(0) = -1$ to (4), obtaining $c_1 + 7c_2 = 9$, $c_1 - 3c_2 = -1$. The solution of this is $c_1 = 2$, $c_2 = 1$. Substituting these values back into (4), we get the desired particular solution

$$x = 2e^{5t} + 7e^{-5t}, \ y = 2e^{5t} - 3e^{-5t}.$$

29. We assume a solution of the form $x = Ae^{\lambda t}$, $y = Be^{\lambda t}$. Substituting into the given system, we obtain

$$\begin{cases} A\lambda e^{\lambda t} = 2Ae^{\lambda t} - 8Be^{\lambda t}, \\ B\lambda e^{\lambda t} = Ae^{\lambda t} + 6Be^{\lambda t}, \end{cases}$$

which leads at once to the algebraic system

$$\begin{cases} (2 - \lambda)A - 8B = 0, \\ A + (6 - \lambda)B = 0. \end{cases} \tag{1}$$

For nontrivial solutions of this, we must have

$$
\begin{vmatrix} 2 - \lambda & -8 \\ 1 & 6 - \lambda \end{vmatrix} = 0.
$$

This leads to the characteristic equation $\lambda^2 - 8\lambda + 20 = 0$, whose roots are $4 \pm 2i$.

Letting $\lambda = 4 + 2i$ in (1), we obtain $(2 - 2i)A - 8B = 0$, $A + (2 - 2i)B = 0$, a nontrivial solution of which is $A = 2(i - 1)$, $B = 1$. With these values of A, B, and λ, we find the complex solution

$$
\begin{cases} x = 2(i - 1)e^{(4+2i)t} \\ y = e^{(4+2i)t}. \end{cases}
$$

Using Euler's formula this takes the form

$$
\begin{cases} x = 2(i - 1)e^{4t}(\cos 2t + i \sin 2t), \\ y = e^{4t}(\cos 2t + i \sin 2t). \end{cases}
$$

Both the real and imaginary parts of this are solutions. Thus we obtain the two linearly independent solutions

$$
\begin{cases} x = -2e^{4t}(\cos 2t + \sin 2t) \\ y = e^{4t}\cos 2t, \end{cases}
$$

and

$$
\begin{cases} x = 2e^{4t}(\cos 2t - \sin 2t), \\ y = e^{4t}\sin 2t. \end{cases}
$$

The G.S. of the system is thus

$$\begin{cases} x = 2e^{4t}[(-c_1 + c_2)\cos 2t - (c_1 + c_2)\sin 2t], \\ y = e^{4t}(c_1 \cos 2t + c_2 \sin 2t). \end{cases} \quad (2)$$

We apply the I.C.'s $x(0) = 4$, $y(0) = 1$ to (2), obtaining $2(-c_1 + c_2) = 4$, $c_1 = 1$. Thus $c_1 = 1$, $c_2 = 3$.

Substituting back into (2), we get the desired particular solution

$$\begin{cases} x = 4e^{4t}(\cos 2t - 2 \sin 2t), \\ y = e^{4t}(\cos 2t + 3 \sin 2t). \end{cases}$$

31. We assume a solution of the form $x = Ae^{\lambda t}$, $y = Be^{\lambda t}$. Substituting into the given system, we obtain

$$\begin{cases} A\lambda e^{\lambda t} = 6Ae^{\lambda t} - 4Be^{\lambda t}, \\ B\lambda e^{\lambda t} = Ae^{\lambda t} + 2Be^{\lambda t}, \end{cases}$$

which leads at once to the algebraic system

$$\begin{cases} (6 - \lambda)A - 4B = 0, \\ A + (2 - \lambda)B = 0. \end{cases} \quad (1)$$

For nontrivial solutions of this, we must have

$$\begin{vmatrix} 6 - \lambda & -4 \\ 1 & 2 - \lambda \end{vmatrix} = 0.$$

This leads to the characteristic equation $\lambda^2 - 8\lambda + 16 = 0$, whose roots are 4, 4 (real and equal).

Setting $\lambda = 4$ in (1), we obtain $2A - 4B = 0$, $A - 2B = 0$, a nontrivial solution of which is $A = 2$, $B = 1$. With these values of A, B, and λ, we find the nontrivial solution

$$x = 2e^{4t}, \quad y = e^{4t}. \tag{2}$$

We now seek a second solution of the form $x = (A_1 t + A_2)e^{4t}$, $y = (B_1 t + B_2)e^{4t}$. Substituting these into the given system, we obtain

$$\begin{cases} (4A_1 t + 4A_2 + A_1)e^{4t} = 6(A_1 t + A_2)e^{4t} - 4(B_1 t + B_2)e^{4t}, \\ (4B_1 t + 4B_2 + B_1)e^{4t} = (A_1 t + A_2)e^{4t} + 2(B_1 t + B_2)e^{4t}. \end{cases}$$

These equations reduce to

$$\begin{cases} (2A_1 - 4B_1)t + (2A_2 - 4B_2 - A_1) = 0, \\ (A_1 - 2B_1)t + (A_2 - 2B_2 - B_1) = 0. \end{cases}$$

Thus we must have

$$\begin{cases} 2A_1 - 4B_1 = 0, \\ A_1 - 2B_1 = 0, \end{cases} \qquad \begin{cases} 2A_2 - 4B_2 - A_1 = 0, \\ A_2 - 2B_2 - B_1 = 0. \end{cases}$$

A simple nontrivial solution of this is $A_1 = 2$, $B_1 = 1$, $A_2 = 1$, $B_2 = 0$. This leads to the solution

$$x = (2t + 1)e^{4t}, \quad y = te^{4t}. \tag{3}$$

The G.S. of the given system is a linear combination of (2) and (3). Thus we find the G.S. in the form

$$x = 2c_1 e^{4t} + c_2(2t + 1)e^{4t}, \quad y = c_1 e^{4t} + c_2 te^{4t}. \tag{4}$$

We apply the I.C.'s $x(0) = 2$, $y(0) = 3$ to (4), obtaining $2c_1 + c_2 = 2$, $c_1 + 0 = 3$. Thus $c_1 = 3$, $c_2 = -4$. Substituting these values back into (4), we get the desired particular solution

$$x = 6e^{4t} - 4(2t + 1)e^{4t}, \quad y = 3e^{4t} - 4te^{4t};$$

or

$$x = 2e^{4t} - 8te^{4t}, \quad y = 3e^{4t} - 4te^{4t}.$$

34. We assume a solution of the form $x = Ae^{\lambda t}$, $y = Be^{\lambda t}$. Substituting into the given system, we obtain

$$\begin{cases} A\lambda e^{\lambda t} = Ae^{\lambda t} - 2Be^{\lambda t}, \\ B\lambda e^{\lambda t} = 8Ae^{\lambda t} - 7Be^{\lambda t}, \end{cases}$$

which leads at once to the algebraic system

$$\begin{cases} (1 - \lambda)A - 2B = 0, \\ 8A + (-7 - \lambda)B = 0. \end{cases} \tag{1}$$

For nontrivial solutions of this, we must have

$$\begin{vmatrix} 1 - \lambda & -2 \\ 8 & -7 - \lambda \end{vmatrix} = 0.$$

This leads to the characteristic equation $\lambda^2 + 6\lambda + 9 = 0$, that is, $(\lambda + 3)^2 = 0$, with roots $\lambda = -3, -3$ (real and equal).

Setting $\lambda = -3$ in (1), we obtain $4A - 2B = 0$, $8A - 4B = 0$, a nontrivial solution of which is $A = 1$, $B = 2$. With these values of A, B, and λ, we find the nontrivial solution

$$x = e^{-3t}, \quad y = 2e^{-3t}. \tag{2}$$

We now seek a second solution of the form $x = (A_1 t + A_2)e^{-3t}$, $y = (B_1 t + B_2)e^{-3t}$. Substituting these into the given system, we obtain

$$\begin{cases} (-3A_1 t - 3A_2 + A_1)e^{-3t} = \\ (A_1 t + A_2)e^{-3t} - 2(B_1 t + B_2)e^{-3t}, \end{cases}$$

$$\begin{cases} (-3B_1 t - 3B_2 + B_1)e^{-3t} = \\ 8(A_1 t + A_2)e^{-3t} - 7(B_1 t + B_2)e^{-3t}. \end{cases}$$

These equations reduce to

$$\begin{cases} (4A_1 - 2B_1)t + (4A_2 - 2B_2 - A_1) = 0, \\ (8A_1 - 4B_1)t + (8A_2 - 4B_2 - B_1) = 0. \end{cases}$$

Thus we must have

$$\begin{cases} 4A_1 - 2B_1 = 0, \\ 8A_1 - 4B_1 = 0, \end{cases} \qquad \begin{cases} 4A_2 - 2B_2 - A_1 = 0, \\ 8A_2 - 4B_2 - B_1 = 0. \end{cases}$$

A simple nontrivial solution of this is $A_1 = 1$, $B_1 = 2$, $A_2 = 1/4$, $B_2 = 0$. This leads to the solution

$$x = (t + 1/4)e^{-3t}, \quad y = 2te^{-3t}. \tag{3}$$

The G.S. of the given system is a linear combination of (2) and (3). Thus we find the G.S.

$$\begin{cases} x = c_1 e^{-3t} + c_2(t + 1/4)e^{-3t}, \\ \\ y = 2c_1 e^{-3t} + 2c_2 te^{-3t}. \end{cases} \tag{4}$$

We apply the I.C.'s $x(0) = 6$, $y(0) = 8$ to (4), obtaining $c_1 + \frac{1}{4}c_2 = 6$, $2c_1 = 8$. Thus $c_1 = 4$, $c_2 = 8$.
Substituting these values back into (4), we get the desired particular solution

$$x = 4e^{-3t} + 8(t + 1/4)e^{-3t}, \quad y = 8e^{-3t} + 16te^{-3t}$$

that is,

$$\begin{cases} x = (8t + 6)e^{-3t}, \\ y = (16t + 8)e^{-3t}. \end{cases}$$

36. We let $t = e^w$, where we assume $t > 0$. Then $w = \ln t$, and $\dfrac{dx}{dt} = \left(\dfrac{dx}{dw}\right)\left(\dfrac{dw}{dt}\right) = \left(\dfrac{1}{t}\right)\dfrac{dx}{dw}$. Thus $t\dfrac{dx}{dt} = \dfrac{dx}{dw}$; and similarily $t\dfrac{dy}{dt} = \dfrac{dy}{dw}$. Substituting into the original system, we obtain the constant coefficient system

$$\begin{cases} \dfrac{dx}{dw} = x + y, \\[2mm] \dfrac{dy}{dw} = -3x + 5y. \end{cases} \tag{1}$$

We assume a solution of (1) of the form $x = Ae^{\lambda w}$, $y = Be^{\lambda w}$. Substituting this into (1), we obtain

$$\begin{cases} A\lambda e^{\lambda w} = Ae^{\lambda w} + Be^{\lambda w}, \\ B\lambda e^{\lambda w} = -3Ae^{\lambda w} + 5Be^{\lambda w}, \end{cases}$$

which leads at once to the algebraic system

$$\begin{cases} (1 - \lambda)A + B = 0, \\ -3A + (5 - \lambda)B = 0. \end{cases} \tag{2}$$

For nontrivial solutions of this, we must have

$$\begin{vmatrix} 1 - \lambda & 1 \\ -3 & 5 - \lambda \end{vmatrix} = 0.$$

This leads to the characteristic equation $\lambda^2 - 6\lambda + 8 = 0$, whose roots are $\lambda_1 = 2$, $\lambda_2 = 4$.

Setting $\lambda = \lambda_1 = 2$ in (2), we obtain $-A + B = 0$, $-3A + 3B = 0$, a nontrivial solution of which is $A = 1$, $B = 1$. With these values of A, B, and λ, we find the nontrivial solution of (1),

$$x = e^{2w}, \quad y = e^{2w}.$$

Setting $\lambda = \lambda_2 = 4$ in (2), we obtain $-3A + B = 0$ (twice), a nontrivial solution of which is $A = 1$, $B = 3$. With these values of A, B, and λ, we find the nontrivial solution of (1),

$$x = e^{4w}, \quad y = 3e^{4w}.$$

The G.S. of (1) is thus $x = c_1 e^{2w} + c_2 e^{4w}$, $y = c_1 e^{2w} + 3c_2 e^{4w}$. We must now return to the original variable t. Since $t = e^w$, we see that the G.S. of the original system is

$$x = c_1 t^2 + c_2 t^4, \quad y = c_1 t^2 + 3c_2 t^4.$$

38. The characteristic equation of the given linear system (S) is $\lambda^2 - (a_1 + b_2)\lambda + (a_1 b_2 - a_2 b_1) = 0$, and its roots are given by $\lambda = \dfrac{a_1 + b_2 \pm \sqrt{(a_1 - b_2)^2 + 4a_2 b_1}}{2}$.

Observe that if $(a_1 - b_2)^2 + 4a_2b_1 > 0$, then

$\sqrt{(a_1 - b_2)^2 + 4a_2b_1}$ is real and positive, so the roots λ_1

and λ_2 of the characteristic equation are real and

distant.   Then by Theorem 7.7, the system has two real

linearly independent solutions of the stated form.

Now note that if $a_2b_1 > 0$, then $(a_1 - b_2)^2 + 4a_2b_1 > 0$

and the reasoning of the previous paragraph follows.   Thus

$a_2'b_1 > 0$ is sufficient for the system to have two

solutions of the stated type and form.

But now observe that there exist systems for which

$a_2b_1 \leq 0$, but $(a_1 - b_2)^2 + 4a_2b_1 > 0$.   For example,

consider the system (S) in which $a_1 = 3$, $b_1 = -1$, $a_2 = 1$,

$b_2 = 0$.   Then $a_2b_1 = -1 < 0$, but $(a_1 - b_2)^2 + 4a_2b_1 = 5 >$

0.   Since $(a_1 - b_2)^2 + 4a_2b_1 > 0$, by the reasoning of the

first paragraph, the system (S) has two solutions of the

stated type and form.   But for the system (S), $a_2b_1 > 0$

does <u>not</u> hold.   Thus the condition $a_2b_1 > 0$ is not

necessary for a system to have two solutions of the stated

type and form.

40.   <u>Suppose</u> there exists a <u>nontrivial</u> solution of the form

$$x = A t e^{\lambda t}, \quad y = B t e^{\lambda t}. \tag{1}$$

If indeed (1) is a solution of the given system, then it

must satisfy the equations of the system.   So,

differentiating (1) and substituting into the system, we must have

$$\begin{cases} (A\lambda t + A)e^{\lambda t} = a_1 \, A \, t \, e^{\lambda t} + b_1 \, B \, t \, e^{\lambda t}, \\[2mm] (B\lambda t + B)e^{\lambda t} = a_2 \, A \, t \, e^{\lambda t} + b_2 \, B \, t \, e^{\lambda t}. \end{cases}$$

These equations quickly reduce to

$$\begin{cases} [(a_1 - \lambda)A + b_1 B]t - A = 0, \\[2mm] [a_2 A + (b_2 - \lambda)B]t - B = 0. \end{cases}$$

Thus we must have

$$\begin{cases} (a_1 - \lambda)A + b_1 B = 0, \\[2mm] a_2 A + (b_2 - \lambda)B = 0, \end{cases} \qquad \begin{cases} A = 0, \\[2mm] B = 0. \end{cases}$$

To satisfy these we must have $A = B = 0$, which means that (1) must be the _trivial_ solution. This is a contradiction. Thus our supposition is invalid, so there exists _no_ nontrivial solution of the stated form.

Section 7.5 A, Page 340.

Here and following, we shall indicate vectors and matrices by placing a bar over the corresponding letter. Thus, for example, we write $\bar{A}$, $\bar{v}$, etc.

1. (b) $\bar{A} + \bar{B} = \begin{pmatrix} 2 + 7 & 1 - 1 & 3 + 6 \\ -1 + 2 & 0 + 4 & 5 - 3 \\ -4 + 5 & 3 - 5 & -2 + 1 \end{pmatrix} = \begin{pmatrix} 9 & 0 & 9 \\ 1 & 4 & 2 \\ 1 & -2 & -1 \end{pmatrix}.$

(c) $\bar{A} + \bar{B} = \begin{pmatrix} -5+7 & 0-2 & 4-3 \\ -2+6 & -1-3 & -3+1 \\ 6-2 & 2+1 & 5-3 \end{pmatrix} = \begin{pmatrix} 2 & -2 & 1 \\ 4 & -4 & -2 \\ 4 & 3 & 2 \end{pmatrix}.$

2. (b) $-4\bar{A} = \begin{pmatrix} -4(1) & -4(-3) & -4(5) \\ -4(6) & -4(-2) & -4(0) \\ -4(3) & -4(1) & -4(2) \end{pmatrix} = \begin{pmatrix} -4 & 12 & -20 \\ -24 & 8 & 0 \\ 12 & -4 & -8 \end{pmatrix}.$

(c) $-3\bar{A} = \begin{pmatrix} -3(5) & -3(-1) & -3(2) \\ -3(4) & -3(-3) & -3(-2) \\ -3(0) & -3(3) & -3(-6) \end{pmatrix} = \begin{pmatrix} -15 & 3 & -6 \\ -12 & 9 & 6 \\ 0 & -9 & 18 \end{pmatrix}.$

3. (b) We have

$$3\bar{x}_1 - 2\bar{x}_2 + 4\bar{x}_4 = 3\begin{bmatrix} 3 \\ 2 \\ -1 \\ 4 \end{bmatrix} - 2\begin{bmatrix} -1 \\ 3 \\ 5 \\ -2 \end{bmatrix} + 4\begin{bmatrix} -1 \\ 2 \\ -3 \\ 5 \end{bmatrix}$$

$$= \begin{bmatrix} 9 \\ 6 \\ -3 \\ 12 \end{bmatrix} + \begin{bmatrix} 2 \\ -6 \\ -10 \\ 4 \end{bmatrix} + \begin{bmatrix} -4 \\ 8 \\ -12 \\ 20 \end{bmatrix} = \begin{bmatrix} 7 \\ 8 \\ -25 \\ 36 \end{bmatrix}.$$

(c) We have

$$-\bar{x}_1 + 5\bar{x}_2 - 2\bar{x}_3 + 3\bar{x}_4 = -\begin{bmatrix} 3 \\ 2 \\ -1 \\ 4 \end{bmatrix} + 5\begin{bmatrix} -1 \\ 3 \\ 5 \\ -2 \end{bmatrix} - 2\begin{bmatrix} 2 \\ 4 \\ 0 \\ 6 \end{bmatrix} + 3\begin{bmatrix} -1 \\ 2 \\ -3 \\ 5 \end{bmatrix}$$

$$= \begin{bmatrix} -3 \\ -2 \\ 1 \\ -4 \end{bmatrix} + \begin{bmatrix} -5 \\ 15 \\ 25 \\ -10 \end{bmatrix} + \begin{bmatrix} -4 \\ -8 \\ 0 \\ -12 \end{bmatrix} + \begin{bmatrix} -3 \\ 6 \\ -9 \\ 15 \end{bmatrix} = \begin{bmatrix} -15 \\ 11 \\ 17 \\ -11 \end{bmatrix}.$$

4. (b) We have

$$\bar{A}\,\bar{x} = \begin{pmatrix} -3 & -5 & 7 \\ 0 & 4 & 1 \\ -2 & 1 & 3 \end{pmatrix} \begin{pmatrix} 2 \\ 3 \\ -2 \end{pmatrix}$$

$$= \begin{pmatrix} -3(2) - 5(3) + 7(-2) \\ 0(2) + 4(3) + 1(-2) \\ -2(2) + 1(3) + 3(-2) \end{pmatrix} = \begin{pmatrix} -35 \\ 10 \\ -7 \end{pmatrix}.$$

(c) We have

$$\bar{A}\,\bar{x} = \begin{pmatrix} 1 & 0 & -3 \\ 2 & -5 & 4 \\ -3 & 1 & 2 \end{pmatrix} \begin{bmatrix} x_1 + x_2 \\ x_1 + 2x_2 \\ x_2 - x_3 \end{bmatrix}$$

$$= \begin{bmatrix} (x_1 + x_2) + 0 - 3(x_2 - x_3) \\ 2(x_1 + x_2) - 5(x_1 + 2x_2) + 4(x_2 - x_3) \\ 3(x_1 + x_2) + (x_1 + 2x_2) + 2(x_2 - x_3) \end{bmatrix}$$

$$= \begin{bmatrix} x_1 - 2x_2 + 3x_3 \\ -3x_1 - 4x_2 - 4x_3 \\ -2x_1 + x_2 - 2x_3 \end{bmatrix}.$$

5. We have

$$\bar{A} = \begin{pmatrix} 3 & -1 & 2 \\ 5 & 4 & -3 \\ -5 & 1 & 2 \end{pmatrix}, \ \bar{x} = \begin{bmatrix} x_1 \\ x_2 \\ x_3 \end{bmatrix}, \ \bar{y} = \begin{bmatrix} y_1 \\ y_2 \\ y_3 \end{bmatrix}, \ c = 4. \quad \text{Then}$$

$$\bar{A}(\bar{x} + \bar{y}) = \begin{pmatrix} 3 & -1 & 2 \\ 5 & 4 & -3 \\ -5 & 1 & 2 \end{pmatrix} \begin{bmatrix} x_1 + y_1 \\ x_2 + y_2 \\ x_3 + y_3 \end{bmatrix}$$

$$= \begin{bmatrix} 3(x_1 + y_1) + (-1)(x_2 + y_2) + 2(x_3 + y_3) \\ 5(x_1 + y_1) + 4(x_2 + y_2) + (-3)(x_3 + y_3) \\ (-5)(x_1 + y_1) + 1(x_2 + y_2) + 2(x_3 + y_3) \end{bmatrix}$$

$$\begin{bmatrix} (3x_1 - x_2 + 2x_3) + (3y_1 - y_2 + 2y_3) \\ (5x_1 + 4x_2 - 3x_3) + (5y_1 + 4y_2 - 3y_3) \\ (-5x_1 + x_2 + 2x_3) + (-5y_1 + y_2 + 2y_3) \end{bmatrix}$$

$$= \begin{bmatrix} 3x_1 - x_2 + 2x_3 \\ 5x_1 + 4x_2 - 3x_3 \\ -5x_1 + x_2 + 2x_3 \end{bmatrix} + \begin{bmatrix} 3y_1 - y_2 + 2y_3 \\ 5y_1 + 4y_2 - 3y_3 \\ -5y_1 + y_2 + 2y_3 \end{bmatrix}$$

$$= \begin{pmatrix} 3 & -1 & 2 \\ 5 & 4 & -3 \\ -5 & 1 & 2 \end{pmatrix} \begin{bmatrix} x_1 \\ x_2 \\ x_3 \end{bmatrix} + \begin{pmatrix} 3 & -1 & 2 \\ 5 & 4 & -3 \\ -5 & 1 & 2 \end{pmatrix} \begin{bmatrix} y_1 \\ y_2 \\ y_3 \end{bmatrix} = \bar{A}\,\bar{x} + \bar{A}\,\bar{y}.$$

Also,

$$\bar{A}(4\bar{x}) = \begin{pmatrix} 3 & -1 & 2 \\ 5 & 4 & -3 \\ -5 & 1 & 2 \end{pmatrix} \begin{bmatrix} 4x_1 \\ 4x_2 \\ 4x_3 \end{bmatrix} = \begin{bmatrix} 12x_1 - 4x_2 + 8x_3 \\ 20x_1 + 16x_2 - 12x_3 \\ -20x_1 + 4x_2 + 8x_3 \end{bmatrix}$$

$$= 4 \begin{bmatrix} 3x_1 - x_2 + 2x_3 \\ 5x_1 + 4x_2 - 3x_3 \\ -5x_1 + x_2 + 2x_3 \end{bmatrix} = 4(\bar{A}\,\bar{x}).$$

Section 7.5B, Page 351.

1. $\bar{A}\,\bar{B} = \begin{pmatrix} 3 & 5 \\ 1 & 7 \end{pmatrix} \begin{pmatrix} 4 & 6 \\ 2 & 1 \end{pmatrix}$

$$= \begin{pmatrix} (3)(4) + (5)(2) & (3)(6) + (5)(1) \\ (1)(4) + (7)(2) & (1)(6) + (7)(1) \end{pmatrix} = \begin{pmatrix} 22 & 23 \\ 18 & 13 \end{pmatrix}.$$

$\bar{B}\,\bar{A} = \begin{pmatrix} 4 & 6 \\ 2 & 1 \end{pmatrix} \begin{pmatrix} 3 & 5 \\ 1 & 7 \end{pmatrix}$

$$= \begin{pmatrix} (4)(3) + (6)(1) & (4)(5) + (6)(7) \\ (2)(3) + (1)(1) & (2)(5) + (1)(7) \end{pmatrix} = \begin{pmatrix} 18 & 62 \\ 7 & 17 \end{pmatrix}.$$

2. $\bar{A}\,\bar{B} \;=\; \begin{pmatrix} 5 & -2 \\ 4 & 3 \end{pmatrix} \begin{pmatrix} -1 & 8 \\ 2 & -5 \end{pmatrix}$

$\qquad = \begin{pmatrix} (5)(-1)+(-2)(2) & (5)(8)+(-2)(-5) \\ (4)(-1)+(3)(2) & (4)(8)+(3)(-5) \end{pmatrix} = \begin{pmatrix} -9 & 50 \\ 2 & 17 \end{pmatrix}.$

$\bar{B}\,\bar{A} \;=\; \begin{pmatrix} -1 & 8 \\ 2 & -5 \end{pmatrix} \begin{pmatrix} 5 & -2 \\ 4 & 3 \end{pmatrix}$

$\qquad = \begin{pmatrix} (-1)(5)+(8)(4) & (-1)(-2)+(8)(3) \\ (2)(5)+(-5)(4) & (2)(-2)+(-5)(3) \end{pmatrix}$

$\qquad = \begin{pmatrix} 27 & 26 \\ -10 & -19 \end{pmatrix}.$

4. $\bar{A}\,\bar{B} \;=\; \begin{pmatrix} 6 & 1 \\ 5 & -2 \end{pmatrix} \begin{pmatrix} 1 & 3 & 2 & 4 \\ 0 & 2 & -1 & -3 \end{pmatrix}$

$\qquad = \begin{pmatrix} (6)(1)+(1)(0) & (6)(3)+(1)(2) \\ (5)(1)+(-2)(0) & (5)(3)+(-2)(2) \end{pmatrix}$

$\qquad\qquad \begin{pmatrix} (6)(2)+(1)(-1) & (6)(4)+(1)(-3) \\ (5)(2)+(-2)(-1) & (5)(4)+(-2)(-3) \end{pmatrix}$

$\qquad = \begin{pmatrix} 6 & 20 & 11 & 21 \\ 5 & 11 & 12 & 26 \end{pmatrix}.$

$\bar{B}\,\bar{A}$ is not defined.

6. $\bar{A}\,\bar{B} = \begin{pmatrix} 3 & 2 & 1 \\ 0 & 1 & 2 \\ 5 & 4 & 3 \end{pmatrix} \begin{pmatrix} 2 & -1 & -3 \\ -6 & 0 & 1 \\ 1 & -3 & 4 \end{pmatrix}$

$= \begin{pmatrix} (3)(2) + (2)(-6) + (1)(1) & (3)(-1) + (2)(0) + (1)(-3) \\ (0)(2) + (1)(-6) + (2)(1) & (0)(-1) + (1)(0) + (2)(-3) \\ (5)(2) + (4)(-6) + (3)(1) & (5)(-1) + (4)(0) + (3)(-3) \end{pmatrix}$

$\begin{pmatrix} (3)(-3) + (2)(1) + (1)(4) \\ (0)(-3) + (1)(1) + (2)(4) \\ (5)(-3) + (4)(1) + (3)(4) \end{pmatrix} = \begin{pmatrix} -5 & -6 & -3 \\ -4 & -6 & 9 \\ -11 & -14 & 1 \end{pmatrix}.$

$\bar{B}\,\bar{A} = \begin{pmatrix} 2 & -1 & -3 \\ -6 & 0 & 1 \\ 1 & -3 & 4 \end{pmatrix} \begin{pmatrix} 3 & 2 & 1 \\ 0 & 1 & 2 \\ 5 & 4 & 3 \end{pmatrix}$

$= \begin{pmatrix} (2)(3)+(-1)(0)+(-3)(5) & (2)(3)+(-1)(1)+(-3)(4) \\ (-6)(3)+ (0)(0)+ (1)(5) & (-6)(2)+ (0)(1)+ (1)(4) \\ (1)(3)+(-3)(0)+ (4)(5) & (1)(2)+(-3)(1)+ (4)(4) \end{pmatrix}$

$\begin{pmatrix} (2)(1)+(-1)(2)+(-3)(3) \\ (-6)(1)+ (0)(2)+ (1)(3) \\ (1)(1)+(-3)(2)+ (4)(3) \end{pmatrix} = \begin{pmatrix} -9 & -9 & -9 \\ -13 & -8 & -3 \\ 23 & 15 & 7 \end{pmatrix}.$

7. $\bar{A}\,\bar{B} = \begin{pmatrix} -2 & 4 & 6 \\ 1 & 3 & 5 \\ 0 & 2 & 0 \end{pmatrix} \begin{pmatrix} 0 & 1 & 2 \\ 3 & -2 & -1 \\ 5 & 4 & 2 \end{pmatrix}$

$$= \begin{pmatrix} (-2)(0)+(4)(3)+(6)(5) & (-2)(1)+(4)(-2)+(6)(4) \\ (1)(0)+(3)(3)+(5)(5) & (1)(1)+(3)(-2)+(5)(4) \\ (0)(0)+(2)(3)+(0)(5) & (0)(1)+(2)(-2)+(0)(4) \end{pmatrix}$$

$$\begin{pmatrix} (-2)(2)+(4)(-1)+(6)(2) \\ (1)(2)+(3)(-1)+(5)(2) \\ (0)(2)+(2)(-1)+(0)(2) \end{pmatrix} = \begin{pmatrix} 42 & 14 & 4 \\ 34 & 15 & 9 \\ 6 & -4 & -2 \end{pmatrix}.$$

$$\bar{B}\,\bar{A} = \begin{pmatrix} 0 & 1 & 2 \\ 3 & -2 & -1 \\ 5 & 4 & 2 \end{pmatrix} \begin{pmatrix} -2 & 4 & 6 \\ 1 & 3 & 5 \\ 0 & 2 & 0 \end{pmatrix}$$

$$= \begin{pmatrix} (0)(-2)+(1)(1)+(2)(0) & (0)(4)+(1)(3)+(2)(2) \\ (3)(-2)+(-2)(1)+(-1)(0) & (3)(4)+(-2)(3)+(-1)(2) \\ (5)(-2)+(4)(1)+(2)(0) & (5)(4)+(4)(3)+(2)(2) \end{pmatrix}$$

$$\begin{pmatrix} (0)(6)+(1)(5)+(2)(0) \\ (3)(6)+(-2)(5)+(-1)(0) \\ (5)(6)+(4)(5)+(2)(0) \end{pmatrix} = \begin{pmatrix} 1 & 7 & 5 \\ -8 & 4 & 8 \\ -6 & 36 & 50 \end{pmatrix}.$$

9. $\bar{A}\,\bar{B} = \begin{pmatrix} 2 & 1 & 0 & 1 \\ 0 & 2 & 3 & -1 \\ 1 & -2 & 1 & 0 \end{pmatrix} \begin{bmatrix} 1 & 2 \\ 0 & 3 \\ -1 & 0 \\ 1 & -2 \end{bmatrix}$

$$= \begin{pmatrix} (2)(1)+(1)(0)+(0)(-1)+(1)(1) \\ (0)(1)+(2)(0)+(3)(-1)+(-1)(1) \\ (1)(1)+(-2)(0)+(1)(-1)+(0)(1) \end{pmatrix}$$

$$\begin{pmatrix} (2)(2)+(1)(3)+(0)(0)+(1)(-2) \\ (0)(2)+(2)(3)+(3)(0)+(-1)(-2) \\ (1)(2)+(-2)(3)+(1)(0)+(0)(-2) \end{pmatrix} = \begin{pmatrix} 3 & 5 \\ -4 & 8 \\ 0 & -4 \end{pmatrix}.$$

Since the number of columns in $\bar{B}$,2, is unequal to the number of rows in $\bar{A}$,3, the product $\bar{B}\,\bar{A}$ is not defined.

11. $\bar{A}^2 = \bar{A}\ \bar{A} = \begin{pmatrix} 1 & 2 & -1 \\ -2 & 1 & 3 \\ 2 & 0 & 1 \end{pmatrix} \begin{pmatrix} 1 & 2 & -1 \\ -2 & 1 & 3 \\ 2 & 0 & 1 \end{pmatrix}$

$= \begin{pmatrix} (1)(1)+(2)(-2)+(-1)(2) & (1)(2)+(2)(1)+(-1)(0) \\ (-2)(1)+(1)(-2)+(3)(2) & (-2)(2)+(1)(1)+(3)(0) \\ (2)(1)+(0)(-2)+(1)(2) & (2)(2)+(0)(1)+(1)(0) \end{pmatrix}$

$\begin{pmatrix} (1)(-1)+(2)(3)+(-1)(1) \\ (-2)(-1)+(1)(3)+(3)(1) \\ (2)(-1)+(0)(3)+(1)(1) \end{pmatrix} = \begin{pmatrix} -5 & 4 & 4 \\ 2 & -3 & 8 \\ 4 & 4 & -1 \end{pmatrix},$

$\bar{A}^3 = \bar{A}\ \bar{A}\ \bar{A} = \bar{A}^2\bar{A} = \begin{pmatrix} -5 & 4 & 4 \\ 2 & -3 & 8 \\ 4 & 4 & -1 \end{pmatrix} \begin{pmatrix} 1 & 2 & -1 \\ -2 & 1 & 3 \\ 2 & 0 & 1 \end{pmatrix}$

$= \begin{pmatrix} (-5)(1)+(4)(-2)+(4)(2) & (-5)(2)+(4)(1)+(4)(0) \\ (2)(1)+(-3)(-2)+(8)(2) & (2)(2)+(-3)(1)+(8)(0) \\ (4)(1)+(4)(-2)+(-1)(2) & (4)(2)+(4)(1)+(-1)(0) \end{pmatrix}$

$\begin{pmatrix} (4)(1)+(4)(-2)+(-1)(2) \\ (4)(2)+(4)(1)+(-1)(0) \\ (4)(-1)+(4)(3)+(-1)(0) \end{pmatrix} = \begin{pmatrix} -5 & -6 & -6 \\ 24 & 1 & 12 \\ -6 & 12 & 8 \end{pmatrix}.$

12. We have $\bar{A}^2 = \bar{A}\bar{A} = \begin{pmatrix} 2 & 3 & 3 \\ 1 & -2 & 1 \\ -3 & -1 & 0 \end{pmatrix} \begin{pmatrix} 2 & 3 & 3 \\ 1 & -2 & 1 \\ -3 & -1 & 0 \end{pmatrix}$

$= \begin{pmatrix} (2)(2)+(3)(1)+(3)(-3) & (2)(3)+(3)(-2)+(3)(-1) \\ (1)(2)+(-2)(1)+(1)(-3) & (1)(3)+(-2)(-2)+(1)(-1) \\ (-3)(2)+(-1)(1)+(0)(-3) & (-3)(3)+(-1)(-2)+(0)(-1) \end{pmatrix}$

$\begin{pmatrix} (2)(3)+(3)(1)+(3)(0) \\ (1)(3)+(-2)(1)+(1)(0) \\ (-3)(3)+(-1)(1)+(0)(0) \end{pmatrix} = \begin{pmatrix} -2 & -3 & 9 \\ -3 & 6 & 1 \\ -7 & -7 & -10 \end{pmatrix}.$

Then

$$\bar{A}^2 + 3\bar{A} + 2\bar{I} = \begin{pmatrix} -2 & -3 & 9 \\ -3 & 6 & 1 \\ -7 & -7 & -10 \end{pmatrix} + \begin{pmatrix} 6 & 9 & 9 \\ 3 & -6 & 3 \\ -9 & -3 & 0 \end{pmatrix}$$

$$+ \begin{pmatrix} 2 & 0 & 0 \\ 0 & 2 & 0 \\ 0 & 0 & 2 \end{pmatrix} = \begin{pmatrix} 6 & 6 & 18 \\ 0 & 2 & 4 \\ -16 & -10 & -8 \end{pmatrix}.$$

13. $\bar{A} = \begin{pmatrix} 1 & 3 \\ 2 & 5 \end{pmatrix}$. We have cof $\bar{A} = \begin{pmatrix} 5 & -2 \\ -3 & 1 \end{pmatrix}$; adj $\bar{A} = (\text{cof } \bar{A})^T =$

$\begin{pmatrix} 5 & -3 \\ -2 & 1 \end{pmatrix}$; and $|\bar{A}| = \begin{vmatrix} 1 & 3 \\ 2 & 5 \end{vmatrix} = -1$. Thus $\bar{A}^{-1} = \dfrac{1}{|\bar{A}|}(\text{adj } \bar{A}) =$

$\begin{pmatrix} -5 & 3 \\ 2 & -1 \end{pmatrix}.$

14. $\bar{A} = \begin{pmatrix} -1 & 5 \\ -2 & 8 \end{pmatrix}$. We have cof $\bar{A} = \begin{pmatrix} 8 & 2 \\ -5 & -1 \end{pmatrix}$; adj $\bar{A} = (\text{cof } \bar{A})^T =$

$\begin{pmatrix} 8 & -5 \\ 2 & -1 \end{pmatrix}$; and $|\bar{A}| = \begin{vmatrix} -1 & 5 \\ -2 & 8 \end{vmatrix} = 2$. Thus $\bar{A}^{-1} = \dfrac{1}{|\bar{A}|}(\text{adj } \bar{A}) =$

$\begin{pmatrix} 4 & -5/2 \\ 1 & -1/2 \end{pmatrix}.$

17. We find

$$
\text{cof } \bar{A} = \left[
\begin{array}{ccc}
\begin{vmatrix} 3 & 2 \\ 1 & 1 \end{vmatrix} & - \begin{vmatrix} 3 & 2 \\ -1 & 1 \end{vmatrix} & \begin{vmatrix} 3 & 3 \\ -1 & 1 \end{vmatrix} \\[12pt]
- \begin{vmatrix} 3 & 1 \\ 1 & 1 \end{vmatrix} & \begin{vmatrix} 4 & 1 \\ -1 & 1 \end{vmatrix} & - \begin{vmatrix} 4 & 3 \\ -1 & 1 \end{vmatrix} \\[12pt]
\begin{vmatrix} 3 & 1 \\ 3 & 2 \end{vmatrix} & - \begin{vmatrix} 4 & 1 \\ 3 & 2 \end{vmatrix} & \begin{vmatrix} 4 & 1 \\ 3 & 2 \end{vmatrix}
\end{array}
\right]
$$

$$
= \begin{pmatrix} 1 & -5 & 6 \\ -2 & 5 & -7 \\ 3 & -5 & 3 \end{pmatrix} ; \quad \text{adj } \bar{A} = (\text{cof } \bar{A})^{T}
$$

$$
= \begin{pmatrix} 1 & -2 & 3 \\ -5 & 5 & -5 \\ 6 & -7 & 3 \end{pmatrix} ; \quad \text{and } |\bar{A}| = \begin{vmatrix} 4 & 3 & 1 \\ 3 & 3 & 2 \\ -1 & 1 & 1 \end{vmatrix} = -5.
$$

Thus

$$
\bar{A}^{-1} = \frac{1}{|\bar{A}|}(\text{adj } \bar{A}) = \begin{pmatrix} -1/5 & 2/5 & -3/5 \\ 1 & -1 & 1 \\ -6/5 & 7/5 & -3/5 \end{pmatrix}.
$$

18. We find

$$\text{cof } \bar{A} = \begin{bmatrix} \begin{vmatrix} 1 & 2 \\ 3 & 4 \end{vmatrix} & -\begin{vmatrix} 1 & 2 \\ -1 & 4 \end{vmatrix} & \begin{vmatrix} 1 & 1 \\ -1 & 3 \end{vmatrix} \\[4mm] -\begin{vmatrix} -1 & 1 \\ 3 & 4 \end{vmatrix} & \begin{vmatrix} 2 & 1 \\ -1 & 4 \end{vmatrix} & -\begin{vmatrix} 2 & -1 \\ -1 & 3 \end{vmatrix} \\[4mm] \begin{vmatrix} -1 & 1 \\ 1 & 2 \end{vmatrix} & -\begin{vmatrix} 2 & 1 \\ 1 & 2 \end{vmatrix} & \begin{vmatrix} 2 & -1 \\ 1 & 1 \end{vmatrix} \end{bmatrix}$$

$$= \begin{pmatrix} -2 & -6 & 4 \\ 7 & 7 & -5 \\ -3 & -3 & 3 \end{pmatrix}.$$

$$\text{adj } \bar{A} = (\text{cof } \bar{A})^T = \begin{pmatrix} -2 & 7 & -3 \\ -6 & 9 & -3 \\ 4 & -5 & 3 \end{pmatrix}. \quad \bar{A} = \begin{vmatrix} 2 & -1 & 1 \\ 1 & 1 & 2 \\ -1 & 3 & 4 \end{vmatrix} = 6.$$

Thus

$$\bar{A}^{-1} = \frac{1}{|\bar{A}|}(\text{adj } \bar{A})$$

$$= \frac{1}{6}\begin{pmatrix} -2 & 7 & -3 \\ -6 & 9 & -3 \\ 4 & -5 & 3 \end{pmatrix} = \begin{pmatrix} -1/3 & 7/6 & -1/2 \\ -1 & 3/2 & -1/2 \\ 2/3 & -5/6 & 1/2 \end{pmatrix}.$$

21. **We find**

$$\text{cof } \bar{A} = \begin{bmatrix} \begin{vmatrix} 4 & 2 \\ 2 & 3 \end{vmatrix} & -\begin{vmatrix} 1 & 2 \\ 2 & 3 \end{vmatrix} & \begin{vmatrix} 1 & 4 \\ 2 & 2 \end{vmatrix} \\[4mm] -\begin{vmatrix} 2 & 4 \\ 2 & 3 \end{vmatrix} & \begin{vmatrix} 3 & 4 \\ 2 & 3 \end{vmatrix} & -\begin{vmatrix} 3 & 2 \\ 2 & 2 \end{vmatrix} \\[4mm] \begin{vmatrix} 2 & 4 \\ 4 & 2 \end{vmatrix} & -\begin{vmatrix} 3 & 4 \\ 1 & 2 \end{vmatrix} & \begin{vmatrix} 3 & 2 \\ 1 & 4 \end{vmatrix} \end{bmatrix} = \begin{pmatrix} 8 & 1 & -6 \\ 2 & 1 & -2 \\ -12 & -2 & -10 \end{pmatrix};$$

$$\text{adj } \bar{A} = (\text{cof } \bar{A})^T = \begin{pmatrix} 8 & 2 & -12 \\ 1 & 1 & -2 \\ -6 & -2 & 10 \end{pmatrix}; \text{ and } \bar{A} = \begin{vmatrix} 3 & 2 & 4 \\ 1 & 4 & 2 \\ 2 & 2 & 3 \end{vmatrix} = 2.$$

Thus $\quad \bar{A}^{-1} = \dfrac{1}{|\bar{A}|}(\text{adj } \bar{A}) = \begin{pmatrix} 4 & 1 & -6 \\ 1/2 & 1/2 & -1 \\ -3 & -1 & 5 \end{pmatrix}$

22. We have

$$\text{cof } \bar{A} = \begin{bmatrix} \begin{vmatrix} 7 & 1 \\ 3 & 3 \end{vmatrix} & -\begin{vmatrix} 3 & 1 \\ 1 & 3 \end{vmatrix} & \begin{vmatrix} 3 & 7 \\ 1 & 3 \end{vmatrix} \\ -\begin{vmatrix} 2 & 0 \\ 3 & 3 \end{vmatrix} & \begin{vmatrix} 1 & 0 \\ 1 & 3 \end{vmatrix} & -\begin{vmatrix} 1 & 2 \\ 1 & 3 \end{vmatrix} \\ \begin{vmatrix} 2 & 0 \\ 7 & 1 \end{vmatrix} & -\begin{vmatrix} 1 & 0 \\ 3 & 1 \end{vmatrix} & \begin{vmatrix} 1 & 2 \\ 3 & 7 \end{vmatrix} \end{bmatrix} = \begin{pmatrix} 18 & -8 & 2 \\ -6 & 3 & -1 \\ 2 & -1 & 1 \end{pmatrix};$$

$$\text{adj } \bar{A} = (\text{cof } \bar{A})^T = \begin{pmatrix} 18 & -6 & 2 \\ -6 & 3 & -1 \\ 2 & -1 & 1 \end{pmatrix}; \text{ and } \bar{A} = \begin{vmatrix} 1 & 2 & 0 \\ 3 & 7 & 1 \\ 1 & 3 & 3 \end{vmatrix} = 2.$$

Thus $\quad \bar{A}^{-1} = \dfrac{1}{|\bar{A}|}(\text{adj } \bar{A}) = \begin{pmatrix} 9 & -3 & 1 \\ -4 & 3/2 & -1/2 \\ 1 & -1/2 & 1/2 \end{pmatrix}.$

Section 7.5C, Page 356.

1. a. We must show there exist numbers c_1, c_2, c_3, not all zero, such that $c_1\bar{v}_1 + c_2\bar{v}_2 + c_3\bar{v}_3 = \bar{0}$. This is

$$c_1\begin{pmatrix} 3 \\ -1 \\ 2 \end{pmatrix} + c_2\begin{pmatrix} 13 \\ 5 \\ -4 \end{pmatrix} + c_3\begin{pmatrix} 2 \\ 4 \\ -5 \end{pmatrix} = \begin{pmatrix} 0 \\ 0 \\ 0 \end{pmatrix}.$$

This is equivalent to the homogeneous linear system

$$\begin{cases} 3c_1 + 13c_2 + 2c_3 = 0, \\ -c_1 + 5c_2 + 4c_3 = 0, \\ 2c_1 - 4c_2 - 5c_3 = 0, \end{cases} \tag{1}$$

the determinant of coefficients of which is

$$\begin{vmatrix} 3 & 13 & 2 \\ -1 & 5 & 4 \\ 2 & -4 & -5 \end{vmatrix} = 0.$$

Hence, by Theorem A (text, page 353), the system (1) has a nontrivial solution for c_1, c_2, c_3. For example, one solution is $c_1 = 3$, $c_2 = -1$, $c_3 = 2$. Thus $\bar{v}_1$, $\bar{v}_2$, $\bar{v}_3$ are linearly dependent.

2. a. Here we must show that if $c_1\bar{v}_1 + c_2\bar{v}_2 + c_3\bar{v}_3 = \bar{0}$, then $c_1 = c_2 = c_3 = 0$. Thus, we suppose

$$c_1\begin{pmatrix} 2 \\ 1 \\ 0 \end{pmatrix} + c_2\begin{pmatrix} 1 \\ 0 \\ 3 \end{pmatrix} + c_3\begin{pmatrix} 0 \\ -1 \\ 1 \end{pmatrix} = \begin{pmatrix} 0 \\ 0 \\ 0 \end{pmatrix}.$$

This is equivalent to the homogeneous linear system

$$\begin{cases} 2c_1 + c_2 = 0, \\ c_1 - c_3 = 0, \\ 3c_2 + c_3 = 0, \end{cases} \tag{1}$$

the determinant of which is

$$\begin{vmatrix} 2 & 1 & 0 \\ 1 & 0 & -1 \\ 0 & 3 & 1 \end{vmatrix} = 5 \neq 0.$$

Thus by Theorem A (text, page 353), the system (1) has only the trivial solution $c_1 = c_2 = c_3 = 0$, and so $\bar{v}_1, \bar{v}_2, \bar{v}_3$ are linearly independent.

3. a.   The given vectors are lineraly dependent if and only if there exist numbers c_1, c_2, c_3, not all zero, such that

$$c_1\bar{v}_1 + c_2\bar{v}_2 + c_3\bar{v}_3 = \bar{0}. \qquad (1)$$

This is

$$c_1\begin{pmatrix} k \\ 2 \\ -1 \end{pmatrix} + c_2\begin{pmatrix} 1 \\ -1 \\ 2 \end{pmatrix} + c_3\begin{pmatrix} 3 \\ 7 \\ -8 \end{pmatrix} = \begin{pmatrix} 0 \\ 0 \\ 0 \end{pmatrix}.$$

This is equivalent to the homogeneous linear system

$$\begin{cases} kc_1 + c_2 + 3c_3 = 0, \\ 2c_1 + c_2 + 7c_3 = 0, \\ -c_1 + 2c_2 - 8c_3 = 0, \end{cases} \qquad (2)$$

the determinant of coefficient of which is

$$\begin{vmatrix} k & 1 & 3 \\ 2 & -1 & 7 \\ -1 & 2 & -8 \end{vmatrix} = -6k + 18. \qquad (3)$$

By Theorem A (text, page 353), the system (2) has a nontrivial solution for c_1, c_2, c_3, and hence there exist c_1, c_2, c_3, not all zero, such that (1) holds, if and only if the determinant (3) is zero. Thus we have $-6k + 18 = 0$, and hence $k = 3$.

4. b. Note that

$$4\bar{\phi}_1(t) - 2\bar{\phi}_2(t) - \bar{\phi}_3(t)$$

$$= \begin{pmatrix} 4\sin t + 4\cos t \\ 8\sin t \\ -4\cos t \end{pmatrix} + \begin{pmatrix} -4\sin t \\ -8\sin t + 2\cos t \\ 2\sin t \end{pmatrix}$$

$$+ \begin{pmatrix} -4\cos t \\ -2\cos t \\ -2\sin t + 4\cos t \end{pmatrix} = \begin{pmatrix} 0 \\ 0 \\ 0 \end{pmatrix}.$$

Thus there exists the set of three numbers 4, -2, -1, none of which are zero, such that $4\bar{\phi}_1(t) + (-2)\bar{\phi}_2(t) + (-1)\bar{\phi}_3(t) = 0$ for all t such that $a \leq t \leq b$, for any a and b. Therefore $\bar{\phi}_1$, $\bar{\phi}_2$, and $\bar{\phi}_3$ are linearly dependent on $a \leq t \leq b$.

5. b. Suppose there exist numbers c_1 and c_2 such that

$c_1 \bar{\phi}_1(t) + c_2 \bar{\phi}_2(t) = 0$ for all t on a $\leq$ t $\leq$ b. Then

$$\begin{cases} 2c_1 e^{2t} + c_2 e^{-t} = 0, \\ \\ -c_1 e^{2t} + 3c_2 e^{-t} = 0, \text{ for all t on a } \leq t \leq b. \end{cases}$$

From this, we have both

$$\begin{cases} 2c_1 + c_2 e^{-3t} = 0, \\ \\ -2c_1 + 6c_2 e^{-3t} = 0, \end{cases} \text{ and } \begin{cases} 6c_1 e^{3t} + 3c_2 = 0, \\ \\ c_1 e^{3t} - 3c_2 = 0. \end{cases} \quad (1)$$

From the former of (1), $7c_2 e^{-3t} = 0$ on a $\leq$ t $\leq$ b, so $c_2 = 0$; and from the latter of (1), $7c_1 e^{3t} = 0$ on a $\leq$ t $\leq$ b, so $c_1 = 0$. Thus if $c_1 \bar{\phi}_1(t) + c_2 \bar{\phi}_2(t) = \bar{0}$ for all t on a $\leq$ t $\leq$ b, we must have $c_1 = c_2 = 0$, and hence ϕ_1 and ϕ_2 are linearly independent on a $\leq$ t $\leq$ b, for any a and b.

Section 7.5D, Page 367.

1. The characteristic equation is

$$\begin{vmatrix} 1 - \lambda & 2 \\ 3 & 2 - \lambda \end{vmatrix} = 0,$$

that is, $\lambda^2 - 3\lambda - 4 = 0$ or $(\lambda + 1)(\lambda - 4) = 0$. Thus, the characteristic values are $\lambda = -1$ and 4.

The characteristic vectors corresponding to $\lambda = -1$ have components x_1 and x_2 such that

$$\begin{pmatrix} 1 & 2 \\ 3 & 2 \end{pmatrix} \begin{pmatrix} x_1 \\ x_2 \end{pmatrix} = (-1) \begin{pmatrix} x_1 \\ x_2 \end{pmatrix}.$$

Thus x_1 and x_2 must satisfy the system

$$\begin{cases} x_1 + 2x_2 = -x_1 \\ 3x_1 + 2x_2 = -x_2 \end{cases} \quad \text{or} \quad \begin{cases} 2x_1 + 2x_2 = 0 \\ 3x_1 + 3x_2 = 0. \end{cases}$$

We find $x_1 = k$, $x_2 = -k$ for every real k.   Hence the characteristic vectors corresponding to $\lambda = -1$ are $\begin{pmatrix} k \\ -k \end{pmatrix}$ for every nonzero real k.

The characteristic vectors corresponding to $\lambda = 4$ have components x_1 and x_2 such that

$$\begin{pmatrix} 1 & 2 \\ 3 & 2 \end{pmatrix} \begin{pmatrix} x_1 \\ x_2 \end{pmatrix} = 4 \begin{pmatrix} x_1 \\ x_2 \end{pmatrix}.$$

Thus x_1 and x_2 must satisfy the system

$$\begin{cases} x_1 + 2x_2 = 4x_1 \\ 3x_1 + 2x_2 = 4x_2 \end{cases} \quad \text{or} \quad \begin{cases} -3x_1 + 2x_2 = 0 \\ 3x_1 - 3x_2 = 0. \end{cases}$$

We find $x_1 = 2k$, $x_2 = 3k$ for every real k. Hence the characteristic vectors corresponding to $\lambda = 4$ are $\begin{pmatrix} 2k \\ 3k \end{pmatrix}$ for every nonzero real k.

2. The characteristic equation is

$$\begin{vmatrix} 3 - \lambda & 2 \\ 6 & -1 - \lambda \end{vmatrix} = 0,$$

that is, $\lambda^2 - 2\lambda - 15 = 0$ or $(\lambda - 5)(\lambda + 3) = 0$. Thus, the characteristic values are $\lambda = 5$ and $\lambda = -3$.

The characteristic vectors corresponding to $\lambda = 5$ have components x_1 and x_2 such that

$$\begin{pmatrix} 3 & 2 \\ 6 & -1 \end{pmatrix} \begin{pmatrix} x_1 \\ x_2 \end{pmatrix} = 5 \begin{pmatrix} x_1 \\ x_2 \end{pmatrix}.$$

Thus x_1 and x_2 must satisfy the system

$$\begin{cases} 3x_1 + 2x_2 = 5x_1 \\ 6x_1 - x_2 = 5x_2 \end{cases} \quad \text{or} \quad \begin{cases} -2x_1 + 2x_2 = 0, \\ x_1 - x_2 = 0. \end{cases}$$

We find $x_1 = x_2 = k$ for every real k. Hence the characteristic vectors corresponding to $\lambda = 5$ are $\begin{pmatrix} k \\ k \end{pmatrix}$ for every nonzero real k.

The characteristic vectors corresponding to $\lambda = -3$ have components x_1 and x_2 such that

$$\begin{pmatrix} 3 & 2 \\ 6 & -1 \end{pmatrix} \begin{pmatrix} x_1 \\ x_2 \end{pmatrix} = (-3) \begin{pmatrix} x_1 \\ x_2 \end{pmatrix}.$$

Thus x_1 and x_2 must satisfy the system

$$\begin{cases} 3x_1 + 2x_2 = -3x_1, \\ 6x_1 - x_2 = -3x_2. \end{cases}$$

From these, $6x_1 + 2x_2 = 0$, so $x_2 = -3x_1$. Thus we find $x_1 = k$, $x_2 = -3k$ for every real k. Hence the characteristic vectors corresponding to $\lambda = -3$ are $\begin{pmatrix} k \\ -3k \end{pmatrix}$ for every nonzero real k.

7. The characteristic equation is

$$\begin{vmatrix} 1 - \lambda & 1 & -1 \\ 2 & 3 - \lambda & -4 \\ 4 & 1 & -4 - \lambda \end{vmatrix} = 0,$$

which reduces to

$$(1 - \lambda) \begin{vmatrix} 3 - \lambda & -4 \\ 1 & 4 - \lambda \end{vmatrix} - \begin{vmatrix} 2 & -4 \\ 4 & -4 - \lambda \end{vmatrix} - \begin{vmatrix} 2 & 3 - \lambda \\ 4 & 1 \end{vmatrix} = 0$$

or $\lambda^3 - 7\lambda + 6 = 0$ or $(\lambda - 1)(\lambda - 2)(\lambda + 3) = 0$. Thus, the characteristic values are $\lambda = 1, 2, -3$.

The characteristic vectors corresponding to $\lambda = 1$ have components x_1, x_2, x_3 such that

$$\begin{pmatrix} 1 & 1 & -1 \\ 2 & 3 & -4 \\ 4 & 1 & -4 \end{pmatrix} \begin{bmatrix} x_1 \\ x_2 \\ x_3 \end{bmatrix} = 1 \begin{bmatrix} x_1 \\ x_2 \\ x_3 \end{bmatrix}.$$

Thus x_1, x_2, x_3 must satisfy the system

$$\begin{cases} x_1 + x_2 - x_3 = x_1, \\ 2x_1 + 3x_2 - 4x_3 = x_2, \\ 4x_1 + x_2 - 4x_3 = x_3, \quad \text{or} \end{cases} \qquad \begin{cases} x_2 - x_3 = 0, \\ 2x_1 + 2x_2 - 4x_3 = 0, \\ 4x_1 + x_2 - 5x_3 = 0. \end{cases}$$

From the first of these, $x_3 = x_2$; and then the second and third become

$$\begin{cases} 2x_1 - 2x_2 = 0, \\ 4x_1 - 4x_2 = 0, \end{cases}$$

a solution of which is $x_1 = x_2 = k$. Thus we find $x_1 = k$, $x_2 = k$, $x_3 = k$ for every real k. Hence the characteristic vectors corresponding to $\lambda = 1$ are $\begin{bmatrix} k \\ k \\ k \end{bmatrix}$ for every nonzero real k.

The characteristic vectors corresponding to $\lambda = 2$ have components x_1, x_2, x_3 such that

$$\begin{pmatrix} 1 & 1 & -1 \\ 2 & 3 & -4 \\ 4 & 1 & -4 \end{pmatrix} \begin{bmatrix} x_1 \\ x_2 \\ x_3 \end{bmatrix} = 2 \begin{bmatrix} x_1 \\ x_2 \\ x_3 \end{bmatrix}.$$

Thus x_1, x_2, x_3 must satisfy the system

$$\begin{cases} x_1 + x_2 - x_3 = 2x_1, \\ 2x_1 + 3x_2 - 4x_3 = 2x_2, \\ 4x_1 + x_2 - 4x_3 = 2x_3, \end{cases} \quad \text{or} \quad \begin{cases} -x_1 + x_2 - x_3 = 0, \\ 2x_1 + x_2 - 4x_3 = 0, \\ 4x_1 + x_2 - 6x_3 = 0. \end{cases}$$

From the first and second, $3x_1 - 3x_3 = 0$, so $x_3 = x_1$, and $3x_2 - 6x_3 = 0$, so $x_2 = 2x_3$. Thus we find $x_1 = k$, $x_2 = 2k$, $x_3 = k$ for every real k. Hence the characteristic vectors corresponding to $\lambda = 2$ are $\begin{bmatrix} k \\ 2k \\ k \end{bmatrix}$ for every nonzero real k.

The characteristic vectors corresponding to $\lambda = -3$ have components x_1, x_2, x_3 such that

$$\begin{pmatrix} 1 & 1 & -1 \\ 2 & 3 & -4 \\ 4 & 1 & -4 \end{pmatrix} \begin{bmatrix} x_1 \\ x_2 \\ x_3 \end{bmatrix} = -3 \begin{bmatrix} x_1 \\ x_2 \\ x_3 \end{bmatrix}.$$

Thus x_1, x_2, x_3 must satisfy the system

$$\begin{cases} x_1 + x_2 - x_3 = -3x_1, \\ 2x_1 + 3x_2 - 4x_3 = -3x_2, \\ 4x_1 + x_2 - 4x_3 = -3x_3, \end{cases} \quad \text{or} \quad \begin{cases} 4x_1 + x_2 - x_3 = 0, \\ 2x_1 + 6x_2 - 4x_3 = 0, \\ 4x_1 + x_2 - x_3 = 0. \end{cases}$$

Regarding the first and second equations as two equations in x_2 and x_3, we find $x_2 = 7x_1$, $x_3 = 11x_1$. Thus we find $x_1 = k$, $x_2 = 7k$, $x_3 = 11k$ for every real k. Hence the characteristic vectors corresponding to $\lambda = -3$ are $\begin{bmatrix} k \\ 7k \\ 11k \end{bmatrix}$

for every nonzero real k.

8. The characteristic equation is

$$\begin{vmatrix} 1 - \lambda & -1 & -1 \\ 1 & 3 - \lambda & 1 \\ -3 & -6 & 6 - \lambda \end{vmatrix} = 0,$$

which reduces to

$$(1 - \lambda) \begin{vmatrix} 3 - \lambda & 1 \\ -6 & 6 - \lambda \end{vmatrix} + \begin{vmatrix} 1 & 1 \\ -3 & 6 - \lambda \end{vmatrix} - \begin{vmatrix} 1 & 3 - \lambda \\ -3 & -6 \end{vmatrix} = 0$$

or $\lambda^3 - 10\lambda^2 + 31\lambda - 30 = 0$ or $(\lambda - 2)(\lambda - 3)(\lambda - 5) = 0$.

The characteristic vectors corresponding to $\lambda = 2$ have components x_1, x_2, x_3 such that

$$\begin{pmatrix} 1 & -1 & -1 \\ 1 & 3 & 1 \\ -3 & -6 & 6 \end{pmatrix} \begin{bmatrix} x_1 \\ x_2 \\ x_3 \end{bmatrix} = 2 \begin{bmatrix} x_1 \\ x_2 \\ x_3 \end{bmatrix}.$$

Thus x_1, x_2, x_3 must satisfy the system

$$\begin{cases} x_1 - x_2 - x_3 = 2x_1, \\ x_1 + 3x_2 + x_3 = 2x_2, \\ -3x_1 - 6x_2 + 6x_3 = 2x_3, \end{cases} \quad \text{or} \quad \begin{cases} -x_1 - x_2 - x_3 = 0, \\ x_1 + x_2 + x_3 = 0, \\ -3x_1 + 6x_2 + 4x_3 = 0. \end{cases}$$

The first two of these are essentially the same. We write the second and third as

$$\begin{cases} x_1 + x_2 = -x_3, \\ 3x_1 + 6x_2 = 4x_3, \end{cases}$$

and solve for x_1 and x_2 in terms of x_3. We find $x_1 = -\dfrac{10}{3}x_3$, $x_2 = \dfrac{7}{3}x_3$. Letting $x_3 = -3k$, we find $x_1 = 10k$, $x_2 = -7k$, $x_3 = -3k$ for every real k. Hence the characteristic vectors corresponding to $\lambda = 2$ are $\begin{bmatrix} 10k \\ -7k \\ -3k \end{bmatrix}$

for every nonzero real k.

The characteristic vectors corresponding to $\lambda = 3$ have components x_1, x_2, x_3 such that

$$\begin{pmatrix} 1 & -1 & -1 \\ 1 & 3 & 1 \\ -3 & -6 & 6 \end{pmatrix} \begin{bmatrix} x_1 \\ x_2 \\ x_3 \end{bmatrix} = 3 \begin{bmatrix} x_1 \\ x_2 \\ x_3 \end{bmatrix}.$$

Thus x_1, x_2, x_3 must satisfy

$$\begin{cases} x_1 - x_2 - x_3 = 3x_1, \\ x_1 + 3x_2 + x_3 = 3x_2, \\ -3x_1 - 6x_2 + 6x_3 = 3x_3, \end{cases} \quad \text{or} \quad \begin{cases} -2x_1 - x_2 - x_3 = 0, \\ x_1 + x_3 = 0, \\ -3x_1 - 6x_2 + 3x_3 = 0. \end{cases}$$

From the second equation, $x_3 = -x_1$; and then the first and third become

$$\begin{cases} -x_1 - x_2 = 0, \\ -6x_1 - 6x_2 = 0. \end{cases}$$

From these $x_2 = -x_1$. Thus we find $x_1 = k$, $x_2 = -k$, $x_3 = -k$, for every real k. Hence the characteristic

vectors to $\lambda = 3$ are $\begin{bmatrix} k \\ -k \\ -k \end{bmatrix}$ for every nonzero real k.

The characteristic vectors corresponding to $\lambda = 5$ have components x_1, x_2, x_3 such that

$$\begin{pmatrix} 1 & -1 & -1 \\ 1 & 3 & 1 \\ -3 & -6 & 6 \end{pmatrix} \begin{bmatrix} x_1 \\ x_2 \\ x_3 \end{bmatrix} = 5 \begin{bmatrix} x_1 \\ x_2 \\ x_3 \end{bmatrix}.$$

Thus x_1, x_2, x_3 must satisfy

$$\begin{cases} x_1 - x_2 - x_3 = 5x_1, \\ x_1 + 3x_2 + x_3 = 5x_2, \\ -3x_1 - 6x_2 + 6x_3 = 5x_3, \end{cases} \quad \text{or} \quad \begin{cases} -4x_1 - x_2 - x_3 = 0, \\ x_1 - 2x_2 + x_3 = 0, \\ -3x_1 - 6x_2 + x_3 = 0. \end{cases}$$

Adding the first two of these, we obtain $-3x_1 - 3x_2 = 0$, from which $x_2 = -x_1$. Then the third equation becomes $3x_1 + x_3 = 0$, from which $x_3 = -3x_1$. Thus we find $x_1 = k$, $x_2 = -k$, $x_3 = -3k$ for every real k. Hence the characteristic vectors corresponding to $\lambda = 5$ are $\begin{bmatrix} k \\ -k \\ -3k \end{bmatrix}$ for every nonzero real k.

10. The characteristic equation is

$$\begin{vmatrix} 1 - \lambda & 1 & 0 \\ 1 & -\lambda & 1 \\ 0 & 1 & 1 - \lambda \end{vmatrix} = 0,$$

which reduces to

$$(1 - \lambda) \begin{vmatrix} -\lambda & 1 \\ 1 & 1 - \lambda \end{vmatrix} - \begin{vmatrix} 1 & 1 \\ 0 & 1 - \lambda \end{vmatrix} = 0$$

or $\lambda^3 - 2\lambda - \lambda + 2 = 0$ or $(\lambda - 1)(\lambda - 2)(\lambda + 1) = 0$.
Thus, the characteristic values are $\lambda = 1, 2, -1$.

The characteristic vectors corresponding to $\lambda = 1$ have components x_1, x_2, x_3 such that

$$\begin{pmatrix} 1 & 1 & 0 \\ 1 & 0 & 1 \\ 0 & 1 & 1 \end{pmatrix} \begin{bmatrix} x_1 \\ x_2 \\ x_3 \end{bmatrix} = 1 \begin{bmatrix} x_1 \\ x_2 \\ x_3 \end{bmatrix}.$$

Thus x_1, x_2, x_3 must satisfy the system

$$\begin{cases} x_1 + x_2 = x_1, \\ x_1 + x_3 = x_2, \\ x_2 + x_3 = x_3, \end{cases} \text{ or } \begin{cases} x_2 = 0, \\ x_1 - x_2 + x_3 = 0, \\ x_2 = 0. \end{cases}$$

From these, we see at once that $x_2 = 0$ and then $x_3 = -x_1$. Thus we find $x_1 = k$, $x_2 = 0$, $x_3 = -k$ for every real k. Hence the characteristic vectors corresponding to $\lambda = 1$

are $\begin{bmatrix} k \\ 0 \\ -k \end{bmatrix}$ for every nonzero real k.

The characteristic vectors corresponding to $\lambda = 2$ have components x_1, x_2, x_3 such that

$$\begin{pmatrix} 1 & 1 & 0 \\ 1 & 0 & 1 \\ 0 & 1 & 1 \end{pmatrix} \begin{bmatrix} x_1 \\ x_2 \\ x_3 \end{bmatrix} = 2 \begin{bmatrix} x_1 \\ x_2 \\ x_3 \end{bmatrix}.$$

Thus x_1, x_2, x_3 must satisfy the system

$$
\begin{cases} x_1 + x_2 = 2x_1, \\ x_1 + x_3 = 2x_2, \\ x_2 + x_3 = 2x_3, \end{cases} \quad \text{or} \quad \begin{cases} x_1 + x_2 = 0 \\ x_1 - 2x_2 + x_3 = 0, \\ x_2 - x_3 = 0. \end{cases}
$$

We find $x_1 = k$, $x_2 = k$, $x_3 = k$ for every real k.   Hence

the characteristic vectors corresponding to $\lambda = 2$ are $\begin{bmatrix} k \\ k \\ k \end{bmatrix}$

for every nonzero real k.

The characteristic vectors corresponding to $\lambda = -1$ have components x_1, x_2, x_3 such that

$$
\begin{pmatrix} 1 & 1 & 0 \\ 1 & 0 & 1 \\ 0 & 1 & 1 \end{pmatrix} \begin{bmatrix} x_1 \\ x_2 \\ x_3 \end{bmatrix} = (-1) \begin{bmatrix} x_1 \\ x_2 \\ x_3 \end{bmatrix}.
$$

Thus x_1, x_2, x_3 must satisfy the system

$$
\begin{cases} x_1 + x_2 = -x_1, \\ x_1 + x_3 = -x_2, \\ x_2 + x_3 = -x_3, \end{cases} \quad \text{or} \quad \begin{cases} 2x_1 + x_2 = 0, \\ x_1 + x_2 + x_3 = 0, \\ x_2 + 2x_3 = 0. \end{cases}
$$

We find $x_1 = k$, $x_2 = -2k$, $x_3 = k$ for every real k.   Hence the characteristic vectors corresponding to $\lambda = -1$ are

$$\begin{bmatrix} k \\ -2k \\ k \end{bmatrix} \text{ for every nonzero real } k.$$

11. The characteristic equation is

$$\begin{vmatrix} 1 - \lambda & 3 & -6 \\ 0 & 2 - \lambda & 2 \\ 0 & -1 & 5 - \lambda \end{vmatrix} = 0,$$

which reduces to

$$(1 - \lambda) \begin{vmatrix} 2 - \lambda & 2 \\ -1 & 5 - \lambda \end{vmatrix} = 0$$

or $\lambda^3 - 8\lambda^2 + 19\lambda - 12 = 0$ or $(\lambda - 1)(\lambda - 3)(\lambda - 4) = 0$.
Thus, the characteristic values are $\lambda = 1, 3, 4$.

The characteristic vectors corresponding to $\lambda = 1$ have components x_1, x_2, x_3 such that

$$\begin{pmatrix} 1 & 3 & -6 \\ 0 & 2 & 2 \\ 0 & -1 & 5 \end{pmatrix} \begin{bmatrix} x_1 \\ x_2 \\ x_3 \end{bmatrix} = (1) \begin{bmatrix} x_1 \\ x_2 \\ x_3 \end{bmatrix}.$$

Thus x_1, x_2, x_3 must satisfy

$$\begin{cases} x_1 + 3x_2 - 6x_3 = x_1, \\ 2x_2 + 2x_3 = x_2, \\ -x_2 + 5x_3 = x_3, \end{cases} \text{ or } \begin{cases} 3x_2 - 6x_3 = 0, \\ x_2 + 2x_3 = 0, \\ -x_2 + 4x_3 = 0. \end{cases}$$

From these we see that x_1 is arbitrary and $x_2 = x_3 = 0$.
Thus we find $x_1 = k$, $x_2 = x_3 = 0$ for every real k.   Hence
the characteristic vectors corresponding to $\lambda = 1$ are $\begin{bmatrix} k \\ 0 \\ 0 \end{bmatrix}$
for every nonzero real k.

The characteristic vectors corresponding to $\lambda = 3$ have
components x_1, x_2, x_3 such that

$$\begin{pmatrix} 1 & 3 & -6 \\ 0 & 2 & 2 \\ 0 & -1 & 5 \end{pmatrix} \begin{bmatrix} x_1 \\ x_2 \\ x_3 \end{bmatrix} = 3 \begin{bmatrix} x_1 \\ x_2 \\ x_3 \end{bmatrix}.$$

Thus x_1, x_2, x_3 must satisfy the system

$$\begin{cases} x_1 + 3x_2 - 6x_3 = 3x_1, \\ 2x_2 + 2x_3 = 3x_2, \\ -x_2 + 5x_3 = 3x_3, \end{cases} \text{ or } \begin{cases} -2x_1 + 3x_2 - 6x_3 = 0, \\ -x_2 + 2x_3 = 0, \\ -x_2 + 2x_3 = 0. \end{cases}$$

The last two equations are identical.   From the second, x_2
$= 2x_3$.Then the first equation becomes $-2x_1 + 6x_3 - 6x_3 =$
0, from which $x_1 = 0$.   Thus we find $x_1 = 0$, $x_2 = 2k$, $x_3 =$
k for every real k.   Hence the characteristic vectors
corresponding to $\lambda = 3$ are $\begin{bmatrix} 0 \\ 2k \\ k \end{bmatrix}$ for every nonzero real k.

The characteristic vectors corresponding to $\lambda = 4$ have components x_1, x_2, x_3 such that

$$\begin{pmatrix} 1 & 3 & -6 \\ 0 & 2 & 2 \\ 0 & -1 & 5 \end{pmatrix} \begin{bmatrix} x_1 \\ x_2 \\ x_3 \end{bmatrix} = 4 \begin{bmatrix} x_1 \\ x_2 \\ x_3 \end{bmatrix}.$$

Thus x_1, x_2, x_3 must satisfy the system

$$\begin{cases} x_1 + 3x_2 - 6x_3 = 4x_1, \\ 2x_2 + 2x_3 = 4x_2, \\ -x_2 + 5x_3 = 4x_3, \quad \text{or} \end{cases} \qquad \begin{cases} -3x_1 + 3x_2 - 6x_3 = 0, \\ -2x_2 + 2x_3 = 0, \\ -x_2 + x_3 = 0. \end{cases}$$

The last two equations are equivalent, and from them, $x_2 = x_3$. Then from the first equation, $x_1 = x_2 - 2x_3 = -x_3$. Thus we find $x_1 = -k$, $x_2 = k$, $x_3 = k$ for every real k. Hence the characteristic vectors corresponding to $\lambda = 4$

are $\begin{bmatrix} -k \\ k \\ k \end{bmatrix}$ for every nonzero real k.

14. The characteristic equation is

$$\begin{vmatrix} -2 - \lambda & 6 & -18 \\ 12 & -23 - \lambda & 66 \\ 5 & -10 & 29 - \lambda \end{vmatrix} = 0,$$

which reduces to

$$(-2 - \lambda) \begin{vmatrix} -23 - \lambda & 66 \\ -10 & 29 - \lambda \end{vmatrix} - 6 \begin{vmatrix} 12 & 66 \\ 5 & 29 - \lambda \end{vmatrix}$$

$$- \begin{vmatrix} 12 & -23 - \lambda \\ 5 & -10 \end{vmatrix} = 0$$

or $\lambda^3 - 4\lambda^2 - \lambda + 4 = 0$ or $(\lambda - 1)(\lambda + 1)(\lambda - 4) = 0$.
Thus the characteristic values are $\lambda = 1, -1, 4$.

The characteristic vectors corresponding to $\lambda = 1$ have
components x_1, x_2, x_3 such that

$$\begin{pmatrix} -2 & 6 & -18 \\ 12 & -23 & 66 \\ 5 & -10 & 29 \end{pmatrix} \begin{bmatrix} x_1 \\ x_2 \\ x_3 \end{bmatrix} = 1 \begin{bmatrix} x_1 \\ x_2 \\ x_3 \end{bmatrix}.$$

Thus x_1, x_2, x_3 must satisfy the system

$$\begin{cases} -2x_1 + 6x_2 - 18x_3 = x_1, \\ 12x_1 - 23x_2 + 66x_3 = x_2, \\ 5x_1 - 10x_2 + 29x_3 = x_3, \end{cases} \text{ or } \begin{cases} -3x_1 + 6x_2 - 18x_3 = 0, \\ 12x_1 - 24x_2 + 66x_3 = 0, \\ 5x_1 - 10x_2 + 28x_3 = 0. \end{cases}$$

Adding four times the first to the second, we find $-6x_3 = 0$, and so $x_3 = 0$. Then the first reduces to $-3x_1 + 6x_2 = 0$, from which $x_1 = 2x_2$. Thus we find $x_1 = 2k$, $x_2 = k$, $x_3 = 0$ for every real k. Hence the characteristic vectors corresponding to $\lambda = 1$ are $\begin{bmatrix} 2k \\ k \\ 0 \end{bmatrix}$ for every nonzero real k.

The characteristic vectors corresponding to $\lambda = -1$ have components x_1, x_2, x_3 such that

$$\begin{pmatrix} -2 & 6 & -18 \\ 12 & -23 & 66 \\ 5 & -10 & 29 \end{pmatrix} \begin{bmatrix} x_1 \\ x_2 \\ x_3 \end{bmatrix} = (-1) \begin{bmatrix} x_1 \\ x_2 \\ x_3 \end{bmatrix}.$$

Thus x_1, x_2, x_3 must satisfy the system

$$\begin{cases} -2x_1 + 6x_2 - 18x_3 = -x_1, \\ 12x_1 - 23x_2 + 66x_3 = -x_2, \\ 5x_1 - 10x_2 + 29x_3 = -x_3, \end{cases} \quad \text{or} \quad \begin{cases} -x_1 + 6x_2 - 18x_3 = 0, \\ 12x_1 - 22x_2 + 66x_3 = 0, \\ 5x_1 - 10x_2 + 30x_3 = 0. \end{cases}$$

Adding 3/5 the third equation to the first gives $2x_1 = 0$, so $x_1 = 0$. Then the first equation becomes $6x_2 - 18x_3 = 0$, from which $x_2 = 3x_3$. Thus we find $x_1 = 0$, $x_2 = 3k$, $x_3 = k$, for every real k. Hence the characteristic vectors corresponding to $\lambda = -1$ are $\begin{bmatrix} 0 \\ 3k \\ k \end{bmatrix}$ for every nonzero real k.

The characteristic vectors corresponding to $\lambda = 4$ have components x_1, x_2, x_3 such that

$$\begin{pmatrix} -2 & 6 & -18 \\ 12 & -23 & 66 \\ 5 & -10 & 29 \end{pmatrix} \begin{bmatrix} x_1 \\ x_2 \\ x_3 \end{bmatrix} = 4 \begin{bmatrix} x_1 \\ x_2 \\ x_3 \end{bmatrix}.$$

Thus x_1, x_2, x_3 must satisfy the system

$$\begin{cases} -2x_1 + 6x_2 - 18x_3 = 4x_1, \\ 12x_1 - 23x_2 + 66x_3 = 4x_2, \\ 5x_1 - 10x_2 + 29x_3 = 4x_3, \end{cases} \quad \text{or} \quad \begin{cases} -6x_1 + 6x_2 - 18x_3 = 0, \\ 12x_1 - 27x_2 + 66x_3 = 0, \\ 5x_1 - 10x_2 + 25x_3 = 0. \end{cases}$$

The first and third reduce to

$$\begin{cases} -x_1 + x_2 - 3x_3 = 0, \\ x_1 - 2x_2 + 5x_3 = 0. \end{cases}$$

Regarding these as two equations in x_1 and x_2, we find $x_1 = -x_3$, $x_2 = 2x_3$. Letting $x_3 = -k$, we find $x_1 = k$, $x_2 = -2k$, $x_3 = -k$ for every real k. Hence the characteristic vectors corresponding to $\lambda = 4$ are $\begin{bmatrix} k \\ -2k \\ -k \end{bmatrix}$ for every nonzero real k.

Section 7.6, Page 377.

1. The characteristic equation of the coefficient matrix

$$\bar{A} = \begin{pmatrix} 5 & -2 \\ 4 & -1 \end{pmatrix} \text{ is}$$

$$|\bar{A} - \bar{\lambda}I| = \begin{vmatrix} 5 - \lambda & -2 \\ 4 & -1 - \lambda \end{vmatrix} = 0,$$

Expanding the determinant and simplifying, this takes the form $\lambda^2 - 4\lambda + 3 = 0$ with roots $\lambda_1 = 1$, $\lambda_2 = 3$. These are the characteristic values of A. They are distinct (and real), and so Theorem 7.10 of the text applies. We use equation (7.118) of the text to find corresponding characteristic vectors.

With $\lambda = \lambda_1 = 1$ and $\bar{a} = \bar{a}^{(1)} = \begin{pmatrix} a_1 \\ a_2 \end{pmatrix}$, (7.118) becomes

$$\begin{pmatrix} 5 & -2 \\ 4 & -1 \end{pmatrix} \begin{pmatrix} a_1 \\ a_2 \end{pmatrix} = 1 \begin{pmatrix} a_1 \\ a_2 \end{pmatrix},$$

from which we find that a_1 and a_2 must satisfy

$$5a_1 - 2a_2 = a_1, \qquad 2a_1 = a_2,$$
$$\text{or}$$
$$4a_1 - a_2 = a_2, \qquad 2a_1 = a_2.$$

A simple nontrivial solution is $a_1 = 1$, $a_2 = 2$, and thus a characteristic vectors corresponding to $\lambda = 1$ is

$\bar{a}^{(1)} = \begin{pmatrix} 1 \\ 2 \end{pmatrix}$. A solution is $\bar{x} = \begin{pmatrix} 1 \\ 2 \end{pmatrix} e^t$, that is,

$$\bar{x} = \begin{pmatrix} e^t \\ 2e^t \end{pmatrix}. \tag{*}$$

With $\lambda = \lambda_2 = 3$ and $\bar{a} = \bar{a}^{(2)} = \begin{pmatrix} a_1 \\ a_2 \end{pmatrix}$, (7.118) becomes

$$\begin{pmatrix} 5 & -2 \\ 4 & -1 \end{pmatrix} \begin{pmatrix} a_1 \\ a_2 \end{pmatrix} = 3 \begin{pmatrix} a_1 \\ a_2 \end{pmatrix}.$$

from which we find that a_1 and a_2 must satisfy

$$\begin{aligned} 5a_1 - 2a_2 &= 3a_1, & a_1 &= a_2, \\ & \text{or} \\ 4a_1 - a_2 &= 3a_2, & a_1 &= a_2. \end{aligned}$$

A simple nontrivial solution is $a_1 = a_2 = 1$, and thus a characteristic vectors corresponding to $\lambda_2 = 3$ is

$\bar{a}^{(2)} = \begin{pmatrix} 1 \\ 1 \end{pmatrix}$. A solution is $\bar{x} = \begin{pmatrix} 1 \\ 1 \end{pmatrix} e^{3t}$, that is,

$$\bar{x} = \begin{pmatrix} e^{3t} \\ e^{3t} \end{pmatrix}. \tag{**}$$

By Theorem 7.10 the solutions (*) and (**) are linearly independent, and a general solution is

$$\bar{x} = c_1 \begin{pmatrix} e^t \\ 2e^t \end{pmatrix} + c_2 \begin{pmatrix} e^{3t} \\ e^{3t} \end{pmatrix},$$

where c_1 and c_2 are arbitrary constants.   In scalar language,

$$x_1 = c_1 e^t + c_2 e^{3t},$$
$$x_2 = 2c_1 e^t + c_2 e^{3t}.$$

5. The characteristic equation of the coefficient matrix
$$\bar{A} = \begin{pmatrix} 3 & 1 \\ 4 & 3 \end{pmatrix} \text{ is}$$

$$|\bar{A} - \bar{\lambda}I| = \begin{vmatrix} 3 - \lambda & 1 \\ 4 & 3 - \lambda \end{vmatrix} = 0,$$

Expanding the determinant and simplifying, this takes the form $\lambda^2 - 6\lambda + 5 = 0$ with roots $\lambda_1 = 1$, $\lambda_2 = 5$.   These are the characteristic values of $\bar{A}$.   They are distinct (and real), and so Theorem 7.10 of the text applies.   We use equation (7.118) of the text to find corresponding characteristic vectors.

With $\lambda = \lambda_1 = 1$ and $\bar{a} = \bar{a}^{(1)} = \begin{pmatrix} a_1 \\ a_2 \end{pmatrix}$, (7.118) becomes

$$\begin{pmatrix} 3 & 1 \\ 4 & 3 \end{pmatrix} \begin{pmatrix} a_1 \\ a_2 \end{pmatrix} = 1 \begin{pmatrix} a_1 \\ a_2 \end{pmatrix},$$

from which we find that a_1 and a_2 must satisfy

$$\begin{cases} 3a_1 + a_2 = a_1, \\ 4a_1 + 3a_2 = a_2, \end{cases} \quad \text{or} \quad \begin{cases} 2a_1 = -a_2, \\ 4a_1 = -2a_2. \end{cases}$$

A simple nontrivial solution is $a_1 = 1$, $a_2 = -2$, and thus a characteristic vector corresponding to $\lambda_1 = 1$ is $\bar{a}^{(1)} = \begin{pmatrix} 1 \\ -2 \end{pmatrix}$. A solution is $\bar{x} = \begin{pmatrix} 1 \\ -2 \end{pmatrix} e^t$, that is,

$$\bar{x} = \begin{pmatrix} e^t \\ -2e^t \end{pmatrix}.$$

With $\lambda = \lambda_2 = 5$ and $\bar{a} = \bar{a}^{(2)} = \begin{pmatrix} a_1 \\ a_2 \end{pmatrix}$, (7.118) becomes

$$\begin{pmatrix} 3 & 1 \\ 4 & 3 \end{pmatrix} \begin{pmatrix} a_1 \\ a_2 \end{pmatrix} = 5 \begin{pmatrix} a_1 \\ a_2 \end{pmatrix},$$

from which we find that a_1 and a_2 must satisfy

$$\begin{cases} 3a_1 + a_2 = 5a_1, \\ 4a_1 + 3a_2 = 5a_2, \end{cases} \quad \text{or} \quad \begin{cases} 2a_1 = a_2, \\ 4a_1 = 2a_2. \end{cases}$$

A simple nontrivial solution is $a_1 = 1$, $a_2 = 2$, and thus a characteristic vector corresponding to $\lambda_2 = 5$ is

$\bar{a}^{(2)} = \begin{pmatrix} 1 \\ 2 \end{pmatrix}$. A solution is $\bar{x} = \begin{pmatrix} 1 \\ 2 \end{pmatrix} e^{5t}$, that is,

$$\bar{x} = \begin{pmatrix} e^{5t} \\ 2e^{5t} \end{pmatrix}. \qquad (**)$$

By Theorem 7.10 the solutions (*) and (**) are linearly independent, and a general solution is

$$\bar{x} = c_1 \begin{pmatrix} e^{t} \\ -2e^{t} \end{pmatrix} + c_2 \begin{pmatrix} e^{5t} \\ 2e^{5t} \end{pmatrix}$$

where c_1 and c_2 are arbitrary constants. In scalar language,

$$\begin{cases} x_1 = c_1 e^{t} + c_2 e^{5t}, \\ x_2 = -2c_1 e^{t} + 2c_2 e^{5t}. \end{cases}$$

6. The characteristic equation of the coefficient matrix

$\bar{A} = \begin{pmatrix} 6 & -1 \\ 3 & 2 \end{pmatrix}$ is

$$|\bar{A} - \lambda \bar{I}| = \begin{vmatrix} 6 - \lambda & -1 \\ 3 & 2 - \lambda \end{vmatrix} = 0,$$

Expanding the determinant and simplifying, this takes the form $\lambda^2 - 8\lambda + 15 = 0$ with roots $\lambda_1 = 3$, $\lambda_2 = 5$. These are the characteristic values of $\bar{A}$. They are distinct (and real), and so Theorem 7.10 of the text applies. We use equation (7.118) of the text to find corresponding characteristic vectors.

With $\lambda = \lambda_1 = 3$ and $\bar{a} = \bar{a}^{(1)} = \begin{pmatrix} a_1 \\ a_2 \end{pmatrix}$, (7.118) becomes

$$\begin{pmatrix} 6 & -1 \\ 3 & 2 \end{pmatrix} \begin{pmatrix} a_1 \\ a_2 \end{pmatrix} = 3 \begin{pmatrix} a_1 \\ a_2 \end{pmatrix},$$

from which we find that a_1 and a_2 must satisfy

$$6a_1 - a_2 = 3a_1, \qquad 3a_1 = a_2,$$

or

$$3a_1 + 2a_2 = 3a_2, \qquad 3a_1 = a_2.$$

A simple nontrivial solution is $a_1 = 1$, $a_2 = 3$, and thus a characteristic vector corresponding to $\lambda_1 = 3$ is $\bar{a}^{(1)} = \begin{pmatrix} 1 \\ 3 \end{pmatrix}$. A solution is $\bar{x} = \begin{pmatrix} 1 \\ 3 \end{pmatrix} e^{3t}$, that is,

$$\bar{x} = \begin{pmatrix} e^{3t} \\ 3e^{3t} \end{pmatrix}. \qquad (*)$$

With $\lambda = \lambda_2 = 5$ and $\bar{a} = \bar{a}^{(2)} = \begin{pmatrix} a_1 \\ a_2 \end{pmatrix}$, (7.118) becomes

$$\begin{pmatrix} 6 & -1 \\ 3 & 2 \end{pmatrix} \begin{pmatrix} a_1 \\ a_2 \end{pmatrix} = 5 \begin{pmatrix} a_1 \\ a_2 \end{pmatrix},$$

from which we find that a_1 and a_2 must satisfy

$$6a_1 - a_2 = 5a_1, \qquad a_1 = a_2,$$
$$\text{or}$$
$$3a_1 + 2a_2 = 5a_2, \qquad a_1 = a_2.$$

A simple nontrivial solution is $a_1 = a_2 = 1$, and thus a characteristic vector corresponding to $\lambda_2 = 5$ is $\bar{a}^{(2)} = \begin{pmatrix} 1 \\ 1 \end{pmatrix}$. A solution is $\bar{x} = \begin{pmatrix} 1 \\ 1 \end{pmatrix} e^{5t}$, that is,

$$\bar{x} = \begin{pmatrix} e^{5t} \\ e^{5t} \end{pmatrix}. \qquad (**)$$

By Theorem 7.10 the solutions (*) and (**) are linearly independent, and a general solution is

$$\bar{x} = c_1 \begin{pmatrix} e^{3t} \\ 3e^{3t} \end{pmatrix} + c_2 \begin{pmatrix} e^{5t} \\ e^{5t} \end{pmatrix},$$

where c_1 and c_2 are arbitrary constants. In scalar language,

$$x_1 = c_1 e^{3t} + c_2 e^{5t},$$
$$x_2 = 3c_1 e^{3t} + c_2 e^{5t}.$$

9. The characteristic equation of the coefficient matrix
$\bar{A} = \begin{pmatrix} 1 & -4 \\ 1 & 1 \end{pmatrix}$ is

$$|\bar{A} - \lambda\bar{I}| = \begin{vmatrix} 1 - \lambda & -4 \\ 1 & 1 - \lambda \end{vmatrix} = 0,$$

Expanding this determinant and simplifying, this takes the form $\lambda^2 - 2\lambda + 5 = 0$ with roots $1 \pm 2i$. These are the characteristic values of $\bar{A}$. They are distinct conjugate complex numbers, and Theorem 7.10 applies. We use equation (7.118) of the text.

With $\lambda = \lambda_1 = 1 + 2i$ and $\bar{a} = \bar{a}^{(1)} = \begin{pmatrix} a_1 \\ a_2 \end{pmatrix}$, (7.118)

becomes

$$\begin{pmatrix} 1 & -4 \\ 1 & 1 \end{pmatrix} \begin{pmatrix} a_1 \\ a_2 \end{pmatrix} = (1 + 2i) \begin{pmatrix} a_1 \\ a_2 \end{pmatrix},$$

from which we find that a_1 and a_2 must satisfy

$$a_1 - 4a_2 = (1 + 2i)a_1, \qquad 2i\, a_1 = -4a_2\, ,$$

$$\text{or}$$

$$a_1 + a_2 = (1 + 2i)a_2, \qquad a_1 = 2i\, a_2\, .$$

A simple nontrivial solution is $a_1 = 2i$, $a_2 = 1$, and thus a characteristic vectors corresponding to $\lambda_1 = 1 + 2i$ is $\bar{a}^{(1)} = \begin{pmatrix} 2i \\ 1 \end{pmatrix}$. A solution is $\bar{x} = \begin{pmatrix} 2i \\ 1 \end{pmatrix} e^{(1+2i)t}$, that is,

$$\bar{x} = \begin{pmatrix} 2i\, e^{(1+2i)t} \\ e^{(1+2i)t} \end{pmatrix}. \qquad (*)$$

We could now let $\lambda = \lambda_2 = 1 - 2i$ in (7118) and find a corresponding characteristic vector $\bar{a}^{(2)}$ and a linearly independent complex solution $\bar{a}^{(2)} e^{(1-2i)t}$. Instead we proceed as in Example 7.18 of Section 7.4C and apply Euler's formula $e^{i\theta} = \cos\theta + i\sin\theta$ to each component of $(*)$. We have

$$\begin{cases} x_1 = 2i\, e^{(1+2i)t} \\[2mm] x_2 = e^{(1+2i)t} \end{cases}$$

and applying the formula and simplifying, this takes the form

$$x_1 = 2e^t(-\sin 2t + i \cos 2t),$$
$$x_2 = e^t(\cos 2t + i \sin 2t).$$

The real and imaginary parts of this solution are themselves solutions, so we obtain

$$x_1 = -2e^t\sin 2t, \qquad x_1 = 2e^t\cos 2t,$$
$$\text{and}$$
$$x_2 = e^t\cos 2t, \qquad x_2 = e^t\sin 2t.$$

These two solutions are linearly independent, so a G.S. may be written

$$x_1 = 2e^t(-c_1 \sin 2t + c_2 \cos 2t),$$
$$x_2 = e^t(c_1 \cos 2t + c_2 \sin 2t).$$

where c_1 and c_2 are arbitrary constants.

10. The characteristic equation of the coefficient matrix

$$\bar{A} = \begin{pmatrix} 2 & -3 \\ 3 & 2 \end{pmatrix} \text{ is }$$

$$|\bar{A} - \lambda \bar{I}| = \begin{vmatrix} 2 - \lambda & -3 \\ 3 & 2 - \lambda \end{vmatrix} = 0,$$

Expanding the determinant and simplifying, this takes the form $\lambda^2 - 4\lambda + 13 = 0$ with roots $\lambda = 2 \pm 3i$. These are the characteristic values of $\bar{A}$. They are distinct conjugate complex numbers, and Theorem 7.10 applies. We use equation (7.118) of the text.

$$\text{With } \lambda = \lambda_1 = 2 + 3i \text{ and } \bar{a} = \bar{a}^{(1)} = \begin{pmatrix} a_1 \\ a_2 \end{pmatrix}, \quad (7.118)$$

becomes

$$\begin{pmatrix} 2 & -3 \\ 3 & 2 \end{pmatrix} \begin{pmatrix} a_1 \\ a_2 \end{pmatrix} = (2 + 3i) \begin{pmatrix} a_1 \\ a_2 \end{pmatrix},$$

from which we find that a_1 and a_2 must satisfy

$$2a_1 - 3a_2 = (2 + 3i)a_1, \qquad -a_2 = i\,a_1,$$
$$\text{or}$$
$$3a_1 + 2a_2 = (2 + 3i)a_2, \qquad a_1 = i\,a_2.$$

A simple nontrivial solution is $a_1 = i$, $a_2 = 1$, and thus a characteristic vector corresponding to $\lambda_1 = 2 + 3i$ is

$$\overline{a}^{(1)} = \begin{pmatrix} i \\ 1 \end{pmatrix}. \quad \text{A solution is } \overline{x} = \begin{pmatrix} i \\ 1 \end{pmatrix} e^{(2+3i)t}, \text{ that is,}$$

$$\overline{x} = \begin{pmatrix} i\, e^{(2+3i)t} \\ e^{(2+3i)t} \end{pmatrix}. \qquad (*)$$

As in our solution to Exercise 9 just preceding this solution, we now proceed as in Example 7.18 of Section 7.4C. We apply Euler's formula $e^{i\,\theta} = \cos\theta + i\sin\theta$ to each component of $(*)$. We have

$$\begin{cases} x_1 = i\, e^{(2+3i)t}, \\ x_2 = e^{(2+3i)t}, \end{cases}$$

and applying the formula and simplifying, this takes the form

$$x_1 = e^{2t}(-\sin 3t + i\cos 3t),$$
$$x_2 = e^{2t}(\cos 3t + i\sin 3t).$$

The real and imaginary parts of this solution are themselves solutions, so we obtain

$$\begin{cases} x_1 = -e^{2t}\sin 3t, \\ x_2 = e^{2t}\cos 3t, \end{cases} \quad \text{and} \quad \begin{cases} x_1 = e^{2t}\cos 3t, \\ x_2 = e^{2t}\sin 3t. \end{cases}$$

These two solutions are linearly independent, so a G.S.
may be written

$$
\begin{cases}
x_1 = e^{2t}(-c_1 \sin 3t + c_2 \cos 3t), \\[2mm]
x_2 = e^{2t}(c_1 \cos 3t + c_2 \sin 3t).
\end{cases}
$$

12. The characteristic equation of the coefficient matrix
$\bar{A} = \begin{pmatrix} 5 & -4 \\ 2 & 1 \end{pmatrix}$ is

$$
|\bar{A} - \lambda\bar{I}| = \begin{vmatrix} 5 - \lambda & -4 \\ 2 & 1 - \lambda \end{vmatrix} = 0,
$$

Expanding this determinant and simplifying, this takes the
form $\lambda^2 - 6\lambda + 13 = 0$ with roots $3 \pm 2i$. These are the
characteristic values of $\bar{A}$. They are distinct conjugate
complex numbers, and Theorem 7.10 applies. We use
equation (7.118) of the text.

$$
\text{With } \lambda = \lambda_1 = 3 + 2i \text{ and } \bar{a} = \bar{a}^{(1)} = \begin{pmatrix} a_1 \\ a_2 \end{pmatrix}, \quad (7.118)
$$

becomes

$$
\begin{pmatrix} 5 & -4 \\ 2 & 1 \end{pmatrix}\begin{pmatrix} a_1 \\ a_2 \end{pmatrix} = (3 + 2i)\begin{pmatrix} a_1 \\ a_2 \end{pmatrix},
$$

from which we find that a_1 and a_2 must satisfy

$$5a_1 - 4a_2 = (3 + 2i)a_1, \qquad (1 - i)a_1 = 2a_2,$$

$$\text{or}$$

$$2a_1 + a_2 = (3 + 2i)a_2, \qquad a_1 = (1 + i)a_2.$$

A simple nontrivial solution is $a_1 = 2$, $a_2 = 1 - i$, and thus a characteristic vector corresponding to $\lambda_1 = 3 + 2i$

is $\bar{a}^{(1)} = \begin{pmatrix} 2 \\ 1 - i \end{pmatrix}$. A solution is $\bar{x} = \begin{pmatrix} 2 \\ 1 - i \end{pmatrix} e^{(3+2i)t}$,

that is,

$$\bar{x} = \begin{pmatrix} 2 e^{(3+2i)t} \\ (1 - i)e^{(3+2i)t} \end{pmatrix}. \qquad (*)$$

As in our solutions to Exercises 9 and 10, we apply Euler's formula $e^{i\theta} = \cos\theta + i\sin\theta$ to each component of $(*)$. We have

$$x_1 = 2e^{(3+2i)t},$$

$$x_2 = (1 - i)e^{(3+2i)t};$$

and applying the formula and simplifying, this takes the form

$$x_1 = 2e^{3t}(\cos 2t + i \sin 2t),$$

$$x_2 = e^{3t}[(\cos 2t + \sin 2t) + i(\sin 2t - \cos 2t)].$$

The real and imaginary parts of this solution are themselves solutions, so we obtain

$$x_1 = 2e^{3t}\cos 2t, \qquad\qquad x_1 = 2e^{3t}\sin 2t,$$

and

$$x_2 = e^{3t}(\cos 2t + \sin 2t), \qquad x_2 = e^{3t}(\sin 2t - \cos 2t).$$

These two solutions are linearly independent, so a G.S. may be written

$$x_1 = 2e^{3t}(c_1 \cos 2t + c_2 \sin 2t),$$
$$x_2 = e^{3t}[c_1(\cos 2t + \sin 2t) + c_2(\sin 2t - \cos 2t),$$

where c_1 and c_2 are arbitrary constants.

14. The characteristic equation of the coefficient matrix

$$\bar{A} = \begin{pmatrix} 4 & -5 \\ 1 & 6 \end{pmatrix} \text{ is}$$

$$|\bar{A} - \lambda\bar{I}| = \begin{vmatrix} 4 - \lambda & -5 \\ 1 & 6 - \lambda \end{vmatrix} = 0,$$

Expanding the determinant and simplifying, this takes the form $\lambda^2 - 10\lambda + 29 = 0$ with roots $5 \pm 2i$. These are the characteristic values of $\bar{A}$. They are distinct conjugate complex numbers, and Theorem 7.10 applies. We use equation (7.118) of the text.

With $\lambda = \lambda_1 = 5 + 2i$ and $\bar{a} = \bar{a}^{(1)} = \begin{pmatrix} a_1 \\ a_2 \end{pmatrix}$, (7.118)

becomes

$$\begin{pmatrix} 4 & -5 \\ 1 & 6 \end{pmatrix}\begin{pmatrix} a_1 \\ a_2 \end{pmatrix} = (5 + 2i)\begin{pmatrix} a_1 \\ a_2 \end{pmatrix},$$

from which we find that a_1 and a_2 must satisfy

$$\begin{cases} 4a_1 - 5a_2 = (5 + 2i)a_1, \\ a_1 + 6a_2 = (5 + 2i)a_2, \end{cases} \quad \text{or} \quad \begin{cases} (1 + 2i)a_1 + 5a_2 = 0, \\ a_1 + (1 - 2i)a_2 = 0. \end{cases}$$

A simple nontrivial solution is $a_1 = 5$, $a_2 = -(1 + 2i)$, and thus a characteristic vector corresponding to $\lambda_1 = 5 + 2i$ is $\bar{a}^{(1)} = \begin{pmatrix} 5 \\ -(1 + 2i) \end{pmatrix}$. A solution is $\bar{x} = \begin{pmatrix} 5 \\ -(1 + 2i) \end{pmatrix} e^{(5+2i)t}$, that is,

$$\bar{x} = \begin{pmatrix} 5\,e^{(5+2i)t} \\ -(1+2i)e^{(5+2i)t} \end{pmatrix}. \tag{*}$$

As in our solution to Exercises 9, 10, and 12 we apply Euler's formula $e^{i\theta} = \cos\theta + i\sin\theta$ to each component of $(*)$. We have

$$\begin{cases} x_1 = 5\,e^{(5+2i)t}, \\[2mm] x_2 = -(1+2i)\,e^{(5+2i)t}, \end{cases}$$

and applying the formula and simplifying, this takes the form

$$\begin{cases} x_1 = 5e^{5t}(\cos 2t + i \sin 2t), \\[2mm] x_2 = -e^{5t}[(\cos 2t - 2 \sin 2t) + i(2 \cos 2t) + \sin 2t)]. \end{cases}$$

The real and imaginary parts of this solution are themselves solutions, so we obtain

$$\begin{cases} x_1 = 5e^{5t}\cos 2t, \\[2mm] x_2 = -e^{5t}(\cos 2t - 2 \sin 2t), \end{cases}$$

and

$$\begin{cases} x_1 = 5e^{5t}\sin 2t, \\[2mm] x_2 = -e^{5t}(2 \cos 2t + \sin 2t). \end{cases}$$

These two solutions are linearly independent, so a G.S. may be written

$$\begin{cases} x_1 = 5e^{5t}(c_1 \cos 2t + c_2 \sin 2t), \\[2mm] x_2 = -e^{5t}[c_1(\cos 2t - 2 \sin 2t) + c_2(2 \cos 2t + \sin 2t)]. \end{cases}$$

15. The characteristic equation of the coefficient matrix

$$\bar{A} = \begin{pmatrix} 3 & -1 \\ 4 & -1 \end{pmatrix} \text{ is}$$

$$|\bar{A} - \bar{\lambda}I| = \begin{vmatrix} 3 - \lambda & -1 \\ 4 & -1 - \lambda \end{vmatrix} = 0,$$

Expanding the determinant and simplifying, this takes the form $\lambda^2 - 2\lambda + 1 = 0$ with double root $\lambda_1 = 1$. That is the characteristic values of A are real and equal and so Theorem 7.11 applies. We use equation (7.118) to find a characteristic vector $\bar{a}$.

With $\lambda = 1$ and $\bar{a} = \begin{pmatrix} a_1 \\ a_2 \end{pmatrix}$, (7.118) becomes

$$\begin{pmatrix} 3 & -1 \\ 4 & -1 \end{pmatrix} \begin{pmatrix} a_1 \\ a_2 \end{pmatrix} = 1 \begin{pmatrix} a_1 \\ a_2 \end{pmatrix},$$

from which we find that a_1 and a_2 must satisfy

$$3a_1 - a_2 = a_1, \qquad 2a_1 = a_2,$$
$$\text{or}$$
$$4a_1 - a_2 = a_2, \qquad 2a_1 = a_2.$$

A simple nontrivial solution is $a_1 = 1$, $a_2 = 2$, and thus a characteristic vector corresponding to $\lambda = 1$ is $\bar{a} = \begin{pmatrix} 1 \\ 2 \end{pmatrix}$.

A solution is $\bar{x} = \begin{pmatrix} 1 \\ 2 \end{pmatrix} e^t$, that is,

$$\bar{x} = \begin{pmatrix} e^t \\ 2e^t \end{pmatrix}. \qquad (*)$$

By Theorem 7.11, a linearly independent solution is of the form $(\bar{a}\, t + \bar{\beta})e^{\lambda t}$, where $\bar{a} = \begin{pmatrix} 1 \\ 2 \end{pmatrix}$, $\lambda = 1$, and $\bar{\beta}$

satisfies $(\bar{A} - \lambda \bar{I})\bar{\beta} = \bar{a}$. Thus $\bar{\beta} = \begin{pmatrix} \beta_1 \\ \beta_2 \end{pmatrix}$ satisfies

$$\left[\begin{pmatrix} 3 & -1 \\ 4 & -1 \end{pmatrix} - \begin{pmatrix} 1 & 0 \\ 0 & 1 \end{pmatrix} \right] \begin{pmatrix} \beta_1 \\ \beta_2 \end{pmatrix} = \begin{pmatrix} 1 \\ 2 \end{pmatrix},$$

which reduces to

$$\begin{pmatrix} 2 & -1 \\ 4 & -2 \end{pmatrix} \begin{pmatrix} \beta_1 \\ \beta_2 \end{pmatrix} = \begin{pmatrix} 1 \\ 2 \end{pmatrix}.$$

From this we find that β_1 and β_2 must satisfy

$$2\beta_1 - \beta_2 = 1,$$
$$4\beta_1 - 2\beta_2 = 2.$$

A simple nontrivial solution is $\beta_1 = 0$, $\beta_2 = -1$. Thus we find $\bar{\beta} = \begin{pmatrix} 0 \\ -1 \end{pmatrix}$, and the "second" solution is

$$\bar{x} = \left[\begin{pmatrix} 1 \\ 2 \end{pmatrix} t + \begin{pmatrix} 0 \\ -1 \end{pmatrix} \right] e^t, \text{ that is,}$$

$$\bar{x} = \begin{pmatrix} te^t \\ (2t - 1)e^t \end{pmatrix}. \tag{**}$$

By Theorem 7.11 the solutions (*)(**) are linearly independent, and a general solution is

$$\bar{x} = c_1 \begin{pmatrix} e^t \\ 2e^t \end{pmatrix} + c_2 \begin{pmatrix} te^t \\ (2t - 1)e^t \end{pmatrix},$$

where c_1 and c_2 are arbitrary constants. In scalar language,

$$x_1 = c_1 e^t + c_2 te^t,$$
$$x_2 = 2c_1 e^t + c_2(2t - 1)e^t.$$

16. The characteristic equation of the coefficient matrix $\bar{A} = \begin{pmatrix} 7 & 4 \\ -1 & 3 \end{pmatrix}$ is

$$|\bar{A} - \bar{\lambda}I| = \begin{vmatrix} 7 - \lambda & 4 \\ -1 & 3 - \lambda \end{vmatrix} = 0,$$

Expanding the determinant and simplifying, this takes the form $\lambda^2 - 10\lambda + 25 = 0$ with double root $\lambda = 5$. That is the characteristic values of $\bar{A}$ are real and equal and so Theorem 7.11 applies. We first use equation (7.118) to find a characteristic vector $\bar{a}$.

With $\lambda = 5$ and $\bar{a} = \begin{pmatrix} a_1 \\ a_2 \end{pmatrix}$, (7.118) becomes

$$\begin{pmatrix} 7 & 4 \\ -1 & 3 \end{pmatrix}\begin{pmatrix} a_1 \\ a_2 \end{pmatrix} = 5\begin{pmatrix} a_1 \\ a_2 \end{pmatrix},$$

from which we find that a_1 and a_2 must satisfy

$$7a_1 + 4a_2 = 5a_1, \qquad a_1 = -2a_2,$$
$$\text{or}$$
$$-a_1 + 3a_2 = 5a_2, \qquad a_1 = -2a_2.$$

A simple nontrivial solution is $a_1 = -2$, $a_2 = 1$, and thus a characteristic vector corresponding to $\lambda - 5$ is

$\bar{a} = \begin{pmatrix} -2 \\ 1 \end{pmatrix}$. A solution is $\bar{x} = \begin{pmatrix} -2 \\ 1 \end{pmatrix}e^{5t}$, that is,

$$\bar{x} = \begin{pmatrix} -2e^{5t} \\ e^{5t} \end{pmatrix}. \qquad\qquad (*)$$

By Theorem 7.11, a linearly independent solution is of the form $(\bar{a}t + \bar{\beta})e^{\lambda t}$, where $\bar{a} = \begin{pmatrix} -2 \\ 1 \end{pmatrix}$, $\lambda = 5$, and $\bar{\beta}$

satisfies $(\bar{A} - \lambda \bar{I})\bar{\beta} = \bar{a}$. Thus $\bar{\beta} = \begin{pmatrix} \beta_1 \\ \beta_2 \end{pmatrix}$ satisfies

$$\left[\begin{pmatrix} 7 & 4 \\ -1 & 3 \end{pmatrix} - 5 \begin{pmatrix} 1 & 0 \\ 0 & 1 \end{pmatrix} \right] \begin{pmatrix} \beta_1 \\ \beta_2 \end{pmatrix} = \begin{pmatrix} -2 \\ 1 \end{pmatrix},$$

which reduces to

$$\begin{pmatrix} 2 & 4 \\ -1 & -2 \end{pmatrix} \begin{pmatrix} \beta_1 \\ \beta_2 \end{pmatrix} = \begin{pmatrix} -2 \\ 1 \end{pmatrix}.$$

From this we find that β_1 and β_2 must satisfy

$$2\beta_1 + 4\beta_2 = -2,$$
$$-\beta_1 - 2\beta_2 = 1.$$

A simple nontrivial solution is $\beta_1 = -1$, $\beta_2 = 0$. Thus we find $\bar{\beta} = \begin{pmatrix} -1 \\ 0 \end{pmatrix}$, and the "second" solution is

$$\bar{x} = \left[\begin{pmatrix} -2 \\ 1 \end{pmatrix} t + \begin{pmatrix} -1 \\ 0 \end{pmatrix} \right] e^{5t}, \text{ that is,}$$

$$\bar{x} = \begin{pmatrix} -(2t + 1)e^{5t} \\ te^{5t} \end{pmatrix}. \tag{**}$$

By Theorem 7.11 the solutions (*) and (**) are linearly independent, and a general solution is

$$\bar{x} = c_1 \begin{pmatrix} -2e^{5t} \\ e^{5t} \end{pmatrix} + c_2 \begin{pmatrix} -(2t + 1)e^{5t} \\ te^{5t} \end{pmatrix},$$

where c_1 and c_2 are arbitrary constants.   In scalar language,

$$x_1 = -2c_1 e^{5t} - c_2(2t + 1)e^{5t},$$
$$x_2 = c_1 e^{5t} + c_2 te^{5t}.$$

19.   The characteristic equation of the coefficient matrix $\bar{A} = \begin{pmatrix} 6 & -4 \\ 1 & 2 \end{pmatrix}$ is

$$|\bar{A} - \lambda \bar{I}| = \begin{vmatrix} 6 - \lambda & -4 \\ 1 & 2 - \lambda \end{vmatrix} = 0,$$

Expanding the determinant and simplifying, this takes the form $\lambda^2 - 8\lambda + 16 = 0$ with double root $\lambda = 4$.   That is, the characteristic values of $\bar{A}$ are real and equal and so Theorem 7.11 applies.   We first use equation (7.118) to find a characteristic vector $\bar{a}$.

With $\lambda = 4$ and $\bar{a} = \begin{pmatrix} a_1 \\ a_2 \end{pmatrix}$, (7.118) becomes

$$\begin{pmatrix} 6 & -4 \\ 1 & 2 \end{pmatrix} \begin{pmatrix} a_1 \\ a_2 \end{pmatrix} = 4 \begin{pmatrix} a_1 \\ a_2 \end{pmatrix},$$

from which we find that a_1 and a_2 must satisfy

$$\begin{cases} 6a_1 - 4a_2 = 4a_1, \\ a_1 + 2a_2 = 4a_2, \end{cases} \quad \text{or} \quad \begin{cases} 2a_1 = 4a_2, \\ a_1 = 2a_2. \end{cases}$$

A simple nontrivial solution is $a_1 = 2$, $a_2 = 1$, and thus a characteristic vector corresponding to $\lambda = 4$ is $\bar{a} = \begin{pmatrix} -2 \\ 1 \end{pmatrix}$.
A solution is $\bar{x} = \begin{pmatrix} 2 \\ 1 \end{pmatrix} e^{4t}$, that is,

$$\bar{x} = \begin{pmatrix} 2e^{4t} \\ e^{4t} \end{pmatrix}. \quad (*)$$

By Theorem 7.11, a linearly independent solution is of the form $(\bar{a}\,t + \bar{\beta})e^{\lambda t}$, where $\bar{a} = \begin{pmatrix} 2 \\ 1 \end{pmatrix}$, $\lambda = 4$, and $\bar{\beta}$

satisfies $(\bar{A} - \lambda\bar{I})\bar{\beta} = \bar{a}$. Thus $\bar{\beta} = \begin{pmatrix} \beta_1 \\ \beta_2 \end{pmatrix}$ satisfies

$$\left[\begin{pmatrix} 6 & -4 \\ 1 & 2 \end{pmatrix} - 4\begin{pmatrix} 1 & 0 \\ 0 & 1 \end{pmatrix} \right]\begin{pmatrix} \beta_1 \\ \beta_2 \end{pmatrix} = \begin{pmatrix} 2 \\ 1 \end{pmatrix},$$

which reduces to

$$\begin{pmatrix} 2 & -4 \\ 1 & -2 \end{pmatrix}\begin{pmatrix} \beta_1 \\ \beta_2 \end{pmatrix} = \begin{pmatrix} 2 \\ 1 \end{pmatrix}.$$

From this we find that β_1 and β_2 must satisfy

$$2\beta_1 - 4\beta_2 = 2,$$
$$\beta_1 - 2\beta_2 = 1.$$

A simple nontrivial solution is $\beta_1 = 1$, $\beta_2 = 0$. Thus we find $\bar{\beta} = \begin{pmatrix} 1 \\ 0 \end{pmatrix}$, and the "second" solution is

$$\bar{x} = \left[\begin{pmatrix} 2 \\ 1 \end{pmatrix}t + \begin{pmatrix} 1 \\ 0 \end{pmatrix} \right]e^{4t}, \text{ that is,}$$

$$\bar{x} = \begin{pmatrix} (2t + 1)e^{4t} \\ te^{4t} \end{pmatrix}. \qquad (**)$$

By Theorem 7.11 the solutions (*) and (**) are linearly independent, and a general solution is

$$\bar{x} = c_1 \begin{pmatrix} 2e^{4t} \\ e^{4t} \end{pmatrix} + c_2 \begin{pmatrix} (2t+1)e^{4t} \\ te^{4t} \end{pmatrix},$$

where c_1 and c_2 are arbitrary constants. In scalar language,

$$x_1 = 2c_1 e^{4t} + c_2(2t+1)e^{4t},$$
$$x_2 = c_1 e^{4t} + c_2 te^{4t}.$$

Section 7.7, Page 400.

1. We assume a solution of the form $\bar{x} = \bar{a}\,e^{\lambda t}$, that is, $x_1 = a_1 e^{\lambda t}$, $x_2 = a_2 e^{\lambda t}$, $x_3 = a_3 e^{\lambda t}$. We know that λ must be a solution of the characteristic equation of the coefficient matrix

$$\bar{A} = \begin{pmatrix} 1 & 1 & -1 \\ 2 & 3 & -4 \\ 4 & 1 & -4 \end{pmatrix}.$$

The characteristic equation is

$$\begin{vmatrix} 1-\lambda & 1 & -1 \\ 2 & 3-\lambda & -4 \\ 4 & 1 & -4-\lambda \end{vmatrix} = 0.$$

Expanding the determinant and simplifying, it reduces to $\lambda^3 - 7\lambda + 6 = 0$ or $(\lambda - 1)(\lambda - 2)(\lambda + 3) = 0$ (see solution of Exercise 7 of Section 7.5D).   Its roots are the characteristic values $\lambda_1 = 1$, $\lambda_2 = 2$, $\lambda_3 = -3$.

We use $\bar{A}\,\bar{a} = \lambda\bar{a}$ to find the corresponding characteristic vectors.   With $\lambda = 1$ this is

$$
\begin{pmatrix} 1 & 1 & -1 \\ 2 & 3 & -4 \\ 4 & 1 & -4 \end{pmatrix}
\begin{bmatrix} a_1 \\ a_2 \\ a_3 \end{bmatrix}
= 1
\begin{bmatrix} a_1 \\ a_2 \\ a_3 \end{bmatrix}.
$$

Thus a_1, a_2, a_3 must satisfy

$$
a_1 - a_3 = 0,
$$
$$
2a_1 + 2a_2 - 4a_3 = 0,
$$
$$
4a_1 + a_2 - 5a_3 = 0.
$$

A nontrivial solution of this is $a_1 = 1$, $a_2 = 1$, $a_3 = 1$ (see solution of Exercise 7 of Section 7.5D, where x's there are used in place of a's here).   Thus a characteristic vector corresponding to $\lambda = 1$ is

$$
\bar{a} = \begin{bmatrix} 1 \\ 1 \\ 1 \end{bmatrix},
$$

and a corresponding solution is

$$
\begin{pmatrix} 1 \\ 1 \\ 1 \end{pmatrix} e^t, \quad \text{that is,} \quad
\begin{bmatrix} e^t \\ e^t \\ e^t \end{bmatrix}. \tag{*}
$$

With $\lambda = 2$, $\bar{A}\bar{a} = \lambda\bar{a}$ is

$$\begin{pmatrix} 1 & 1 & -1 \\ 2 & 3 & -4 \\ 4 & 1 & -4 \end{pmatrix} \begin{bmatrix} a_1 \\ a_2 \\ a_3 \end{bmatrix} = 2 \begin{bmatrix} a_1 \\ a_2 \\ a_3 \end{bmatrix}.$$

Thus a_1, a_2, a_3 must satisfy

$$-a_1 + a_2 - a_3 = 0,$$

$$2a_1 + a_2 - 4a_3 = 0,$$

$$4a_1 + a_2 - 6a_3 = 0.$$

A nontrivial solution of this is $a_1 = 1$, $a_2 = 2$, $a_3 = 1$. Thus a characteristic vector corresponding to $\lambda = 2$ is

$$\bar{a} = \begin{bmatrix} 1 \\ 2 \\ 1 \end{bmatrix},$$

and a corresponding solution is

$$\begin{pmatrix} 1 \\ 2 \\ 1 \end{pmatrix} e^{2t}, \quad \text{that is,} \quad \begin{bmatrix} e^{2t} \\ 2e^{2t} \\ e^{2t} \end{bmatrix}. \qquad (**)$$

With $\lambda = -3$, $\bar{A}\bar{a} = \lambda\bar{a}$ is

$$\begin{pmatrix} 1 & 1 & -1 \\ 2 & 3 & -4 \\ 4 & 1 & -4 \end{pmatrix} \begin{bmatrix} a_1 \\ a_2 \\ a_3 \end{bmatrix} = -3 \begin{bmatrix} a_1 \\ a_2 \\ a_3 \end{bmatrix}.$$

Thus a_1, a_2, a_3 must satisfy

$$4a_1 + a_2 - a_3 = 0,$$

$$2a_1 + 6a_2 - 4a_3 = 0,$$

$$4a_1 + a_2 - a_3 = 0.$$

A nontrivial solution of this is $a_1 = 1$, $a_2 = 7$, $a_3 = 11$.
Thus a characteristic vector corresponding to $\lambda = -3$ is

$$\bar{a} = \begin{bmatrix} 1 \\ 7 \\ 11 \end{bmatrix},$$

and a corresponding solution is

$$\begin{pmatrix} 1 \\ 7 \\ 11 \end{pmatrix} e^{-3t}, \quad \text{that is,} \quad \begin{bmatrix} e^{-3t} \\ 7e^{-3t} \\ 11e^{-3t} \end{bmatrix}. \qquad (***)$$

By Theorem 7.16, the solutions (*), (**), and (***)
are linearly independent, and a G.S. is

$$\bar{x} = c_1 \begin{bmatrix} e^t \\ e^t \\ e^t \end{bmatrix} + c_2 \begin{bmatrix} e^{2t} \\ 2e^{2t} \\ e^{2t} \end{bmatrix} + c_3 \begin{bmatrix} e^{-3t} \\ 7e^{-3t} \\ 11e^{-3t} \end{bmatrix},$$

where c_1, c_2, c_3 are arbitrary constants. In component form, this is

$$x_1 = c_1e^t + c_2e^{2t} + c_3e^{-3t},$$
$$x_2 = c_1e^t + 2c_2e^{2t} + 7c_3e^{-3t},$$
$$x_3 = c_1e^t + c_2e^{2t} + 11c_3e^{-3t}.$$

2. We assume a solution of the form $\bar{x} = \bar{a}\,e^{\lambda t}$, that is,
$x_1 = a_1e^{\lambda t}$, $x_2 = a_2e^{\lambda t}$, $x_3 = a_3e^{\lambda t}$. We know that λ must
be a solution of the characteristic equation of the
coefficient matrix

$$\bar{A} = \begin{pmatrix} 1 & -1 & -1 \\ 1 & 3 & 1 \\ -3 & -6 & 6 \end{pmatrix}.$$

The characteristic equation is

$$\begin{vmatrix} 1 - \lambda & -1 & -1 \\ 1 & 3 - \lambda & 1 \\ -3 & -6 & 6 - \lambda \end{vmatrix} = 0. \qquad (*)$$

We evaluate the determinant in $(*)$ by expanding by
cofactors along its first row. We obtain

$$(1 - \lambda)\begin{vmatrix} 3 - \lambda & 1 \\ -6 & 6 - \lambda \end{vmatrix} - (-1)\begin{vmatrix} 1 & 1 \\ -3 & 6 - \lambda \end{vmatrix}$$

$$+ (-1)\begin{vmatrix} 1 & 3 - \lambda \\ -3 & -6 \end{vmatrix} = 0.$$

This reduces to

$$(1 - \lambda)[(3 - \lambda)(6 - \lambda) + 6] + [6 - \lambda + 3]$$
$$- [-6 + 9 - 3\lambda] = 0$$

which simplifies to

$$(\lambda - 1)(\lambda^2 - 9\lambda + 24) - (-\lambda + 9) + (-3\lambda + 3) = 0$$

and hence to

$$\lambda^3 - 10\lambda^2 + 31\lambda - 30 = 0.$$

Thus the characteristic equation of $\bar{A}$ reduces to the cubic equation $\lambda^3 - 10\lambda^2 + 31\lambda - 30 = 0$. We see by inspection that $\lambda = 2$ is a root of this, and then by synthetic division we obtain $(\lambda - 2)(\lambda^2 - 8\lambda + 15) = 0$. Finally, then, we obtain the characteristic equation in the factored form $(\lambda - 2)(\lambda - 3)(\lambda - 5) = 0$. The roots of this are the characteristic values $\lambda_1 = 2$, $\lambda_2 = 3$, $\lambda_3 = 5$.

We have gone through the preceeding evaluation and solution in some detail. In third-order problems it is essential that we carry out the evaluation and solution of the characteristic equation carefully and correctly. Making an error here can be very time-consuming and frustrating. We shall omit the corresponding details from future solutions in this manual. We refer the reader in need of help to Appendices 1 and 2 of the text.

Going on with the present solution, we use $\bar{A}\bar{a} = \lambda\bar{a}$ to find the corresponding characteristic vectors. With $\lambda = 2$, this is

$$\begin{pmatrix} 1 & -1 & -1 \\ 1 & 3 & 1 \\ -3 & -6 & 6 \end{pmatrix} \begin{bmatrix} a_1 \\ a_2 \\ a_3 \end{bmatrix} = 2 \begin{bmatrix} a_1 \\ a_2 \\ a_3 \end{bmatrix}.$$

Thus a_1, a_2, a_3 must satisfy

$$-a_1 - a_2 - a_3 = 0,$$

$$a_1 + a_2 + a_3 = 0,$$

$$-3a_1 - 6a_2 + 4a_3 = 0.$$

The first two equations are equivalent. Solving the last two for a_1 and a_2 in terms of a_3, we find $a_1 = -(10/3)a_3$, $a_2 = (7/3)a_3$. Thus a nontrivial solution of the algebraic system is $a_1 = 10$, $a_2 = -7$, $a_3 = -3$. Thus a character-istic vector corresponding to $\lambda = 2$ is

$$\bar{a} = \begin{bmatrix} 10 \\ -7 \\ -3 \end{bmatrix},$$

and a corresponding solution is

$$\begin{pmatrix} 10 \\ -7 \\ -3 \end{pmatrix} e^{2t}, \quad \text{that is,} \quad \begin{bmatrix} 10e^{2t} \\ -7e^{2t} \\ -3e^{2t} \end{bmatrix}. \qquad (**)$$

With $\lambda = 3$, $\bar{A}\bar{a} = \lambda\bar{a}$ is

$$\begin{pmatrix} 1 & -1 & -1 \\ 1 & 3 & 1 \\ -3 & -6 & 6 \end{pmatrix}\begin{bmatrix} a_1 \\ a_2 \\ a_3 \end{bmatrix} = 3\begin{bmatrix} a_1 \\ a_2 \\ a_3 \end{bmatrix}.$$

Thus a_1, a_2, a_3 must satisfy

$$-2a_1 - a_2 - a_3 = 0,$$
$$a_1 + a_3 = 0,$$
$$-3a_1 - 6a_2 + 3a_3 = 0.$$

A nontrivial solution of this is $a_1 = 1$, $a_2 = -1$, $a_3 = -1$.
Thus a characteristic vector corresponding to $\lambda = 3$ is

$$\bar{a} = \begin{bmatrix} 1 \\ -1 \\ -1 \end{bmatrix},$$

and a corresponding solution is

$$\begin{pmatrix} 1 \\ -1 \\ -1 \end{pmatrix}e^{3t}, \quad \text{that is,} \quad \begin{bmatrix} e^{3t} \\ -e^{3t} \\ -e^{3t} \end{bmatrix}. \qquad (***)$$

With $\lambda = 5$, $\bar{A}\bar{a} = \lambda\bar{a}$ is

$$\begin{pmatrix} 1 & -1 & -1 \\ 1 & 3 & 1 \\ -3 & -6 & 6 \end{pmatrix} \begin{bmatrix} a_1 \\ a_2 \\ a_3 \end{bmatrix} = 5 \begin{bmatrix} a_1 \\ a_2 \\ a_3 \end{bmatrix}.$$

Thus a_1, a_2, a_3 must satisfy

$$-4a_1 - a_2 - a_3 = 0,$$

$$a_1 - 2a_2 + a_3 = 0,$$

$$-3a_1 - 6a_2 + a_3 = 0.$$

Adding the first two equations, we obtain $-3a_1 - 3a_2 = 0$, from which $a_2 = -a_1$. Then the third equation gives $a_3 = 3a_1 + 6a_2 = -3a_1$. So a nontrivial solution of the algebraic system is $a_1 = 1$, $a_2 = -1$, $a_3 = -3$. Thus a characteristic vector corresponding to $\lambda = 5$ is

$$\bar{a} = \begin{bmatrix} 1 \\ -1 \\ -3 \end{bmatrix},$$

and a corresponding solution is

$$\begin{pmatrix} 1 \\ -1 \\ -3 \end{pmatrix} e^{5t}, \quad \text{that is,} \quad \begin{bmatrix} e^{5t} \\ -e^{5t} \\ -3e^{5t} \end{bmatrix} \qquad (****)$$

By Theorem 7.16, the solutions (**), (***), and (****) are linearly independent, and a G.S. is

$$\bar{x} = c_1 \begin{bmatrix} 10e^{2t} \\ -7e^{2t} \\ -3e^{2t} \end{bmatrix} + c_2 \begin{bmatrix} e^{3t} \\ -e^{3t} \\ -e^{3t} \end{bmatrix} + c_3 \begin{bmatrix} e^{5t} \\ -e^{5t} \\ -3e^{5t} \end{bmatrix},$$

where c_1, c_2, c_3 are arbitrary constants. In component form, this is

$$x_1 = 10c_1 e^{2t} + c_2 e^{3t} + c_3 e^{5t},$$
$$x_2 = -7c_1 e^{2t} - c_2 e^{3t} - c_3 e^{5t},$$
$$x_3 = -3c_1 e^{2t} - c_2 e^{3t} - 3c_3 e^{5t}.$$

6. We assume a solution of the form $\bar{x} = \bar{a}\,e^{\lambda t}$, that is, $x_1 = a_1 e^{\lambda t}$, $x_2 = a_2 e^{\lambda t}$, $x_3 = a_3 e^{\lambda t}$. We know that λ must be a solution of the characteristic equation of the coefficient matrix

$$\bar{A} = \begin{pmatrix} 1 & 1 & 0 \\ 1 & 0 & 1 \\ 0 & 1 & 1 \end{pmatrix}.$$

The characteristic equation is

$$\begin{vmatrix} 1 - \lambda & 1 & 0 \\ 1 & -\lambda & 1 \\ 0 & 1 & 1 - \lambda \end{vmatrix} = 0.$$

Expanding the determinant and simplifying, it reduces to $\lambda^3 - 2\lambda^2 - \lambda + 2 = 0$ or $(\lambda - 1)(\lambda - 2)(\lambda + 1) = 0$. Its roots are the characteristic values $\lambda_1 = 1$, $\lambda_2 = 2$, $\lambda_3 = -1$.

We use $\bar{A}\,\bar{a} = \lambda a$ to find the corresponding characteristic vectors. With $\lambda = 1$ this is

$$\begin{pmatrix} 1 & 1 & 0 \\ 1 & 0 & 1 \\ 0 & 1 & 1 \end{pmatrix} \begin{bmatrix} a_1 \\ a_2 \\ a_3 \end{bmatrix} = 1 \begin{bmatrix} a_1 \\ a_2 \\ a_3 \end{bmatrix}.$$

Thus a_1, a_2, a_3 must satisfy

$$a_1 = 0,$$
$$a_1 - a_2 + a_3 = 0,$$
$$a_1 = 0.$$

A nontrivial solution of this is $a_1 = 1$, $a_2 = 0$, $a_3 = -1$. Thus a characteristic vector corresponding to $\lambda = 1$ is

$$\bar{a} = \begin{bmatrix} 1 \\ 0 \\ -1 \end{bmatrix},$$

and a corresponding solution is

$$\begin{pmatrix} 1 \\ 0 \\ -1 \end{pmatrix} e^t, \quad \text{that is,} \quad \begin{bmatrix} e^t \\ 0 \\ -e^t \end{bmatrix}. \qquad (*)$$

With $\lambda = 2$, $\bar{A}\bar{a} = \lambda\bar{a}$ is

$$\begin{pmatrix} 1 & 1 & 0 \\ 1 & 0 & 1 \\ 0 & 1 & 1 \end{pmatrix} \begin{bmatrix} a_1 \\ a_2 \\ a_3 \end{bmatrix} = 2 \begin{bmatrix} a_1 \\ a_2 \\ a_3 \end{bmatrix}.$$

Thus a_1, a_2, a_3 must satisfy

$$-a_1 + a_2 = 0,$$

$$a_1 - 2a_2 + a_3 = 0,$$

$$a_1 - a_3 = 0.$$

A nontrivial solution of this is $a_1 = 1$, $a_2 = 1$, $a_3 = 1$.
Thus a characteristic vector corresponding to $\lambda = 2$ is

$$\bar{a} = \begin{bmatrix} 1 \\ 1 \\ 1 \end{bmatrix},$$

and a corresponding solution is

$$\begin{pmatrix} 1 \\ 1 \\ 1 \end{pmatrix} e^{2t}, \quad \text{that is,} \quad \begin{bmatrix} e^{2t} \\ e^{2t} \\ e^{2t} \end{bmatrix}. \qquad (**)$$

With $\lambda = -1$, $\bar{A}\bar{a} = \lambda\bar{a}$ is

$$\begin{pmatrix} 1 & 1 & 0 \\ 1 & 0 & 1 \\ 0 & 1 & 1 \end{pmatrix} \begin{bmatrix} a_1 \\ a_2 \\ a_3 \end{bmatrix} = -1 \begin{bmatrix} a_1 \\ a_2 \\ a_3 \end{bmatrix}.$$

Thus a_1, a_2, a_3 must satisfy

$$2a_1 + a_2 = 0,$$

$$a_1 + a_2 + a_3 = 0,$$

$$a_2 + 2a_3 = 0.$$

A nontrivial solution of this is $a_1 = 1$, $a_2 = -2$, $a_3 = 1$. Thus a characteristic vector corresponding to $\lambda = -1$ is

$$\bar{a} = \begin{bmatrix} 1 \\ -2 \\ 1 \end{bmatrix},$$

and a corresponding solution is

$$\begin{pmatrix} 1 \\ -2 \\ 1 \end{pmatrix} e^{-t}, \quad \text{that is,} \quad \begin{bmatrix} e^{-t} \\ -2e^{-t} \\ e^{-t} \end{bmatrix} \qquad (***)$$

By Theorem 7.16, the solutions (*), (**), and (***) are linearly independent, and a G.S. is

$$\bar{x} = c_1 \begin{bmatrix} e^t \\ 0 \\ -e^t \end{bmatrix} + c_2 \begin{bmatrix} e^{2t} \\ e^{2t} \\ e^{2t} \end{bmatrix} + c_3 \begin{bmatrix} e^{-t} \\ -2e^{-t} \\ e^{-t} \end{bmatrix},$$

where c_1, c_2, c_3 are arbitrary constants. In component form, this is

$$x_1 = c_1 e^t + c_2 e^{2t} + c_3 e^{-t},$$
$$x_2 = c_2 e^{2t} - 2c_3 e^{-t},$$
$$x_3 = c_1 e^t + c_2 e^{2t} + c_3 e^{-t}.$$

7. We assume a solution of the form $\bar{x} = \bar{a} e^{\lambda t}$, that is, $x_1 = a_1 e^{\lambda t}$, $x_2 = a_2 e^{\lambda t}$, $x_3 = a_3 e^{\lambda t}$. We know that λ must be a solution of the characteristic equation of the coefficient matrix

$$\bar{A} = \begin{pmatrix} 1 & -2 & 0 \\ -2 & 3 & 0 \\ 0 & 0 & 2 \end{pmatrix}.$$

The characteristic equation is

$$\begin{vmatrix} 1-\lambda & -2 & 0 \\ -2 & 3-\lambda & 0 \\ 0 & 0 & 2-\lambda \end{vmatrix} = 0.$$

Expanding the determinant and simplifying, it reduces to $\lambda^3 - 6\lambda^2 + 7\lambda + 2 = 0$ or $(\lambda - 2)(\lambda^2 - 4\lambda - 1) = 0$. Its roots are the characteristic values $\lambda_1 = 2 + \sqrt{5}$, $\lambda_2 = 2 - \sqrt{5}$, $\lambda_3 = 2$.

We use $\bar{A}\,\bar{a} = \lambda\bar{a}$ to find the corresponding characteristic vectors. With $\lambda = 2 + \sqrt{5}$ this is

$$\begin{pmatrix} 1 & -2 & 0 \\ -2 & 3 & 0 \\ 0 & 0 & 2 \end{pmatrix} \begin{bmatrix} a_1 \\ a_2 \\ a_3 \end{bmatrix} = (2 + \sqrt{5}) \begin{bmatrix} a_1 \\ a_2 \\ a_3 \end{bmatrix} .$$

Thus a_1, a_2, a_3 must satisfy

$$(-1 - \sqrt{5})a_1 - 2a_2 = 0,$$

$$-2a_1 + (1 - \sqrt{5})a_2 = 0,$$

$$- \sqrt{5}\,a_3 = 0.$$

A nontrivial solution of this is $a_1 = -2$, $a_2 = 1 + \sqrt{5}$, $a_3 = 0$. Thus a characteristic vector corresponding to $\lambda = 2 + \sqrt{5}$ is

$$\bar{a} = \begin{pmatrix} -2 \\ 1 + \sqrt{5} \\ 0 \end{pmatrix},$$

and a corresponding solution is

$$\begin{pmatrix} -2 \\ 1 + \sqrt{5} \\ 0 \end{pmatrix} e^{(2+\sqrt{5})t}, \text{ that is, } \begin{bmatrix} -2e^{(2+\sqrt{5})t} \\ (1 + \sqrt{5})e^{(2+\sqrt{5})t} \\ 0 \end{bmatrix} \quad (*)$$

With $\lambda = 2 - \sqrt{5}$, $A\bar{a} = \lambda\bar{a}$ is

$$\begin{pmatrix} 1 & -2 & 0 \\ -2 & 3 & 0 \\ 0 & 0 & 2 \end{pmatrix} \begin{bmatrix} a_1 \\ a_2 \\ a_3 \end{bmatrix} = (2 - \sqrt{5}) \begin{bmatrix} a_1 \\ a_2 \\ a_3 \end{bmatrix}.$$

Thus a_1, a_2, a_3 must satisfy

$$(-1 + \sqrt{5})a_1 - 2a_2 = 0,$$
$$-2a_1 + (1 + \sqrt{5})a_2 = 0,$$
$$\sqrt{5}\, a_3 = 0.$$

A nontrivial solution of this is $a_1 = 2$, $a_2 = -1 + \sqrt{5}$, $a_3 = 0$. Thus a characteristic vector corresponding to $\lambda = 2 - \sqrt{5}$ is

$$\begin{pmatrix} 2 \\ -1 + \sqrt{5} \\ 0 \end{pmatrix},$$

and a corresponding solution is

$$\begin{pmatrix} 2 \\ 1 + \sqrt{5} \\ 0 \end{pmatrix} e^{(2-\sqrt{5})t}, \text{ that is, } \begin{bmatrix} 2e^{(2-\sqrt{5})t} \\ (-1 + \sqrt{5})e^{(2-\sqrt{5})t} \\ 0 \end{bmatrix} \quad (**)$$

With $\lambda = 2$, $\bar{A}\bar{a} = \lambda\bar{a}$ is

$$\begin{pmatrix} 1 & -2 & 0 \\ -2 & 3 & 0 \\ 0 & 0 & 2 \end{pmatrix} \begin{bmatrix} a_1 \\ a_2 \\ a_3 \end{bmatrix} = 2 \begin{bmatrix} a_1 \\ a_2 \\ a_3 \end{bmatrix}.$$

Thus a_1, a_2, a_3 must satisfy

$$-a_1 - 2a_2 = 0,$$
$$-2a_1 + a_2 = 0,$$
$$0a_3 = 0.$$

A nontrivial solution of this is $a_1 = 0$, $a_2 = 0$, $a_3 = 1$. Thus a characteristic vector corresponding to $\lambda = 2$ is

$$\bar{a} = \begin{bmatrix} 0 \\ 0 \\ 1 \end{bmatrix},$$

and a corresponding solution is

$$\begin{pmatrix} 0 \\ 0 \\ 1 \end{pmatrix} e^{2t}, \quad \text{that is,} \quad \begin{bmatrix} 0 \\ 0 \\ e^{2t} \end{bmatrix}. \qquad (***)$$

By Theorem 7.16, the solutions (*), (**), and (***) are linearly independent, and a G.S. is

$$
\bar{x} = c_1 \begin{bmatrix} -2e^{(2+\sqrt{5})t} \\ (1 + \sqrt{5})e^{(2+\sqrt{5})t} \\ 0 \end{bmatrix} + c_2 \begin{bmatrix} 2e^{(2-\sqrt{5})t} \\ (-1 + \sqrt{5})e^{(2-\sqrt{5})t} \\ 0 \end{bmatrix}
$$

$$
+ c_3 \begin{bmatrix} 0 \\ 0 \\ e^{2t} \end{bmatrix}.
$$

where c_1, c_2, c_3 are arbitrary constants. In component form, this is

$$
x_1 = -2c_1 e^{(2+\sqrt{5})t} + 2c_2 e^{(2-\sqrt{5})t},
$$

$$
x_2 = (1 + \sqrt{5})c_1 e^{(2+\sqrt{5})t} + (-1 + \sqrt{5})c_2 e^{(2-\sqrt{5})t},
$$

$$
x_3 = c_3 e^{2t}.
$$

10. We assume a solution of the form $\bar{x} = \bar{a}\,e^{\lambda t}$, that is, $x_1 = a_1 e^{\lambda t}$, $x_2 = a_2 e^{\lambda t}$, $x_3 = a_3 e^{\lambda t}$. We know that λ must be a solution of the characteristic equation of the coefficient matrix

$$
\bar{A} = \begin{pmatrix} 3 & 7 & -3 \\ 1 & 2 & -2 \\ 1 & 6 & -2 \end{pmatrix}.
$$

The characteristic equation is

$$
\begin{vmatrix} 3 - \lambda & 7 & -3 \\ 1 & 2 - \lambda & -2 \\ 1 & 6 & -2 - \lambda \end{vmatrix} = 0.
$$

Expanding the determinant and simplifying, it reduces to $\lambda^3 - 3\lambda^2 + 4\lambda - 12 = 0$, or $(\lambda - 3)(\lambda^2 + 4) = 0$. Its roots are the characteristic values $\lambda_1 = 3$, $\lambda_2 = 2i$, $\lambda_3 = -2i$.

We use $\bar{A}\bar{a} = \lambda\bar{a}$ to find the corresponding characteristic vectors. With $\lambda = 3$ this is

$$\begin{pmatrix} 3 & 7 & -3 \\ 1 & 2 & -2 \\ 1 & 6 & -2 \end{pmatrix} \begin{bmatrix} a_1 \\ a_2 \\ a_3 \end{bmatrix} = 3 \begin{bmatrix} a_1 \\ a_2 \\ a_3 \end{bmatrix}.$$

Thus a_1, a_2, a_3 must satisfy

$$7a_2 - 3a_3 = 0,$$
$$a_1 - a_2 - 2a_3 = 0,$$
$$a_1 + 6a_2 - 5a_3 = 0.$$

A nontrivial solution of this is $a_1 = 17$, $a_2 = 3$, $a_3 = 7$. Thus a characteristic vector corresponding to $\lambda = 3$ is

$$\bar{a} = \begin{bmatrix} 17 \\ 3 \\ 7 \end{bmatrix},$$

and a corresponding solution is

$$\begin{pmatrix} 17 \\ 3 \\ 7 \end{pmatrix} e^{3t}, \quad \text{that is,} \quad \begin{bmatrix} 17e^{3t} \\ 3e^{3t} \\ 7e^{3t} \end{bmatrix} \quad (*)$$

With $\lambda = 2i$, $\bar{A}\,\bar{a} = \lambda\bar{a}$ is

$$\begin{pmatrix} 3 & 7 & -3 \\ 1 & 2 & -2 \\ 1 & 6 & -2 \end{pmatrix} \begin{bmatrix} a_1 \\ a_2 \\ a_3 \end{bmatrix} = 2i \begin{bmatrix} a_1 \\ a_2 \\ a_3 \end{bmatrix}.$$

Thus a_1, a_2, a_3 must satisfy

$$(3 - 2i)a_1 + 7a_2 - 3a_3 = 0,$$

$$a_1 + (2 - 2i)a_2 - 2a_3 = 0,$$

$$a_1 + 6a_2 + (-2 - 2i)a_3 = 0.$$

Subtracting the second equation from the third, we obtain

$$(4 + 2i)a_2 - 2i\, a_3 = 0$$

and hence $(2 + i)a_2 = i\, a_3$. A nontrivial solution of this is $a_2 = 1$, $a_3 = 1 - 2i$. Then $a_1 = (-2 + 2i)a_2 + 2a_3 = (-2 + 2i)(1) + 2(1 - 2i) = -2i$. Thus a characteristic vector corresponding to $\lambda = 2i$ is

$$\bar{a} = \begin{bmatrix} -2i \\ 1 \\ 1 - 2i \end{bmatrix},$$

and a corresponding solution is

$$
\begin{pmatrix} -2i \\ 1 \\ 1 - 2i \end{pmatrix} e^{2i\,t}, \quad \text{that is,} \quad \begin{bmatrix} -2ie^{2i\,t} \\ e^{2i\,t} \\ (1-2)e^{2i\,t} \end{bmatrix}.
$$

Proceeding as in Section 7.6, we apply Euler's formula $e^{i\theta} = \cos\theta + i\sin\theta$ to each component. We obtain the solution

$$
\begin{pmatrix} -2i(\cos 2t + i \sin 2t) \\ \cos 2t + i \sin 2t \\ (1 - 2i)(\cos 2t + i \sin 2t) \end{pmatrix},
$$

which simplifies to

$$
\begin{pmatrix} 2 \sin 2t + i(-2 \cos 2t) \\ \cos 2t + i \sin 2t \\ \cos 2t + 2 \sin 2t + i(-2 \cos 2t + \sin 2t) \end{pmatrix}.
$$

The real and imaginary parts of this solution are themselves solutions, so we obtain the solutions

$$
\begin{pmatrix} 2 \sin 2t \\ \cos 2t \\ \cos 2t + 2 \sin 2t \end{pmatrix} \quad \text{and} \quad \begin{pmatrix} -2 \cos 2t \\ \sin 2t \\ -2 \cos 2t + \sin 2t \end{pmatrix}. \quad (**)
$$

The solution (*) and the two solutions (**) are linearly independent, so a G.S. is

$$\bar{x} = c_1 \begin{bmatrix} 17e^{3t} \\ 3e^{3t} \\ 7e^{3t} \end{bmatrix} + c_2 \begin{pmatrix} 2 \sin 2t \\ \cos 2t \\ \cos 2t + 2 \sin 2t \end{pmatrix}$$

$$+ c_3 \begin{pmatrix} -2 \cos 2t \\ \sin 2t \\ \sin 2t - 2 \cos 2t \end{pmatrix}.$$

where c_1, c_2, c_3 are arbitrary constants. In component form, this is

$$\begin{cases} x_1 = 17c_1 e^{3t} + 2c_2 \sin 2t - 2c_3 \cos 2t, \\[2mm] x_2 = 3c_1 e^{3t} + c_2 \cos 2t + c_3 \sin 2t, \\[2mm] x_3 = 7c_1 e^{3t} + c_2(\cos 2t + 2 \sin 2t) \\[2mm] \qquad + c_3(\sin 2t - 2 \cos 2t). \end{cases}$$

13.   We assume a solution of the form $\bar{x} = \bar{a}\, e^{\lambda t}$, that is, $x_1 = a_1 e^{\lambda t}$, $x_2 = a_2 e^{\lambda t}$, $x_3 = a_3 e^{\lambda t}$. We know that λ must be a solution of the characteristic equation of the coefficient matrix

$$\bar{A} = \begin{pmatrix} 1 & -3 & 9 \\ 0 & -5 & 18 \\ 0 & -3 & 10 \end{pmatrix}.$$

The characteristic equation is

$$\begin{vmatrix} 1 - \lambda & -3 & 9 \\ 0 & -5 - \lambda & 18 \\ 0 & -3 & 10 - \lambda \end{vmatrix} = 0.$$

Expanding this determinant and simplifying, it reduces to $(\lambda - 1)(\lambda^2 - 5\lambda + 4) = 0$ or $(\lambda - 1)^2(\lambda - 4) = 0$. Its roots are the characteristic vectors $\lambda_1 = 4$, $\lambda_2 = \lambda_3 = 1$.

We use $\bar{A}\,\bar{a} = \lambda\bar{a}$ to find the corresponding characteristic vectors. With $\lambda = 4$ this is

$$\begin{pmatrix} 1 & -3 & 9 \\ 0 & -5 & 18 \\ 0 & -3 & 10 \end{pmatrix} \begin{bmatrix} a_1 \\ a_2 \\ a_3 \end{bmatrix} = 4 \begin{bmatrix} a_1 \\ a_2 \\ a_3 \end{bmatrix}.$$

Thus a_1, a_2, a_3 must satisfy

$$-3a_1 - 3a_2 + 9a_3 = 0,$$
$$- 9a_2 + 18a_3 = 0,$$
$$- 3a_2 + 6a_3 = 0.$$

From either the second or third equation $a_2 = 2a_3$. Then the first equation gives $a_1 = -a_2 + 3a_3 = a_3$. A non-trivial solution of this is $a_1 = 1$, $a_2 = 2$, $a_3 = 1$. Thus a characteristic vector corresponding to $\lambda = 4$ is

$$\bar{a} = \begin{bmatrix} 1 \\ 2 \\ 1 \end{bmatrix},$$

and a corresponding solution is

$$\begin{pmatrix} 1 \\ 2 \\ 1 \end{pmatrix} e^{4t}, \quad \text{that is,} \quad \begin{bmatrix} e^{4t} \\ 2e^{4t} \\ e^{4t} \end{bmatrix}. \qquad (*)$$

With $\lambda = 1$, $\bar{A}\bar{a} = \lambda\bar{a}$ is

$$\begin{pmatrix} 1 & -3 & 9 \\ 0 & -5 & 18 \\ 0 & -3 & 10 \end{pmatrix} \begin{bmatrix} a_1 \\ a_2 \\ a_3 \end{bmatrix} = (1) \begin{bmatrix} a_1 \\ a_2 \\ a_3 \end{bmatrix}.$$

Thus a_1, a_2, a_3 must satisfy

$$-3a_2 + 9a_3 = 0,$$

$$-6a_2 + 18a_3 = 0,$$

$$-3a_2 + 9a_3 = 0.$$

Each of these three relationships is equivalent to the one relation

$$a_2 = 3a_3, \qquad (**)$$

and this the only relationship which a_1, a_2, a_3 must satisfy. In particular, observe that a_1 is arbitrary. Two linearly independent solutions of $(**)$ are $a_1 = 0$, $a_2 = 3$, $a_3 = 1$ and $a_1 = 1$, $a_2 = 3$, $a_3 = 1$. That is, corresponding to the double characteristic value $\lambda = 1$, we have the two linearly independent characteristic vectors

$$\begin{pmatrix} 0 \\ 3 \\ 1 \end{pmatrix} \quad \text{and} \quad \begin{pmatrix} 1 \\ 3 \\ 1 \end{pmatrix}.$$

Respective corresponding solutions are

$$\begin{pmatrix} 0 \\ 3 \\ 1 \end{pmatrix} e^t, \quad \text{and} \quad \begin{pmatrix} 1 \\ 3 \\ 1 \end{pmatrix} e^t,$$

that is,

$$\begin{bmatrix} 0 \\ 3e^t \\ e^t \end{bmatrix} \quad \text{and} \quad \begin{bmatrix} e^t \\ 3e^t \\ e^t \end{bmatrix}. \qquad (\ast\ast\ast)$$

The three solutions given by $(\ast)$ and $(\ast\ast\ast)$ are linearly independent, and a G.S. is

$$\bar{x} = c_1 \begin{bmatrix} e^{4t} \\ 2e^{4t} \\ e^{4t} \end{bmatrix} + c_2 \begin{bmatrix} e^t \\ 3e^t \\ e^t \end{bmatrix} + c_3 \begin{bmatrix} 0 \\ 3e^t \\ e^t \end{bmatrix},$$

where c_1, c_2, c_3 are arbitrary constants. In component form, this is

$$x_1 = c_1 e^{4t} + c_2 e^t,$$

$$x_2 = 2c_1 e^{4t} + 3c_2 e^t + 3c_3 e^t,$$

$$x_3 = c_1 e^{4t} + c_2 e^t + c_3 e^t.$$

15. We assume a solution of the form $\bar{x} = \bar{a}\, e^{\lambda t}$, that is, $x_1 = a_1 e^{\lambda t}$, $x_2 = a_2 e^{\lambda t}$, $x_3 = a_3 e^{\lambda t}$. We know that λ must be a solution of the characteristic equation of the coefficient matrix

$$\bar{A} = \begin{pmatrix} 11 & 6 & 18 \\ 9 & 8 & 18 \\ -9 & -6 & -16 \end{pmatrix}.$$

The characteristic equation is

$$\begin{vmatrix} 11 - \lambda & 6 & 18 \\ 9 & 8 - \lambda & 18 \\ -9 & -6 & -16 - \lambda \end{vmatrix} = 0.$$

Expanding the determinant and simplifying, it reduces to $\lambda^3 - 3\lambda^2 + 4 = 0$ or $(\lambda + 1)(\lambda - 2)^2 = 0$. Its roots are the characteristic vectors $\lambda_1 = -1$, $\lambda_2 = \lambda_3 = 2$.

We use $\bar{A}\,\bar{a} = \lambda \bar{a}$ to find the corresponding characteristic vectors. With $\lambda = -1$ this is

$$\begin{pmatrix} 11 & 6 & 18 \\ 9 & 8 & 18 \\ -9 & -6 & -16 \end{pmatrix} \begin{bmatrix} a_1 \\ a_2 \\ a_3 \end{bmatrix} = (-1) \begin{bmatrix} a_1 \\ a_2 \\ a_3 \end{bmatrix}.$$

Thus a_1, a_2, a_3 must satisfy

$$12a_1 + 6a_2 + 18a_3 = 0,$$
$$9a_1 + 9a_2 + 18a_3 = 0,$$
$$-9a_1 - 6a_2 - 15a_3 = 0.$$

These simplify to

$$2a_1 + a_2 + 3a_3 = 0,$$
$$a_1 + a_2 + 2a_3 = 0,$$
$$3a_1 + 2a_2 + 5a_3 = 0.$$

A nontrivial solution of this is $a_1 = 1$, $a_2 = 1$, $a_3 = -1$. Thus a characteristic vector corresponding to $\lambda = -1$ is

$$\bar{a} = \begin{bmatrix} 1 \\ 1 \\ -1 \end{bmatrix},$$

and a corresponding solution is

$$\begin{pmatrix} 1 \\ 1 \\ -1 \end{pmatrix} e^{-t}, \quad \text{that is,} \quad \begin{bmatrix} e^{-t} \\ e^{-t} \\ -e^{-t} \end{bmatrix}. \qquad (*)$$

With $\lambda = 2$, $\bar{A}\bar{a} = \lambda\bar{a}$ is

$$
\begin{pmatrix} 11 & 6 & 18 \\ 9 & 8 & 18 \\ -9 & -6 & -16 \end{pmatrix} \begin{bmatrix} a_1 \\ a_2 \\ a_3 \end{bmatrix} = 2 \begin{bmatrix} a_1 \\ a_2 \\ a_3 \end{bmatrix}.
$$

Thus a_1, a_2, a_3 must satisfy

$$9a_1 + 6a_2 + 18a_3 = 0,$$

$$9a_1 + 6a_2 + 18a_3 = 0,$$

$$-9a_1 - 6a_2 - 18a_3 = 0.$$

Each of these three relationships is equivalent to the one relation

$$3a_1 + 2a_2 + 6a_3 = 0,$$

and this the only relationship which a_1, a_2, a_3 must satisfy. Two linearly independent solutions of it are $a_1 = 2$, $a_2 = 0$, $a_3 = -1$ and $a_1 = 0$, $a_2 = 3$, $a_3 = -1$. That is, corresponding to the double characteristic value $\lambda = 2$, we have the two linearly independent characteristic vectors

$$
\begin{bmatrix} 2 \\ 0 \\ -1 \end{bmatrix} \quad \text{and} \quad \begin{bmatrix} 0 \\ 3 \\ -1 \end{bmatrix}.
$$

Respective corresponding solutions are

$$\begin{pmatrix} 2 \\ 0 \\ -1 \end{pmatrix} e^{2t}, \quad \text{and} \quad \begin{pmatrix} 0 \\ 3 \\ -1 \end{pmatrix} e^{2t},$$

that is,

$$\begin{bmatrix} 2e^{2t} \\ 0 \\ -e^{2t} \end{bmatrix} \quad \text{and} \quad \begin{bmatrix} 0 \\ 3e^{2t} \\ -e^{2t} \end{bmatrix}. \qquad (**)$$

The three solutions given by (*) and (**) are linearly independent, and a G.S. is

$$\bar{x} = c_1 \begin{bmatrix} e^{-t} \\ e^{-t} \\ -e^{-t} \end{bmatrix} + c_2 \begin{bmatrix} 2e^{2t} \\ 0 \\ -e^{2t} \end{bmatrix} + c_3 \begin{bmatrix} 0 \\ 3e^{2t} \\ -e^{2t} \end{bmatrix},$$

where c_1, c_2, c_3 are arbitrary constants. In component form, this is

$$x_1 = c_1 e^{-t} + 2c_2 e^{2t},$$
$$x_2 = c_1 e^{-t} + 3c_3 e^{2t},$$
$$x_3 = -c_1 e^{-t} + c_2 e^{2t} - c_3 e^{2t}.$$

16. We assume a solution of the form $\bar{x} = \bar{a} e^{\lambda t}$, that is,
$x_1 = a_1 e^{\lambda t}$, $x_2 = a_2 e^{\lambda t}$, $x_3 = a_3 e^{\lambda t}$. We know that λ must

be a solution of the characteristic equation of the coefficient matrix

$$\bar{A} = \begin{pmatrix} 1 & 9 & 8 \\ 0 & 19 & 18 \\ 0 & 9 & 10 \end{pmatrix}.$$

The characteristic equation is

$$\begin{vmatrix} 1 - \lambda & 9 & 9 \\ 0 & 19 - \lambda & 18 \\ 0 & 9 & 10 - \lambda \end{vmatrix} = 0.$$

Expanding the determinant and simplifying, it reduces to $\lambda^3 - 30\lambda^2 + 57\lambda - 28 = 0$ or $(\lambda - 1)^2(\lambda - 28) = 0$. Its roots are the characteristic vectors $\lambda_1 = 28$, $\lambda_2 = \lambda_3 = 1$.

We use $\bar{A}\,\bar{a} = \lambda\bar{a}$ to find the corresponding characteristic vectors. With $\lambda = 28$ this is

$$\begin{pmatrix} 1 & 9 & 9 \\ 0 & 19 & 18 \\ 0 & 9 & 10 \end{pmatrix} \begin{bmatrix} a_1 \\ a_2 \\ a_3 \end{bmatrix} = 28 \begin{bmatrix} a_1 \\ a_2 \\ a_3 \end{bmatrix}.$$

Thus a_1, a_2, a_3 must satisfy

$$-27a_1 + 9a_2 + 9a_3 = 0,$$
$$-9a_2 + 18a_3 = 0,$$
$$9a_2 - 18a_3 = 0.$$

A nontrivial solution of this is $a_1 = 1$, $a_2 = 2$, $a_3 = 1$.
Thus a characteristic vector corresponding to $\lambda = 28$ is

$$\bar{a} = \begin{bmatrix} 1 \\ 2 \\ 1 \end{bmatrix},$$

and a corresponding solution is

$$\begin{pmatrix} 1 \\ 2 \\ 1 \end{pmatrix} e^{28t}, \quad \text{that is,} \quad \begin{bmatrix} e^{28t} \\ 2e^{28t} \\ e^{28t} \end{bmatrix} \qquad (*)$$

With $\lambda = 1$, $\bar{A}\,\bar{a} = \lambda\bar{a}$ is

$$\begin{pmatrix} 1 & 9 & 9 \\ 0 & 19 & 18 \\ 0 & 9 & 10 \end{pmatrix} \begin{bmatrix} a_1 \\ a_2 \\ a_3 \end{bmatrix} = 1 \begin{bmatrix} a_1 \\ a_2 \\ a_3 \end{bmatrix}.$$

Thus a_1, a_2, a_3 must satisfy

$$9a_2 + 9a_3 = 0,$$
$$18a_2 + 18a_3 = 0,$$
$$9a_2 + 9a_3 = 0.$$

Each of these three relationships is equivalent to the one
relation

$$a_2 + a_3 = 0,$$

and this the only relationship which a_1, a_2, a_3 must satisfy. Note that it allows a_1 to be arbitrary. Two linearly independent solutions of it are $a_1 = 1$, $a_2 = a_3 = 0$, and $a_1 = 0$, $a_2 = 1$, $a_3 = -1$. That is, corresponding to the double characteristic value $\lambda = 1$, we have the two linearly independent characteristic vectors

$$\begin{bmatrix} 1 \\ 0 \\ 0 \end{bmatrix} \quad \text{and} \quad \begin{bmatrix} 0 \\ 1 \\ -1 \end{bmatrix} .$$

Respective corresponding solutions are

$$\begin{bmatrix} 1 \\ 0 \\ 0 \end{bmatrix} e^t, \quad \text{and} \quad \begin{bmatrix} 0 \\ 1 \\ -1 \end{bmatrix} e^t ,$$

that is,

$$\begin{bmatrix} e^t \\ 0 \\ 0 \end{bmatrix} \quad \text{and} \quad \begin{bmatrix} 0 \\ e^t \\ -e^t \end{bmatrix} . \qquad (**)$$

The three solutions given by (*) and (**) are linearly independent, and a C.S. is

$$\bar{x} = c_1 \begin{bmatrix} e^{28t} \\ 2e^{28t} \\ e^{28t} \end{bmatrix} + c_2 \begin{bmatrix} e^t \\ 0 \\ 0 \end{bmatrix} + c_3 \begin{bmatrix} 0 \\ e^t \\ -e^t \end{bmatrix} ,$$

where c_1, c_2, c_3 are arbitrary constants. In component
form, this is

$$x_1 = c_1 e^{28t} + c_2 e^t,$$
$$x_2 = 2c_1 e^{28t} + c_3 e^t,$$
$$x_3 = c_1 e^{28t} - c_3 e^t.$$

17. We assume a solution of the form $\bar{x} = \bar{a} e^{\lambda t}$, that is,
$x_1 = a_1 e^{\lambda t}$, $x_2 = a_2 e^{\lambda t}$, $x_3 = a_3 e^{\lambda t}$. We know that λ must
be a solution of the characteristic equation of the
coefficient matrix

$$\bar{A} = \begin{pmatrix} -5 & -12 & 6 \\ 1 & 5 & -1 \\ -7 & -10 & 8 \end{pmatrix}.$$

The characteristic equation is

$$\begin{vmatrix} -5 - \lambda & -12 & 6 \\ 1 & 5 - \lambda & -1 \\ -7 & -10 & 8 - \lambda \end{vmatrix} = 0.$$

This reduces to $\lambda^3 - 8\lambda^2 + 19\lambda - 12 = 0$, that is,
$(\lambda - 1)(\lambda - 3)(\lambda - 4) = 0$. Its roots are $\lambda_1 = 1$, $\lambda_2 = 3$,
$\lambda_3 = 4$.

We use $\bar{A}\bar{a} = \lambda\bar{a}$ to find the corresponding characteristic vectors. With $\lambda = 1$ this is

$$\begin{pmatrix} -5 & -12 & 6 \\ 1 & 5 & -1 \\ -7 & -10 & 8 \end{pmatrix} \begin{bmatrix} a_1 \\ a_2 \\ a_3 \end{bmatrix} = (1) \begin{bmatrix} a_1 \\ a_2 \\ a_3 \end{bmatrix}$$

Thus a_1, a_2, a_3 must satisfy

$$\begin{cases} -6a_1 - 12a_2 + 6a_3 = 0, \\ a_1 + 4a_2 - a_3 = 0, \\ -7a_1 - 10a_2 + 7a_3 = 0. \end{cases}$$

A nontrivial solution of this is $a_1 = 1$, $a_2 = 0$, $a_3 = 1$.
Thus a characteristic vector corresponding to $\lambda = 1$ is

$$\bar{a} = \begin{bmatrix} 1 \\ 0 \\ 1 \end{bmatrix},$$

and a corresponding solution is

$$\begin{pmatrix} 1 \\ 0 \\ 1 \end{pmatrix} e^t, \quad \text{that is,} \quad \begin{bmatrix} e^t \\ 0 \\ e^t \end{bmatrix} \qquad (*)$$

With $\lambda = 3$, $\bar{A}\bar{a} = \lambda\bar{a}$ is

$$\begin{pmatrix} -5 & -12 & 6 \\ 1 & 5 & -1 \\ -7 & -10 & 8 \end{pmatrix} \begin{bmatrix} a_1 \\ a_2 \\ a_3 \end{bmatrix} = 3 \begin{bmatrix} a_1 \\ a_2 \\ a_3 \end{bmatrix}.$$

Thus a_1, a_2, a_3 must satisfy

$$\begin{cases} -8a_1 - 12a_2 + 6a_3 = 0, \\ a_1 + 2a_2 - a_3 = 0, \\ -7a_1 - 10a_2 + 5a_3 = 0. \end{cases}$$

A nontrivial solution of this is $a_1 = 0$, $a_2 = 1$, $a_3 = 2$.
Thus a characteristic vector corresponding to $\lambda = 3$ is

$$\bar{a} = \begin{pmatrix} 0 \\ 1 \\ 2 \end{pmatrix}$$

and a corresponding solution is

$$\begin{bmatrix} 0 \\ 1 \\ 2 \end{bmatrix} e^{3t}, \quad \text{that is,} \quad \begin{bmatrix} 0 \\ e^{3t} \\ 2e^{3t} \end{bmatrix}. \tag{**}$$

With $\lambda = 4$, $\bar{A}\bar{a} = \lambda\bar{a}$ is

$$\begin{pmatrix} -5 & -12 & 6 \\ 1 & 5 & -1 \\ -7 & -10 & 8 \end{pmatrix} \begin{bmatrix} a_1 \\ a_2 \\ a_3 \end{bmatrix} = 4 \begin{bmatrix} a_1 \\ a_2 \\ a_3 \end{bmatrix}.$$

Thus a_1, a_2, a_3 must satisfy

$$\begin{cases} -9a_1 - 12a_2 + 6a_3 = 0, \\ a_1 + a_2 - a_3 = 0, \\ -7a_1 - 10a_2 + 4a_3 = 0. \end{cases}$$

A nontrivial solution of this is $a_1 = 2$, $a_2 = -1$, $a_3 = 1$. Thus a characteristic vector corresponding to $\lambda = 4$ is

$$\bar{a} = \begin{pmatrix} 2 \\ -1 \\ 1 \end{pmatrix}$$

and a corresponding solution is

$$\begin{bmatrix} 2 \\ -1 \\ 1 \end{bmatrix} e^{4t}, \quad \text{that is,} \quad \begin{bmatrix} 2e^{4t} \\ -e^{4t} \\ e^{4t} \end{bmatrix}. \qquad (***)$$

By Theorem 7.16, the solutions (*), (**), and (**) are linearly independent, and a G.S. is

$$\bar{x} = c_1 \begin{bmatrix} e^t \\ 0 \\ e^t \end{bmatrix} + c_2 \begin{bmatrix} 0 \\ e^{3t} \\ 2e^{3t} \end{bmatrix} + c_3 \begin{bmatrix} 2e^{4t} \\ -e^{4t} \\ e^{4t} \end{bmatrix},$$

where c_1, c_2, c_3 are arbitrary constants. In component form, this is

$$x_1 = c_1 e^t + 2c_3 e^{4t},$$

$$x_2 = c_2 e^{3t} - c_3 e^{4t},$$

$$x_3 = c_1 e^t + 2c_2 e^{3t} + c_3 e^{4t}.$$

19. We assume a solution of the form $\bar{x} = \bar{a} e^{\lambda t}$, that is, $x_1 = a_1 e^{\lambda t}$, $x_2 = a_2 e^{\lambda t}$, $x_3 = a_3 e^{\lambda t}$. We know that λ must be a solution of the characteristic equation of the coefficient matrix

$$\bar{A} = \begin{pmatrix} -5 & -3 & -3 \\ 8 & 5 & 7 \\ -2 & -1 & -3 \end{pmatrix}.$$

The characteristic equation is

$$\begin{vmatrix} -5 - \lambda & -3 & -3 \\ 8 & 5 - \lambda & 7 \\ -2 & -1 & -3 - \lambda \end{vmatrix} = 0.$$

Expanding the determinant and simplifying, it reduces to $\lambda^3 + 3\lambda^2 - 4 = 0$, or $(\lambda - 1)(\lambda + 2)^2 = 0$. Its roots are the characteristic values $\lambda_1 = 1$, $\lambda_2 = \lambda_3 = -2$.

We use $\bar{A}\bar{a} = \lambda\bar{a}$ to find the corresponding characteristic vectors. With $\lambda = 1$ this is

$$\begin{pmatrix} -5 & -3 & -3 \\ 8 & 5 & 7 \\ -2 & -1 & -3 \end{pmatrix} \begin{bmatrix} a_1 \\ a_2 \\ a_3 \end{bmatrix} = (1) \begin{bmatrix} a_1 \\ a_2 \\ a_3 \end{bmatrix}.$$

Thus a_1, a_2, a_3 must satisfy

$$\begin{cases} -6a_1 - 3a_2 - 3a_3 = 0, \\ 8a_1 + 4a_2 + 7a_3 = 0, \\ -2a_1 - a_2 - 4a_3 = 0. \end{cases}$$

Adding the second equation to four times the third equation, we find $-9a_3 = 0$, from which $a_3 = 0$. With $a_3 = 0$, each equation is equivalent to $a_3 = -2a_1$. Hence a nontrivial solution of the algebraic system is $a_1 = 1$, $a_2 = -2$, $a_3 = 0$. Thus a characteristic vector corresponding to $\lambda = 1$ is

$$\bar{a} = \begin{bmatrix} 1 \\ -2 \\ 0 \end{bmatrix},$$

and a corresponding solution is

$$\begin{pmatrix} 1 \\ -2 \\ 0 \end{pmatrix} e^t, \quad \text{that is,} \quad \begin{bmatrix} e^t \\ -2e^t \\ 0 \end{bmatrix}. \tag{*}$$

With $\lambda = -2$, $\bar{A}\bar{a} = \lambda\bar{a}$ is

$$\begin{pmatrix} -5 & -3 & -3 \\ 8 & 5 & 7 \\ -2 & -1 & -3 \end{pmatrix} \begin{bmatrix} a_1 \\ a_2 \\ a_3 \end{bmatrix} = -2 \begin{bmatrix} a_1 \\ a_2 \\ a_3 \end{bmatrix}.$$

Thus a_1, a_2, a_3 must satisfy

$$\begin{cases} -3a_1 - 3a_2 - 3a_3 = 0, \\ 8a_1 + 7a_2 + 7a_3 = 0, \\ -2a_1 - a_2 - a_3 = 0. \end{cases}$$

Adding the first equation to (-3) times the third, we find $3a_1 = 0$, from which $a_1 = 0$. With $a_1 = 0$, each equation is equivalent to $a_3 = -a_2$. Hence a nontrivial solution of the algebraic system is $a_1 = 0$, $a_2 = 1$, $a_3 = -1$. Thus a characteristic vector corresponding to $\lambda = -2$ is

$$\bar{a} = \begin{pmatrix} 0 \\ 1 \\ -1 \end{pmatrix}$$

and a corresponding solution is

$$\begin{bmatrix} 0 \\ 1 \\ -1 \end{bmatrix} e^{-2t}, \quad \text{that is,} \quad \begin{bmatrix} 0 \\ e^{-2t} \\ -e^{-2t} \end{bmatrix}. \qquad (**)$$

A second solution corresponding to $\lambda = -2$ is of the form $(\bar{a}t + \bar{\beta})e^{-2t}$, where $\bar{\beta}$ satisfies $(\bar{A} + 2\bar{I})\bar{\beta} = \bar{a}$. This is

$$\begin{pmatrix} -3 & -3 & -3 \\ 8 & 7 & 7 \\ -2 & -1 & -1 \end{pmatrix} \begin{bmatrix} \beta_1 \\ \beta_2 \\ \beta_3 \end{bmatrix} = \begin{pmatrix} 0 \\ 1 \\ -1 \end{pmatrix}.$$

Multiplying this out yields

$$\begin{cases} -3\beta_1 - 3\beta_2 - 3\beta_3 = 0, \\ 8\beta_1 + 7\beta_2 + 7\beta_3 = 1, \\ -2\beta_1 - \beta_2 - \beta_3 = -1. \end{cases}$$

a solution of which is $\beta_1 = 1$, $\beta_2 = -1$, $\beta_3 = 0$. Thus we obtain

$$\bar{\beta} = \begin{pmatrix} 1 \\ -1 \\ 0 \end{pmatrix}$$

and the second solution $(\bar{\alpha}t + \bar{\beta})e^{-2t}$ is

$$\left[\begin{bmatrix} 0 \\ 1 \\ -1 \end{bmatrix} + \begin{bmatrix} 1 \\ -1 \\ 0 \end{bmatrix} \right] e^{-2t}, \quad \text{that is,} \quad \begin{bmatrix} e^{-2t} \\ (t-1)e^{-2t} \\ -te^{-2t} \end{bmatrix}. \qquad (\ast\ast\ast)$$

The three solutions $(\ast)$, $(\ast\ast)$, and $(\ast\ast)$ are linearly independent. Thus a G.S. is

$$\bar{x} = c_1 \begin{bmatrix} e^t \\ -2e^t \\ 0 \end{bmatrix} + c_2 \begin{bmatrix} 0 \\ e^{-2t} \\ -2e^{-2t} \end{bmatrix} + c_3 \begin{bmatrix} e^{-2t} \\ (t-1)e^{-2t} \\ -te^{-2t} \end{bmatrix},$$

where c_1, c_2, c_3 are arbitrary constants. In component form, this is

$$\begin{cases} x_1 = c_1 e^t + c_3 e^{-2t}, \\ x_2 = -2c_1 e^t + c_2 e^{-2t} + c_3 (t - 1) e^{-2t}, \\ x_3 = -2c_2 e^{-2t} - c_3 t e^{-2t}. \end{cases}$$

22. We assume a solution of the form $\bar{x} = \bar{a} e^{\lambda t}$, that is, $x_1 = a_1 e^{\lambda t}$, $x_2 = a_2 e^{\lambda t}$, $x_3 = a_3 e^{\lambda t}$. We know that λ must be a solution of the characteristic equation of the coefficient matrix

$$\bar{A} = \begin{pmatrix} 3 & -2 & -1 \\ -4 & 2 & 4 \\ 5 & -3 & -3 \end{pmatrix}.$$

The characteristic equation is

$$\begin{vmatrix} 3 - \lambda & -2 & -1 \\ -4 & 2 - \lambda & 4 \\ 5 & -3 & -3 - \lambda \end{vmatrix} = 0.$$

It reduces to $\lambda^3 - 2\lambda^2 = 0$. Its roots are the characteristic values $\lambda_1 = 2$, $\lambda_2 = \lambda_3 = 0$.

We use $\bar{A}\bar{a} = \lambda\bar{a}$ to find the corresponding characteristic vectors. With $\lambda = 2$ this is

$$\begin{pmatrix} 3 & -2 & -1 \\ -4 & 2 & 4 \\ 5 & -3 & -3 \end{pmatrix} \begin{bmatrix} a_1 \\ a_2 \\ a_3 \end{bmatrix} = 2 \begin{bmatrix} a_1 \\ a_2 \\ a_3 \end{bmatrix}.$$

Thus a_1, a_2, a_3 must satisfy

$$\begin{cases} a_1 - 2a_2 - a_3 = 0, \\ \quad\;\; -4a_1 + 4a_3 = 0, \\ 5a_1 - 3a_2 - 5a_3 = 0. \end{cases}$$

A nontrivial solution of this is $a_1 = 1$, $a_2 = 0$, $a_3 = 1$. Thus a characteristic vector corresponding to $\lambda = 2$ is

$$\bar{a} = \begin{bmatrix} 1 \\ 0 \\ 1 \end{bmatrix},$$

and a corresponding solution is

$$\begin{pmatrix} 1 \\ 0 \\ 1 \end{pmatrix} e^{2t}, \quad \text{that is,} \quad \begin{bmatrix} e^{2t} \\ 0 \\ e^{2t} \end{bmatrix}. \qquad (*)$$

We now consider the repeated characteristic value $\lambda_2 = \lambda_3 = 0$. With $\lambda = 0$, $\bar{A}\,\bar{a} = \lambda\bar{a}$ becomes

$$\begin{pmatrix} 3 & -2 & -1 \\ -4 & 2 & 4 \\ 5 & -3 & -3 \end{pmatrix} \begin{bmatrix} a_1 \\ a_2 \\ a_3 \end{bmatrix} = 0 \begin{bmatrix} a_1 \\ a_2 \\ a_3 \end{bmatrix}.$$

Thus a_1, a_2, a_3 must satisfy

$$\begin{cases} 3a_1 - 2a_2 - a_3 = 0, \\ -4a_1 + 2a_2 + 4a_3 = 0, \\ 5a_1 - 3a_2 - 3a_3 = 0. \end{cases}$$

Adding the first two equations, we obtain $-a_1, + 3a_3 = 0$. From this $a_1 = 3a_3$. Then $2a_2 = 3a_1 - a_3 = 8a_3$, so $a_2 = 4a_3$. Thus a simple nontrivial solution is $a_1 = 3$, $a_2 = 4$, $a_3 = 1$. A characteristic vector corresponding to $\lambda = 0$ is

$$\bar{a} = \begin{pmatrix} 3 \\ 4 \\ 1 \end{pmatrix}$$

and the corresponding solution is

$$\begin{bmatrix} 3 \\ 4 \\ 1 \end{bmatrix} e^{0t}, \quad \text{that is,} \quad \begin{bmatrix} 3 \\ 4 \\ 1 \end{bmatrix}. \tag{**}$$

A second solution corresponding to $\lambda = 0$ is of the form $(\bar{a}t + \bar{\beta})e^{0t} = \bar{a}t + \bar{\beta}$, where $\bar{\beta}$ satisfies $(\bar{A} - \lambda\bar{I})\bar{\beta} = \bar{a}$, that is $\bar{A}\bar{\beta} = \bar{a}$. This is

$$\begin{pmatrix} 3 & -2 & -1 \\ -4 & 2 & 4 \\ 5 & -3 & -3 \end{pmatrix} \begin{bmatrix} \beta_1 \\ \beta_2 \\ \beta_3 \end{bmatrix} = \begin{pmatrix} 3 \\ 4 \\ 1 \end{pmatrix}.$$

Multiplying this out yields

$$\begin{cases} 3\beta_1 - 2\beta_2 - \beta_3 = 3, \\ -4\beta_1 + 2\beta_2 + 4\beta_3 = 4, \\ 5\beta_1 - 3\beta_2 - 3\beta_3 = 1. \end{cases}$$

Multiplying the first equation by 4, the second by 3, and adding, we find $-2\beta_2 + 8\beta_3 = 24$ and so $\beta_2 = 4\beta_3 - 12$. The first equation gives $\beta_1 = \frac{2}{3}\beta_2 + \frac{1}{3}\beta_3 + 1$. Then substituting $\beta_2 = 4\beta_3 - 12$ into this, we have $\beta_1 = 3\beta_3 - 7$. One quickly checks that the pair $\beta_1 = 3\beta_3 - 7$, $\beta_2 = 4\beta_3 - 12$ satisfies the third equation for arbitrary β_3. Choosing $\beta_3 - 3$, we have the simple nontrivial solution $\beta_1 = 2$, $\beta_2 = 0$, $\beta_3 = 3$. Thus we obtain

$$\bar{\beta} = \begin{pmatrix} 2 \\ 0 \\ 3 \end{pmatrix},$$

and the second solution $\bar{\alpha}t + \bar{\beta}$ is

$$\left(\begin{bmatrix} 3 \\ 4 \\ 1 \end{bmatrix} t + \begin{bmatrix} 2 \\ 0 \\ 3 \end{bmatrix} \right), \quad \text{that is,} \quad \begin{bmatrix} 3t + 2 \\ 4t \\ t + 3 \end{bmatrix}. \qquad (***)$$

The three solutions (*), (**), and (**) are linearly independent. Thus a G.S. is

$$\bar{x} = c_1 \begin{bmatrix} e^{2t} \\ 0 \\ e^{2t} \end{bmatrix} + c_2 \begin{bmatrix} 3 \\ 4 \\ 1 \end{bmatrix} + c_3 \begin{bmatrix} 3t + 2 \\ 4t \\ t + 3 \end{bmatrix}$$

where c_1, c_2, c_3 are arbitrary constants. In component form, this is

$$\begin{cases} x_1 = c_1 e^{2t} + 3c_2 + c_3(3t + 2), \\ x_2 = 4c_2 + 4c_2 t, \\ x_3 = c_1 e^{2t} + c_2 + c_3(t + 3). \end{cases}$$

23. We assume a solution of the form $\bar{x} = \bar{a}\, e^{\lambda t}$, that is, $x_1 = a_1 e^{\lambda t}$, $x_2 = a_2 e^{\lambda t}$, $x_3 = a_3 e^{\lambda t}$. We know that λ must be a solution of the characteristic equation of the coefficient matrix

$$\bar{A} = \begin{pmatrix} 7 & 4 & 4 \\ -6 & -4 & -7 \\ -2 & -1 & 2 \end{pmatrix}.$$

The characteristic equation is

$$\begin{vmatrix} 7 - \lambda & 4 & 4 \\ -6 & -4 - \lambda & -7 \\ -2 & -1 & 2 - \lambda \end{vmatrix} = 0.$$

It reduces to $\lambda^3 - 5\lambda^2 + 3\lambda + 9 = 0$, that is, $(\lambda + 1)(\lambda - 3)^2 = 0$. Its roots are the characteristic values $\lambda_1 = -1$, $\lambda_2 = \lambda_3 = 3$.

We use $\bar{A}\,\bar{a} = \lambda\bar{a}$ to find the corresponding characteristic vectors. With $\lambda = -1$ this is

$$\begin{pmatrix} 7 & 4 & 4 \\ -6 & -4 & -7 \\ -2 & -1 & 2 \end{pmatrix} \begin{bmatrix} a_1 \\ a_2 \\ a_3 \end{bmatrix} = -1 \begin{bmatrix} a_1 \\ a_2 \\ a_3 \end{bmatrix}.$$

Thus a_1, a_2, a_3 must satisfy

$$\begin{cases} 8a_1 + 4a_2 + 4a_3 = 0, \\ -6a_1 - 3a_2 - 7a_3 = 0, \\ -2a_1 - a_2 + 3a_3 = 0. \end{cases}$$

A nontrivial solution of this is $a_1 = 1$, $a_2 = -2$, $a_3 = 0$. Thus a characteristic vector corresponding to $\lambda = -1$ is

$$\bar{a} = \begin{bmatrix} 1 \\ -2 \\ 0 \end{bmatrix},$$

and the corresponding solution is

$$\begin{pmatrix} 1 \\ -2 \\ 0 \end{pmatrix} e^{-t}, \quad \text{that is,} \quad \begin{bmatrix} e^{-t} \\ -2e^{-t} \\ 0 \end{bmatrix}. \qquad (*)$$

We now consider the repeated characteristic value $\lambda_2 = \lambda_3 = 3$. With $\lambda = 3$, $\bar{A}\,\bar{a} = \lambda\bar{a}$ becomes

$$\begin{pmatrix} 7 & 4 & 4 \\ -6 & -4 & -7 \\ -2 & -1 & 2 \end{pmatrix} \begin{bmatrix} a_1 \\ a_2 \\ a_3 \end{bmatrix} = 3 \begin{bmatrix} a_1 \\ a_2 \\ a_3 \end{bmatrix}.$$

Thus a_1, a_2, a_3 must satisfy

$$4a_1 + 4a_2 + 4a_3 = 0,$$

$$-6a_1 - 7a_2 - 7a_3 = 0,$$

$$-2a_1 - a_2 - a_3 = 0.$$

A solution of this is $a_1 = 0$, $a_2 = 1$, $a_3 = -1$.   Thus a characteristic vector corresponding to $\lambda = 3$ is

$$\bar{a} = \begin{pmatrix} 0 \\ 1 \\ -1 \end{pmatrix},$$

and the corresponding solution is

$$\begin{bmatrix} 0 \\ 1 \\ -1 \end{bmatrix} e^{3t}, \quad \text{that is,} \quad \begin{bmatrix} 0 \\ e^{3t} \\ -e^{3t} \end{bmatrix}. \qquad (**)$$

A second solution corresponding to $\lambda = 3$ is of the form $(\bar{a}t + \bar{\beta})e^{3t}$, where $\bar{\beta}$ satisfies $(\bar{A} - 3\bar{I})\bar{\beta} = \bar{a}$.   This is

$$\begin{pmatrix} 4 & 4 & 4 \\ -6 & -7 & -7 \\ -2 & -1 & -1 \end{pmatrix} \begin{bmatrix} \beta_1 \\ \beta_2 \\ \beta_3 \end{bmatrix} = \begin{pmatrix} 0 \\ 1 \\ -1 \end{pmatrix}.$$

Multiplying this out yields

$$4\beta_1 + 4\beta_2 + 4\beta_3 = 0,$$

$$-6\beta_1 - 7\beta_2 - 7\beta_3 = 1,$$

$$-2\beta_1 - \beta_2 - \beta_3 = -1.$$

a solution of which is $\beta_1 = 1$, $\beta_2 = -1$, $\beta_3 = 0$. Thus we obtain

$$\bar{\beta} = \begin{pmatrix} 1 \\ -1 \\ 0 \end{pmatrix},$$

and the second solution $(\bar{a}t + \bar{\beta})e^{3t}$ is

$$\left[\begin{bmatrix} 0 \\ 1 \\ -1 \end{bmatrix} t + \begin{bmatrix} 1 \\ -1 \\ 0 \end{bmatrix} \right] e^{3t}, \quad \text{that is,} \quad \begin{bmatrix} e^{3t} \\ (t-1)e^{3t} \\ -te^{3t} \end{bmatrix}. \qquad (***)$$

The three solutions (*), (**), and (***) are linearly independent. Thus a G.S. is

$$\bar{x} = c_1 \begin{bmatrix} e^{-t} \\ -2e^{-t} \\ 0 \end{bmatrix} + c_2 \begin{bmatrix} 0 \\ e^{3t} \\ -e^{3t} \end{bmatrix} + c_3 \begin{bmatrix} e^{3t} \\ (t-1)e^{3t} \\ -te^{3t} \end{bmatrix},$$

where c_1, c_2, and c_3 are arbitrary constants. In component form, this is

$$x_1 = c_1 e^{-t} + c_3 e^{3t},$$

$$x_2 = -2c_1 e^{-t} + c_2 e^{3t} + c_3(t - 1)e^{3t},$$

$$x_3 = -c_2 e^{3t} - c_3 t e^{3t}.$$

26. We assume a solution of the form $\bar{x} = \bar{a}\, e^{\lambda t}$, that is, $x_1 = a_1 e^{\lambda t}$, $x_2 = a_2 e^{\lambda t}$, $x_3 = a_3 e^{\lambda t}$. We know that λ must be a solution of the characteristic equation of the coefficient matrix

$$\bar{A} = \begin{pmatrix} 4 & -1 & -1 \\ 2 & 1 & -1 \\ 2 & -1 & 1 \end{pmatrix}.$$

The characteristic equation is

$$\begin{vmatrix} 4 - \lambda & -1 & -1 \\ 2 & 1 - \lambda & -1 \\ 2 & -1 & 1 - \lambda \end{vmatrix} = 0.$$

It reduces to $\lambda^3 - 6\lambda^2 + 12\lambda - 8 = 0$, that is, $(\lambda - 2)^3 = 0$. Its roots are the characteristic values $\lambda_1 = \lambda_2 = \lambda_3 = 2$.

We use $\bar{A}\,\bar{a} = \lambda\bar{a}$ to find the corresponding characteristic vectors. With $\lambda = 2$ this is

$$\begin{pmatrix} 4 & -1 & -1 \\ 2 & 1 & -1 \\ 2 & -1 & 1 \end{pmatrix} \begin{bmatrix} a_1 \\ a_2 \\ a_3 \end{bmatrix} = 2 \begin{bmatrix} a_1 \\ a_2 \\ a_3 \end{bmatrix}.$$

Each of the three resulting relationships in a_1, a_2, a_3 is $2a_1 - a_2 - a_3 = 0$. Both $a_1 = 1$, $a_2 = 2$, $a_3 = 0$ and $a_1 = 1$, $a_2 = 0$, $a_3 = 2$ are distinct solutions of this. Thus we obtain the characteristic vectors

$$\bar{a}^{(1)} = \begin{pmatrix} 1 \\ 2 \\ 0 \end{pmatrix} \quad \text{and} \quad \bar{a}^{(2)} = \begin{pmatrix} 1 \\ 0 \\ 2 \end{pmatrix}$$

and the corresponding solutions $\bar{a}^{(1)} e^{2t}$ and $\bar{a}^{(2)} e^{2t}$, that is

$$\begin{bmatrix} e^{2t} \\ 2e^{2t} \\ 0 \end{bmatrix} \quad \text{and} \quad \begin{bmatrix} e^{2t} \\ 0 \\ 2e^{2t} \end{bmatrix}. \qquad (*)$$

A third solution corresponding to $\lambda = 2$ is of the form $(\bar{a}t + \bar{\beta})e^{2t}$, where $\bar{a}$ satisfies $(\bar{A} - 2\bar{I})\bar{a} = \bar{0}$ and $\bar{\beta}$ satisfies $(\bar{A} - 2\bar{I})\bar{\beta} = \bar{a}$. We apply this last equation with

$$\bar{a} = k_1 \bar{a}^{(1)} + k_2 \bar{a}^{(2)} = k_1 \begin{pmatrix} 1 \\ 2 \\ 0 \end{pmatrix} + k_2 \begin{pmatrix} 1 \\ 0 \\ 2 \end{pmatrix} = \begin{bmatrix} k_1 + k_2 \\ 2k_1 \\ 2k_2 \end{bmatrix}.$$

Thus we have

$$
\begin{pmatrix} 2 & -1 & -1 \\ 2 & -1 & -1 \\ 2 & -1 & -1 \end{pmatrix} \begin{bmatrix} \beta_1 \\ \beta_2 \\ \beta_3 \end{bmatrix} = \begin{bmatrix} k_1 + k_2 \\ 2k_1 \\ 2k_2 \end{bmatrix}
$$

Multiplying this out yields

$$
\begin{aligned}
2\beta_1 - \beta_2 - \beta_3 &= k_1 + k_2, \\
2\beta_1 - \beta_2 - \beta_3 &= 2k_1, \qquad\qquad (**)\\
2\beta_1 - \beta_2 - \beta_3 &= 2k_2.
\end{aligned}
$$

From these $k_1 + k_2 = 2k_1 = 2k_2$, a nontrivial solution of which is $k_1 = k_2 = 1$. Thus

$$
\bar{a} = \begin{pmatrix} 2 \\ 2 \\ 2 \end{pmatrix}
$$

and relations (**) each become $2\beta_1 - \beta_2 - \beta_3 = 2$. A solution of this is $\beta_1 = 1$, $\beta_2 = \beta_3 = 0$; and we obtain

$$
\bar{\beta} = \begin{pmatrix} 1 \\ 0 \\ 0 \end{pmatrix}.
$$

Thus the third solution $(\bar{\alpha}t + \bar{\beta})e^{2t}$ is

$$\left(\begin{bmatrix}2\\2\\2\end{bmatrix}t + \begin{bmatrix}1\\0\\0\end{bmatrix}\right)e^{2t}, \qquad \text{that is,} \qquad \begin{bmatrix}(2t+1)e^{2t}\\2t\,e^{2t}\\2t\,e^{2t}\end{bmatrix}.$$

This solution and the two solutions (*) are linearly independent, and a G.S. is the linear combination

$$\bar{x} = c_1\begin{bmatrix}e^{2t}\\2e^{2t}\\0\end{bmatrix} + c_2\begin{bmatrix}e^{2t}\\0\\2e^{2t}\end{bmatrix} + c_3\begin{bmatrix}(2t+1)e^{2t}\\2t\,e^{2t}\\2t\,e^{2t}\end{bmatrix},$$

where c_1, c_2, c_3 are arbitrary constants. In component form, this is

$$x_1 = c_1 e^{2t} + c_2 e^{2t} + c_3(2t+1)e^{2t},$$
$$x_2 = 2c_1 e^{2t} + 2c_3 te^{2t},$$
$$x_3 = 2c_2 e^{2t} + 2c_3 t\,e^{2t}.$$

28. We assume a solution of the form $\bar{x} = \bar{a}\,e^{\lambda t}$, that is, $x_1 = a_1 e^{\lambda t}$, $x_2 = a_2 e^{\lambda t}$, $x_3 = a_3 e^{\lambda t}$. We know that λ must be a solution of the characteristic equation of the coefficient matrix

$$\bar{A} = \begin{pmatrix}4 & 6 & -1\\-1 & -2 & 1\\-2 & -8 & 4\end{pmatrix}.$$

The characteristic equation is

$$\begin{vmatrix} 4 - \lambda & 6 & -1 \\ -1 & -2 - \lambda & 1 \\ -2 & -8 & 4 - \lambda \end{vmatrix} = 0.$$

It reduces to $\lambda^3 - 6\lambda^2 + 12\lambda - 8 = 0$, that is, $(\lambda - 2)^3 = 0$. Its roots are the characteristic values $\lambda_1 = \lambda_2 = \lambda_3 = 2$.

We use $\bar{A}\,\bar{a} = \lambda\bar{a}$ to find the corresponding characteristic vectors. With $\lambda = 2$ this is

$$\begin{pmatrix} 4 & 6 & -1 \\ -1 & -2 & 1 \\ -2 & -8 & 4 \end{pmatrix} \begin{bmatrix} a_1 \\ a_2 \\ a_3 \end{bmatrix} = 2 \begin{bmatrix} a_1 \\ a_2 \\ a_3 \end{bmatrix}.$$

The resulting relationships in a_1, a_2, a_3 are

$$2a_1 + 6a_2 - a_3 = 0,$$
$$-a_1 - 4a_2 + a_3 = 0,$$
$$-2a_1 - 8a_2 + 2a_3 = 0.$$

We find that $a_1 = 2$, $a_2 = -1$, $a_3 = -2$ is a nontrivial solution of these. Thus we obtain the characteristic vector

$$\bar{a} = \begin{pmatrix} 2 \\ -1 \\ -2 \end{pmatrix}.$$

and the corresponding solution $\bar{a}e^{2t}$, that is,

$$\begin{bmatrix} 2e^{2t} \\ -e^{2t} \\ -2e^{2t} \end{bmatrix}. \qquad (*)$$

A second solution corresponding to $\lambda = 2$ is of the form $(\bar{a}t + \bar{\beta})e^{2t}$, where $\bar{\beta}$ satisfies $(\bar{A} - 2\bar{I})\beta = \bar{a}$. This is

$$\begin{pmatrix} 2 & 6 & -1 \\ -1 & -4 & 1 \\ -2 & -8 & 2 \end{pmatrix} \begin{bmatrix} \beta_1 \\ \beta_2 \\ \beta_3 \end{bmatrix} = \begin{pmatrix} 2 \\ -1 \\ -2 \end{pmatrix}.$$

Multiplying this out yields

$$2\beta_1 + 6\beta_2 - \beta_3 = 2,$$
$$-\beta_1 - 4\beta_2 + \beta_3 = -1,$$
$$-2\beta_1 - 8\beta_2 + 2\beta_3 = -2,$$

a solution of which is $\beta_1 = 1$, $\beta_2 = \beta_3 = 0$. Thus we obtain

$$\bar{\beta} = \begin{pmatrix} 1 \\ 0 \\ 0 \end{pmatrix},$$

and the second solution $(\bar{a}t + \bar{\beta})e^{2t}$ is

$$\left(\begin{bmatrix} 2 \\ -1 \\ -2 \end{bmatrix} t + \begin{bmatrix} 1 \\ 0 \\ 0 \end{bmatrix}\right) e^{2t}, \quad \text{that is,} \quad \begin{bmatrix} (2t + 1)e^{2t} \\ -t\, e^{2t} \\ -2t\, e^{2t} \end{bmatrix}. \qquad (**)$$

A third solution if of the form $\left(\dfrac{\bar{a}t^2}{2} + \bar{\beta}t + \bar{\gamma}\right)e^{2t}$,

where $\bar{\gamma}$ satisfies $(\bar{A} - 2\bar{I})\bar{\gamma} = \bar{\beta}$. This is

$$\begin{pmatrix} 2 & 6 & -1 \\ -1 & -4 & 1 \\ -2 & -8 & 2 \end{pmatrix} \begin{bmatrix} \gamma_1 \\ \gamma_2 \\ \gamma_3 \end{bmatrix} = \begin{pmatrix} 1 \\ 0 \\ 0 \end{pmatrix}.$$

Multiplying this out yields

$$2\gamma_1 + 6\gamma_2 - \gamma_3 = 1,$$
$$-\gamma_1 - 4\gamma_2 + \gamma_3 = 0,$$
$$-2\gamma_1 - 8\gamma_2 + 2\gamma_3 = 0.$$

A solution of this is $\gamma_1 = 1$, $\gamma_2 = 0$, $\gamma_3 = 1$; and we obtain

$$\bar{\gamma} = \begin{pmatrix} 1 \\ 0 \\ 1 \end{pmatrix}.$$

Thus the third solution $\left(\dfrac{\bar{a}t^2}{2} + \bar{\beta}t + \bar{\gamma}\right)e^{2t}$ is

$$\left[\begin{bmatrix} 2 \\ -1 \\ -2 \end{bmatrix}\frac{t^2}{2} + \begin{bmatrix} 1 \\ 0 \\ 0 \end{bmatrix}t + \begin{bmatrix} 1 \\ 0 \\ 1 \end{bmatrix}\right]e^{2t}, \quad \text{that is,} \quad \begin{bmatrix} (t^2 + t + 1)e^{2t} \\ (-1/2)t^2 e^{2t} \\ (-t^2 + 1)e^{2t} \end{bmatrix}.$$

This and the two solutions (*) and (**) are linearly independent, and a general solution is the linear combination

$$\bar{x} = c_1\begin{bmatrix} 2e^{2t} \\ -e^{2t} \\ -2e^{2t} \end{bmatrix} + c_2\begin{bmatrix} (2t + 1)e^{2t} \\ -t\,e^{2t} \\ -2t\,e^{2t} \end{bmatrix} + c_3\begin{bmatrix} (t^2 + t + 1)e^{2t} \\ (-1/2)t^2 e^{2t} \\ (-t^2 + 1)e^{2t} \end{bmatrix},$$

where c_1, c_2, c_3 are arbitrary constants. In component form, this is

$$x_1 = 2c_1 e^{2t} + c_2(2t + 1)e^{2t} + c_3(t^2 + t + 1)e^{2t},$$
$$x_2 = -c_1 e^{2t} - c_2 t e^{2t} - (c_3/2)t^2 e^{2t},$$
$$x_3 = -2c_1 e^{2t} - 2c_2 t e^{2t} + c_3(-t^2 + 1)e^{2t}.$$

Chapter 8

Section 8.1B, Page 427

Exercise 2.

We separate the solution into four graphs. The differential equation y′ = xy determines the family of isoclines c = xy; these are the hyperbolas y = c/x. Our first graph is the result of drawing these hyperbolas for c = ±1/10, ±3/10, ±1/2, ±1, ±2, ±3, ±4, ±6, ±8, and ±10. We note that if c = 0, the corresponding isoclines are the x and y axes. See figure 8.1B-2A.

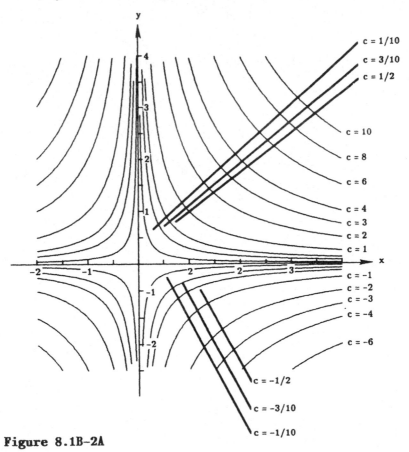

Figure 8.1B-2A

Now, on each of the isoclines drawn in our first
graph, we draw several line elements having the
appropriate slope, or, angle of inclination. The slope of
each line element on the fixed isocline $y = c_0/x$ is c_0;

the corresponding angle of inclination is then
$a_0 = \arctan c_0$. We obtain the graph found in Figure
8.1B-2B.

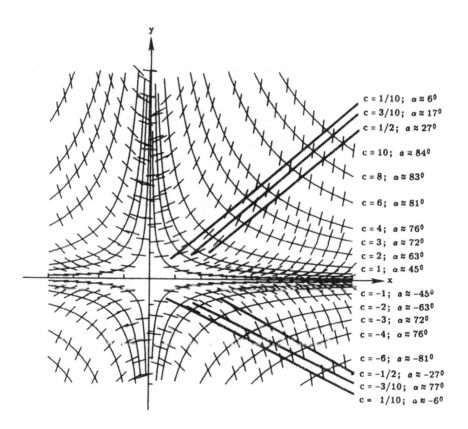

c = 1/10; $a \approx 6^0$
c = 3/10; $a \approx 17^0$
c = 1/2; $a \approx 27^0$

c = 10; $a \approx 84^0$

c = 8; $a \approx 83^0$

c = 6; $a \approx 81^0$

c = 4; $a \approx 76^0$
c = 3; $a \approx 72^0$
c = 2; $a \approx 63^0$
c = 1; $a \approx 45^0$

c = -1; $a \approx -45^0$
c = -2; $a \approx -63^0$
c = -3; $a \approx 72^0$
c = -4; $a \approx 76^0$

c = -6; $a \approx -81^0$
c = -1/2; $a \approx -27^0$
c = -3/10; $a \approx 77^0$
c = -1/10; $a \approx -6^0$

Figure 8.1B-2B

The line element configuration seen in the second
graph indicates how to draw the approximate integral
curves. We sketch several. In particular, we show those
curves which have y intercepts -1, -0.1, -0.01, -0.001,
1/16, 1/8, 1/4, 1/2, 1, and 2, respectively. See Figure
8.1B-2C.

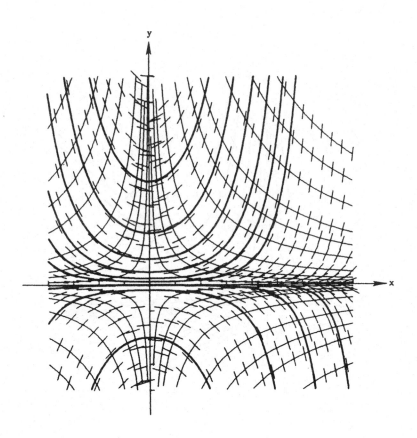

Figure 8.1B-2C

Our last graph just shows how the integral curves
appear without the background of isoclines and the
corresponding line elements. See Figure 8.1B-2D.

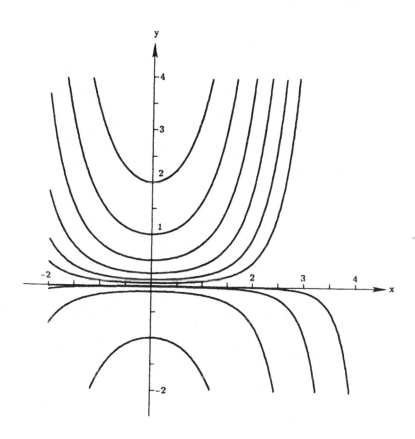

Figure 8.1B-2D

Exercise 11.

We separate the solution into four graphs. The differential equation $y' = y \sin x$ determines the family of isoclines $c = y \sin x$, or equivalently, $y = c \csc x$. Our first graph is the result of drawing these isoclines for $c = \pm 1/10,\ \pm 1/2,\ \pm 1,\ \pm 2,\ \pm 3,\ \pm 4,$ and -5. See Figure 8.1B–11A.

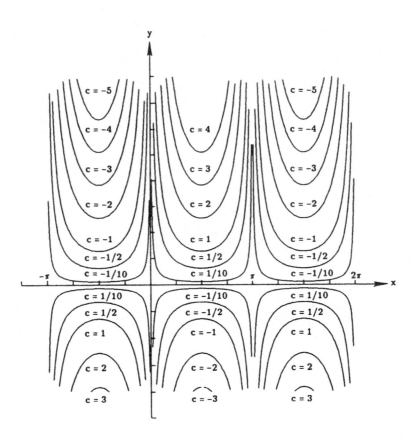

Figure 8.1B–11A

Now, on each of the isoclines drawn in our first
graph, we draw several line elements having the
appropriate slope, or, angle of inclination.   The slope of
each line element on the fixed isocline $y = c_0 \csc x$ is c_0;
the corresponding angle of inclination is then
$a_0 = \arctan c_0$.   We obtain the graph found in Figure
8.1B-11B.

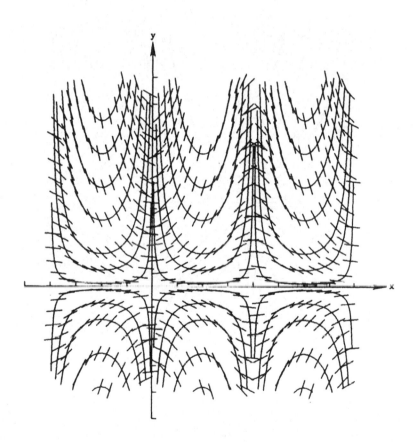

Figure 8.1B-11B

The line element configuration seen in the second graph indicates how to draw the approximate integral curves. We sketch several, including one whose y intercept is π. See Figure 8.1B–11C.

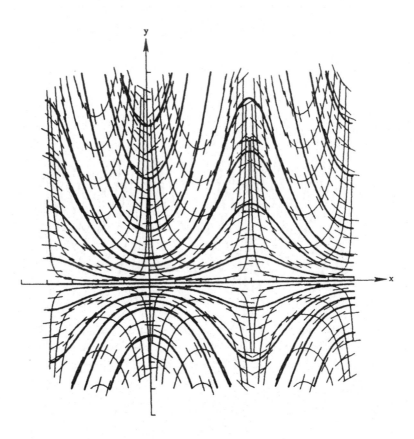

Figure 8.1B–11C

Our last graph just shows how the integral curves
appear without the background of isoclines and the
corresponding line elements.   See Figure 8.1B-11D.

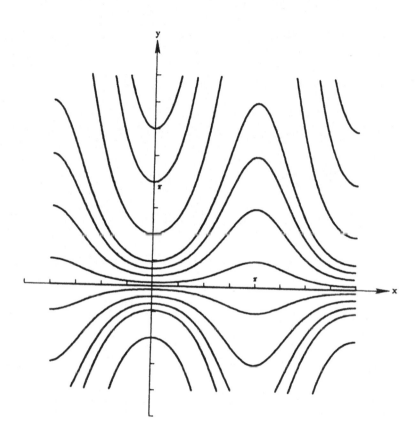

Figure 8.1B-11D

Section 8.2, Page 434.

3.   a.   We have

$$y(x) = y(0) + y'(0)x + y''(0)\frac{x^2}{2!} + y'''(0)\frac{x^3}{3!}$$

$$+ y^{iv}(0)\frac{x^4}{4!} + \cdots. \qquad (1)$$

The I.C. states $y(0) = 2$, and the D.E. gives $y'(0) = 1 + (0)(2)^2 = 1$. Differentiating the D.E., we obtain

$$y'' = 2xyy' + y^2, \qquad (2)$$

$$y''' = 2xyy'' + 2x(y')^2 + 4yy', \qquad (3)$$

$$y^{iv} = 2xyy''' + 6xy'y'' + 6yy'' + 6(y')^2. \qquad (4)$$

Substituting $x = 0$, $y = 2$, $y' = 1$ into (2), we obtain

$$y''(0) = 2(0)(2)(1) + (2)^2 = 4.$$

Substituting $x = 0$, $y = 2$, $y' = 1$, $y'' = 4$ into (3), we obtain

$$y'''(0) = (2)(0)(2)(4) + (2)(0)(1)^2 + (4)(2)(1) = 8.$$

Substituting $x = 0$, $y = 2$, $y' = 1$, $y'' = 4$, $y''' = 8$ into (4), we obtain

$$y^{iv}(0) = (2)(0)(1)(8) + (6)(0)(1)(4) + (6)(2)(4)$$

$$+ (6)(1)^2 = 54.$$

Now substituting the values of $y(0)$, $y'(0)$, $y''(0)$, $y'''(0)$, and $y^{iv}(0)$ so determined into (1), we obtain

$$y = 2 + 1x + \frac{4x^2}{2!} + \frac{8x^3}{3!} + \frac{54x^4}{4!} + \cdots$$

$$= 2 + x + 2x^2 + \frac{4x^3}{3} + \frac{9x^4}{4} + \cdots.$$

b. We assume

$$y = c_0 + c_1 x + c_2 x^2 + c_3 x^3 + c_4 x^4 + \cdots.$$

To satisfy the I.C. $y(0) = 2$, we must have $c_0 = 2$, and hence

$$y = 2 + c_1 x + c_2 x^2 + c_3 x^3 + c_4 x^4 + \cdots. \qquad (5)$$

Differentiating this, we find

$$y' = c_1 + 2c_2 x + 3c_3 x^2 + 4c_4 x^3 + \cdots. \qquad (6)$$

Since the initial y value is 2, we must express y^2 in the D.E. in powers of $y - 2$. Then the D.E. takes the form

$$y' = 1 + x[(y - 2)^2 + 4(y - 2) + 4].$$

Now substituting (5) and (6) into this, we obtain

$$c_1 + 2c_2x + 3c_3x^2 + 4c_4x^3 + \cdots$$

$$= 1 + x[(c_1x + c_2x^2 + \cdots)^2$$

$$+ 4(c_1x + c_2x^2 + c_3x^3 + \cdots) + 4]$$

or

$$c_1 + 2c_2x + 3c_3x^2 + 4c_4x^3 + \cdots$$

$$= 1 + 4x + 4c_1x^2 + (c_1^2 + 4c_2)x^3 + \cdots$$

From this, $c_1 = 1$, $2c_2 = 4$, $3c_3 = 4c_1$, $4c_4 = c_1^2 + 4c_2$;

and from these, $c_1 = 1$, $c_2 = 2$, $c_3 = \dfrac{4c_1}{3} = \dfrac{4}{3}$,

$c_4 = \dfrac{c_1^2 + 4c_2}{4} = \dfrac{9}{4}$.   Substituting these into (5) we again

obtain

$$y = 2 + x + 2x^2 + \frac{4x^3}{3} + \frac{9x^4}{4} + \cdots.$$

4. a. We have

$$y(x) = y(0) + y'(0)x + y''(0)\frac{x^2}{2!} + y'''(0)\frac{x^3}{3!}$$

$$+ y^{iv}(0)\frac{x^4}{4!} + \cdots. \tag{1}$$

The I.C. states that $y(0) = 3$, and the D.E. gives $y'(0) = (0)^3 + (3)^3 = 27$. Differentiating the D.E., we obtain

$$y'' = 3x^2 + 3y^2 y', \tag{2}$$

$$y''' = 6x + 3y^2 y'' + 6y(y')^2, \tag{3}$$

$$y^{iv} = 6 + 3y^2 y''' + 18yy'y'' + 6(y')^3 \tag{4}$$

Substituting $x = 0$, $y = 3$, $y' = 27$ into (1), we obtain

$$y''(0) = 3(0)^2 + 3(3)^2(27) = 729.$$

Substituting $x = 0$, $y = 3$, $y' = 27$, $y'' = 729$ into (3), we obtain

$$y'''(0) = 6(0) + 3(3)^2(729) + 6(3)(27)^2 = 32,805.$$

Substituting $y = 3$, $y' = 27$, $y'' = 729$, $y''' = 32,805$ into (4), we obtain

$$y^{iv}(0) = 6 + 3(3)^2(32,805) + (18)(3)(27)(729)$$
$$+ 6(27)^3 = 2,066,721.$$

Now substituting the values of $y(0)$, $y'(0)$, $y''(0)$, $y'''(0)$ and $y^{iv}(0)$ so determined into (1), we obtain

$$y = 3 + 27x + \frac{729x^2}{2} + \frac{32,805x^3}{6} + \frac{2,066,721x^4}{24} + \cdots$$

$$= 3 + 27x + \frac{729x^2}{2} + \frac{10,935x^3}{2} + \frac{688,907x^4}{8} + \cdots$$

b.   We assume $y = c_0 + c_1x + c_2x^2 + c_3x^3 + c_4x^4 + \cdots$.   To satisfy the I.C. $y(0) = 3$, we must have $c_0 = 3$, and hence

$$y = 3 + c_1x + c_2x^2 + c_3x^3 + c_4x^4 + \cdots \qquad (5)$$

Differentiating this, we find

$$y' = c_1 + 2c_2x + 3c_3x^2 + 4c_4x^3 + \cdots. \qquad (6)$$

Since the initial y value is 3, we must express y^3 in the D.E. in powers of $y - 3$.   Then the D.E. takes the form

$$y' = x^3 + (y - 3)^3 + 9(y - 3)^2 + 27(y - 3) + 27.$$

Now substituting (5) and (6) into this, we obtain

$$c_1 + 2c_2x + 3c_3x^2 + 4c_4x^3 + \cdots$$

$$= x^3 + (c_1x + c_2x^2 + \cdots)^3 + 9(c_1x + c_2x^2 + \cdots)^2$$

$$+ 27(c_1x + c_2x^2 + c_3x^3 + \cdots) + 27$$

or

$$c_1 + 2c_2x + 3c_3x^2 + 4c_4x^3 + \cdots$$

$$= 27 + 27c_1x + (9c_1^2 + 27c_2)x^2$$

$$+ (1 + c_1^3 + 18c_1c_2 + 27c_3)x^3 + \cdots.$$

From this, $c_1 = 27$, $2c_2 = 27c_1$, $3c_3 = 9c_1^2 + 27c_2$,

$$4c_4 = 1 + c_1^3 + 18c_1c_2 + 27c_3; \text{ and from these, } c_1 = 27,$$

$$c_2 = \frac{27c_1}{2} = \frac{729}{2}, \; c_3 = 3c_1^2 + 9c_2 = 3(27)^2 + 9\frac{729}{2} = \frac{10,935}{2},$$

$$c_4 = \frac{1 + c_1^3 + 18c_1c_2 + 27c_3}{4}$$

$$= \frac{1 + 19,683 + 177,147 + \frac{295,245}{2}}{4} = \frac{688,907}{8}.$$

Substituting these into (5), we again obtain

$$y = 3 + 27x + \frac{729x^2}{2} + \frac{10,935x^3}{2} + \frac{688,907x^4}{8} + \cdots.$$

5. a. We have

$$y(x) = y(0) + y'(0)x + \frac{y''(0)}{2!}x^2 + \frac{y'''(0)}{3!}x^3$$

$$+ \frac{y^{iv}(0)}{4!}x^4 + \frac{y^{v}(0)}{5!}x^5 + \cdots.$$

The I.C. states $y(0) = 0$, and the D.E. gives $y'(0) = 0 + \sin 0 = 0$. Differentiating the D.E., we obtain

$$y'' = 1 + (\cos y)y',$$

$$y''' = (\cos y)y'' - (\sin y)(y')^2.$$

$$y^{iv} = (\cos y)y''' - 3(\sin y)y'y'' - (\cos y)(y')^3,$$

$$y^v = (\cos y)y^{iv} - 4(\sin y)y'y''' - 3(\sin y)(y'')^2$$
$$- 6(\cos y)(y')^2 y'' + (\sin y)(y')^4.$$

Substituting $x = 0$, $y = 0$, $y' = 0$ into these, we find

$$y''(0) = 1, \quad y'''(0) = 1, \quad y^{iv}(0) = 1, \quad y^v(0) = 1.$$

Thus the assumed Taylor Series solution is

$$y = \frac{1}{2} x^2 + \frac{1}{6} x^3 + \frac{1}{24} x^4 + \frac{1}{120} x^5 + \cdots.$$

b.   We assume

$$y = c_0 + c_1 x + c_2 x^2 + c_3 x^3 + c_4 x^4 + c_5 x^5 + \cdots.$$

The I.C. $y(0) = 0$ gives $c_0 = 0$, so

$$y = c_1 x + c_2 x^2 + c_3 x^3 + c_4 x^4 + c_5 x^5 + \cdots. \tag{1}$$

Differentiating, we find

$$y' = c_1 + 2c_2 x + 3c_3 x^2 + 4c_4 x^3 + 5c_5 x^4 + \cdots. \tag{2}$$

Also, the Maclaurin Series for $\sin y$ is

$$\sin y = y - \frac{1}{6} y^3 + \cdots.$$

Thus the D.E. takes the form

$$y' = x + y - \frac{1}{6} y^3 + \cdots .$$

Now substituting (1) and (2) into this, we obtain

$$c_1 + 2c_2 x + 3c_3 x^2 + 4c_4 x^3 + 5c_5 x^4 + \cdots$$

$$= x + [c_1 x + c_2 x^2 + c_3 x^3 + c_4 x^4 + \cdots]$$

$$- \frac{1}{6} [c_1^3 x^3 + 3c_1^2 c_2 x^4 + \cdots] + \cdots .$$

or

$$c_1 + 2c_2 x + 3c_3 x^2 + 4c_4 x^3 + 5c_5 x^4 + \cdots$$

$$= (1 + c_1)x + c_2 x^2 + (c_3 - \frac{1}{6} c_1^3)x^3$$

$$+ (c_4 - \frac{1}{2} c_1^2 c_2)x^4 + \cdots .$$

From this, $c_1 = 0$, $2c_2 = 1 + c_1$, $3c_3 = c_2$,

$4c_4 = c_3 - \frac{1}{6} c_1^3$, $5c_5 = c_4 - \frac{1}{2} c_1^2 c_2$, $\cdots$.

and from these, $c_1 = 0$, $c_2 = \frac{1}{2}$, $c_3 = \frac{1}{3} c_2 = \frac{1}{6}$,

$c_4 = \frac{1}{4} c_3 = \frac{1}{24}$; $c_5 = \frac{1}{5} c_4 = \frac{1}{120}$.

Substituting this into (1), we again obtain

$$y = \frac{1}{2} x^2 + \frac{1}{6} x^3 + \frac{1}{24} x^4 + \frac{1}{120} x^5 + \cdots .$$

8. a.   We have

$$y(x) = y(0) + y'(0)x + \frac{y''(0)}{2!} x^2 + \frac{y'''(0)}{3!} x^3$$

$$+ \frac{y^{iv}(0)}{4!} x^4 + \cdots .$$

The I.C. states that $y(0) = 1$, and the D.E. gives $y'(0) = 1$. Differentiating the D.E., we obtain

$$y'' = 4x^3 + 4y^3 y' ,$$

$$y''' = 12x^2 + 4y^3 y'' + 12y^2 (y')^2 ,$$

$$y^{iv} = 24x + 4y^3 y''' + 36y^2 y' y'' + 24y(y')^3 .$$

Substituting $x = 0$, $y = 1$, $y' = 1$ into these, we find

$$y''(0) = 4, \qquad y'''(0) = 28, \qquad y^{iv}(0) = 280 .$$

Thus the assumed Taylor Series is

$$y = 1 + x + 2x^2 + \frac{14}{3} x^3 + \frac{35}{3} x^4 + \cdots .$$

b.   We assume

$$y = c_0 + c_1 x + c_2 x^2 + c_3 x^3 + c_4 x^4 + \cdots .$$

The I.C. $y(0) = 1$ gives $c_0 = 1$, so

$$y = 1 + c_1 x + c_2 x^2 + c_3 x^3 + c_4 x^4 + \cdots . \tag{1}$$

Differentiating we find

$$y' = c_1 + 2c_2x + 3c_3x^2 + 4c_4x^3 + \cdots \qquad (2)$$

Since the initial y value is 1, we express y^4 in the D.E. in powers of $y - 1$. Then the D.E. takes the form

$$y' = x^4 + 1 + 4(y - 1) + 6(y - 1)^2 + 4(y - 1)^3$$
$$+ (y - 1)^4.$$

Substituting (1) and (2) into this, we obtain

$$c_1 + 2c_2x + 3c_3x^2 + 4c_4x^3 + 5c_5x^4 + \cdots$$

$$= x^4 + 1 + 4(c_1x + c_2x^2 + c_3x^3 + c_4x^4 + \cdots)$$

$$+ 6[c_1^2x^2 + 2c_1c_2x^3 + (2c_1c_3 + c_2^2)x^4 + \cdots]$$

$$+ 4[c_1^3x^3 + 3c_1^2c_2x^4 + \cdots] + [c_1^4x^4 + \cdots]$$

or

$$c_1 + 2c_2x + 3c_3x^2 + 4c_4x^3 + 5c_5x^4 + \cdots$$

$$= 1 + 4c_1x + (4c_2 + 6c_1^2)x^2 + (4c_3 + 12c_1c_2 + 4c_1^3)x^3$$

$$+ (1 + 4c_4 + 12c_1c_3 + 6c_2^2 + 12c_1^2c_2 + c_1^4)x^4 + \cdots.$$

From these, $c_1 = 1$, $2c_2 = 4c_1$, $3c_3 = 4c_2 + 6c_1^2$,

$$4c_4 = 4c_3 + 12c_1c_2 + 4c_1^3, \cdots;$$

and from these, $c_1 = 1$, $c_2 = 2c_1 = 2$,

$$c_3 = \frac{4}{3} c_2 + 2c_1^2 = \frac{14}{3}, \quad c_4 = c_3 + 3c_1c_2 + c_1^3 + \frac{35}{3}, \quad \cdots.$$

Substituting these into (1), we again obtain

$$y = 1 + x + 2x^2 + \frac{14}{3} x^3 + \frac{35}{3} x^4 + \cdots.$$

12. a. We have

$$y(x) = y(1) + y'(1)(x - 1) + y''(1) \frac{(x - 1)^2}{2!}$$

$$+ y'''(1) \frac{(x - 1)^3}{3!} + \cdots$$

$$= \sum_{n = 0}^{\infty} y^{(n)}(1) \frac{(x - 1)^n}{n!}. \tag{1}$$

The I.C. states that $y(1) = \pi$, and the D.E. gives
$y'(1) = 1 + \cos \pi = 0$. Differentiating the D.E., we
obtain

$$y'' = 1 - (\sin y)y', \tag{2}$$

$$y''' = -(\sin y)y'' - (\cos y)(y')^2, \tag{3}$$

$$y^{iv} = -(\sin y)y''' - 3(\cos y)y''y'$$
$$+ (\sin y)(y')^2, \tag{4}$$

$$y^v = -(\sin y)y^{iv} - 4(\cos y)y'''y'$$
$$- 3(\cos y)(y'')^2 + 3(\sin y)y''(y')^2$$
$$+ 2(\sin y)y'y'' + (\cos y)(y')^3. \tag{5}$$

Substituting $x = 1$, $y = \pi$, $y' = 0$ into (2),
we obtain $y''(1) = 1 - (\sin \pi)(0) = 1$.
Substituting $y = \pi$, $y' = 0$, $y'' = 1$ into (3),
we obtain $y'''(1) = -(\sin \pi)(1) - (\cos \pi)(0)^2 = 0$.
Substituting $y = \pi$, $y' = 0$, $y'' = 1$, $y''' = 0$ into (4),
we obtain

$$y^{iv}(1) = -(\sin \pi)(0) - (3 \cos \pi)(1)(0)$$
$$+ (\sin \pi)(0)^2 = 0.$$

Substituting $y = \pi$, $y' = 0$, $y'' = 1$, $y''' = 0$, $y^{iv} = 0$ into
(5), we obtain

$$y^v(1) = -(\sin \pi)(0) - 4(\cos \pi)(0)(0) - 3(\cos \pi)(1)^2$$
$$+ 3(\sin \pi)(1)(0)^2 + 2(\sin \pi)(0)(1)$$
$$+ (\cos \pi)(0)^3 = 3.$$

Now substituting the values of $y(1)$, $y'(1)$, $\cdots$, $y^v(1)$ so
determined into (1), we obtain

$$y = \pi + \frac{(x - 1)^2}{2!} + 3\frac{(x - 1)^5}{5!} + \cdots$$

$$= \pi + \frac{(x - 1)^2}{2} + \frac{(x - 1)^5}{40} + \cdots$$

(b) We assume

$$y = c_0 + c_1(x - 1) + c_2(x - 1)^2 + c_3(x - 1)^3$$

$$+ c_4(x - 1)^4 + c_5(x - 1)^5 + \cdots$$

To satisfy the I.C. $y(1) = \pi$, we must have $c_0 = \pi$, and hence

$$y = \pi + c_1(x - 1) + c_2(x - 1)^2 + c_3(x - 1)^3$$

$$+ c_4(x - 1)^4 + c_5(x - 1)^5 + \cdots. \qquad (6)$$

Differentiating this, we find

$$y' = c_1 + 2c_2(x - 1) + 3c_3(x - 1)^2 + 4c_4(x - 1)^3$$

$$+ 5c_5(x - 1)^4 + \cdots. \qquad (7)$$

Since the initial y value is π, we must express $\cos y$ in the D.E. in powers of $y - \pi$. By Taylor's Theorem,

$$\phi(y) = \sum_{n = 0}^{\infty} \frac{\phi^{(n)}(\pi)}{n!} (y - \pi)^n$$

with $\phi(y) = \cos y$, we obtain

$$\cos y = -1 + \frac{(y - \pi)^2}{2!} - \frac{(y - \pi)^4}{4!} + \cdots$$

We also express x in powers of x - 1, and then the D.E. takes the form

$$y' = [(x - 1) + 1] + \left[-1 + \frac{(y - \pi)^2}{2!} - \frac{(y - \pi)^4}{4!} + \cdots \right]$$

or

$$y' = x - 1 + \frac{(y - \pi)^2}{2} - \frac{(y - \pi)^4}{24} + \cdots .$$

Now substituting (6) and (7) into this, we obtain

$$c_1 + 2c_2(x - 1) + 3c_3(x - 1)^2 + 4c_4(x - 1)^3$$

$$+ 5c_5(x - 1)^4 + \cdots$$

$$= x - 1 + \frac{[c_1(x - 1) + c_2(x - 1)^2 + \cdots]^2}{2}$$

$$- \frac{[c_1(x - 1) + c_2(x - 1)^2 + \cdots]^4}{24} + \cdots$$

$$= (x - 1) + \left(\frac{c_1^2}{2} \right)(x - 1)^2 + c_1 c_2 (x - 1)^3$$

$$+ \left(c_1 c_3 + \frac{c_2^2}{2} + \frac{c_1^4}{24} \right)(x - 1)^4 + \cdots .$$

From this, $c_1 = 0$, $2c_2 = 1$, $3c_3 = \dfrac{c_1^2}{2}$, $4c_4 = c_1 c_2$,

$5c_5 = c_1 c_3 + \dfrac{c_2^2}{2} + \dfrac{c_1^2}{24}$; and from these $c_1 = 0$, $c_2 = \dfrac{1}{2}$,

$$c_3 = 0, \ c_4 = 0, \ c_5 = \frac{0 + \frac{1}{8} + 0}{5} = \frac{1}{40} \ . \quad \text{Substituting these}$$

into (6), we again obtain

$$y = x + \frac{(x - 1)^2}{2} + \frac{(x - 1)^5}{40} + \cdots .$$

14. (a) We have

$$y(x) = y(1) + y'(1)(x - 1) + \frac{y''(1)}{2!}(x - 1)^2$$

$$+ \frac{y'''(1)}{3!}(x - 1)^3 + \cdots .$$

The I.C. states that $y(1) = 2$, and the D.E. gives $y'(1) = e + 2$. Differentiating the D.E., we obtain

$$y'' = y' + (x + 1)e^x.$$

$$y''' = y'' + (x + 2)e^x,$$

$$y^{iv} = y''' + (x + 3)e^x.$$

Substituting $x = 1$, $y = 2$, $y' = e + 2$ into these, we find

$$y''(1) = 3e + 2, \quad y'''(1) = 6e + 2, \quad y^{iv}(1) = 10e + 2.$$

Thus the assumed Taylor Series is

$$y = 2 + (e + 2)(x - 1) + \left(\frac{3}{2}e + 1\right)(x - 1)^2$$

$$+ \left(e + \frac{1}{3}\right)(x - 1)^3 + \left(\frac{5}{12}e + \frac{1}{12}\right)(x - 1)^4 + \cdots .$$

(b) We assume

$$y = c_0 + c_1(x - 1) + c_2(x - 1)^2 + c_3(x - 1)^3$$
$$+ c_4(x - 1)^4 + \cdots.$$

The I.C. $y(1) = 2$ gives $c_0 = 2$, so

$$y = 2 + c_1(x - 1) + c_2(x - 1)^2 + c_3(x - 1)^3$$
$$+ c_4(x - 1)^4 + \cdots. \tag{1}$$

Differentiating, we find

$$y' = c_1 + 2c_2(x - 1) + 3c_3(x - 1)^2$$
$$+ 4c_4(x - 1)^3 + \cdots. \tag{2}$$

We also need to expand xe^x in powers of $x - 1$. We have

$$xe^x = e[1 + (x - 1)]e^{(x - 1)}$$
$$= e[1 + (x - 1)][1 + (x - 1) + \frac{1}{2!}(x - 1)^2$$
$$+ \frac{1}{3!}(x - 1)^3 + \cdots].$$

Thus the D.E. takes the form

$$y' = y + e[1 + (x - 1)][1 + (x - 1) + \frac{1}{2}(x - 1)^2$$
$$+ \frac{1}{6}(x - 1)^3 + \cdots]$$

Now substituting (1) and (2) into this and multiplying the bracketed x factors together, we obtain

$$c_1 + 2c_2(x - 1) + 3c_3(x - 1)^2 + 4c_4(x - 1)^3 + \cdots$$

$$= 2 + c_1(x - 1) + c_2(x - 1)^2 + c_3(x - 1)^3 + \cdots$$

$$+e + 2e(x - 1) + \frac{3}{2}e(x - 1)^2 + \frac{2}{3}e(x - 1)^3 + \cdots$$

$$= (e + 2) + (2e + c_1)(x - 1) + \left(\frac{3}{2}e + c_2\right)(x - 1)^2$$

$$+ \left(\frac{2}{3}e + c_3\right)(x - 1)^3 + \cdots.$$

From this, $c_1 = e + 2$, $2c_2 = 2e + c_1$,

$3c_3 = \frac{3}{2}e + c_2$, $4c_4 = \frac{2}{3}e + c_3$;

and from these, $c_1 = e + 2$, $c_2 = e + \frac{1}{2}c_1 = \frac{3}{2}e + 1$,

$c_3 = \frac{1}{2}e + \frac{1}{3}c_2 = e + \frac{1}{3}$, $c_4 = \frac{1}{6}e + \frac{1}{4}c_3 = \frac{5}{12}e + \frac{1}{12}$.

Substituting these into (1), we again obtain

$$y = 2 + (e + 2)(x - 1) + \left(\frac{3}{2}e + 1\right)(x - 1)^2$$

$$+ \left(e + \frac{1}{3}\right)(x - 1)^3 + \left(\frac{5}{12}e + \frac{1}{12}\right)(x - 1)^4 + \cdots.$$

Section 8.3, Page 438.

4. Since the initial y value is 0, we choose the zeroth approximation ϕ_0 to be the function defined by $\phi_0(x) = 0$ for all x. The nth approximation ϕ_n for $n \geq 1$ is given by

$$\phi_n(x) = y_0 + \int_{x_0}^x f[t, \ \phi_{n-1}(t)] \ dt$$

$$= 0 + \int_0^x \{1 + t[\phi_{n-1}(t)^2\} \ dt, \ n \geq 1.$$

Using this formula for n = 1, 2, 3, $\cdots$, we obtain successively

$$\phi_1(x) = \int_0^x \{1 + t[\phi_0(t)]^2\} \ dt = \int_0^x (1 + 0) \ dt = x,$$

$$\phi_2(x) = \int_0^x \{1 + t[\phi_1(t)]^2\} \ dt = \int_0^x (1 + t^3) \ dt$$

$$= x + \frac{x^4}{4},$$

$$\phi_3(x) = \int_0^x \{1 + t[\phi_2(t)]^2\} \ dt$$

$$= \int_0^x \left[1 + t\left(t + \frac{t^4}{4}\right)^2\right] \ dt$$

$$= \int_0^x \left(1 + t^3 + \frac{t^6}{2} + \frac{t^9}{16}\right) \ dt$$

$$= x + \frac{x^4}{4} + \frac{x^7}{14} + \frac{x^{10}}{160}.$$

5. Since the initial y value is 0, we choose the zeroth
    approximation ϕ_0 to be the function defined by $\phi_0(x) = 0$
    for all x. The nth approximation ϕ_n for $n \geq 1$ is given by

$$\phi_n(x) = y_0 + \int_{x_0}^{x} f[t, \phi_{n-1}(t)] \, dt$$

$$= 0 + \int_{0}^{x} \{e^t + [\phi_{n-1}(t)]^2\} \, dt, \quad n \geq 1.$$

Using this formula for $n = 1, 2, 3, \cdots$, we obtain
successively

$$\phi_1(x) = \int_{0}^{x} (e^t + 0) \, dt = e^t \Big|_{0}^{x} = e^x - 1.$$

$$\phi_2(x) = \int_{0}^{x} [e^t + (e^t - 1)^2] \, dt$$

$$= \int_{0}^{x} (e^{2t} - e^t + 1) \, dt$$

$$= \frac{e^{2t}}{2} - e^t + t \Big|_{0}^{x} = \frac{e^{2x}}{2} - e^x + x + \frac{1}{2},$$

$$\phi_3(x) = \int_{0}^{x} \left[e^t + \left(\frac{e^{2t}}{2} - e^t + t + \frac{1}{2} \right)^2 \right] \, dt$$

$$= \int_0^x \left[\frac{e^{4t}}{4} - e^{3t} + te^{2t} + \frac{3e^{2t}}{2}\right.$$

$$\left. - 2te^t + t^2 + t + \frac{1}{4}\right] dt$$

$$= \left[\frac{e^{4t}}{16} - \frac{e^{3t}}{3} + e^{2t}\left(\frac{2t-1}{4}\right) + \frac{3e^{2t}}{4}\right.$$

$$\left. - 2e^t(t-1) + \frac{t^3}{3} + \frac{t^2}{2} + \frac{t}{4}\right]_0^x$$

$$= \frac{e^{4x}}{16} - \frac{e^{3x}}{3} + \frac{xe^{2x}}{2} + \frac{e^{2x}}{2}$$

$$- 2xe^x + 2e^x + \frac{x^3}{3} + \frac{x^2}{2} + \frac{x}{4} - \frac{107}{48}.$$

7. As in the solutions of Exercises 4 and 5, we choose $\phi_0(x) = 0$ for all x. For $n \geq 1$, $\phi_n(x)$ is given by

$$\phi_n(x) = y_0 + \int_{x_0}^x f[t, \phi_{n-1}(t)] dt$$

$$= 0 + \int_0^x \{2t + [\phi_{n-1}(t)]^2\} dt.$$

Using this formula for $n = 1, 2, 3, \cdots$, we obtain successively

$$\phi_1(x) = \int_0^x [2t + (0)^3] \, dt = x^2,$$

$$\phi_2(x) = \int_0^x [2t + (t^2)^3] \, dt = x^2 + \frac{x^7}{7},$$

$$\phi_3(x) = \int_0^x \left[2t + \left(t^2 + \frac{t^7}{7} \right)^3 \right] dt$$

$$= \int_0^x \left(2t + t^6 + \frac{3t^{11}}{7} + \frac{3t^{16}}{49} + \frac{t^{21}}{343} \right) dt$$

$$= x^2 + \frac{x^7}{7} + \frac{x^{12}}{28} + \frac{3x^{17}}{833} + \frac{x^{22}}{7546}.$$

NOTE   In the solutions of Sections 8.4 through 8.7, the exact solutions of the problems in Exercises 1, 4, 5, 8, 9, and 12 are required. The D.E.'s in Exercises 1 and 4 are linear (see Section 2.3A), the ones in Exercises 9 and 12 are separable (see Section 2.2A), and the ones in Exercises 5 and 8 are both linear and separable.

We list the exact solution of each of these problems here:

1. $y = \frac{1}{4} (2x - 1 + 5e^{-2x}).$

4. $y = \frac{1}{4} (-2x - 1 + 3e^2 e^{2x}).$

5. $y = e^{x^2/2 - 2x + 2}.$

8.　$y = \frac{1}{2} e^{1 - \cos x}$.

9.　$y = \frac{1}{5} \sqrt{25x^2 + 1}$.

12.　$y = \sqrt{3 - 2 \cos x}$.

Section 8.4, Page 447.

General Information:　For each problem below, we show the first three calculations in detail, while summarizing calculations at each value of x_n in a table.　The problems were done maintaining 11 to 12 digit accuracy from step to step; however, we've rounded the results to fewer places (usually four to six) as we tabulate them below.

1.　Let $h = 0.2$ and $f(x,y) = x - 2y$ in (8.53); we have $x_0 = 0$ and $y_0 = 1$.

(a) $x_1 = x_0 + h = 0.0 + 0.2 = 0.2$.

To find y_1, we use (8.53) with $n = 0$:　$y_1 = y_0 + h$

$f(x_0, y_0) = 1.0000 + (0.2)f(0.0, 1.0000) = 1.0000 +$

$(0.2)(0.0 - 2(1.0000)) = 0.6000$.

(b) $x_2 = x_1 + h = 0.2 + 0.2 = 0.4$.

To find y_2, we use (8.53) with $n = 1$:　$y_2 = y_1 +$

$hf(x_1,y_1) = 0.6000 + (0.2)f(0.2, 0.6000) = 0.6000 +$

$(0.2)(0.2 - 2(0.6000)) = 0.4000$.

(c) $x_3 = x_2 + h = 0.4 + 0.2 = 0.6$.

To find y_3, we use (8.53) with n = 2: $y_3 = y_2 +$

$hf(x_2,y_2) = 0.4000 + (0.2)f(0.4, 0.4000) = 0.4000 +$

$(0.2)(0.4 - 2(0.4000)) = 0.3200$.

Proceeding in this manner, using (8.53) with n = 3 and n = 4, we successively obtain $y_4 = 0.3120$ corresponding to $x_4 = 0.8$, and $y_5 = 0.3472$ corresponding to $x_5 = 1.0$.

All results and errors are summarized in Table 8.4.1. The exact solution of the differential equation is given in the pages immediately preceding this section of this manual.

TABLE 8.4.1   Euler Method for $y' = x - 2y$, $y(0) = 1$,
with h = 0.2.

x_n	Exact Solution	Euler Method	Error	% Rel Error
0.2	0.687900	0.600000	0.087900	12.778027
0.4	0.511661	0.400000	0.111661	21.823270
0.6	0.426493	0.320000	0.106493	24.969419
0.8	0.402371	0.312000	0.090371	22.459553
1.0	0.419169	0.347200	0.071969	17.169468

4. Let h = 0.2 and $f(x,y) = x + 2y$ in (8.53);
we have $x_0 = -1$ and $y_0 = 1$.

(a) $x_1 = x_0 + h = -1.0 + 0.2 = -0.8$.

To find y_1, we use (8.53) with n = 0: $y_1 = y_0 +$

$hf(x_0,y_0) = 1.0000 + (0.2)f(-1.0, 1.0000) = 1.0000 +$

$(0.2)(-1.0 + 2(1.0000)) = 1.2000$.

(b) $x_2 = x_1 + h = -0.8 + 0.2 = -0.6$.

To find y_2, we use (8.53) with n = 1: $y_2 = y_1 +$

$hf(x_1, y_1) = 1.2000 + (0.2)f(-0.8, 1.2000) = 1.2000 +$

$(0.2)(-0.8 + 2(1.2000)) = 1.5200$.

(c) $x_3 = x_2 + h = -0.6 + 0.2 = -0.4$.

To find y_3, we use (8.53) with n = 2: $y_3 = y_2 +$

$hf(x_2, y_2) = 1.5200 + (0.2)f(-0.6, 1.5200) = 1.5200 +$

$(0.2)(-0.6 + 2(1.5200)) = 2.0080$.

Proceeding in this manner, using (8.53) with n = 3 we

successively obtain $y_4 = 2.7312$ corresponding to

$x_4 = -0.2$, and $y_5 = 3.7837$ corresponding to $x_5 = 0.0$.

All results and errors are summarized in Table 8.4.4. The

exact solution of the differential equation is given in

the pages immediately preceding this section of this

manual.

TABLE 8.4.4   Euler Method for $y' = x + 2y$, $y(-1) = 1$,
             with h = 0.2.

x_n	Exact Solution	Euler Method	Error	% Rel Error
-0.8	1.268869	1.200000	0.068869	5.427554
-0.6	1.719156	1.520000	0.199156	11.584506
-0.4	2.440088	2.008000	0.432088	17.707876
-0.2	3.564774	2.731200	0.833574	23.383649
0.0	5.291792	3.783680	1.508112	28.499080

5.   Let $h = 0.1$ and $f(x,y) = xy - 2y$ in (8.53);
     we have $x_0 = 2$ and $y_0 = 1$.

(a) $x_1 = x_0 + h = 2.0 + 0.1 = 2.1$.

     To find y_1, we use (8.53) with $n = 0$: $y_1 = y_0 +$
     $hf(x_0,y_0) = 1.0000 + (0.1)f(2.0, 1.0000) = 1.0000 +$
     $(0.1)((2.0)(1.0000) - 2(1.0000)) = 1.0000 +$
     $(0.1)(0.0000) = 1.0000$.

(b) $x_2 = x_1 + h = 2.1 + 0.1 = 2.2$.

     To find y_2, we use (8.53) with $n = 1$: $y_2 = y_1 +$
     $hf(x_1,y_1) = 1.0000 + (0.1)f(2.1,1.0000) = 1.0000 +$
     $(0.1)((2.1)(1.0000) - 2(1.0000)) = 1.0000 +$
     $(0.1)(0.1000) = 1.0100$.

(c) $x_3 = x_2 + h = 2.2 + 0.1 = 2.3$.

     To find y_3, we use (8.53) with $n = 2$: $y_3 = y_2 +$
     $hf(x_2,y_2) = 1.0100 + (0.1)f(2.2, 1.0100) = 1.0100 +$
     $(0.1)((2.2)(1.0100) - 2(1.0100)) = 1.0100 +$
     $(0.1)(0.2020) = 1.0302$.

Proceeding in this manner, using (8.53) with $n = 3$ and
$n = 4$, we successively obtain $y_4 = 1.0611$ corresponding to
$x_4 = 2.4$, and $y_5 = 1.1036$ corresponding to $x_5 = 2.5$.   All
results and errors are summarized in Table 8.4.5.   The
exact solution of the differential equation is given in
the pages immediately preceding this section of this
manual.

TABLE 8.4.5 Euler Method for $y' = xy - 2y$, $y(2) = 1$, with $h = 0.1$

x_n	Exact Solution	Euler Method	Error	% Rel Error
2.1	1.005013	1.000000	0.005013	0.498752
2.2	1.020201	1.010000	0.010201	0.999934
2.3	1.046028	1.030200	0.015828	1.513139
2.4	1.083287	1.061106	0.022181	2.047571
2.5	1.133148	1.103550	0.029598	2.612033

8. Let $h = 0.2$ and $f(x,y) = y \sin x$ in (8.53); we have $x_0 = 0$ and $y_0 = 0.5$.

(a) $x_1 = x_0 + h = 0.0 + 0.2 = 0.2$.

To find y_1, we use (8.53) with $n = 0$: $y_1 = y_0 + hf(x_0, y_0) = 0.5000 + (0.2)f(0.0, 0.5000) = 0.5000 + (0.2)(0.5000 \sin (0.0)) = 0.5000 + (0.2)(0.0000) = 0.5000$.

(b) $x_2 = x_1 + h = 0.2 + 0.2 = 0.4$.

To find y_2, we use (8.53) with $n = 1$: $y_2 = y_1 + hf(x_1, y_1) = 0.5000 + (0.2)f(0.2, 0.5000) = 0.5000 + (0.2)(0.5000 \sin (0.2)) = 0.5000 + (0.2)(0.0993) = 0.5199$.

(c) $x_3 = x_2 + h = 0.4 + 0.2 = 0.6$.

To find y_3, we use (8.53) with $n = 2$: $y_3 = y_2 + hf(x_2, y_2) = 0.5199 + (0.2)f(0.4, 0.5199) = 0.5199 + (0.2)(0.5199 \sin (0.4)) = 0.5199 + (0.2)(0.2024) = 0.5604$.

Proceeding in this manner, using (8.53) with n = 3 and
n = 4, we successively obtain y_4 = 0.6236 corresponding to
x_4 = 0.8, and y_5 = 0.7131 corresponding to x_5 = 1.0. All
results and errors are summarized in Table 8.4.8. The
exact solution of the differential equation is given in
the pages immediately preceding this section of this
manual.

TABLE 8.4.8   Euler Method for y' = y sin x, y(0) = 0.5,
with h = 0.2.

x_n	Exact Solution	Euler Method	Error	% Rel Error
0.2	0.510067	0.500000	0.010067	1.973607
0.4	0.541069	0.519867	0.021202	3.918579
0.6	0.595423	0.560356	0.035067	5.889452
0.8	0.677156	0.623636	0.053520	7.903582
1.0	0.791798	0.713110	0.078687	9.937829

9. Let h = 0.2 and f(x,y) = x/y in (8.53);
we have x_0 = 0 and y_0 = 0.2.

(a) x_1 = x_0 + h = 0.0 + 0.2 = 0.2.

To find y_1, we use (8.53) with n = 0: y_1 = y_0 +
hf(x_0,y_0) = 0.2000 + (0.2)f(0.0, 0.2000) = 0.2000 +
(0.2)(0.0/0.2000) = 0.2000.

(b) x_2 = x_1 + h = 0.2 + 0.2 = 0.4.

To find y_2, we use (8.53) with n = 1: y_2 = y_1 +
hf(x_1,y_1) = 0.2000 + (0.2)f(0.2, 0.2000) = 0.2000 +
(0.2)(0.2/0.2000) = 0.4000.

(c) $x_3 = x_2 + h = 0.4 + 0.2 = 0.6$.

To find y_3, we use (8.53) with $n = 2$: $y_3 = y_2 +$

$hf(x_2,y_2) = 0.4000 + (0.2)f(0.4,0.4000) = 0.4000 +$

$(0.2)(0.4/0.4000) = 0.6000$.

Proceeding in this manner, using (8.53) with $n = 3$,
$n = 4$, ..., $n = 9$, we successively obtain $y_4 = 0.8000$

corresponding to $x_4 = 0.8$, $y_5 = 1.000$ corresponding to

$x_5 = 1.0$, ..., and finally $y_{10} = 2.0000$ corresponding

to $x_{10} = 2.0$. Results for all x_n and the errors are

summarized in Table 8.4.9. The exact solution of the
differential equation is given in the pages immediately
preceding this section of this manual.

TABLE 8.4.9   Euler Method for $y' = x/y$, $y(0) = 0.2$,
with $h = 0.2$.

x_n	Exact Solution	Euler Method	Error	% Rel Error
0.2	0.282843	0.200000	0.082843	29.289322
0.4	0.447214	0.400000	0.047214	10.557281
0.6	0.632456	0.600000	0.032456	5.131670
0.8	0.824621	0.800000	0.024621	2.985750
1.0	1.019804	1.000000	0.019804	1.941932
1.2	1.216553	1.200000	0.016553	1.360608
1.4	1.414214	1.400000	0.014214	1.005051
1.6	1.612452	1.600000	0.012452	0.772212
1.8	1.811077	1.800000	0.011077	0.611627
2.0	2.009975	2.000000	0.009975	0.496281

12.   Let $h = 0.2$ and $f(x,y) = (\sin x)/y$ in (8.53);
      we have $x_0 = 0$ and $y_0 = 1$.

(a) $x_1 = x_0 + h = 0.0 + 0.2 = 0.2$.

      To find y_1, we use (8.53) with $n = 0$: $y_1 = y_0 +$
      $hf(x_0,y_0) = 1.0000 + (0.2)f(0.0, 1.0000) = 1.0000 +$
      $(0.2)(\sin(0.0)/1.0000) = 1.0000 + (0.2)(0.0000) =$
      1.0000.

(b) $x_2 = x_1 + h = 0.2 + 0.2 = 0.4$.

      To find y_2, we use (8.53) with $n = 1$: $y_2 = y_1 +$
      $hf(x_1,y_1) = 1.0000 + (0.2)f(0.2, 1.0000) = 1.0000 +$
      $(0.2)(\sin(0.2)/1.0000) = 1.0000 + (0.2)(0.1987) =$
      1.0397.

(c) $x_3 = x_2 + h = 0.4 + 0.2 = 0.6$.

      To find y_3, we use (8.53) with $n = 2$: $y_3 = y_2 +$
      $hf(x_2,y_2) = 1.0397 + (0.2)f(0.4, 1.0397) = 1.0397 +$
      $(0.2)(\sin(0.4)/1.0397) = 1.0397 + (0.2)(0.3745) =$
      1.1146.

Proceeding in this manner, using (8.53) with $n = 3$,
$n = 4$, ..., $n = 9$, we successively obtain $y_4 = 1.2160$
corresponding to $x_4 = 0.8$, $y_5 = 1.3339$ corresponding
to $x_5 = 1.0$, ..., and $y_{10} = 1.9352$ corresponding to
$x_{10} = 2.0$. Results for all x_n and corresponding errors

are summarized in Table 8.4.12. The exact solution of the differential equation is given in the pages immediately preceding this section of this manual.

TABLE 8.4.12   Euler Method for $y' = (\sin x)/y$, $y(0) = 1$, with $h = 0.2$.

x_n	Exact Solution	Euler Method	Error	% Rel Error
0.2	1.019739	1.000000	0.019739	1.935654
0.4	1.076047	1.039734	0.036314	3.374715
0.6	1.161606	1.114641	0.046965	4.043104
0.8	1.267512	1.215955	0.051557	4.067577
1.0	1.385422	1.333946	0.051477	3.715614
1.2	1.508405	1.460108	0.048296	3.201819
1.4	1.630971	1.587775	0.043195	2.648449
1.6	1.748828	1.711905	0.036923	2.111296
1.8	1.858603	1.828684	0.029919	1.609740
2.0	1.957624	1.935192	0.022432	1.145899

Section 8.5, Page 454.

General Information: For each problem below, we show the first three calculations in detail, while summarizing calculations at each value of x_n in a table. The problems were done maintaining 11 to 12 digit accuracy from step to step; however, we've rounded the results to fewer places (usually four to six) as we tabulate them below.

1. Let $h = 0.2$ and $f(x,y) = x - 2y$ in (8.59) and (8.60); we have $x_0 = 0$ and $y_0 = 1$.

(a) $x_1 = x_0 + h = 0.0 + 0.2 = 0.2$.

To find y_1, we use (8.59) and (8.60) with $n = 0$.

First using (8.59), we find

$$\hat{y}_1 = y_0 + hf(x_0,y_0) = 1.0000 + (0.2)f(0.0, 1.0000)$$

$$= 1.0000 + (0.2)(-2.0000) = 0.60000.$$

Now using (8.60) with this value of $\hat{y}_1$ we obtain

$$y_1 = y_0 + h \frac{f(x_0,y_0) + f(x_1,\hat{y}_1)}{2}$$

$$= 1.0000 + (0.2) \frac{f(0.0, 1.0000) + f(0.2, 0.6000)}{2}$$

$$= 1.0000 + (0.2) \frac{-2.0000 - 1.0000}{2} = 0.7000.$$

(b) $x_2 = x_2 + h = 0.2 + 0.2 = 0.4.$

To find y_2, we use (8.59) and (8.60) with $n = 1$.

First using (8.59), we find

$$\hat{y}_2 = y_1 + hf(x_1,y_1) = 0.7000 + (0.2)f(0.2, 0.7000)$$

$$= 0.7000 + (0.2)(-1.2000) = 0.4600.$$

Now using (8.60) with this value of $\hat{y}_2$ we obtain

$$y_2 = y_1 + h \frac{f(x_1,y_1) + f(x_2,\hat{y}_2)}{2}$$

$$= 0.7000 + (0.2) \frac{f(0.2, 0.7000) + f(0.4, 0.4600)}{2}$$

$$= 0.7000 + (0.2) \frac{-1.2000 - 0.5200}{2} = 0.5280.$$

(c) $x_3 = x_2 + h = 0.4 + 0.2 = 0.6.$

To find y_3, we use (8.59) and (8.60) with n = 2.

First using (8.59), we find

$$\hat{y}_3 = y_2 + hf(x_2, y_2) = 0.5280 + (0.2)f(0.4, 0.5280)$$

$$= 0.5280 + (0.2)(-0.6560) = 0.3968.$$

Now using (8.60) with this value of $\hat{y}_3$ we obtain

$$y_3 = y_2 + h\frac{f(x_2, y_2) + f(x_3, \hat{y}_3)}{2}$$

$$= 0.5280 + (0.2)\frac{f(0.4, 0.5280) + f(0.6, 0.3968)}{2}$$

$$= 0.5280 + (0.2)\frac{-0.6560 - 0.1936}{2} = 0.4430.$$

These results and those for the remaining x_n's, as well as the errors are summarized in Table 8.5.1.   The exact solution of the differential equation is given in the pages immediately preceding Section 8.4 of this manual.

TABLE 8.5.1   Improved Euler Method for y' = x - 2y, y(0) = 1, with h = 0.2.

x_n	Exact Solution	Improved Euler	Error	% Rel Error
0.2	0.687900	0.700000	0.012100	1.7590
0.4	0.511661	0.528000	0.016339	3.1933
0.6	0.426493	0.443040	0.016547	3.8798
0.8	0.402371	0.417267	0.014897	3.7022
1.0	0.419169	0.431742	0.012573	2.9994

4. Let $h = 0.2$ and $f(x,y) = x + 2y$ in (8.59) and (8.60);
    we have $x_0 = -1$ and $y_0 = 1$.

    (a) $x_1 = x_0 + h = -1.0 + 0.2 = -0.8$.

    To find y_1, we use (8.59) and (8.60) with $n = 0$.

    First using (8.59), we find

    $$\hat{y}_1 = y_0 + hf(x_0,y_0) = 1.0000 + (0.2)f(-1.0, 1.0000)$$

    $$= 1.0000 + (0.2)(1.0000) = 1.2000.$$

    Now using (8.60) with this value of $\hat{y}_1$ we obtain

    $$y_1 = y_0 + h \frac{f(x_0,y_0) + f(x_1,\hat{y}_1)}{2}$$

    $$= 1.0000 + (0.2) \frac{f(-1.0, 1.0000) + f(-0.8, 1.2000)}{2}$$

    $$= 1.0000 + (0.2) \frac{1.0000 + 1.6000}{2} = 1.2600.$$

    (b) $x_2 = x_1 + h = -0.8 + 0.2 = -0.6$.

    To find y_2, we use (8.59) and (8.60) with $n = 1$.

    First using (8.59), we find

    $$\hat{y}_2 = y_1 + hf(x_1,y_1) = 1.2600 + (0.2)f(-0.8, 1.2600)$$

    $$= 1.2600 + (0.2)(1.7200) = 1.6040.$$

    Now using (8.60) with this value of $\hat{y}_2$ we obtain

$$y_2 = y_1 + h \frac{f(x_1,y_1) + f(x_2,\hat{y}_2)}{2}$$

$$= 1.2600 + (0.2) \frac{f(-0.8,\ 1.2600) + f(-0.6,\ 1.6040)}{2}$$

$$= 1.2600 + (0.2) \frac{1.7200 + 2.6080}{2} = 1.6928.$$

(c) $x_3 = x_2 + h = -0.6 + 0.2 = -0.4.$

To find y_3, we use (8.59) and (8.60) with $n = 2$.

First using (8.59), we find

$$\hat{y}_3 = y_2 + hf(x_2,y_2) = 1.6928 + (0.2)f(-0.6,\ 1.6928)$$

$$= 1.6928 + (0.2)(2.7856) = 2.2499.$$

Now using (8.60) with this value of $\hat{y}_3$ we obtain

$$y_3 = y_2 + h \frac{f(x_2,y_2) + f(x_3,\hat{y}_3)}{2}$$

$$= 1.6928 + (0.2) \frac{f(-0.6,\ 1.6928) + f(-0.4,\ 2.2499)}{2}$$

$$= 1.6928 + (0.2) \frac{2.7856 + 4.0998}{2} = 2.3813.$$

These results and those for the remaining x_n's, as well as the errors are summarized in Table 8.5.4. The exact solution of the differential equation is given in the pages immediately preceding Section 8.4 of this manual.

TABLE 8.5.4   Improved Euler Method for $y' = x + 2y$, $y(-1) = 1$, with $h = 0.2$.

x_n	Exact Solution	Improved Euler	Error	% Rel Error
-0.8	1.268869	1.260000	0.008869	0.6989
-0.6	1.719156	1.692800	0.026356	1.5331
-0.4	2.440088	2.381344	0.058744	2.4074
-0.2	3.564774	3.448389	0.116385	3.2649
0.0	5.291792	5.075616	0.216176	4.0851

5. Let $h = 0.1$ and $f(x,y) = xy - 2y$ in (8.59) and (8.60); we have $x_0 = 2$ and $y_0 = 1$.

(a) $x_1 = x_0 + h = 2.0 + 0.1 = 2.1$.

To find y_1, we use (8.59) and (8.60) with $n = 0$.

First using (8.59), we find

$$\hat{y}_1 = y_0 + hf(x_0, y_0) = 1.0000 + (0.1)f(2.0, 1.0000)$$
$$= 1.0000 + (0.1)(0.0000) = 1.0000.$$

Now using (8.60) with this value of $\hat{y}_1$ we obtain

$$y_1 = y_0 + h \frac{f(x_0, y_0) + f(x_1, \hat{y}_1)}{2}$$

$$= 1.0000 + (0.1) \frac{f(2.0, 1.0000) + f(2.1, 1.0000)}{2}$$

$$= 1.0000 + (0.1) \frac{0.0000 + 0.1000}{2} = 1.0050.$$

(b) $x_2 = x_1 + h = 2.1 + 0.1 = 2.2$.

To find y_2, we use (8.59) and (8.60) with $n = 1$.

First using (8.59), we find

$$\hat{y}_2 = y_1 + hf(x_1,y_1) = 1.0050 + (0.1)f(2.1, 1.0050)$$

$$= 1.0050 + (0.1)(1.1005) = 1.0150.$$

Now using (8.60) with this value of $\hat{y}_2$ we obtain

$$y_2 = y_1 + h\ \frac{f(x_1,y_1) + f(x_2,\hat{y}_2)}{2}$$

$$= 1.0050 + (0.1)\ \frac{f(2.1, 1.0050) + f(2.2, 1.0150)}{2}$$

$$= 1.0050 + (0.1)\ \frac{0.1005 + 0.2030}{2} = 1.0202.$$

(c) $x_3 = x_2 + h = 2.2 + 0.1 = 2.3.$

To find y_3, we use (8.59) and (8.60) with $n = 2$.

First using (8.59), we find

$$\hat{y}_3 = y_2 + hf(x_2,y_2) = 1.0202 + (0.1)f(2.2, 1.0202)$$

$$= 1.0202 + (0.1)(0.2040) = 1.0406.$$

Now using (8.60) with this value of $\hat{y}_3$ we obtain

$$y_3 = y_2 + h\ \frac{f(x_2,y_2) + f(x_3,\hat{y}_3)}{2}$$

$$= 1.0202 + (0.1)\ \frac{f(2.2, 1.0202) + f(2.3, 1.0406)}{2}$$

$$= 1.0202 + (0.1)\ \frac{0.2040 + 0.3122}{2} = 1.0460.$$

These results and those for the remaining x_n's, as well as the errors are summarized in Table 8.5.5. The exact solution of the differential equation is given in the pages immediately preceding Section 8.4 of this manual.

TABLE 8.5.5   Improved Euler Method for $y' = xy - 2y$, $y(2) = 1$, with $h = 0.1$.

x_n	Exact Solution	Improved Euler	Error	% Rel Error
2.1	1.005013	1.005000	0.000013	0.0012
2.2	1.020201	1.020175	0.000026	0.0025
2.3	1.046028	1.045986	0.000042	0.0040
2.4	1.083287	1.083223	0.000064	0.0059
2.5	1.133148	1.133051	0.000097	0.0086

8. Let $h = 0.2$ and $f(x,y) = y \sin x$ in (8.59) and (8.60); we have $x_0 = 0$ and $y_0 = 0.5$.

(a) $x_1 = x_0 + h = 0.0 + 0.2 = 0.2$.

To find y_1, we use (8.59) and (8.60) with $n = 0$.

First using (8.59), we find

$$\hat{y}_1 = y_0 + hf(x_0, y_0) = 0.5000 + (0.2)f(0.0, 0.5000)$$
$$= 0.5000 + (0.2)(0.0000) = 0.5000.$$

Now using (8.60) with this value of $\hat{y}_1$ we obtain

$$y_1 = y_0 + h \; \frac{f(x_0, y_0) + f(x_1, \hat{y}_1)}{2}$$

$$= 0.5000 + (0.2) \; \frac{f(0.0, \; 0.5000) + f(0.2, \; 0.5000)}{2}$$

$$= 0.5000 + (0.2) \; \frac{0.0000 + 0.0993}{2} = 0.5099.$$

(b) $x_2 = x_1 + h = 0.2 + 0.2 = 0.4.$

To find y_2, we use (8.59) and (8.60) with $n = 1$.

First using (8.59), we find

$$\hat{y}_2 = y_1 + hf(x_1, y_1) = 0.5099 + (0.2)f(0.2, \; 0.5099)$$

$$= 0.5099 + (0.2)(0.1013) = 0.5302.$$

Now using (8.60) with this value of $\hat{y}_2$ we obtain

$$y_2 = y_1 + h \; \frac{f(x_1, y_1) + f(x_2, \hat{y}_2)}{2}$$

$$= 0.5099 + (0.2) \; \frac{f(0.2, \; 0.5099) + f(0.4, \; 0.5302)}{2}$$

$$= 0.5099 + (0.2) \; \frac{0.1013 + 0.2065}{2} = 0.5407.$$

(c) $x_3 = x_2 + h = 0.4 + 0.2 = 0.6.$

To find y_3, we use (8.59) and (8.60) with $n = 2$.

First using (8.59), we find

$$\hat{y}_3 = y_2 + hf(x_2, y_2) = 0.5407 + (0.2)f(0.4, \; 0.5407)$$

$$= 0.5407 + (0.2)(0.2106) = 0.5828.$$

Now using (8.60) with this value of $\hat{y}_3$ we obtain

$$y_3 = y_2 + h \frac{f(x_2,y_2) + f(x_3,\hat{y}_3)}{2}$$

$$= 0.5407 + (0.2) \frac{f(0.4,\ 0.5407) + f(0.6,\ 0.5828)}{2}$$

$$= 0.5407 + (0.2) \frac{0.2106 + 0.3291}{2} = 0.5947.$$

These results and those for the remaining x_n's, as well as the errors are summarized in Table 8.5.8. The exact solution of the differential equation is given in the pages immediately preceding Section 8.4 of this manual.

TABLE 8.5.8   Improved Euler Method for $y' = y \sin x$, $y(0) = 0.5$, with $h = 0.2$.

x_n	Exact Solution	Improved Euler	Error	% Rel Error
0.2	0.510067	0.509933	0.000133	0.0261
0.4	0.541069	0.540711	0.000358	0.0662
0.6	0.595423	0.594676	0.000747	0.1255
0.8	0.677156	0.675731	0.001425	0.2104
1.0	0.791798	0.789224	0.002574	0.3251

9. Let $h = 0.2$ and $f(x,y) = x/y$ in (8.59) and (8.60); we have $x_0 = 0$ and $y_0 = 0.2$.

(a) $x_1 = x_0 + h = 0.0 + 0.2 = 0.2.$

To find y_1, we use (8.59) and (8.60) with $n = 0$.

First using (8.59), we find

$$\hat{y}_1 = y_0 + hf(x_0, y_0) = 0.2000 + (0.2)f(0.0, 0.2000)$$

$$= 0.2000 + (0.2)(0.0000) = 0.2000.$$

Now using (8.60) with this value of $\hat{y}_1$ we obtain

$$y_1 = y_0 + h \frac{f(x_0, y_0) + f(x_1, \hat{y}_1)}{2}$$

$$= 0.2000 + (0.2) \frac{f(0.0, 0.2000) + f(0.2, 0.2000)}{2}$$

$$= 0.2000 + (0.2) \frac{0.0000 + 1.0000}{2} = 0.3000.$$

(b) $x_2 = x_1 + h = 0.2 + 0.2 = 0.4.$

To find y_2, we use (8.59) and (8.60) with $n = 1$.

First using (8.59), we find

$$\hat{y}_2 = y_1 + hf(x_1, y_1) = 0.3000 + (0.2)f(0.2, 0.3000)$$

$$= 0.3000 + (0.2)(0.6667) = 0.4333.$$

Now using (8.60) with this value of $\hat{y}_2$ we obtain

$$y_2 = y_1 + h \frac{f(x_1, y_1) + f(x_2, \hat{y}_2)}{2}$$

$$= 0.3000 + (0.2) \frac{f(0.2, 0.3000) + f(0.4, 0.4333)}{2}$$

$$= 0.3000 + (0.2) \frac{0.6667 + 0.9231}{2} = 0.4590.$$

(c) $x_3 = x_2 + h = 0.4 + 0.2 = 0.6.$

To find y_3, we use (8.59) and (8.60) with $n = 2$.

First using (8.59), we find

$$\hat{y}_3 = y_2 + hf(x_2,y_2) = 0.4590 + (0.2)f(0.4, 0.4590)$$

$$= 0.4590 + (0.2)(0.8715) = 0.6333.$$

Now using (8.60) with this value of $\hat{y}_3$ we obtain

$$y_3 = y_2 + h \frac{f(x_2,y_2) + f(x_3,\hat{y}_3)}{2}$$

$$= 0.4590 + (0.2) \frac{f(0.4, 0.4590) + f(0.6, 0.6333)}{2}$$

$$= 0.4590 + (0.2) \frac{0.8715 + 0.9475}{2} = 0.6409.$$

These results and those for the remaining x_n's, as well as the errors are summarized in Table 8.5.9. The exact solution of the differential equation is given in the pages immediately preceding Section 8.4 of this manual.

TABLE 8.5.9   Improved Euler Method for $y' = x/y$, $y(0) = 0.2$, with $h = 0.2$.

x_n	Exact Solution	Improved Euler	Error	% Rel Error
0.2	0.282843	0.300000	0.017157	6.0660
0.4	0.447214	0.458974	0.011761	2.6298
0.6	0.632456	0.640871	0.008415	1.3305
0.8	0.824621	0.831098	0.006477	0.7854
1.0	1.019804	1.025049	0.005245	0.5144
1.2	1.216553	1.220953	0.004401	0.3617
1.4	1.414214	1.418001	0.003787	0.2678
1.6	1.612452	1.615774	0.003323	0.2061
1.8	1.811077	1.814036	0.002959	0.1634
2.0	2.009975	2.012642	0.002667	0.1327

12. Let $h = 0.2$ and $f(x,y) = (\sin x)/y$ in (8.59) and (8.60); we have $x_0 = 0$ and $y_0 = 1$.

(a) $x_1 = x_0 + h = 0.0 + 0.2 = 0.2$.

To find y_1, we use (8.59) and (8.60) with $n = 0$.

First using (8.59), we find

$$\hat{y}_1 = y_0 + hf(x_0, y_0) = 1.0000 + (0.2)f(0.0, 1.0000)$$

$$= 1.0000 + (0.2)(0.0000) = 1.0000.$$

Now using (8.60) with this value of $\hat{y}_1$ we obtain

$$y_1 = y_0 + h\,\frac{f(x_0, y_0) + f(x_1, \hat{y}_1)}{2}$$

$$= 1.0000 + (0.2)\,\frac{f(0.0, 1.0000) + f(0.2, 1.0000)}{2}$$

$$= 1.0000 + (0.2)\,\frac{0.0000 + 0.1987}{2} = 1.0199.$$

(b) $x_2 = x_1 + h = 0.2 + 0.2 = 0.4$.

To find y_2, we use (8.59) and (8.60) with $n = 1$.

First using (8.59), we find

$$\hat{y}_2 = y_1 + hf(x_1,y_1) = 1.0199 + (0.2)f(0.2, 1.0199)$$

$$= 1.0199 + (0.2)(0.1948) = 1.0588.$$

Now using (8.60) with this value of $\hat{y}_2$ we obtain

$$y_2 = y_1 + h\ \frac{f(x_1,y_1) + f(x_2,\hat{y}_2)}{2}$$

$$= 1.0199 + (0.2)\ \frac{f(0.2,\ 1.0199) + f(0.4,\ 1.0588)}{2}$$

$$= 1.0199 + (0.2)\ \frac{0.1948 + 0.3678}{2} = 1.0761.$$

(c) $x_3 = x_2 + h = 0.4 + 0.2 = 0.6$.

To find y_3, we use (8.59) and (8.60) with $n = 2$.

First using (8.59), we find

$$\hat{y}_3 = y_2 + hf(x_2,y_2) = 1.0761 + (0.2)f(0.4,\ 1.0761)$$

$$= 1.0761 + (0.2)(0.3619) = 1.1485.$$

Now using (8.60) with this value of $\hat{y}_3$ we obtain

$$y_3 = y_2 + h \frac{f(x_2, y_2) + f(x_3, \hat{y}_3)}{2}$$

$$= 1.0761 + (0.2) \frac{f(0.4, \ 1.0761) + f(0.6, \ 1.1485)}{2}$$

$$= 1.0761 + (0.2) \frac{0.3619 + 0.4916}{2} = 1.1615.$$

These results and those for the remaining x_n's, as well as the errors are summarized in Table 8.5.12. The exact solution of the differential equation is given in the pages immediately preceding Section 8.4 of this manual.

TABLE 8.5.12   Improved Euler Method for $y' = (\sin x)/y$, $y(0) = 1$, with $h = 0.2$.

x_n	Exact Solution	Improved Euler	Error	% Rel Error
0.2	1.019739	1.019867	0.000128	0.0126
0.4	1.076047	1.076125	0.000078	0.0072
0.6	1.161606	1.161476	0.000130	0.0112
0.8	1.267512	1.267082	0.000430	0.0340
1.0	1.385422	1.384659	0.000764	0.0551
1.2	1.508405	1.507310	0.001095	0.0726
1.4	1.630971	1.629565	0.001405	0.0862
1.6	1.748828	1.747140	0.001688	0.0965
1.8	1.858603	1.856666	0.001937	0.1042
2.0	1.957624	1.955473	0.002152	0.1099

Section 8.6, Page 462.

General Information: For each problem below, we show the first two calculations in detail, while summarizing calculations at each value of x_n in a table. The problems were done maintaining 11 to 12 digit accuracy from step to step; however, we've rounded the results to fewer places as we tabulate them below.

1.   Let $h = 0.2$ and $f(x,y) = x - 2y$ in (8.59) and (8.60); we have $x_0 = 0$ and $y_0 = 1$.

$$k_1 = hf(x_0,y_0) = 0.2f(0.0,\ 1.0000000) = 0.2(-2.0000000)$$
$$= -0.4000000,$$

$$k_2 = hf(x_0 + h/2,\ y_0 + k_1/2) = 0.2f(0.1,\ 0.8000000)$$
$$= 0.2(-1.5000000) = -0.3000000,$$

$$k_3 = hf(x_0 + h/2,\ y_0 + k_2/2) = 0.2f(0.1,\ 0.8500000)$$
$$= 0.2(-1.6000000) = -0.3200000,$$

$$k_4 = hf(x_0 + h,\ y_0 + k_3) = 0.2f(0.2,\ 0.6800000)$$
$$= 0.2(-1.1600000) = -0.2320000.$$

Therefore

$$K = (k_1 + 2k_2 + 2k_3 + k_4)/6$$
$$= (-0.4000000 - 0.6000000 - 0.6400000 - 0.2320000)/6$$
$$= -0.3120000,$$

and so the approximate value of the solution at $x_1 = 0.2$ is

$$y_1 = 1.0000000 - 0.3120000 = 0.6880000.$$

Now using the above value y_1, we calculate successively *new* k_1, k_2, k_3, k_4, and then K. We first find

$$k_1 = hf(x_1, y_1) = 0.2f(0.2, 0.6880000) = 0.2(-1.1760000)$$
$$= -0.2352000,$$

$$k_2 = hf(x_1 + h/2, y_1 + k_1/2) = 0.2f(0.3, 0.5704000)$$
$$= 0.2(-0.8408000) = -0.1681600,$$

$$k_3 = hf(x_1 + h/2, y_1 + k_2/2) = 0.2f(0.3, 0.6039200)$$
$$= 0.2(-0.9078400) = -0.1815680,$$

$$k_4 = hf(x_1 + h, y_1 + k_3) = 0.2f(0.4, 0.5064320)$$
$$= 0.2(-0.6128640) = -0.1225728.$$

Therefore

$$K = (k_1 + 2k_2 + 2k_3 + k_4)/6$$
$$= (-0.2352000 - 0.3363200 - 0.3631360 - 0.1225728)/6$$
$$= -0.1762048,$$

and so the approximate value of the solution at $x_2 = 0.4$ is

$$y_2 = 0.6880000 - 0.1762048 = 0.5117952.$$

We summarize the calculations for these and the remaining y_n's in Table 8.6.1. The exact solution of the

differential equation is given in the pages immediately
preceding Section 8.4 of this manual.

TABLE 8.6.1   Runge–Kutta Method for $x - 2y$,
$y(0) = 1$, with $h = 0.2$.

x_n	k_1	k_2	k_3	k_4	K	Exact Solution	Runge-Kutta	Error	% Rel Error
0.2	-0.4000	-0.3000	-0.3200	-0.2320	-0.3120	0.687900	0.688000	0.000100	0.014529
0.4	-0.2352	-0.1682	-0.1816	-0.1226	-0.1762	0.511661	0.511795	0.000134	0.026188
0.6	-0.1247	-0.0798	-0.0888	-0.0492	-0.0852	0.426493	0.426628	0.000135	0.031592
0.8	-0.0507	-0.0205	-0.0265	0.0000	-0.0241	0.402371	0.402491	0.000120	0.029930
1.0	-0.0010	0.0192	0.0152	0.0329	0.0168	0.419169	0.419270	0.000101	0.024075

4. Let $h = 0.2$ and $f(x,y) = x + 2y$ in (8.59) and (8.60);
    we have $x_0 = -1$ and $y_0 = 1$.

$k_1 = hf(x_0, y_0) = 0.2f(-1.0, 1.0000000)$

$\quad = 0.2(1.0000000) = 0.2000000,$

$k_2 = hf(x_0 + h/2, y_0 + k_1/2) = 0.2f(-0.9, 1.1000000)$

$\quad = 0.2(1.3000000) = 0.2600000,$

$k_3 = hf(x_0 + h/2, y_0 + k_2/2) = 0.2f(-0.9, 1.1300000)$

$\quad = 0.2(1.3600000) = 0.2720000,$

$k_4 = hf(x_0 + h, y_0 + k_3) = 0.2f(-0.8, 1.2720000)$

$\quad = 0.2(1.7440000) = 0.3488000.$

Therefore

$$K = (k_1 + 2k_2 + 2k_3 + k_4)/6$$
$$= (0.2000000 + 0.5200000 + 0.5440000 + 0.3488000)/6$$
$$= 0.2688000,$$

and so the approximate value of the solution at $x_1 = -0.8$ is

$$y_1 = 1.0000000 + 0.2688000 = 1.2688000.$$

Now using the above value y_1, we calculate successively *new* k_1, k_2, k_3, k_4, and then K. We first find

$$k_1 = hf(x_1,y_1) = 0.2f(-0.8, 1.2688000) = 0.2(1.7376000)$$
$$= 0.3475200,$$

$$k_2 = hf(x_1 + h/2, y_1 + k_1/2) = 0.2f(-0.7, 1.4425600)$$
$$= 0.2(2.1851200) = 0.4370240,$$

$$k_3 = hf(x_1 + h/2, y_1 + k_2/2) = 0.2f(-0.7, 1.4873120)$$
$$= 0.2(2.2746240) = 0.4549248,$$

$$k_4 = hf(x_1 + h, y_1 + k_3) = 0.2f(-0.6, 1.7237248)$$
$$= 0.2(2.8474496) = 0.5694899.$$

Therefore

$$K = (k_1 + 2k_2 + 2k_3 + k_4)/6$$

$$= (0.3475200 + 0.8740480 + 0.9098496 + 0.5694899)/6$$

$$= 0.4501513,$$

and so the approximate value of the solution at $x_2 = -0.6$ is

$$y_2 = 1.2688000 + 0.4501513 = 1.7189513.$$

We summarize the calculations for these and the remaining y_n's in Table 8.6.4. The exact solution of the differential equation is given in the pages immediately preceding Section 8.4 of this manual.

TABLE 8.6.4   Runge-Kutta Method for $y' = x + 2y$, $y(-1) = 1$, with $h = 0.2$.

x_n	k_1	k_2	k_3	k_4	K	Exact Solution	Runge-Kutta	Error	% Rel Error
-0.8	0.2000	0.2600	0.2720	0.3488	0.2688	1.268869	1.268800	0.000069	0.005400
-0.6	0.3475	0.4370	0.4549	0.5695	0.4502	1.719156	1.718951	0.000204	0.011892
-0.4	0.5676	0.7011	0.7278	0.8987	0.7207	2.440088	2.439630	0.000457	0.018748
-0.2	0.8959	1.0950	1.1349	1.3898	1.1242	3.564774	3.563864	0.000910	0.025526
0.0	1.3855	1.6827	1.7421	2.1224	1.7262	5.291792	5.290095	0.001697	0.032064

5. Let $h = 0.1$ and $f(x,y) = xy - 2y$ in (8.59) and (8.60); we have $x_0 = 2$ and $y_0 = 1$.

$$k_1 = hf(x_0,y_0) = 0.1f(2.0,\ 1.0000000) = 0.1(0.0000000)$$
$$= 0.0000000,$$

$$k_2 = hf(x_0 + h/2,\ y_0 + k_1/2) = 0.1f(2.05,\ 1.00000000)$$
$$= 0.1(0.0500000) = 0.0050000,$$

$$k_3 = hf(x_0 + h/2,\ y_0 + k_2/2) = 0.1f(2.05,\ 1.0025000)$$
$$= 0.1(0.0501250) = 0.0050125,$$

$$k_4 = hf(x_0 + h,\ y_0 + k_3) = 0.1f(2.10,\ 1.0050125)$$
$$= 0.1(0.1005012) = 0.0100501.$$

Therefore

$$K = (k_1 + 2k_2 + 2k_3 + k_4)/6$$
$$= (0.0000000 + 0.0100000 + 0.0100250 + 0.0100501)/6$$
$$= 0.0050125,$$

and so the approximate value of the solution at $x_1 - 2.1$ is

$$y_1 = 1.0000000 + 0.0050125 = 1.0050125.$$

Now using the above value y_1, we calculate successively
new k_1, k_2, k_3, k_4, and then K. We first find

$$k_1 = hf(x_1, y_1) = 0.1f(2.1, \ 1.0050125 = 0.1(0.1005013)$$
$$= 0.0100501,$$

$$k_2 = hf(x_1 + h/2, \ y_1 + k_1/2) = 0.1f(2.15, \ 1.0100376)$$
$$= 0.1(0.1515056) = 0.0151506,$$

$$k_3 = hf(x_1 + h/2, \ y_1 + k_2/2) = 0.1f(2.15, \ 1.0125878)$$
$$= 0.1(0.1518882) = 0.0151888,$$

$$k_4 = hf(x_1 + h, \ y_1 + k_3) = 0.1f(2.20, \ 1.0202013)$$
$$= 0.1(0.2040403) = 0.0204040.$$

Therefore

$$K = (k_1 + 2k_2 + 2k_3 + k_4)/6$$
$$= (0.0100501 + 0.0303011 + 0.0303776 + 0.0204040)/6$$
$$= 0.0151888,$$

and so the approximate value of the solution at $x_2 = 2.2$
is

$$y_2 = 1.0050125 + 0.0151888 = 1.0202013.$$

We summarize the calculations for these and the remaining y_n's in Table 8.6.5. The exact solution of the differential equation is given in the pages immediately preceding Section 8.4 of this manual.

TABLE 8.6.5   Runge-Kutta Method for $y' = xy - 2y$, $y(2) = 1$, with $h = 0.1$.

x_n	k_1	k_2	k_3	k_4	K	Exact Solution	Runge-Kutta	Error	% Rel Error
2.1	0.0000	0.0050	0.0050	0.0101	0.0050	1.005013	1.005013	0.000000	0.000000
2.2	0.0101	0.0152	0.0152	0.0204	0.0152	1.020201	1.020201	0.000000	0.000000
2.3	0.0204	0.0258	0.0258	0.0314	0.0258	1.046028	1.046028	0.000000	0.000000
2.4	0.0314	0.0372	0.0373	0.0433	0.0373	1.083287	1.083287	0.000000	0.000000
2.5	0.0433	0.0497	0.0499	0.0567	0.0499	1.133148	1.133148	0.000000	0.000001

8. Let $h = 0.2$ and $f(x,y) = y \sin x$ in (8.59) and (8.60); we have $x_0 = 0$ and $y_0 = 0.5$.

$$k_1 = hf(x_0, y_0) = 0.2f(0.0, 0.5000000) = 0.2(0.0000000)$$
$$= 0.0000000,$$

$$k_2 = hf(x_0 + h/2, y_0 + k_1/2) = 0.2f(0.1, 0.5000000)$$
$$= 0.2(0.0499167) = 0.0099833,$$

$$k_3 = hf(x_0 + h/2, y_0 + k_2/2) = 0.2f(0.1, 0.5049917)$$
$$= 0.2(0.0504150) = 0.0100830,$$

$$k_4 = hf(x_0 + h, y_0 + k_3) = 0.2f(0.2, 0.5100830)$$
$$= 0.2(0.1013379) = 0.0202676.$$

Therefore

$$K = (k_1 + 2k_2 + 2k_3 + k_4)/6$$
$$= (0.0000000 + 0.0199667 + 0.0201660 + 0.0202676)/6$$
$$= 0.0100667,$$

and so the approximate value of the solution at $x_1 = 0.2$ is

$$y_1 = 0.5000000 + 0.0100667 = 0.5100667.$$

Now using the above value y_1, we calculate successively *new* k_1, k_2, k_3, k_4, and then K. We first find

$$k_1 = hf(x_1,y_1) = 0.2f(0.2, 0.5100667) = 0.2(0.1013346)$$
$$= 0.0202669,$$

$$k_2 = hf(x_1 + h/2, y_1 + k_1/2) = 0.2f(0.3, 0.5202002)$$
$$= 0.2(0.1537297) = 0.0307459,$$

$$k_3 = hf(x_1 + h/2, y_1 + k_2/2) = 0.2f(0.3, 0.5254397)$$
$$= 0.2(0.1552780) = 0.0310556,$$

$$k_4 = hf(x_1 + h, y_1 + k_3) = 0.2f(0.4, 0.5411223)$$
$$= 0.2(0.2107230) = 0.0421446.$$

Therefore

$$K = (k_1 + 2k_2 + 2k_3 + k_4)/6$$

$$= (0.0202669 + 0.0614919 + 0.0621112 + 0.0421446)/6$$

$$= 0.0310024,$$

and so the approximate value of the solution at $x_2 = 0.4$ is

$$y_2 = 0.5100667 + 0.0310024 = 0.5410691.$$

We summarize the calculations for these and the remaining y_n's in Table 8.6.8. The exact solution of the differential equation is given in the pages immediately preceding Section 8.4 of this manual.

TABLE 8.6.8   Runge–Kutta Method for $y' = y \sin x$, $y(0) = 0.5$, with $h = 0.2$.

x_n	k_1	k_2	k_3	k_4	K	Exact Solution	Runge-Kutta	Error	% Rel Error
0.2	0.0000	0.0100	0.0101	0.0203	0.0101	0.510067	0.510067	0.000000	0.000000
0.4	0.0203	0.0307	0.0311	0.0421	0.0310	0.541069	0.541069	0.000000	0.000003
0.6	0.0421	0.0539	0.0545	0.0673	0.0544	0.595423	0.595423	0.000000	0.000016
0.8	0.0672	0.0810	0.0819	0.0972	0.0817	0.677156	0.677155	0.000000	0.000055
1.0	0.0972	0.1137	0.1150	0.1333	0.1146	0.791798	0.791796	0.000001	0.000139

9. Let $h = 0.2$ and $f(x,y) = x/y$ in (8.59) and (8.60); we have $x_0 = 0$ and $y_0 = 0.2$.

$k_1 = hf(x_0, y_0) = 0.2f(0.0, 0.2000000) = 0.2(0.0000000)$

$\quad = 0.0000000$,

$k_2 = hf(x_0 + h/2, y_0 + k_1/2) = 0.2f(0.1, 0.2000000)$

$\quad = 0.2(0.5000000) = 0.1000000$,

$k_3 = hf(x_0 + h/2, y_0 + k_2/2) = 0.2f(0.1, 0.2500000)$

$\quad = 0.2(0.4000000) = 0.0800000$,

$k_4 = hf(x_0 + h, y_0 + k_3) = 0.2f(0.2, 0.2800000)$

$\quad = 0.2(0.7142857) = 0.1428571$.

Therefore

$K = (k_1 + 2k_2 + 2k_3 + k_4)/6$

$\quad = (0.0000000 + 0.2000000 + 0.1600000 + 0.1428571)/6$

$\quad = 0.0838095$,

and so the approximate value of the solution at $x_1 = 0.2$ is

$$y_2 = 0.2000000 + 0.0838095 = 0.2838095.$$

Now using the above value y_1, we calculate successively *new* k_1, k_2, k_3, k_4, and then K. We first find

$$k_1 = hf(x_1, y_1) = 0.2f(0.2,\ 0.2838095) = 0.2(0.7046980)$$

$$= 0.1409396,$$

$$k_2 = hf(x_1 + h/2,\ y_1 + k_1/2) = 0.2f(0.3,\ 0.3542793)$$

$$= 0.2(0.8467895) = 0.1693579,$$

$$k_3 = hf(x_1 + h/2,\ y_1 + k_2/2) = 0.2f(0.3,\ 0.3684885)$$

$$= 0.2(0.8141367) = 0.1628273,$$

$$k_4 = hf(x_1 + h,\ y_1 + k_3) = 0.2f(0.4,\ 0.4466369)$$

$$= 0.2(0.8955821) = 0.1791164.$$

Therefore

$$K = (k_1 + 2k_2 + 2k_3 + k_4)/6$$

$$= (0.1409396 + 0.3387158 + 0.3256547 + 0.1791164)/6$$

$$= 0.1640711,$$

and so the approximate value of the solution at $x_2 = 0.4$ is

$$y_2 = 0.2838095 + 0.1640711 = 0.4478806.$$

We summarize the calculations for these and the remaining y_n's in Table 8.6.9. The exact solution of the differential equation is given in the pages immediately preceding Section 8.4 of this manual.

TABLE 8.6.9   Runge-Kutta Method for $y' = x/y$,
$y(0) = 0.2$, with $h = 0.2$.

x_n	k_1	k_2	k_3	k_4	K	Exact Solution	Runge-Kutta	Error	% Rel Error
0.2	0.0000	0.1000	0.0800	0.1429	0.0838	0.282843	0.283810	0.000967	0.341819
0.4	0.1409	0.1694	0.1628	0.1791	0.1641	0.447214	0.447881	0.000667	0.149149
0.6	0.1786	0.1862	0.1849	0.1897	0.1850	0.632456	0.632930	0.000474	0.074943
0.8	0.1896	0.1924	0.1920	0.1940	0.1921	0.824621	0.824985	0.000364	0.044113
1.0	0.1939	0.1952	0.1951	0.1961	0.1951	1.019804	1.020098	0.000294	0.028848
1.2	0.1961	0.1968	0.1967	0.1972	0.1967	1.216553	1.216799	0.000247	0.020273
1.4	0.1972	0.1977	0.1976	0.1980	0.1976	1.414214	1.414426	0.000212	0.015002
1.6	0.1980	0.1982	0.1982	0.1984	0.1982	1.612452	1.612638	0.000186	0.011541
1.8	0.1984	0.1986	0.1986	0.1988	0.1986	1.811077	1.811243	0.000166	0.009148
2.0	0.1988	0.1989	0.1989	0.1990	0.1989	2.009975	2.010124	0.000149	0.007427

12. Let $h = 0.2$ and $f(x,y) = (\sin x)/y$ in (8.59) and (8.60);
     we have $x_0 = 0$ and $y_0 = 1$.

$k_1 = hf(x_0, y_0) = 0.2f(0.0, 1.0000000) = 0.2(0.0000000)$

$\quad = 0.0000000,$

$k_2 = hf(x_0 + h/2, y_0 + k_1/2) = 0.2f(0.1, 1.0000000)$

$\quad = 0.2(0.0998334) = 0.0199667,$

$k_3 = hf(x_0 + h/2, y_0 + k_2/2) = 0.2f(0.1, 1.0099833)$

$\quad = 0.2(0.0988466) = 0.0197693,$

$k_4 = hf(x_0 + h, y_0 + k_3) = 0.2f(0.2, 1.0197693)$

$\quad = 0.2(0.1948179) = 0.0389636.$

Therefore

$$K = (k_1 + 2k_2 + 2k_3 + k_4)/6$$
$$= (0.0000000 + 0.0399334 + 0.0395386 + 0.0389636)/6$$
$$= 0.0197393,$$

and so the approximate value of the solution at $x_1 = 0.2$ is

$$y_1 = 1.0000000 + 0.0197393 = 1.0197393.$$

Now using the above value y_1, we calculate successively *new* k_1, k_2, k_3, k_4, and then K. We first find

$$k_1 = hf(x_1,y_1) = 0.2f(0.2, 1.0197393) = 0.2(0.1948237)$$
$$= 0.0389647,$$

$$k_2 = hf(x_1 + h/2, y_1 + k_1/2) = 0.2f(0.3, 1.0392216)$$
$$= 0.2(0.2843669) = 0.0568734,$$

$$k_3 = hf(x_1 + h/2, y_1 + k_2/2) = 0.2f(0.3, 1.0481760)$$
$$= 0.2(0.2819376) = 0.0563875,$$

$$k_4 = hf(x_1 + h, y_1 + k_3) = 0.2f(0.4, 1.0761268)$$
$$= 0.2(0.3618703) = 0.0723741.$$

Therefore

$$K = (k_1 + 2k_2 + 2k_3 + k_4)/6$$

$$= (0.0389647 + 0.1137467 + 0.1127750 + 0.0723741)/6$$

$$= 0.0563101,$$

and so the approximate value of the solution at $x_2 = 0.4$ is

$$y_2 = 1.0197393 + 0.0563101 = 1.0760494.$$

We summarize the calculations for these and the remaining y_n's in Table 8.6.12. The exact solution of the differential equation is given in the pages immediately preceding Section 8.4 of this manual.

TABLE 8.6.12   Runge-Kutta Method for $y' = (\sin x)/y$, $y(0) = 1$, with $h = 0.2$.

x_n	k_1	k_2	k_3	k_4	K	Exact Solution	Runge-Kutta	Error	% Rel Error
0.2	0.0000	0.0200	0.0198	0.0390	0.0197	1.019739	1.019739	0.000001	0.000064
0.4	0.0390	0.0569	0.0564	0.0724	0.0563	1.076047	1.076049	0.000002	0.000182
0.6	0.0724	0.0862	0.0857	0.0972	0.0856	1.161606	1.161609	0.000003	0.000248
0.8	0.0972	0.1065	0.1061	0.1132	0.1059	1.267512	1.267515	0.000003	0.000251
1.0	0.1132	0.1183	0.1181	0.1215	0.1179	1.385422	1.385426	0.000003	0.000225
1.2	0.1215	0.1233	0.1232	0.1236	0.1230	1.508405	1.508408	0.000003	0.000196
1.4	0.1236	0.1227	0.1228	0.1208	0.1226	1.630971	1.630974	0.000003	0.000173
1.6	0.1208	0.1179	0.1181	0.1143	0.1179	1.748828	1.748831	0.000003	0.000156
1.8	0.1143	0.1098	0.1100	0.1048	0.1098	1.858603	1.858605	0.000003	0.000144
2.0	0.1048	0.0990	0.0992	0.0929	0.0990	1.957624	1.957627	0.000003	0.000137

Section 8.7, Page 468.

General Information: For each problem below, we show the first two calculations in detail, while summarizing calculations at each value of x_n in a table. The problems were done maintaining 11 to 12 digit accuracy from step to step; however, we've rounded the results to fewer places as we tabulate them below.

1. Let $h = 0.2$ and $f(x,y) = x - 2y$ in (8.67), (8.68) and (8.69); we have $x_0 = 0$ and $y_0 = 1$.

   Before we can begin using the ABAM method, we need to have values for y_0, y_1, y_2, and y_3. The first of these is given by the initial condition $y(0) = 1.0$, while values for the other three are supplied by the text and were found using the Runge-Kutta method. We have

$$x_0 = 0.0, \qquad y_0 = 1.0000000000,$$
$$x_1 = 0.2, \qquad y_1 = 0.6880000000,$$
$$x_2 = 0.4, \qquad y_2 = 0.5117952000,$$
$$x_3 = 0.6, \qquad y_3 = 0.4266275021,$$

and we set $x_4 = 0.8$. Now we find

$$y_0' = f(x_0, y_0) = f(0.0,\ 1.0000000) = -2.0000000,$$
$$y_1' = f(x_1, y_1) = f(0.2,\ 0.6880000) = -1.1760000,$$
$$y_2' = f(x_2, y_2) = f(0.4,\ 0.5117952) = -0.6235904,$$
$$y_3' = f(x_3, y_3) = f(0.6,\ 0.4266275) = -0.2532550.$$

We now use (8.67) with n = 3 and h = 0.2 to determine $\hat{y}_4$. We have

$$\hat{y}_4 = y_3 + \frac{0.2}{24} (55y_3' - 59y_2' + 37y_1' - 9y_0')$$

$$= 0.4266275$$

$$+ \frac{(-13.9290252 + 36.7918336 - 43.5120000 + 18.0000000)}{120.0}$$

$$= 0.4045509.$$

Having thus determined $\hat{y}_4$, we use (8.68) with n = 3 to find $\hat{y}_4'$. We obtain

$$\hat{y}_4' = f(x_4, \hat{y}_4) = f(0.8, 0.4045509) = -0.0091018.$$

We use this value of $\hat{y}_4'$ in (8.69) with n = 3 and h = 0.2 to finally obtain y_4 as follows:

$$y_4 = y_3 + \frac{0.2}{24} (9\hat{y}_4' + 19y_3' - 5y_2' + y_1')$$

$$= 0.4266275$$

$$+ \frac{(-0.0819163 - 4.8118451 + 3.1179520 - 1.1760000)}{120.0}$$

$$= 0.4020291.$$

Now we set n = 4 in order to calculate $y_{n+1} = y_5$. Using the value we just found for y_4, we first calculate

$$y_4' = f(x_4, y_4) = f(0.80, 0.4020291) = -0.0040582.$$

We next use (8.67) with n = 4 and h = 0.20 to determine $\hat{y}_5$. We find

$$\hat{y}_5 = y_4 + \frac{0.2}{24} (55y_4' - 59y_3' + 37y_2' - 9y_1')$$

$$= 0.4020291$$

$$+ \frac{(-0.2232000 + 14.9420452 - 23.0728448 + 10.5840000)}{120.0}$$

$$= 0.4206124.$$

Having thus determined $\hat{y}_5$, we use (8.68) with n = 4 to find $\hat{y}_5'$. We obtain

$$\hat{y}_5' = f(x_5, \hat{y}_5) = f(1.0, 0.4206124) = 0.1587751.$$

We use this value of $\hat{y}_5'$ in (8.69) with n = 4 and h = 0.2 to finally obtain y_5 as follows:

$$y_5 = y_4 + \frac{0.2}{24} (9\hat{y}_5' + 19y_4' - 5y_3' + y_2')$$

$$= 0.4020291$$

$$+ \frac{(1.4289763 - 0.0771054 + 1.2662750 - 0.6235904)}{120.0}$$

$$= 0.4186504.$$

The exact solution of the differential equation is given in the pages immediately preceding Section 8.4 of this manual.

TABLE 8.7.1A   ABAM Method (Calculations Summary) for
$y' = x - 2y$, $y(0) = 1$, with $h = 0.2$.

n	x_n	y_n	y'_n	x_{n+1}	$\hat{y}_{n+1}$	$\hat{y}'_{n+1}$	y_{n+1}
0	0.0	1.000000	-2.000000	–	—	—	—
1	0.2	0.688000	-1.176000	–	—	—	—
2	0.4	0.511795	-0.623590	–	—	—	—
3	0.6	0.426628	-0.253255	0.8	0.404551	-0.009102	0.402029
4	0.8	0.402029	-0.004058	1.0	0.420612	0.158775	0.418650
5	1.0	0.418650					

TABLE 8.7.1B   ABAM Method (Comparisons and Errors) for
$y' = x - 2y$, $y(0) = 1$, with $h = 0.2$.

x_n	Exact Solution	ABAM Method	Error	% Rel Error
0.8	0.402371	0.402029	0.000342	0.084886
1.0	0.419169	0.418650	0.000519	0.123749

4. Let $h = 0.2$ and $f(x,y) = x + 2y$ in (8.67), (8.68) and
(8.69); we have $x_0 = -1$ and $y_0 = 1$.

Before we can begin using the ABAM method, we need to
have values for y_0, y_1, y_2, and y_3. The first of these is
given by the initial condition $y(-1) = 1.0$, while values
for the other three are supplied by the text and were
found using the Runge–Kutta method. We have

$$x_0 = -1.0, \qquad y_0 = 1.0000000000,$$

$$x_1 = -0.8, \qquad y_1 = 1.2688000000,$$

$$x_2 = -0.6, \qquad y_2 = 1.7189512533,$$

$$x_3 = -0.4, \qquad y_3 = 2.4396302163,$$

and we set $x_4 = -0.2$. Now we find

$$y_0' = f(x_0, y_0) = f(-1.0, \ 1.0000000) = 1.0000000,$$

$$y_1' = f(x_1, y_1) = f(-0.8, \ 1.2688000) = 1.7376000,$$

$$y_2' = f(x_2, y_2) = f(-0.6, \ 1.7189513) = 2.8379025$$

$$y_3' = f(x_3, y_3) = f(-0.4, \ 2.4396302) = 4.4792604.$$

We now use (8.67) with n = 3 and h = 0.2 to determine $\hat{y}_4$.
We have

$$\hat{y}_4 = y_3 + \frac{0.2}{24} \ (55y_3' - 59y_2' + 37y_1' - 9y_0')$$

$$= 2.4396302$$

$$+ \frac{(246.3593238 - 167.4362479 + 64.2912000 - 9.0000000)}{120.0}$$

$$= 3.5580825.$$

Having thus determined $\hat{y}_4$, we use (8.68) with n = 3 to
find $\hat{y}_4'$. We obtain

$$\hat{y}_4' = f(x_4, \hat{y}_4) = f(-0.2, \ 3.5580825) = 6.9161650.$$

We use this value of $\hat{y}_4'$ in (8.69) with n = 3 and h = 0.2
to finally obtain y_4 as follows:

$$y_4 = y_3 + \frac{0.2}{24}(9\hat{y}_4' + 19y_3' - 5y_2' + y_1')$$

$$= 2.4396302$$

$$+ \frac{(62.2454853 + 85.1059482 - 14.1895125 + 1.7376000)}{120.0}$$

$$= 3.5637929.$$

Now we set n = 4 in order to calculate $y_{n+1} = y_5$. Using the value we just found for y_4, we first calculate

$$y_4' = f(x_4,y_4) = f(-0.20, 3.5637929) = 6.9275858.$$

We next use (8.67) with n = 4 and h = 0.20 to determine $\hat{y}_5$. We find

$$\hat{y}_5 = y_4 + \frac{0.2}{24}(55y_4' - 59y_3' + 37y_2' - 9y_1')$$

$$= 3.5637929$$

$$+ \frac{(381.0172180 - 264.2763655 + 105.0023927 - 15.6384000}{120.0}$$

$$= 5.2813333,$$

Having thus determined $\hat{y}_5$ we use (8.68) with n = 4 to fin· $\hat{y}_5'$. We obtain

$$\hat{y}_5' = f(x_5,\hat{y}_5) = f(0.0, 5.2813333) = 10.5626665.$$

We use this value of $\hat{y}_5'$ in (8.69) with n = 4 and h = 0.2 to finally obtain y_5 as follows:

$$y_5 = y_4 + \frac{0.2}{24} (9\hat{y}'_5 + 19y'_4 - 5y'_3 + y'_2)$$

$$= 3.5637929$$

$$+ \frac{(95.0639988 + 131.6241299 - 22.3963022 + 2.8379025)}{120.0}$$

$$= 5.2898740.$$

The exact solution of the differential equation is given in the pages immediately preceding Section 8.4 of this manual.

TABLE 8.7.4A   ABAM Method (Calculations Summary) for
$y' = x + 2y$, $y(-1) = 1$, with $h = 0.2$.

n	x_n	y_n	y'_n	x_{n+1}	$\hat{y}_{n+1}$	$\hat{y}'_{n+1}$	y_{n+1}
0	-1.0	1.000000	1.000000	–	—	—	—
1	-0.8	1.268800	1.737600	–	—	—	—
2	-0.6	1.718951	2.837903	–	—	—	—
3	-0.4	2.439630	4.479260	-0.2	3.558083	6.916165	3.563793
4	-0.2	3.563793	6.927586	0.0	5.281333	10.562667	5.289874
5	0.0	5.289874					

TABLE 8.7.4B   ABAM Method (Comparisons and Errors) for
$y' = x + 2y$, $y(-1) = 1$, with $h = 0.2$.

x_n	Exact Solution	ABAM Method	Error	% Rel Error
-0.2	3.564774	3.563793	0.000981	0.027531
0.0	5.291792	5.289874	0.001918	0.036247

5.   Let $h = 0.1$ and $f(x,y) = xy - 2y$ in (8.67), (8.68) and
(8.69); we have $x_0 = 2$ and $y_0 = 1$.

Before we can begin using the ABAM method, we need to have values for y_0, y_1, y_2, and y_3. The first of these is given by the initial condition $y(2) = 1.0$, while values for the other three are supplied by the text and were found using the Runge-Kutta method. We have

$$x_0 = 2.0, \qquad y_0 = 1.0000000000,$$

$$x_1 = 2.1, \qquad y_1 = 1.0050125208,$$

$$x_2 = 2.2, \qquad y_2 = 1.0202013398,$$

$$x_3 = 2.3, \qquad y_3 = 1.0460278589,$$

and we set $x_4 = 2.4$. Now we find

$$y_0' = f(x_0, y_0) = f(2.0, \ 1.0000000) = 0.0000000,$$

$$y_1' = f(x_1, y_1) = f(2.1, \ 1.0050125) = 0.1005013,$$

$$y_2' = f(x_2, y_2) = f(2.2, \ 1.0202013) = 0.2040403,$$

$$y_3' = f(x_3, y_3) = f(2.3, \ 1.0460279) = 0.3138084.$$

We now use (8.67) with $n = 3$ and $h = 0.1$ to determine $\hat{y}_4$.
We have

$$\hat{y}_4 = y_3 + \frac{0.1}{24} \ (55y_3' - 59y_2' + 37y_1' - 9y_0')$$

$$= 1.0460279$$

$$+ \ \frac{(17.2594597 - 12.0383758 + 3.7185463 - 0.0000000)}{240.0}$$

$$= 1.0832763.$$

Having thus determined $\hat{y}_4$, we use (8.68) with $n = 3$ to find $\hat{y}_4'$. We obtain

$$\hat{y}_4' = f(x_4, \hat{y}_4) = f(2.4, \ 1.0832763) = 0.4333105.$$

We use this value of $\hat{y}_4'$ in (8.69) with n = 3 and h = 0.1 to finally obtain y_4 as follows:

$$y_4 = y_3 + \frac{0.1}{24} \ (9\hat{y}_4' + 19y_3' - 5y_2' + y_1')$$

$$= 1.0460279$$

$$+ \ \frac{(3.8997947 + 5.9623588 - 1.0202013 + 0.1005013)}{240.0}$$

$$= 1.0832881.$$

Now we set n = 4 in order to calculate $y_{n+1} = y_5$. Using the value we just found for y_4, we first calculate

$$y_4' = f(x_4, y_4) = f(2.40, \ 1.0832881) = 0.4333152.$$

We next use (8.67) with n = 4 and h = 0.10 to determine $\hat{y}_5$. We find

$$\hat{y}_5 = y_4 + \frac{0.1}{24} \ (55y_4' - 59y_3' + 37y_2' - 9y_1')$$

$$= 1.0832881$$

$$+ \ \frac{(23.8323378 - 18.5146931 + 7.5494899 - 0.9045113)}{240.0}$$

$$= 1.1331323.$$

Having thus determined $\hat{y}_5$, we use (8.68) with n = 4 to find $\hat{y}_5'$. We obtain

$$\hat{y}_5' = f(x_5, \hat{y}_5) = f(2.5,\ 1.1331323) = 0.5665662.$$

We use this value of $\hat{y}_5'$ in (8.69) with $n = 4$ and $h = 0.1$ to finally obtain y_5 as follows:

$$y_5 = y_4 + \frac{0.1}{24}\ (9\hat{y}_5' + 19y_4' - 5y_3' + y_2')$$

$$= 1.0832881$$

$$+\ \frac{(5.0990956 + 8.2329894 - 1.5690418 + 0.2040403)}{240.0}$$

$$= 1.1331509.$$

The exact solution of the differential equation is given in the pages immediately preceding Section 8.4 of this manual.

TABLE 8.7.5A   ABAM Method (Calculations Summary) for $y' = xy - 2y$, $y(2) = 1$, with $h = 0.1$.

n	x_n	y_n	y'_n	x_{n+1}	$\hat{y}_{n+1}$	$\hat{y}'_{n+1}$	y_{n+1}
0	2.0	1.000000	0.000000	–	—	—	—
1	2.1	1.005013	0.100501	–	—	—	—
2	2.2	1.020201	0.204040	–	—	—	—
3	2.3	1.046028	0.313808	2.4	1.083276	0.433311	1.083288
4	2.4	1.083288	0.433315	2.5	1.133132	0.566566	1.133151
5	2.5	1.133151					

TABLE 8.7.5B   ABAM Method (Comparisons and Errors) for $y' = xy - 2y$, $y(2) = 1$, with $h = 0.1$.

x_n	Exact Solution	ABAM Method	Error	% Rel Error
2.4	1.083287	1.083288	0.000001	0.000094
2.5	1.133148	1.133151	0.000002	0.000219

8. Let $h = 0.2$ and $f(x,y) = y \sin x$ in (8.67), (8.68) and (8.69); we have $x_0 = 0$ and $y_0 = 0.5$.

Before we can begin using the ABAM method, we need to have values for y_0, y_1, y_2, and y_3. The first of these is given by the initial condition $y(0) = 0.5$, while values for the other three are supplied by the text and were found using the Runge-Kutta Method. We have

$$x_0 = 0.0, \qquad y_0 = 0.5000000000,$$

$$x_1 = 0.2, \qquad y_1 = 0.5100667118,$$

$$x_2 = 0.4, \qquad y_2 = 0.5410691444,$$

$$x_3 = 0.6, \qquad y_3 = 0.5954231442,$$

and we set $x_4 = 0.8$. Now we find

$$y_0' = f(x_0,y_0) = f(0.0,\ 0.5000000) = 0.0000000,$$

$$y_1' = f(x_1,y_1) = f(0.2,\ 0.5100667) = 0.1013346,$$

$$y_2' = f(x_2,y_2) = f(0.4,\ 0.5410691) = 0.2107022,$$

$$y_3' = f(x_3,y_3) = f(0.6,\ 0.5954231) = 0.3362012.$$

We now use (8.67) with $n = 3$ and $h = 0.2$ to determine $\hat{y}_4$. We have

$$\hat{y}_4 = y_3 + \frac{0.2}{24}\,(55y_3' - 59y_2' + 37y_1' - 9y_0')$$

$$= 0.5954231$$

$$+\ \frac{(18.4910658 - 12.4314327 + 3.7493807 - 0.0000000)}{120.0}$$

$$= 0.6771649.$$

Having thus determined $\hat{y}_4$, we use (8.68) with $n = 3$ to find $\hat{y}_4'$. We obtain

$$\hat{y}_4' = f(x_4,\hat{y}_4) = f(0.8,\ 0.6771649) = 0.4857684.$$

We use this value of $\hat{y}_4'$ in (8.69) with $n = 3$ and $h = 0.2$ to finally obtain y_4 as follows:

$$y_4 = y_3 + \frac{0.2}{24}\,(9\hat{y}_4' + 19y_3' - 5y_2' + y_1')$$

$$= 0.5954231$$

$$+\ \frac{(4.3719155 + 6.3878227 - 1.0535112 + 0.1013346)}{120.0}$$

$$= 0.6771528.$$

Now we set $n = 4$ in order to calculate $y_{n+1} = y_5$. Using the value we just found for y_4, we first calculate

$$y_4' = f(x_4,y_4) = f(0.80,\ 0.6771528) = 0.4857597.$$

We next use (8.67) with $n = 4$ and $h = 0.20$ to determine $\hat{y}_5$. We find

$$\hat{y}_5 = y_4 + \frac{0.2}{24} (55y_4' - 59y_3' + 37y_2' - 9y_1')$$

$$= 0.6771528$$

$$+ \frac{(26.7167836 - 19.8358706 + 7.7959832 - 0.9120115)}{120.0}$$

$$= 0.7918602.$$

Having thus determined $\hat{y}_5$, we use (8.68) with n = 4 to find $\hat{y}_5'$. We obtain

$$\hat{y}_5' = f(x_5, \hat{y}_5) = f(1.0, \ 0.7918602) = 0.6663274.$$

We use this value of $\hat{y}_5'$ in (8.69) with n = 4 and h = 0.2 to finally obtain y_5 as follows:

$$y_5 = y_4 + \frac{0.2}{24} (9\hat{y}_5' + 19y_4' - 5y_3' + y_2')$$

$$= 0.6771528$$

$$+ \frac{(5.9969464 + 9.2294344 - 1.6810060 + 0.2107022)}{120.0}$$

$$= 0.7917868.$$

The exact solution of the differential equation is given in the pages immediately preceding Section 8.4 of this manual.

TABLE 8.7.8A   ABAM Method (Calculations Summary) for
$y' = y \sin x$, $y(0) = 0.5$, with $h = 0.2$.

n	x_n	y_n	y'_n	x_{n+1}	$\hat{y}_{n+1}$	$\hat{y}'_{n+1}$	y_{n+1}
0	0.0	0.500000	0.000000	–	—	—	—
1	0.2	0.510067	0.101335	–	—	—	—
2	0.4	0.541069	0.210702	–	—	—	—
3	0.6	0.595423	0.336201	0.8	0.677165	0.485768	0.677153
4	0.8	0.677153	0.485760	1.0	0.791860	0.666327	0.791787
5	1.0	0.791787					

TABLE 8.7.8B   ABAM Method (Comparisons and Errors) for
$y' = y \sin x$, $y(0) = 0.5$, with $h = 0.2$.

x_n	Exact Solution	ABAM Method	Error	% Rel Error
0.8	0.677156	0.677153	0.000003	0.000440
1.0	0.791798	0.791787	0.000011	0.001363

9.   Let $h = 0.2$ and $f(x,y) = x/y$ in (8.67), (8.68) and (8.69);
we have $x_0 = 0$ and $y_0 = 0.2$.

Before we can begin using the ABAM method, we need to
have values for y_0, y_1, y_2, and y_3.   The first of these is
given by the initial condition $y(0) = 0.2$, while values
for the other three are supplied by the text and were
found using the Runge–Kutta method.   We have

$$x_0 = 0.0, \qquad y_0 = 0.2000000000,$$

$$x_1 = 0.2, \qquad y_1 = 0.2838095238,$$

$$x_2 = 0.4, \qquad y_2 = 0.4478806080,$$

$$x_3 = 0.6, \qquad y_3 = 0.6329295155,$$

and we set $x_4 = 0.8$.   Now we find

$$y_0' = f(x_0, y_0) = f(0.0, \ 0.2000000) = 0.0000000,$$

$$y_1' = f(x_1, y_1) = f(0.2, \ 0.2838095) = 0.7046980,$$

$$y_2' = f(x_2, y_2) = f(0.4, \ 0.4478806) = 0.8930952,$$

$$y_3' = f(x_3, y_3) = f(0.6, \ 0.6329295) = 0.9479729.$$

We now use (8.67) with $n = 3$ and $h = 0.2$ to determine $\hat{y}_4$.
We have

$$\hat{y}_4 = y_3 + \frac{0.2}{24} \ (55y_3' - 59y_2' + 37y_1' - 9y_0')$$

$$= 0.6329295$$

$$+ \ \frac{(52.1385070 - 52.6926140 + 26.0738255 - 0.0000000)}{120.0}$$

$$= 0.8455938.$$

Having thus determined $\hat{y}_4$, we use (8.68) with $n = 3$ to
find $\hat{y}_4'$.   We obtain

$$\hat{y}_4' = f(x_4, \hat{y}_4) = f(0.8, \ 0.8455938) = 0.9460807.$$

We use this value of $\hat{y}_4'$ in (8.69) with $n = 3$ and $h = 0.2$
to finally obtain y_4 as follows:

$$y_4 = y_3 + \frac{0.2}{24} \left(9\hat{y}_4' + 19y_3' - 5y_2' + y_1' \right)$$

$$= 0.6329295$$

$$+ \frac{(8.5147262 + 18.0114843 - 4.4654758 + 0.7046980)}{120.0}$$

$$= 0.8226415.$$

Now we set n = 4 in order to calculate $y_{n+1} = y_5$. Using the value we just found for y_4, we first calculate

$$y_4' = f(x_4, y_4) = f(0.80, 0.8226415) = 0.9724771.$$

We next use (8.67) with n = 4 and h = 0.20 to determine $\hat{y}_5$. We find

$$\hat{y}_5 = y_4 + \frac{0.2}{24} \left(55y_4' - 59y_3' + 37y_2' - 9y_1' \right)$$

$$= 0.8226415$$

$$+ \frac{(53.4862421 - 55.9303985 + 33.0445207 - 6.3422819)}{120.0}$$

$$= 1.0247921.$$

Having thus determined $\hat{y}_5$, we use (8.68) with n = 4 to find $\hat{y}_5'$. We obtain

$$\hat{y}_5' = f(x_5, \hat{y}_5) = f(1.0, 1.0247921) = 0.9758076.$$

We use this value of $\hat{y}_5'$ in (8.69) with n = 4 and h = 0.2 to finally obtain y_5 as follows:

$$y_5 = y_4 + \frac{0.2}{24} \left(9\hat{y}_5' + 19y_4' - 5y_3' + y_2'\right)$$

$$= 0.8226415$$

$$+ \frac{(8.7822688 + 18.4770655 - 4.7398643 + 0.8930952)}{120.0}$$

$$= 1.0177462.$$

The exact solution of the differential equation is given in the pages immediately preceding Section 8.4 of this manual.

TABLE 8.7.9A   ABAM Method (Calculations Summary) for $y' = x/y$, $y(0) = 0.2$, with $h = 0.2$.

n	x_n	y_n	y_n'	x_{n+1}	$\hat{y}_{n+1}$	$\hat{y}_{n+1}'$	y_{n+1}
0	0.0	0.200000	0.000000	–	—	—	—
1	0.2	0.283810	0.704698	–	—	—	—
2	0.4	0.447881	0.893095	–	—	—	—
3	0.6	0.632930	0.947973	0.8	0.845594	0.946081	0.822641
4	0.8	0.822641	0.972477	1.0	1.024792	0.975808	1.017746
5	1.0	1.017746	0.982563	1.2	1.215263	0.987441	1.214757
6	1.2	1.214757	0.987852	1.4	1.413178	0.990675	1.412631
7	1.4	1.412631	0.991058	1.6	1.611193	0.993053	1.611055
8	1.6	1.611055	0.993138	1.8	1.809868	0.994547	1.809831
9	1.8	1.809831	0.994568	2.0	2.008869	0.995585	2.008851
10	2.0	2.008851					

TABLE 8.7.9B   ABAM Method (Comparisons and Errors) for $y' = x/y$, $y(0) = 0.2$, with $h = 0.2$.

x_n	Exact Solution	ABAM Method	Error	% Rel Error
0.8	0.824621	0.822641	0.001980	0.240070
1.0	1.019804	1.017746	0.002058	0.201778
1.2	1.216553	1.214757	0.001796	0.147619
1.4	1.414214	1.412631	0.001583	0.111901
1.6	1.612452	1.611055	0.001396	0.086603
1.8	1.811077	1.809831	0.001246	0.068800
2.0	2.009975	2.008851	0.001124	0.055916

12.   Let $h = 0.2$ and $f(x,y) = (\sin x)/y$ in (8.67), (8.68) and (8.69); we have $x_0 = 0$ and $y_0 = 1$.

Before we can begin using the ABAM method, we need to have values for y_0, y_1, y_2, and y_3. The first of these is given by the initial condition $y(0) = 1.0$, while values for the other three are supplied by the text and were found using the Runge-Kutta method. We have

$$x_0 = 0.0, \qquad y_0 = 1.0000000000,$$

$$x_1 = 0.2, \qquad y_1 = 1.0197392647,$$

$$x_2 = 0.4, \qquad y_2 = 1.0760493618,$$

$$x_3 = 0.6, \qquad y_3 = 1.1616089997,$$

and we set $x_4 = 0.8$. Now we find

$$y_0' = f(x_0,y_0) = f(0.0,\ 1.0000000) = 0.0000000,$$

$$y_1' = f(x_1,y_1) = f(0.2,\ 1.0197393) = 0.1948237,$$

$$y_2' = f(x_2,y_2) = f(0.4,\ 1.0760494) = 0.3618964,$$

$$y_3' = f(x_3,y_3) = f(0.6,\ 1.1616090) = 0.4860865.$$

We now use (8.67) with $n = 3$ and $h = 0.2$ to determine $\hat{y}_4$. We have

$$\hat{y}_4 = y_3 + \frac{0.2}{24}(55y_3' - 59y_2' + 37y_1' - 9y_0')$$

$$= 1.1616090$$

$$+ \frac{(26.7347585 - 21.3518850 + 7.2084752 - 0.0000000)}{120.0}$$

$$= 1.2665369.$$

Having thus determined $\hat{y}_4$, we use (8.68) with n = 3 to find $\hat{y}_4'$. We obtain

$$\hat{y}_4' = f(x_4, \hat{y}_4) = f(0.8, 1.2665369) = 0.5663918.$$

We use this value of $\hat{y}_4'$ in (8.69) with n = 3 and h = 0.2 to finally obtain y_4 as follows:

$$y_4 = y_3 + \frac{0.2}{24}(9\hat{y}_4' + 19y_3' - 5y_2' + y_1')$$

$$= 1.1616090$$

$$+ \frac{(5.0975260 + 9.2356438 - 1.8094818 + 0.1948237)}{120.0}$$

$$= 1.2675966.$$

Now we set n = 4 in order to calculate $y_{n+1} = y_5$. Using the value we just found for y_4, we first calculate

$$y_4' = f(x_4, y_4) = f(0.80, 1.2675966) = 0.5659183.$$

We next use (8.67) with n = 4 and h = 0.20 to determine $\hat{y}_5$. We find

$$\hat{y}_5 = y_4 + \frac{0.2}{24} (55y_4' - 59y_3' + 37y_2' - 9y_1')$$

$$= 1.2675966$$

$$+ \frac{(31.1255056 - 28.6791045 + 13.3901652 - 1.7534129)}{120.0}$$

$$= 1.3849562.$$

Having thus determined $\hat{y}_5$, we use (8.68) with n = 4 to find $\hat{y}_5$. We obtain

$$\hat{y}_5' = f(x_5, \hat{y}_5) = f(1.0, 1.3849562) = 0.6075795.$$

We use this value of $\hat{y}_5'$ in (8.69) with n = 4 and h = 0.2 to finally obtain y_5 as follows:

$$y_5 = y_4 + \frac{0.2}{24} (9\hat{y}_5' + 19y_4' - 5y_3' + y_2')$$

$$= 1.2675966$$

$$+ \frac{(5.4682154 + 10.7524474 - 2.4304326 + 0.3618964)}{120.0}$$

$$= 1.3855310.$$

The exact solution of the differential equation is given in the pages immediately preceding Section 8.4 of this manual.

TABLE 8.7.12A   ABAM Method (Calculations Summary) for $y' = (\sin x)/y$, $y(0) = 1$, with $h = 0.2$.

n	x_n	y_n	y'_n	x_{n+1}	$\hat{y}_{n+1}$	$\hat{y}'_{n+1}$	y_{n+1}
0	0.0	1.000000	0.000000	–	—	—	—
1	0.2	1.019739	0.194824	–	—	—	—
2	0.4	1.076049	0.361896	–	—	—	—
3	0.6	1.161609	0.486087	0.8	1.266537	0.566392	1.267597
4	0.8	1.267597	0.565918	1.0	1.384956	0.607579	1.385531
5	1.0	1.385531	0.607327	1.2	1.508381	0.617907	1.508505
6	1.2	1.508505	0.617856	1.4	1.631121	0.604155	1.631054
7	1.4	1.631054	0.604180	1.6	1.749006	0.571509	1.748897
8	1.6	1.748897	0.571545	1.8	1.858756	0.523924	1.858660
9	1.8	1.858660	0.523951	2.0	1.957744	0.464462	1.957674
10	2.0	1.957674					

TABLE 8.7.12B   ABAM Method (Comparisons and Errors) for $y' = (\sin x)/y$, $y(0) = 1$, with $h = 0.2$.

x_n	Exact Solution	ABAM Method	Error	% Rel Error
0.8	1.267512	1.267597	0.000085	0.006677
1.0	1.385422	1.385531	0.000109	0.007833
1.2	1.508405	1.508505	0.000100	0.006654
1.4	1.630971	1.631054	0.000084	0.005132
1.6	1.748828	1.748897	0.000069	0.003923
1.8	1.858603	1.858660	0.000057	0.003090
2.0	1.957624	1.957674	0.000050	0.002540

Section 8.8, Page 477.

General Information: For each problem below, we show in detail the first two steps for the Euler Method, and the first for the Runge-Kutta method. All calculations for each value of t_n are summarized in tables. The problems were done maintaining 11 to 12 digit accuracy from step to step; however, we've rounded the results to fewer places (usually four to seven) as we tabulate them below.

1. The exact solution is $x = 2e^t$, $y = 4e^t$. First we find the solution by the Euler method. We have $f(t,x,y) = 5x - 2y$, $g(t,x,y) = 4x - y$, $t_0 = 0.0$, $x_0 = 2.0000$, $y_0 = 4.0000$, and $h = 0.1$. To find the approximate values at $t_1 = 0.0 + 0.1 = 0.1$, we use Formulas (8.74). We have

$$x_1 = x_0 + hf(t_0, x_0, y_0)$$
$$= 2.0000 + 0.1f(0.0, 2.0000, 4.0000)$$
$$= 2.0000 + (0.1)2.0000 = 2.2000,$$

$$y_1 = y_0 + hg(t_0, x_0, y_0)$$
$$= 4.0000 + 0.1g(0.0, 2.0000, 4.0000)$$
$$= 4.0000 + (0.1)4.0000 = 4.4000.$$

Now to find the approximate values at $t_2 = 0.1 + 0.1 = 0.2$, we use Formulas 8.75 with $n = 1$. We have

$$x_2 = x_1 + hf(t_1, x_1, y_1)$$
$$= 2.2000 + 0.1f(0.1, 2.2000, 4.4000)$$
$$= 2.2000 + (0.1)2.2000 = 2.4200,$$

$$y_2 = y_1 + hg(t_1, x_1, y_1)$$
$$= 4.4000 + 0.1g(0.1, 2.2000, 4.4000)$$
$$= 4.4000 + (0.1)4.4000 = 4.8400.$$

The results for all t_n and the corresponding errors have been summarized in Table 8.8.1-Euler.

Now we solve the problem by the Runge-Kutta method. Again, $f(t,x,y) = 5x - 2y$, $g(t,x,y) = 4x - y$, $t_0 = 0.0$, $x_0 = 2.0000000$, $y_0 = 4.0000000$, and $h = 0.1$. To find the approximate values at $t_1 = 0.0 + 0.1 = 0.1$, we use Formulas (8.78-80) with $n = 0$. By (8.80) we have

$$k_1 = hf(t_0, x_0, y_0)$$
$$= (0.1)f(0.0,\ 2.0000000,\ 4.0000000)$$
$$= (0.1)2.0000000 = 0.2000000,$$

$$m_1 = hg(t_0, x_0, y_0)$$
$$= (0.1)g(0.0,\ 2.0000000,\ 4.0000000)$$
$$= (0.1)4.0000000 = 0.4000000.$$

Using these values for k_1 and m_1 we now find

$$k_2 = hf\left(t_0 + \frac{h}{2},\ x_0 + \frac{k_1}{2},\ y_0 + \frac{m_1}{2}\right)$$
$$= (0.1)f(0.05,\ 2.1000000,\ 4.2000000)$$
$$= (0.1)2.1000000 = 0.2100000,$$

$$m_2 = hg\left(t_0 + \frac{h}{2},\ x_0 + \frac{k_1}{2},\ y_0 + \frac{m_1}{2}\right)$$
$$= (0.1)g(0.05,\ 2.1000000,\ 4.2000000)$$
$$= (0.1)4.2000000 = 0.4200000.$$

Next, the above values for k_2 and m_2 are used to obtain

$$k_3 = hf\left(t_0 + \frac{h}{2},\ x_0 + \frac{k_2}{2},\ y_0 + \frac{m_2}{2}\right)$$
$$= (0.1)f(0.05,\ 2.1050000,\ 4.2100000)$$
$$= (0.1)2.1050000 = 0.2105000,$$

$$m_3 = hg\left(t_0 + \frac{h}{2}, \ x_0 + \frac{k_2}{2}, \ y_0 + \frac{m_2}{2}\right)$$
$$= (0.1)g(0.05, \ 2.1050000, \ 4.2100000)$$
$$= (0.1)4.2100000 = 0.4210000.$$

With these values for k_3 and m_3 we find

$$k_4 = hf(t_0 + h, \ x_0 + k_3, \ y_0 + m_3)$$
$$= (0.1)f(0.1, \ 2.2105000, \ 4.4210000)$$
$$= (0.1)2.2105000 = 0.2210500,$$

$$m_4 = hg(t_0 + h, \ x_0 + k_3, \ y_0 + m_3)$$
$$= (0.1)g(0.1, \ 2.2105000, \ 4.4210000)$$
$$= (0.1)4.4210000 = 0.4421000.$$

Then from (8.79) we set

$$K = \frac{1}{6} \ (0.2000000 + 2(0.2100000) + 2(0.2105000)$$
$$+ \ 0.2210500)$$
$$= \frac{1}{6} \ (1.2620500) = 0.2103417,$$

$$M = \frac{1}{6} \ (0.4000000 + 2(0.4200000) + 2(0.4210000)$$
$$+ \ 0.4421000)$$
$$= \frac{1}{6} \ (2.5241000) = 0.4206833.$$

Finally from (8.78) we obtain

$$x_1 = 2.0000000 + 0.2103417 = 2.2103417,$$
$$y_1 = 4,0000000 + 0.4206833 = 4.4206833.$$

These calculations and those for the remaining t_n are summarized in Table 8.8.1-RK.

TABLE 8.8.1-Euler   Summary: Euler Method Solution (with h = 0.1) of the System

$$x' = 5x - 2y, \quad y' = 4x - y, \quad x(0.0) = 2.0, \quad y(0.0) = 4.0.$$

t_n	Exact $\begin{cases} x(t_n) \\ y(t_n) \end{cases}$	Euler $\begin{cases} x_n \\ y_n \end{cases}$	Errors	% Rel Errors
0.1	2.21034	2.20000	0.01034	0.4679
	4.42068	4.40000	0.02068	0.4679
0.2	2.44281	2.42000	0.02281	0.9336
	4.88561	4.84000	0.04561	0.9336
0.3	2.69972	2.66200	0.03772	1.3971
	5.39944	5.32400	0.07544	1.3971
0.4	2.98365	2.92820	0.05545	1.8584
	5.96730	5.85640	0.11090	1.8584
0.5	3.29744	3.22102	0.07642	2.3176
	6.59489	6.44204	0.15285	2.3176

TABLE 8.8.1-R-K   Summary: Runge-Kutta Solution
(with h = 0.1) of the System

$$x' = 5x - 2y, \quad y' = 4x - y, \quad x(0.0) = 2.0, \quad y(0.0) = 4.0.$$

t_n k	k_1, m_1	k_2, m_2	k_3, m_3	k_4, m_4	K, M	Exact $\{x(t_n)/y(t_n)\}$	R-K $\{x_n/y_n\}$	Errors	% Rel Errors
0.1	0.2000	0.2100	0.2105	0.2210	0.2103	2.2103418	2.2103417	0.0000002	0.00001
	0.4000	0.4200	0.4210	0.4421	0.4207	4.4206837	4.4206833	0.0000003	0.00001
0.2	0.2210	0.2321	0.2326	0.2443	0.2325	2.4428055	2.4428051	0.0000004	0.00002
	0.4421	0.4642	0.4653	0.4886	0.4649	4.8856110	4.8856103	0.0000007	0.00002
0.3	0.2443	0.2565	0.2571	0.2700	0.2569	2.6997176	2.6997170	0.0000006	0.00002
	0.4886	0.5130	0.5142	0.5400	0.5138	5.3994352	5.3994340	0.0000012	0.00002
0.4	0.2700	0.2835	0.2841	0.2984	0.2839	2.9836494	2.9836485	0.0000009	0.00003
	0.5399	0.5669	0.5683	0.5968	0.5679	5.9672988	5.9672970	0.0000018	0.00003
0.5	0.2984	0.3133	0.3140	0.3298	0.3138	3.2974425	3.2974413	0.0000013	0.00004
	0.5967	0.6266	0.6281	0.6595	0.6276	6.5948851	6.5948826	0.0000025	0.00004

4. The exact solution is $x = e^{-t}$, $y = -e^{-t}$. First we find the solution by the Euler method. We have $f(t,x,y) = x + 2y$, $g(t,x,y) = 3x + 2y$, $t_0 = 0.0$, $x_0 = 1.0000$, $y_0 = -1.0000$, and $h = 0.1$. To find the approximate values at $t_1 = 0.0 + 0.1 = 0.1$, we use Formulas (8.74). We have

$$x_1 = x_0 + hf(t_0, x_0, y_0)$$
$$= 1.0000 + 0.1f(0.0, 1.0000, -1.0000)$$
$$= 1.0000 + (0.1)(-1.0000) = 0.9000,$$

$$y_1 = y_0 + hg(t_0, x_0, y_0)$$
$$= -1.0000 + 0.1g(0.0, 1.0000, -1.0000)$$
$$= -1.0000 + (0.1)1.0000 = -0.9000.$$

Now to find the approximate values at $t_2 = 0.1 + 0.1 = 0.2$, we use Formulas 8.75 with $n = 1$. We have

$$x_2 = x_1 + hf(t_1, x_1, y_1)$$
$$= 0.9000 + 0.1f(0.1, 0.9000, -0.9000)$$
$$= 0.9000 + (0.1)(-0.9000) = 0.8100,$$

$$y_2 + y_1 + hg(t_1, x_1, y_1)$$
$$= -0.9000 + 0.1g(0.1, 0.9000, -0.9000)$$
$$= -0.9000 + (0.1)0.9000 = -0.8100.$$

The results for all t_n and the corresponding errors have been summarized in Table 8.8.4-Euler.

Now we solve the problem by the Runge-Kutta method. Again, $f(t,x,y) = x + 2y$, $g(t,x,y) = 3x + 2y$, $t_0 = 0.0$, $x_0 = 1.0000000$, $y_0 = -1.0000000$, and $h = 0.1$. To find the approximate values at $t_1 = 0.0 + 0.1 = 0.1$, we use Formulas (8.78-80) with $n = 0$. By (8.80) we have

$$k_1 = hf(t_0,x_0,y_0)$$
$$= (0.1)f(0.0, \ 1.0000000, \ -1.0000000)$$
$$= (0.1)(-1.0000000) = -0.1000000,$$

$$m_1 = hg(t_0,x_0,y_0)$$
$$= (0.1)g(0.0, \ 1.0000000, \ -1.0000000)$$
$$= (0.1)1.0000000 = 0.1000000.$$

Using these values for k_1 and m_1 we now find

$$k_2 = hf\left(t_0 + \frac{h}{2}, \ x_0 + \frac{k_1}{2}, \ y_0 + \frac{m_1}{2}\right)$$
$$= (0.1)f(0.05, \ 0.9500000, \ -0.9500000)$$
$$= (0.1)(-0.9500000) = -0.0950000,$$

$$m_2 = hg\left(t_0 + \frac{h}{2}, \ x_0 + \frac{k_1}{2}, \ y_0 + \frac{m_1}{2}\right)$$
$$= (0.1)g(0.05, \ 0.9500000, \ -0.9500000)$$
$$= (0.1)0.9500000 = 0.0950000.$$

Next, the above values for k_2 and m_2 are used to obtain

$$k_3 = hf\left(t_0 + \frac{h}{2}, \; x_0 + \frac{k_2}{2}, \; y_0 + \frac{m_2}{2}\right)$$
$$= (0.1)f(0.05, \; 0.9525000, \; -0.9525000)$$
$$= (0.1)(-0.9525000) = -0.0952500,$$

$$m_3 = hg\left(t_0 + \frac{h}{2}, \; x_0 + \frac{k_2}{2}, \; y_0 + \frac{m_2}{2}\right)$$
$$= (0.1)g(0.05, \; 0.9525000, \; -0.9525000)$$
$$= (0.1)0.9525000 = 0.0952500.$$

With these values for k_3 and m_3 we find

$$k_4 = hf(t_0 + h, \; x_0 + k_3, \; y_0 + m_3)$$
$$= (0.1)f(0.1, \; 0.9047500, \; -0.9047500)$$
$$= (0.1)(-0.9047500) = -0.0904750,$$

$$m_4 = hg(t_0 + h, \; x_0 + k_3, \; y_0 + m_3)$$
$$= (0.1)g(0.1, \; 0.9047500, \; -0.9047500)$$
$$= (0.1)0.9047500 = 0.0904750.$$

Then from (8.79) we set

$$K = \frac{1}{6} \left(-0.1000000 + 2(0.0950000) + 2(-0.0952500)\right.$$
$$\left. + -0.0904750\right)$$
$$= \frac{1}{6} \left(-0.5709750\right) = -0.0951625,$$

$$M = \frac{1}{6} \; (0.1000000 + 2(0.0950000) + 2(0.0952500)$$

$$+ \; 0.0904750)$$

$$= \frac{1}{6} \; (0.5709750) = 0.0951625.$$

Finally from (8.78) we obtain

$$x_1 = 1.0000000 + -0.0951625 = 0.9048375,$$

$$y_1 = -1.0000000 + 0.0951625 = -0.9048375.$$

These calculations and those for the remaining t_n are summarized in Table 8.8.4–RK.

TABLE 8.8.4–Euler   Summary: Euler Method Solution
(with h = 0.1) of the System

$$x' = x + 2y, \; y' = 3x + 2y, \; x(0.0) = 1.0, \; y(0.0) = -1.0.$$

t_n	Exact $\begin{cases} x(t_n) \\ y(t_n) \end{cases}$	Euler $\begin{cases} x_n \\ y_n \end{cases}$	Errors	% Rel Errors
0.1	0.90484	0.90000	0.00484	0.5346
	-0.90484	-0.90000	0.00484	0.5346
0.2	0.81873	0.81000	0.00873	1.0664
	-0.81873	-0.81000	0.00873	1.0664
0.3	0.74082	0.72900	0.01182	1.5953
	-0.74082	-0.72900	0.01182	1.5953
0.4	0.67032	0.65610	0.01422	2.1214
	-0.67032	-0.65610	0.01422	2.1214
0.5	0.60653	0.59049	0.01604	2.6447
	-0.60653	-0.59049	0.01604	2.6447

TABLE 8.8.4-R-K Summary: Runge-Kutta Solution (with h = 0.1) of the System

$$x' = x + 2y, \quad y' = 3x + 2y, \quad x(0.0) = 1.0, \quad y(0.0) = -1.0.$$

t_n	$k_1,$ m_1	$k_2,$ m_2	$k_3,$ m_3	$k_4,$ m_4	$K,$ M	Exact $\begin{cases} x(t_n) \\ y(t_n) \end{cases}$	R-K $\begin{cases} x_n \\ y_n \end{cases}$	Errors	% Rel Errors
0.1	-0.1000	-0.0950	-0.0952	-0.0905	-0.0952	0.9048374	0.9048375	0.0000001	0.00001
	0.1000	0.0950	0.0952	0.0905	0.0952	-0.9048374	-0.9048375	0.0000001	0.00001
0.2	-0.0905	-0.0860	-0.0862	-0.0819	-0.0861	0.8187308	0.8187309	0.0000001	0.00002
	0.0905	0.0860	0.0862	0.0819	0.0861	-0.8187308	-0.8187309	0.0000001	0.00002
0.3	-0.0819	-0.0778	-0.0780	-0.0741	-0.0779	0.7408182	0.7408184	0.0000002	0.00003
	0.0819	0.0778	0.0780	0.0741	0.0779	-0.7408182	-0.7408184	0.0000002	0.00003
0.4	-0.0741	-0.0704	-0.0706	-0.0670	-0.0705	0.6703200	0.6703203	0.0000002	0.00004
	0.0741	0.0704	0.0706	0.0670	0.0705	-0.6703200	-0.6703203	0.0000002	0.00004
0.5	-0.0670	-0.0637	-0.0638	-0.0606	-0.0638	0.6065307	0.6065309	0.0000003	0.00005
	0.0670	0.0637	0.0638	0.0606	0.0638	-0.6065307	-0.6065309	0.0000003	0.00005

7.   The exact solution is $x = -e^{3t} + 3e^{4t} + 2e^{2t}$, $y = -e^{3t} + 2e^{4t} + 3e^{2t}$.   First we find the solution by the Euler method.   We have $f(t,x,y) = 6x - 3y + e^{2t}$, $g(t,x,y) = 2x + y - e^{2t}$, $t_0 = 0.0$, $x_0 = 4.0000$, $y_0 = 4.0000$ and $h = 0.1$.

To find the approximate values at $t_1 = 0.0 + 0.1 = 0.1$, we use Formulas (8.74).   We have

$$x_1 = x_0 + hf(t_0,x_0,y_0)$$
$$= 4.0000 + 0.1f(0.0,\ 4.0000,\ 4.0000)$$
$$= 4.0000 + (0.1)13.0000 = 5.3000,$$

$$y_1 = y_0 + hg(t_0,\ x_0,\ y_0)$$
$$= 4.0000 + 0.1g(0.0,\ 4.0000,\ 4.0000)$$
$$= 4.0000 + (0.1)11.0000 = 5.1000.$$

Now to find the approximate values at $t_2 = 0.1 + 0.1 = 0.2$, we use Formulas 8.75 with $n = 1$.   We have

$$x_2 = x_1 + hf(t_1,x_1,y_1)$$
$$= 5.3000 + 0.1f(0.1,\ 5.3000,\ 5.1000)$$
$$= 5.3000 + (0.1)17.7214 = 7.0721,$$

$$y_2 = y_1 + hg(t_1,x_1,y_1)$$
$$= 5.1000 + 0.1g(0.1,\ 5.3000,\ 5.1000)$$
$$= 5.1000 + (0.1)14.4786 = 6.5479.$$

The results for all t_n and the corresponding errors have been summarized in Table 8.8.7-Euler.

Now we solve the problem by the Runge-Kutta method. Again, $f(t,x,y) = 6x - 3y + e^{2t}$, $g(t,x,y) = 2x + y - e^{2t}$, $t_0 = 0.0$, $x_0 = 4.0000000$, $y_0 = 4.0000000$, and $h = 0.1$. To find the approximate values at $t_1 = 0.0 + 0.1 = 0.1$, we use Formulas (8.78-80) with $n = 0$. By (8.80) we have

$$k_1 = hf(t_0, x_0, y_0)$$
$$= (0.1)f(0.0, 4.0000000, 4.0000000)$$
$$= (0.1)13.0000000 = 1.3000000,$$

$$m_1 = hg(t_0, x_0, y_0)$$
$$= (0.1)g(0.0, 4.0000000, 4.0000000)$$
$$= (0.1)11.0000000 = 1.1000000.$$

Using these values for k_1 and m_1 we now find

$$k_2 = hf\left(t_0 + \frac{h}{2},\ x_0 + \frac{k_1}{2},\ y_0 + \frac{m_1}{2}\right)$$
$$= (0.1)f(0.05, 4.6500000, 4.5500000)$$
$$= (0.1)15.3551709 = 1.5355171,$$

$$m_2 = hg\left(t_0 + \frac{h}{2},\ x_0 + \frac{k_1}{2},\ y_0 + \frac{m_1}{2}\right)$$
$$= (0.1)g(0.05, 4.6500000, 4.5500000)$$
$$= (0.1)12.7448291 = 1.2744829.$$

Next, the above values for k_2 and m_2 are used to obtain

$$k_3 = hf\left(t_0 + \frac{h}{2},\ x_0 + \frac{k_2}{2},\ y_0 + \frac{m_2}{2}\right)$$
$$= (0.1)f(0.05,\ 4.7677585,\ 4.6372415)$$
$$= (0.1)15.7999978 = 1.5799998,$$

$$m_3 = hg\left(t_0 + \frac{h}{2},\ x_0 + \frac{k_2}{2},\ y_0 + \frac{m_2}{2}\right)$$
$$= (0.1)g(0.05,\ 4.7677585,\ 4.6372415)$$
$$= (0.1)13.0675876 = 1.3067588.$$

With these values for k_3 and m_3 we find

$$k_4 = hf(t_0 + h,\ x_0 + k_3,\ y_0 + m_3)$$
$$= (0.1)f(0.1,\ 5.5799998,\ 5.3067588)$$
$$= (0.1)18.7811252 = 1.8781125,$$

$$m_4 = hg(t_0 + h,\ x_0 + k_3,\ y_0 + m_3)$$
$$= (0.1)g(0.1,\ 5.5799998,\ 5.3067588)$$
$$= (0.1)15.2453556 = 1.5245356.$$

Then from (8.79) we set

$$K = \frac{1}{6}\ (1.3000000 + 2(1.5355171) + 2(1.5799998)$$
$$+ 1.8781125)$$

$$= \frac{1}{6}\ (9.4091463) = 1.5681910,$$

$$M = \frac{1}{6} (1.1000000 + 2(1.2744829) + 2(1.3067588)$$
$$+ 1.5245356)$$

$$= \frac{1}{6} (7.7870189) = 1.2978365.$$

Finally from (8.78) we obtain

$$x_1 = 4.0000000 + 1.5681910 = 5.5681910,$$

$$y_1 = 4.0000000 + 1.2978365 = 5.2978365.$$

These calculations and those for the remaining t_n are summarized in Table 8.8.7-RK.

TABLE 8.8.7-Euler   Summary: Euler Method Solution (with h = 0.1) of the System

$x' = 6x - 3y + e^{2t}$, $y' = 2x + y - e^{2t}$, $x(0.0) = 4.0$, $y(0.0) = 4.0$.

t_n	Exact $\begin{cases} x(t_n) \\ y(t_n) \end{cases}$	Euler $\begin{cases} x_n \\ y_n \end{cases}$	Errors	% Rel Errors
0.1	5.56842	5.30000	0.26842	4.8204
	5.29800	5.10000	0.19800	3.7372
0.2	7.83815	7.07214	0.76601	9.7729
	7.10444	6.54786	0.55658	7.8342
0.3	11.14499	9.50025	1.64474	14.758
	9.64699	8.46789	1.17910	12.222
0.4	15.99006	12.84224	3.14782	19.686
	13.26257	11.03252	2.23005	16.815
0.5	23.12204	17.46039	5.66166	24.486
	18.45127	14.48166	3.96960	21.514

TABLE 8.8.7-R-K   Summary: Runge–Kutta Solution (with $h = 0.1$) of the System

$$x' = 6x - 3y + e^{2t}, \quad y' = 2x + y - e^{2t}, \quad x(0.0) = 4.0,$$
$$y(0.0) = 4.0.$$

t_n, k	k_1, m_1	k_2, m_2	k_3, m_3	k_4, m_4	K, M	Exact $\begin{Bmatrix} x(t_n) \\ y(t_n) \end{Bmatrix}$	R-K $\begin{Bmatrix} x_n \\ y_n \end{Bmatrix}$	Errors	% Rel Errors
0.1	1.3000	1.5355	1.5800	1.8781	1.5682	5.5684208	5.5681910	0.0002298	0.00413
	1.1000	1.2745	1.3068	1.5245	1.2978	5.2979989	5.2978365	0.0001624	0.00306
0.2	1.8737	2.2205	2.2869	2.7272	2.2693	7.8381534	7.8374626	0.0006908	0.00881
	1.5213	1.7719	1.8191	2.1335	1.8061	7.1044371	7.1039538	0.0004833	0.00680
0.3	2.7205	3.2330	3.3323	3.9846	3.3060	11.1449853	11.1434276	0.0015577	0.01398
	2.1287	2.4915	2.5609	3.0182	2.5420	9.6469871	9.6459066	0.0010806	0.01120
0.4	3.9745	4.7343	4.8830	5.8519	4.8435	15.9900622	15.9869409	0.0031213	0.01952
	3.0111	3.5399	3.6423	4.3116	3.6145	13.2625707	13.2604207	0.0021500	0.01621
0.5	5.8366	6.9658	7.1883	8.6305	7.1292	23.1220429	23.1161805	0.0058624	0.02535
	4.3009	5.0762	5.2279	6.2121	5.1868	18.4512686	18.4472544	0.0040143	0.02176

9. Before we begin using the Euler method, we convert the given second order D.E. into a system, following the procedure described in section 7.1A. We have $f(t,x,y) = y$, $g(t,x,y) = -6x + 5y$, $t_0 = 0.0$, $x_0 = 1.0000$, $y_0 = 2.5000$, and $h = 0.1$. To find the approximate values at $t_1 = 0.0 + 0.1 = 0.1$, we use Formulas (8.74). We have

$$x_1 = x_0 + hf(t_0,x_0,y_0)$$
$$= 1.0000 + 0.1f(0.0,\ 1.0000,\ 2.5000)$$
$$= 1.0000 + (0.1)2.5000 = 1.2500,$$

$$y_1 = y_0 + hg(t_0,x_0,y_0)$$
$$= 2.5000 + 0.1g(0.0,\ 1.0000,\ 2.5000)$$
$$= 2.5000 + (0.1)6.5000 = 3.1500.$$

Now to find the approximate values at $t_2 = 0.1 + 0.1 = 0.2$, we use Formulas 8.75 with $n = 1$. We have

$$x_2 = x_1 + hf(t_1,x_1,y_1)$$
$$= 1.2500 + 0.1f(0.1,\ 1.2500,\ 3.1500)$$
$$= 1.2500 + (0.1)3.1500 = 1.5650,$$

$$y_2 = y_1 + hg(t_1,x_1,y_1)$$
$$= 3.1500 + 0.1g(0.1,\ 1.2500,\ 3.1500)$$
$$= 3.1500 + (0.1)8.2500 = 3.9750.$$

The results for all t_n and the corresponding errors have been summarized in Table 8.8.9-Euler.

Now we solve the problem by the Runge-Kutta method. Again, $f(t,x,y) = y$, $g(t,x,y) = -6x + 5y$, $t_0 = 0.0$, $x_0 = 1.0000000$, $y_0 = 2.50000000$, and $h = 0.1$. To find the approximate values at $t_1 = 0.0 + 0.1 = 0.1$, we use Formulas (8.78-80) with $n = 0$. By (8.80) we have

$$k_1 = hf(t_0,x_0,y_0)$$
$$= (0.1)f(0.0, \ 1.0000000, \ 2.5000000)$$
$$= (0.1)2.5000000 = 0.2500000,$$

$$m_1 = hg(t_0,x_0,y_0)$$
$$= (0.1)g(0.0, \ 1.0000000, \ 2.5000000)$$
$$= (0.1)6.5000000 = 0.6500000.$$

Using these values for k_1 and m_1 we now find

$$k_2 = hf\left(t_0 + \frac{h}{2}, \ x_0 + \frac{k_1}{2}, \ y_0 + \frac{m_1}{2}\right)$$
$$= (0.1)f(0.05, \ 1.1250000, \ 2.8250000)$$
$$= (0.1)2.8250000 = 0.2825000,$$

$$m_2 = hg\left(t_0 + \frac{h}{2}, \ x_0 + \frac{k_1}{2}, \ y_0 + \frac{m_1}{2}\right)$$
$$= (0.1)g(0.05, \ 1.1250000, \ 2.8250000)$$
$$= (0.1)7.3750000 = 0.7375000.$$

Next, the above values for k_2 and m_2 are used to obtain

$$k_3 = hf\left(t_0 + \frac{h}{2},\ x_0 + \frac{k_2}{2},\ y_0 + \frac{m_2}{2}\right)$$

$$= (0.1)f(0.05,\ 1.1412500,\ 2.8687500)$$

$$= (0.1)2.8687500 = 0.2868750,$$

$$m_3 = hg\left(t_0 + \frac{h}{2},\ x_0 + \frac{k_2}{2},\ y_0 + \frac{m_2}{2}\right)$$

$$= (0.1)g(0.05,\ 1.1412500,\ 2.8687500)$$

$$= (0.1)7.4962500 = 0.7496250.$$

With these values for k_3 and m_3 we find

$$k_4 = hf(t_0 + h,\ x_0 + k_3,\ y_0 + m_3)$$

$$= (0.1)f(0.1,\ 1.2868750,\ 3.2496250)$$

$$= (0.1)3.2496250 = 0.3249625,$$

$$m_4 = hg(t_0 + h,\ x_0 + k_3,\ y_0 + m_3)$$

$$= (0.1)g(0.1,\ 1.2868750,\ 3.2496250$$

$$= (0.1)8.5268750 = 0.8526875.$$

Then from (8.79) we set

$$K = \frac{1}{6}\ (0.2500000 + 2(0.2825000) + 2(0.2868750)$$

$$+ 0.3249625)$$

$$= \frac{1}{6}\ (1.7137125) = 0.2856187,$$

$$M = \frac{1}{6} (0.6500000 + 2(0.7375000) + 2(0.7496250)$$
$$+ 0.8526875)$$

$$= \frac{1}{6} (4.4769375) = 0.7461562.$$

Finally from (8.78) we obtain

$$x_1 = 1.0000000 + 0.2856187 = 1.2856187,$$

$$y_1 = 2.5000000 + 0.7461562 = 3.2461562.$$

These calculations and those for the remaining t_n are summarized in Table 8.8.9-R.K. The exact solution to the stated problem is $x = \frac{1}{2} (e^{2t} + e^{3t})$; we have also $y = x' = \frac{1}{2} (2e^{2t} + 3e^{3t})$.

TABLE 8.8.9-Euler   Summary: Euler Method Solution (with $h = 0.1$) of the System

$$x' = y, \quad y' = -6x + 5y, \quad x(0.0) = 1.0, \quad y(0.0) = 2.5.$$

t_n	Exact $\begin{cases} x(t_n) \\ y(t_n) \end{cases}$	Euler $\begin{cases} x_n \\ y_n \end{cases}$	Errors	% Rel Errors
0.1	1.28563	1.25000	0.03563	2.7715
	3.24619	3.15000	0.09619	2.9632
0.2	1.65697	1.56500	0.09197	5.5506
	4.22500	3.97500	0.25000	5.9172
0.3	2.14086	1.96250	0.17836	8.3313
	5.51152	5.02350	0.48802	8.8546
0.4	2.77283	2.46485	0.30798	11.107
	7.20572	6.35775	0.84797	11.768
0.5	3.59999	3.10062	0.49936	13.871
	9.44082	8.05771	1.38310	14.650

TABLE 8.8.9-R-K Summary: Runge-Kutta Solution
(with $h = 0.1$) of the System

$$x' = y, \quad y' = -6x + 5y, \quad x(0.0) = 1.0, \quad y(0.0) = 2.5.$$

t_n,k	k_1, m_1	k_2, m_2	k_3, m_3	k_4, m_4	K, M	Exact $\begin{Bmatrix} x(t_n) \\ y(t_n) \end{Bmatrix}$	R-K $\begin{Bmatrix} x_n \\ y_n \end{Bmatrix}$	Errors	% Rel Errors
0.1	0.2500	0.2825	0.2869	0.3250	0.2856	1.2856308	1.2856187	0.0000120	0.00094
	0.6500	0.7375	0.7496	0.8527	0.7462	3.2461910	3.2461562	0.0000347	0.00107
0.2	0.3246	0.3672	0.3730	0.4230	0.3713	1.6569717	1.6569396	0.0000321	0.00194
	0.8517	0.9672	0.9834	1.1196	0.9788	4.2250029	4.2249099	0.0000930	0.00220
0.3	0.4225	0.4784	0.4860	0.5517	0.4839	2.1408610	2.1407965	0.0000644	0.00301
	1.1183	1.2711	1.2925	1.4729	1.2864	5.5115235	5.5113364	0.0001871	0.00339
0.4	0.5511	0.6247	0.6348	0.7214	0.6319	2.7728289	2.7727141	0.0001149	0.00414
	1.4712	1.6736	1.7022	1.9414	1.6940	7.2057163	7.2053818	0.0003345	0.00464
0.5	0.7205	0.8175	0.8309	0.9451	0.8271	3.5999854	3.5997932	0.0001922	0.00534
	1.9391	2.2077	2.2457	2.5634	2.2349	9.4408154	9.4402542	0.0005613	0.00594

12.   Before we begin using the Euler method, we convert the given second order D.E. into a system, following the procedure described in section 7.1A.   We have $f(t,x,y) = y$, $g(t,x,y) = 3y - 2x + 4t^2$, $t_0 = 0.0$, $x_0 = 8.0000$, $y_0 = 8.0000$, and $h = 0.1$.   To find the approximate values at $t_1 = 0.0 + 0.1 = 0.1$, we use Formulas (8.74).   We have

$$x_1 = x_0 + hf(t_0,x_0,y_0)$$
$$= 8.0000 + 0.1f(0.0,\ 8.0000,\ 8.0000)$$
$$= 8.0000 + (0.1)8.0000 = 8.8000,$$

$$y_1 = y_0 + hg(t_0,x_0,y_0)$$
$$= 8.0000 + 0.1g(0.0,\ 8.0000,\ 8.0000)$$
$$= 8.0000 + (0.1)8.0000 = 8.8000.$$

Now to find the approximate values at $t_2 = 0.1 + 0.1 = 0.2$, we use Formulas 8.75 with $n = 1$.   We have

$$x_2 = x_1 + hf(t_1,x_1,y_1)$$
$$= 8.8000 + 0.1f(0.1,\ 8.8000,\ 8.8000)$$
$$= 8.8000 + (0.1)8.8000 = 9.6800,$$

$$y_2 = y_1 + hg(t_1,x_1,y_1)$$
$$= 8.8000 + 0.1g(0.1,\ 8.8000,\ 8.8000)$$
$$= 8.8000 + (0.1)8.8400 = 9.6840.$$

The results for all t_n and the corresponding errors have been summarized in Table 8.8.12-Euler.

Now we solve the problem by the Runge-Kutta method.
Again, $f(t,x,y) = y$, $g(t,x,y) = 3y - 2x + 4t^2$, $t_0 = 0.0$,
$x_0 = 8.0000000$, $y_0 = 8.0000000$, and $h = 0.1$. To find the
approximate values at $t_1 = 0.0 + 0.1 = 0.1$, we use
Formulas (8.78-80) with $n = 0$. By (8.80) we have

$$k_1 = hf(t_0, x_0, y_0)$$
$$= (0.1)f(0.0,\ 8.0000000,\ 8.0000000)$$
$$= (0.1)8.0000000 = 0.8000000,$$

$$m_1 = hg(t_0,\ x_0,\ y_0)$$
$$= (0.1)g(0.0,\ 8.0000000,\ 8.0000000)$$
$$= (0.1)8.0000000 = 0.8000000.$$

Using these values for k_1 and m_1 we now find

$$k_2 = hf\left(t_0 + \frac{h}{2},\ x_0 + \frac{k_1}{2},\ y_0 + \frac{m_1}{2}\right)$$
$$= (0.1)f(0.05,\ 8.4000000,\ 8.4000000)$$
$$= (0.1)8.4000000 = 0.8400000,$$

$$m_2 = hg\left(t_0 + \frac{h}{2},\ x_0 + \frac{k_1}{2},\ y_0 + \frac{m_1}{2}\right)$$
$$= (0.1)g(0.05,\ 8.4000000,\ 8.4000000)$$
$$= (0.1)8.4100000 = 0.8410000.$$

Next, the above values for k_2 and m_2 are used to obtain

$$k_3 = hf\left(t_0 + \frac{h}{2},\ x_0 + \frac{k_2}{2},\ y_0 + \frac{m_2}{2}\right)$$
$$= (0.1)f(0.05,\ 8.4200000,\ 8.4205000)$$
$$= (0.2)8.4205000 = 0.8420500,$$

$$m_3 = hg\left(t_0 + \frac{h}{2},\ x_0 + \frac{k_2}{2},\ y_0 + \frac{m_2}{2}\right)$$
$$= (0.1)g(0.05,\ 8.4200000,\ 8.4205000)$$
$$= (0.1)8.4315000 = 0.8431500.$$

With these values for k_3 and m_3 we find

$$k_4 = hf(t_0 + h,\ x_0 + k_3,\ y_0 + m_3)$$
$$= (0.1)f(0.1,\ 8.8420500,\ 8.8431500)$$
$$= (0.1)8.8431500 = 0.8843150,$$

$$m_4 = hg(t_0 + h,\ x_0 + k_3,\ y_0 + m_3)$$
$$= (0.1)g(0.1,\ 8.8420500,\ 8.8431500)$$
$$= (0.1)8.8853500 = 0.8885350.$$

Then from (8.79) we set

$$K = \frac{1}{6}\ (0.8000000 + 2(0.8400000) + 2(0.8420500)$$
$$+ 0.8843150)$$

$$= \frac{1}{6}\ (5.0484150) = 0.8414025,$$

$$M = \frac{1}{6} (0.8000000 + 2(0.8410000) + 2(0.8431500)$$

$$+ 0.8885350)$$

$$= \frac{1}{6} (5.0568350) = 0.8428058.$$

Finally from (8.78) we obtain

$$x_1 = 8.0000000 + 0.8414025 = 8.8414025,$$

$$y_1 = 8.0000000 + 0.8428058 = 8.8428058.$$

These calculations and those for the remaining t_n are summarized in Table 8.8.12-R-K.

TABLE 8.8.12-Euler   Summary: Euler Method Solution
(with $h = 0.1$) of the System

$$x' = y, \ y' = 3y - 2x + 4t^2, \ x(0.0) = 8.0, \ y(0.0) = 8.0.$$

t_n	Exact $\begin{cases} x(t_n) \\ y(t_n) \end{cases}$	Euler $\begin{cases} x_n \\ y_n \end{cases}$	Errors	% Rel Errors
0.1	8.84140	8.80000	0.04140	0.4683
	8.84281	8.80000	0.04281	0.4841
0.2	9.77182	9.68000	0.09182	0.9397
	9.78365	9.68400	0.09965	1.0185
0.3	10.80212	10.64840	0.15372	1.4230
	10.84424	10.66920	0.17504	1.6141
0.4	11.94554	11.71532	0.23022	1.9273
	12.05106	11.77628	0.27480	2.2803
0.5	13.21828	12.89295	0.32533	2.4612
	13.43656	13.03010	0.40646	3.0251

TABLE 8.8.12-R-K   Summary: Runge-Kutta Solution (with h = 0.1) of the System

$$x' = y, \quad y' = 3y - 2x + 4t^2, \quad x(0.0) = 8.0, \quad y(0.0) = 8.0.$$

t_n, k	$k_1,$ m_1	$k_2,$ m_2	$k_3,$ m_3	$k_4,$ m_4	$K,$ M	Exact $\left\{\begin{array}{l}x(t_n)\\y(t_n)\end{array}\right.$	R-K $\left\{\begin{array}{l}x_n\\y_n\end{array}\right.$	Errors	% Rel Errors
0.1	0.8000	0.8400	0.8420	0.8843	0.8414	8.8414028	8.8414025	0.0000003	0.00000
	0.8000	0.8410	0.8431	0.8885	0.8428	8.8428055	8.8428058	0.0000003	0.00000
0.2	0.8843	0.9287	0.9312	0.9784	0.9304	9.7718247	9.7718236	0.0000011	0.00001
	0.8886	0.9384	0.9415	0.9968	0.9408	9.7836494	9.7836490	0.0000004	0.00000
0.3	0.9784	1.0282	1.0312	1.0845	1.0303	10.8021188	10.8021161	0.0000027	0.00003
	0.9967	1.0574	1.0615	1.1289	1.0606	10.8442376	10.8442349	0.0000027	0.00002
0.4	1.0844	1.1409	1.1446	1.2052	1.1434	11.9455409	11.9455354	0.0000055	0.00005
	1.1288	1.2027	1.2082	1.2904	1.2068	12.0510819	12.0510747	0.0000072	0.00006
0.5	1.2051	1.2696	1.2741	1.3438	1.2727	13.2182818	13.2182719	0.0000100	0.00008
	1.2902	1.3802	1.3873	1.4876	1.3855	13.4365637	13.4365489	0.0000148	0.00011

The exact solution to the stated problem is $x = e^{2t} + 2t^2 + 6t + 7$; we also have $y = x' = 2e^{2t} + 4t + 6$.

Chapter 9

1. We have

$$\mathcal{L}\{f(t)\} = \int_0^\infty e^{-st}f(t)dt = \int_0^\infty t^2 e^{-st}dt$$

$$= \lim_{R\to\infty}\left[-\frac{e^{-st}}{s^3}(s^2t^2 + 2st + 2)\right]\Bigg|_0^R$$

$$= \lim_{R\to\infty}\left[-\frac{e^{-sR}}{s^3}(s^2R^2 + 2sR + 2)\right] + \frac{2}{s^3}$$

$$= 0 + \frac{2}{s^3} = \frac{2}{s^3}, \text{ where } s > 0.$$

4. We have

$$\mathcal{L}\{f(t)\} = \int_0^\infty e^{-st}f(t)dt = \int_0^3 4e^{-st}dt + \int_3^\infty 2e^{-st}dt$$

$$= \frac{-4e^{-st}}{s}\Bigg|_0^3 + \lim_{R\to\infty}\left[\frac{-2e^{-st}}{s}\right]\Bigg|_3^R$$

$$= \frac{-4e^{-3s}}{s} + \frac{4}{s} + \lim_{R\to\infty}\left[\frac{-2e^{-sR}}{s} + \frac{2e^{-3s}}{s}\right]$$

$$= \frac{-4e^{-3s}}{s} + \frac{4}{s} + 0 + \frac{2e^{-3s}}{s}$$

$$= \frac{2(2 - e^{-3s})}{s}, \text{ where } s > 0.$$

5.   We have

$$\mathcal{L}\{f(t)\} = \int_0^\infty e^{-st} f(t) dt = \int_0^2 te^{-st} dt + \int_2^\infty 3e^{-st} dt$$

$$= - \left. \frac{e^{-st}(st + 1)}{s^2} \right|_0^2 + \lim_{R \to \infty} \left[\frac{-3e^{-st}}{s} \right] \Bigg|_2^R$$

$$= - \frac{e^{-2s}(2s + 1)}{s^2} + \frac{1}{s^2} + \lim_{R \to \infty} \left[\frac{-3e^{-sR}}{s} + \frac{3e^{-2s}}{s} \right]$$

$$= \frac{-e^{-2s}(2s + 1)}{s^2} + \frac{1}{s^2} + 0 + \frac{3e^{-2s}}{s}$$

$$= \frac{1}{s^2} + \frac{e^{-2s}}{s} - \frac{e^{-2s}}{s^2}, \text{ where } s > 0.$$

8.   We have

$$\mathcal{L}\{f(t)\} = \int_0^\infty e^{-st} f(t) dt = \int_0^1 2te^{-st} dt + \int_1^3 2e^{-st} dt$$

$$+ \int_3^\infty (8 - 2t)e^{-st} dt$$

$$= \left. \frac{2e^{-st}}{s^2} (-st - 1) \right|_0^1 - \left. \frac{2e^{-st}}{s} \right|_1^3$$

$$+ \lim_{R \to \infty} \left[-\frac{8e^{-st}}{s} - \frac{2e^{-st}}{s^2} (-st - 1) \right] \Bigg|_3^R$$

$$= - \frac{2e^{-s}}{s^2} (s + 1) + \frac{2}{s^2} - \frac{2e^{-3s}}{s} + \frac{2e^{-s}}{s}$$

$$+ \lim_{R \to \infty} \left[-\frac{8e^{-sR}}{s} + \frac{2e^{-sR}}{s^2} (sR + 1) \right]$$

$$+ \left[\frac{8e^{-3s}}{s} - \frac{2e^{-3s}}{s^2} (3s + 1) \right]$$

$$= -\frac{2e^{-s}}{s} - \frac{2e^{-s}}{s^2} + \frac{2}{s^2} - \frac{2e^{-3s}}{s} + \frac{2e^{-s}}{s}$$

$$+ 0 + \frac{8e^{-3s}}{s} - \frac{6e^{-3s}}{s} - \frac{2e^{-3s}}{s^2}$$

$$= \frac{2}{s^2} - \frac{2e^{-s}}{s^2} - \frac{2e^{-3s}}{s^2}$$

$$= \frac{2}{s^2} (1 - e^{-s} - e^{-3s}), \text{ where } s > 0.$$

Section 9.1B, Page 496.

2.   We use the identity

$$\sin at \sin bt = \frac{1}{2} \cos(a - b)t - \frac{1}{2} \cos(a + b)t.$$

Applying Theorem 9.2, we have

$$\mathcal{L}\{\sin at \sin bt\} = \frac{1}{2} \mathcal{L}\{\cos(a - b)t\} - \frac{1}{2} \mathcal{L}\{\cos(a + b)t\}.$$

By (9.6), $\mathcal{L}\{\cos ct\} = \dfrac{s}{s^2 + c^2}$, $(s > 0)$, where c is a

constant.   Applying this with $c = a - b$ and $c = a + b$, we

have

$$\mathcal{L}\{\sin at \sin bt\} = \frac{1}{2} \cdot \frac{s}{s^2 + (a - b)^2} - \frac{1}{2} \cdot \frac{s}{s^2 + (a + b)^2}$$

$$= \frac{s}{2} \cdot \frac{[s^2 + (a + b)^2] - [s^2 + (a - b)^2]}{[s^2 + (a - b)^2][s^2 + (a + b)^2]}$$

$$= \frac{2abs}{[s^2 + (a - b)^2][s^2 + (a + b)^2]}.$$

3. We use the identity $\sin^3 at = \dfrac{3 \sin at - \sin 3at}{4}$. Applying Theorem 9.2, we have

$$\mathcal{L}\{\sin^3 at\} = \frac{3\mathcal{L}\{\sin at\}}{4} - \frac{\mathcal{L}\{\sin 3at\}}{4}.$$

By (9.5), $\mathcal{L}\{\sin at\} = \dfrac{a}{s^2 + a^2}$ and $\mathcal{L}\{\sin 3at\} = \dfrac{3a}{s^2 + 9a^2}$.

Thus

$$\mathcal{L}\{\sin^3 at\} = \frac{3a}{4(s^2 + a^2)} - \frac{3a}{4(s^2 + 9a^2)}$$

$$= \frac{6a^3}{(s^2 + a^2)(s^2 + 9a^2)}.$$

Now note that $(\sin^3 at)' = 3a \sin^2 at \cos at$. Then by Theorem 9.3, $\mathcal{L}\{f'(t)\} = s\mathcal{L}\{f(t)\} - f(0)$, so with $f(t) = \sin^3 at$ we have, $\mathcal{L}\{3a \sin^2 at \cos at\} = s\mathcal{L}\{\sin^3 at\}$ - 0. Thus $\mathcal{L}\{\sin^2 at \cos at\} = \dfrac{s\mathcal{L}\{\sin^3 at\}}{3a}$

$$= \frac{2a^2 s}{(s^2 + a^2)(s^2 + 9a^2)}.$$

6. We apply Theorem 9.4 with n = 2. We have
$$\mathcal{L}\{f''(t)\} = s^2\mathcal{L}\{f(t)\} - sf(0) - f'(0). \text{ Letting } f(t) = t^4,$$

$f'(t) = 4t^3$, $f''(t) = 12t^2$, this becomes $\mathcal{L}\{12t^2\} = s^2\mathcal{L}\{t^4\}$
$- 0 - 0$, so $\mathcal{L}\{t^4\} = \dfrac{\mathcal{L}\{12t^2\}}{s^2} = \dfrac{24}{s^5}.$

7. We apply Theorem 9.2 to obtain $\mathcal{L}\{f''(t)\} + 3\mathcal{L}\{f'(t)\}$
    $+ 2\mathcal{L}\{f(t)\} = \mathcal{L}\{0\}$. Then, using (9.11) and (9.13), we have
    $[s^2\mathcal{L}\{f(t)\} - sf(0) - f'(0)] + 3[s\mathcal{L}\{f(t)\} - f(0)]$
    $+ 2\mathcal{L}\{f(t)\} = 0$. Now applying the conditions $f(0) = 1$,
    $f'(0) = 2$, this becomes

    $$s^2\mathcal{L}\{f(t)\} - s - 2 + 3s\mathcal{L}\{f(t)\} - 3 + 2\mathcal{L}\{f(t)\} = 0,$$

    or $(s^2 + 3s + 2)\mathcal{L}\{f(t)\} = s + 5$. Hence

    $$\mathcal{L}\{f(t)\} = \frac{s + 5}{s^2 + 3s + 2}.$$

9. We take the Laplace Transform of both sides of
    $f'''(t) = f'(t)$ to obtain

    $$\mathcal{L}\{f'''(t)\} = \mathcal{L}\{f'(t)\}. \tag{1}$$

    By (9.17) with $n = 3$, we have

    $$\mathcal{L}\{f'''(t)\} = s^3\mathcal{L}\{f(t)\} - s^2f(0) - sf'(0) - f''(0).$$

    By (9.11), $\mathcal{L}\{f'(t)\} = s\mathcal{L}\{f(t)\} - f(0)$. Using these
    expressions, (1) becomes

    $$s^3\mathcal{L}\{ft)\} - s^2f(0) - sf'(0) - f''(0) = s\mathcal{L}\{f(t)\} - f(0).$$

    Now apply the conditions $f(0) = 0$, $f'(0) = 1$, $f''(0) = 2$ to
    obtain

    $$s^3\mathcal{L}\{f(t)\} - s - 2 = s\mathcal{L}\{f(t)\}.$$

From this,

$$(s^3 - s)\mathcal{L}\{f(t)\} = s + 2$$

and hence

$$\mathcal{L}\{f(t)\} = \frac{s + 2}{s^3 - s}.$$

10. We take the Laplace Transform of both sides of
     $f^{iv}(t) = f''(t)$, to obtain

$$\mathcal{L}\{f^{iv}(t)\} = \mathcal{L}\{f''(t)\}. \qquad (1)$$

By (9.17), with n = 4, we have

$$\mathcal{L}\{f^{iv}(t)\} = s^4\mathcal{L}\{f(t)\} - s^3 f(0) - s^2 f'(0)$$
$$- sf''(0) - f'''(0).$$

By (9.18), $\mathcal{L}\{f''(t)\} = s^2\mathcal{L}\{f(t)\} - sf(0) - f'(0)$. Using
these expressions, (1) becomes

$$s^4\mathcal{L}\{f(t)\} - s^3 f(0) - s^2 f'(0) - sf''(0) - f'''(0)$$
$$= s^2\mathcal{L}\{f(t)\} - sf(0) - f'(0).$$

Now apply the conditions $f(0) = -1$, $f'(0) = 0$, $f''(0) = 0$,
$f'''(0) = 1$ to obtain

$$s^4\mathcal{L}\{f(t)\} + s^3 - 1 = s^2\mathcal{L}\{f(t)\} + s.$$

From this, $(s^4 - s^2)\mathcal{L}\{f(t)\} = -s^3 + s + 1$, and hence

$$\mathcal{L}\{f(t)\} = \frac{-s^3 + s + 1}{s^4 - s^2}.$$

12. We apply Theorem 9.2 to obtain

$$3\mathcal{L}\{f''(t)\} - 5\mathcal{L}\{f'(t)\} + 7\mathcal{L}\{f(t)\} = \mathcal{L}\{\sin 2t\}. \qquad (1)$$

Then by formula (9.18), $\mathcal{L}\{f''(t)\} = s^2\mathcal{L}\{f(t)\} - sf(0)$
$- f'(0)$; and by formula (9.11), $\mathcal{L}\{f'(t)\} = s\mathcal{L}\{f(t)\}$
$- f(0)$. By the given conditions, these become $\mathcal{L}\{f''(t)\}$
$= s^2\mathcal{L}\{f(t)\} - 4s - 6$ and $\mathcal{L}\{f'(t)\} = s\mathcal{L}\{f(t)\} - 4$.
Substituting these expressions into the left member of
(1), the left member becomes $3s^2\mathcal{L}\{f(t)\} - 12s - 18$
$- 5s\mathcal{L}\{f(t)\} + 20 + 7\mathcal{L}\{f(t)\}$ or $(3s^2 - 5s + 7)\mathcal{L}\{f(t)\}$
$- 12s + 2$. By Example 9.4, $\mathcal{L}\{\sin 2t\} = \dfrac{2}{s^2 + 4}$; and so the
right member of (1) becomes $\dfrac{2}{s^2 + 4}$. Thus (1) becomes

$$(3s^2 - 5s + 7)\mathcal{L}\{f(t)\} - 12s + 2 = \frac{2}{s^2 + 4}.$$

From this,

$$(3s^2 - 5s + 7)\mathcal{L}\{f(t)\} = 12s - 2 + \frac{2}{s^2 + 4}$$

and hence

$$\mathcal{L}\{f(t)\} = \frac{12s - 2}{3s^2 - 5s + 7} + \frac{2}{(3s^2 - 5s + 7)(s^2 + 4)}.$$

13. We let $f(t) = t^2$. Then, using information provided in
Exercises 5 or 6, $F(s) = \mathcal{L}\{f(t)\} = \mathcal{L}\{t^2\} = \dfrac{2}{s^3}$. Then by
Theorem 9.5, we have

$$\mathcal{L}\{e^{at}t^2\} = \mathcal{L}\{e^{at}f(t)\} = F(s - a) = \frac{2}{(s - a)^3}.$$

16. We let $f(t) = \sin bt$. Then $F(s) = \mathcal{L}\{f(t)\} = \mathcal{L}\{\sin bt\}$
$= \dfrac{b}{s^2 + b^2}$. Then by Theorem 9.6,

$$\mathcal{L}\{t^3 \sin bt\} = (-1)^3 \frac{d^3}{ds^3} [F(s)]. \tag{1}$$

As in Example 9.18, $\dfrac{d^2}{ds^2} [F(s)] = \dfrac{6bs^2 - 2b^3}{(s^2 + b^2)^3}$. From this,

$$\frac{d^3}{ds^3} [F(s)] = \frac{24bs(b^2 - s^2)}{(s^2 + b^2)^4}. \quad \text{Hence by (1)},$$

$$\mathcal{L}\{t^3 \sin bt\} = \frac{24bs(s^2 - b^2)}{(s^2 + b^2)^4}.$$

Section 9.2A, Page 504.

1. $\mathcal{L}^{-1}\left\{\dfrac{2}{s} + \dfrac{3}{s-5}\right\} = 2 + 3e^{5t}$, where we used Table 9.1, number

1, and number 2 with a = 5, respectively.

4. $\mathcal{L}^{-1}\left\{\dfrac{2s}{s^2 + 9}\right\} = 2\mathcal{L}^{-1}\left\{\dfrac{s}{s^2 + 3^2}\right\} = 2\cos 3t$, when we used Table

9.1, number 4, with b = 3.

6. $\mathcal{L}^{-1}\left\{\dfrac{5s + 6}{s^3}\right\} = 5\mathcal{L}^{-1}\left\{\dfrac{1}{s^2}\right\} + 3\mathcal{L}^{-1}\left\{\dfrac{2}{s^3}\right\} = 5t + 3t^2$, where we

used Table 9.1, number 7, with n = 1 and n = 2,
respectively.

7. We write

$$F(s) = \frac{s + 2}{s^2 + 4s + 7} = \frac{s + 2}{(s + 2)^2 + (\sqrt{3})^2}.$$

Then using Table 9.1, number 12, with a = 2, b = $\sqrt{3}$, we have

$$\mathcal{L}^{-1}\{F(s)\} = \mathcal{L}^{-1}\left\{\frac{s + 2}{(s + 2)^2 + (\sqrt{3})^2}\right\} = e^{-2t}\cos\sqrt{3}t.$$

10. We write $\dfrac{2s + 3}{s^2 - 4} = 2\left(\dfrac{s}{s^2 - 4}\right) + \dfrac{3}{2}\left(\dfrac{2}{s^2 - 4}\right)$. By Table 9.1, number 6, $\mathcal{L}^{-1}\left\{\dfrac{s}{s^2 - 4}\right\} = \cosh 2t$; and by number 5,

$$\mathcal{L}^{-1}\left\{\frac{2}{s^2 - 4}\right\} = \sinh 2t. \quad \text{Thus } \mathcal{L}^{-1}\left\{\frac{2s + 3}{s^2 - 4}\right\} = 2\mathcal{L}^{-1}\left\{\frac{s}{s^2 - 4}\right\}$$

$$+ \frac{3}{2}\mathcal{L}^{-1}\left\{\frac{2}{s^2 - 4}\right\} = 2\mathcal{L}^{-1}\cosh 2t + \frac{3}{2}\sinh 2t.$$

12. We employ partial fractions. We have

$$\frac{2s + 6}{8s^2 - 2s - 3} = \frac{A}{4s - 3} + \frac{B}{2s + 1}$$

and hence 2s + 6 = A(2s + 1) + B(4s - 3). Letting s = $-\dfrac{1}{2}$, we find B = -1; and letting s = $\dfrac{3}{4}$. we find A = 3. Thus we have the partial fractions decomposition

$$\frac{2s + 6}{8s^2 - 2s - 3} = \frac{3}{4s - 3} - \frac{1}{2s + 1}$$

$$= \frac{3}{4\left(s - \frac{3}{4}\right)} - \frac{1}{2\left(s + \frac{1}{2}\right)}.$$

Then, using Table 9.1, number 2, we find

$$\mathcal{L}^{-1}\{F(s)\} = \frac{3e^{3t/4}}{4} - \frac{e^{-t/2}}{2}.$$

13.   We express F(s) as follows:

$$F(s) = \frac{5s}{s^2 + 4s + 4} = \frac{5(s + 2) - 10}{(s + 2)^2} = \frac{5}{s + 2} - \frac{10}{(s + 2)^2}.$$

Now, using Table 9.1, number 2 with a = -2, and number 8 with a = -2 and n = 1, we find

$$\mathcal{L}^{-1}\{F(s)\} = 5\mathcal{L}^{-1}\left\{\frac{1}{s + 2}\right\} - 10\mathcal{L}^{-1}\left\{\frac{1}{(s + 2)^2}\right\}$$

$$= 5e^{-2t} - 10te^{-2t} = 5e^{-2t}(1 - 2t).$$

14.   We first employ partial fractions. We have $\dfrac{s + 1}{s^3 + 2s}$

$$= \frac{A}{s} + \frac{Bs + c}{s^2 + 2} \text{ and hence } s + 1 = (A + B)s^2 + Cs + 2A. \quad \text{Thus}$$

$A + B = 0$, $C = 1$, $2A = 1$.   Hence $A = \dfrac{1}{2}$, $B = -\dfrac{1}{2}$, $C = 1$;

and we have the partial fractions decomposition

$$\frac{s + 1}{s^3 + 2s} = \frac{1}{2s} - \frac{s}{2(s^2 + 2)} + \frac{1}{s^2 + 2};$$

so

$$\mathcal{L}^{-1}\left\{\frac{s + 1}{s^3 + 2s}\right\} = \frac{1}{2} \mathcal{L}^{-1}\left\{\frac{1}{s}\right\} - \frac{1}{2} \mathcal{L}^{-1}\left\{\frac{s}{s^2 + 2}\right\}$$

$$+ \frac{1}{\sqrt{2}} \mathcal{L}^{-1}\left\{\frac{\sqrt{2}}{s^2 + 2}\right\}.$$

Then using Table 9.1, numbers 1, 4, and 3, respectively, we find

$$\mathcal{L}^{-1}\left\{\frac{s + 1}{s^2 + 2s}\right\} = \frac{1}{2} - \frac{1}{2} \cos\sqrt{2}t + \frac{1}{\sqrt{2}} \sin\sqrt{2}t.$$

16.   We write

$$F(s) = \frac{2s + 7}{(s + 3)^4} = \frac{2(s + 3) + 1}{(s + 3)^4} = \frac{2}{(s + 3)^3} + \frac{1}{(s + 3)^4}.$$

Then

$$\mathcal{L}^{-1}\{F(s)\} = \mathcal{L}^{-1}\left\{\frac{2!}{(s + 3)^3}\right\} + \frac{1}{6}\,\mathcal{L}^{-1}\left\{\frac{3!}{(s + 3)^4}\right\}.$$

Using Table 9.1, number 8, with a = -3 and n = 2 and 3, respectively, we find

$$\mathcal{L}^{-1}\{F(s)\} = t^2 e^{-3t} + \frac{t^3 e^{-3t}}{6} = t^2 e^{-3t}\left(1 + \frac{1}{6}\right).$$

18.   We write $\dfrac{8(s + 1)}{(2s + 1)^3} = \dfrac{8\left(s + \frac{1}{2}\right) + 4}{\left[2\left(s + \frac{1}{2}\right)\right]^3} = \dfrac{1}{\left(s + \frac{1}{2}\right)^2}$

$+ \dfrac{1}{4} \cdot \dfrac{2}{\left(s + \frac{1}{2}\right)^3}.$ By Table 9.1, number 8, with n = 1,

$$\mathcal{L}^{-1}\left\{\frac{1}{\left(s + \frac{1}{2}\right)^2}\right\} = te^{-t/2}; \text{ and by number 8, with } n = 2,$$

$$\mathcal{L}^{-1}\left\{\frac{2}{\left(s + \frac{1}{2}\right)^3}\right\} = t^2 e^{-t/2}. \text{ Thus } \mathcal{L}^{-1}\left\{\frac{8(s + 1)}{(2s - 1)^3}\right\} = te^{-t/2}$$

$\mid \dfrac{1}{4} t^2 e^{-t/2}.$

20.   Write

$$\frac{s^2 - 4s - 4}{(s^2 + 4)^2} = \frac{s^2 - 4}{(s^2 + 4)^2} - \frac{4s}{(s^2 + 4)^2}.$$

By Table 9.1, number 10, with b = 2,

$$\mathcal{L}^{-1}\left\{\frac{s^2 - 4}{(s^2 + 4)^2}\right\} = t\cos 2t.$$

By number 9, with b = 2,

$$\mathcal{L}^{-1}\left\{\frac{4s}{(s^2 + 4)^2}\right\} = t\sin 2t.$$

Thus $\mathcal{L}^{-1}\left\{\dfrac{s^2 - 4s - 4}{(s^2 + 4)^2}\right\} = t\cos 2t - t\sin 2t$

$= t(\cos 2t - \sin 2t).$

22. Write

$$\frac{5s + 17}{s^2 + 4s + 13} = \frac{5(s + 2) + \frac{7}{3}(3)}{(s + 2)^2 + (3)^2}$$

$$= 5\left[\frac{s + 2}{(s + 2)^2 + (3)^2}\right] + \frac{7}{3}\left[\frac{3}{(s + 2)^2 + (3)^2}\right].$$

By Table 9.1, number 12, with a = 2, b = 3,

$$\mathcal{L}^{-1}\left\{\frac{s + 2}{(s + 2)^2 + (3)^2}\right\} = e^{-2t}\cos 3t.$$

By number 11, with a = 2, b = 3,

$$\mathcal{L}^{-1}\left\{\frac{3}{(s + 2)^2 + (3)^2}\right\} = e^{-2t}\sin 3t.$$

Thus $\mathcal{L}^{-1}\left\{\dfrac{5s + 17}{s^2 + 4s + 13}\right\} = 5e^{-2t}\cos 3t + \dfrac{7}{3}e^{-2t}\sin 3t$

$= e^{-2t}\left[5\cos 3t + \dfrac{7}{3}\sin 3t\right].$

23. We first employ partial fractions,

$$\frac{10s + 23}{s^2 + 7s + 12} = \frac{A}{s + 3} + \frac{B}{s + 4}.$$

and hence $10s + 23 = A(s + 4) + B(s + 3)$. Letting $s = -3$, we find $A = -7$; and letting $s = -4$, we find $B = 17$. Thus

$$\mathcal{L}^{-1}\left\{\frac{10s + 23}{s^2 + 7s + 12}\right\} = -7\mathcal{L}^{-1}\left\{\frac{1}{s + 3}\right\} + 17\mathcal{L}^{-1}\left\{\frac{1}{s + 4}\right\}$$

$$= -7e^{-3t} + 17e^{-4t},$$

when we used Table 9.1, number 2, with $a = -3$ and $a = -4$, respectively.

25. We first employ partial fractions

$$\frac{1}{s^3 + 4s^2 + 3s} = \frac{1}{s(s + 1)(s + 3)} = \frac{A}{s} + \frac{B}{s + 1} + \frac{C}{s + 3}.$$

Then, clearing fractions,

$$1 = A(s + 1)(s + 3) + Bs(s + 3) + Cs(s + 1).$$

Letting $s = 0$, we find $A = 1/3$; letting $s = -1$, we find $B = -1/2$; and letting $s = -3$, we find $C = 1/6$. Thus

$$\mathcal{L}^{-1}\left\{\frac{1}{s^3 + 4s^2 + 3s}\right\} = \frac{1}{3}\mathcal{L}^{-1}\left\{\frac{1}{s}\right\} - \frac{1}{2}\mathcal{L}^{-1}\left\{\frac{1}{s + 1}\right\}$$

$$+ \frac{1}{6}\mathcal{L}^{-1}\left\{\frac{1}{s + 3}\right\}.$$

By Table 9.1, number 1, $\mathcal{L}^{-1}\left\{\frac{1}{s}\right\} = 1$. By number 2,

$$\mathcal{L}^{-1}\left\{\frac{1}{s + 1}\right\} = e^{-t} \text{ and } \mathcal{L}^{-1}\left\{\frac{1}{s + 3}\right\} = e^{-3t}. \text{ Thus}$$

$$\mathcal{L}^{-1}\left\{\frac{1}{s^3 + 4s^2 + 3s}\right\} = \frac{1}{3} - \frac{1}{2}e^{-t} + \frac{1}{6}e^{-3t}.$$

26.   We first employ partial fractions.   We have

$$\frac{s + 5}{s^4 + 3s^3 + 2s^2} = \frac{s + 5}{s^2(s + 1)(s + 2)}$$

$$= \frac{A}{s} + \frac{B}{s^2} + \frac{C}{s + 1} + \frac{D}{s + 2}.$$

and hence $s + 5 = As(s + 1)(s + 2) + B(s + 1)(s + 2)$
$+ Cs^2(s + 2) + Ds^2(s + 1)$. Letting $s = 0$, we find $2B = 5$
so $B = \frac{5}{2}$;   letting $s = -1$, we find $C = 4$; letting $s = -2$,
we find $-4D = 3$, so $D = -\frac{3}{4}$; and letting $s = 1$, we have
$6A + 6B + 3C + 2D = 6$ so $A = 1 - B - \frac{C}{2} - \frac{D}{3} = -\frac{13}{4}$.   Thus
we have the partial fractions decomposition

$$\frac{s + 5}{s^4 + 3s^3 + 2s^2} = -\frac{13}{4}\left(\frac{1}{s}\right) + \frac{5}{2}\left(\frac{1}{s^2}\right) + 4\left(\frac{1}{s + 1}\right)$$

$$- \frac{3}{4}\left(\frac{1}{s + 2}\right).$$

Then using Table 9.1, numbers 1, 7, 2, and 2 respectively,
we find

$$\mathcal{L}^{-1}\{F(s)\} = -\frac{13}{4} + \frac{5t}{2} + 4e^{-t} - \frac{3e^{-2t}}{4}.$$

27.   We first employ partial fractions,

$$\frac{7s^2 + 8s + 8}{s^3 + 4s} = \frac{7s^2 + 8s + 8}{s(s^2 + 4)} = \frac{A}{s} + \frac{Bs + C}{s^2 + 4}.$$

Clearing fractions, we have

$$7s^2 + 8s + 8 = A(s^2 + 4) + s(Bs + C)$$

$$= (A + B)s^2 + Cs + 4A.$$

Equating coefficients of like powers of s, we have

A + B = 7, C = 8, 4A = 8, and hence A = 2, B = 5, C = 8.

Thus

$$\mathcal{L}^{-1}\left\{\frac{7s^2 + 8s + 8}{s^3 + 4s}\right\} = \mathcal{L}^{-1}\left\{\frac{2}{s} + \frac{5s + 8}{s^2 + 4}\right\}$$

$$= 2\mathcal{L}^{-1}\left\{\frac{1}{s}\right\} + 5\mathcal{L}^{-1}\left\{\frac{s}{s^2 + 4}\right\} + 4\mathcal{L}^{-1}\left\{\frac{2}{s^2 + 4}\right\}.$$

By Table 9.1, number 1, $\mathcal{L}^{-1}\left\{\frac{1}{s}\right\} = 1$.   By number 4, with

$$b = 2, \; \mathcal{L}^{-1}\left\{\frac{s}{s^2 + 4}\right\} = \cos 2t. \quad \text{By number 3, with}$$

$$b = 2, \; \mathcal{L}^{-1}\left\{\frac{2}{s^2 + 4}\right\} = \sin 2t. \quad \text{Thus}$$

$$\mathcal{L}^{-1}\left\{\frac{7s^2 + 8s + 8}{s^2 + 4s}\right\} = 2 + 5\cos 2t + 4\sin 2t.$$

28. We can first employ partial fractions,

$$\frac{3s^3 + 4s^2 - 16s + 16}{s^3(s - 2)^2} = \frac{A}{s} + \frac{B}{s^2} + \frac{C}{s^3} + \frac{D}{s - 2} + \frac{E}{(s - 2)^2}.$$

Clearing fractions, expanding and multiplying, and then rearranging terms, we find

$$3s^3 + 4s^2 - 16s + 16 = As^2(s - 2)^2 + Bs(s - 2)^2$$
$$+ C(s - 2)^2 + Ds^3(s - 2) + Es^3$$
$$= (A + D)s^4 + (-4A + B - 2D + E)s^3$$
$$+ (4A - 4B + C)s^2 + (4B - 4C)s + (4C). \qquad (*)$$

Letting s = 0, one quickly obtains 16 = 4C, so C = 4.
Letting s = 2, one easily finds 24 = 8E, so E = 3.   Then
equating coefficients of s in the extreme members of (*),

one obtains 4B – 4C = –16, from which B = C – 4. But
C = 4, so B = 0. Next equating coefficients of s^2 in (*)
one has 4A – 4B + C = 4; but B = 0 and C = 4, so this
reduces to 4A + 4 = 4, so A = 0. Finally, equating
coefficients of s^4 in (*) gives A + D = 0; and since
A = 0, D = 0 also. Thus

$$\frac{3s^3 + 4s^2 - 16s + 16}{s^3(s - 2)^2} = \frac{4}{s^3} + \frac{3}{(s - 2)^2}.$$

Alternatively, we could have done this by careful
observation. For, we have

$$\frac{3s^3 + 4s^2 - 16s + 16}{s^3(s - 2)^2} = \frac{3s^3}{s^3(s - 2)^2} + \frac{4(s^2 - 4s + 4)}{s^3(s - 2)^2}$$

$$= \frac{3}{(s - 2)^2} + \frac{4}{s^3}.$$

Thus

$$\mathcal{L}^{-1}\left\{\frac{3s^3 + 4s^2 - 16s + 16}{s^3(s - 2)^2}\right\} = \mathcal{L}^{-1}\left\{\frac{4}{s^3}\right\} + \mathcal{L}^{-1}\left\{\frac{3}{(s - 2)^2}\right\}.$$

By Table 9.1, number 7, with n = 2, $\mathcal{L}^{-1}\left\{\dfrac{4}{s^3}\right\} = 2\mathcal{L}^{-1}\left\{\dfrac{2}{s^3}\right\}$

$= 2t^2$. By number 8, with n = 1 and a = 2, $\mathcal{L}^{-1}\left\{\dfrac{3}{(s - 2)^2}\right\}$

$= 3\mathcal{L}^{-1}\left\{\dfrac{1}{(s - 2)^2}\right\} = 3te^{2t}$. Thus, finally,

$$\mathcal{L}^{-1}\left\{\frac{3s^3 + 4s^2 - 16s + 16}{s^3(s - 2)^2}\right\} = 2t^2 + 3te^{2t}.$$

Section 9.2B, Page 509.

4. We write $\dfrac{1}{s(s^2 + 4s + 13)}$ as the product $F(s)G(s)$, where

$F(s) = \dfrac{1}{s}$ and $G(s) = \dfrac{1}{s^2 + 4s + 13}$. By Table 9.1, number

1, $f(t) = \mathcal{L}^{-1}\{F(s)\} = \mathcal{L}^{-1}\left\{\dfrac{1}{s}\right\} = 1$, and by number 11,

$$g(t) = \mathcal{L}^{-1}\{G(s)\} = \mathcal{L}^{-1}\left\{\dfrac{1}{s^2 + 4s + 13}\right\} = \dfrac{1}{3}\,\mathcal{L}^{-1}\left\{\dfrac{3}{(s + 2)^2 + 9}\right\}$$

$= \dfrac{1}{3}\,e^{-2t}\sin 3t$.   Then

$$\mathcal{L}^{-1}\{H(s)\} = \mathcal{L}^{-1}\{F(s)G(s)\} = f(t)*g(t)$$

$$= \int_0^t f(\tau)g(t - \tau)\,d\tau$$

$$= \int_0^t [1]\left[\dfrac{1}{3}\,e^{-2(t-\tau)}\sin 3(t - \tau)\right]d\tau$$

or

$$\mathcal{L}^{-1}\{H(s)\} = \mathcal{L}^{-1}\{G(s)F(s)\} = g(t)*f(t)$$

$$= \int_0^t g(\tau)f(t - \tau)\,d\tau$$

$$= \int_0^t \left[\dfrac{1}{3}\,e^{-2\tau}\sin 3\tau\right][1]\,d\tau.$$

We evaluate the latter of these two integral expressions.

$$\mathcal{L}^{-1}\{H(s)\} = \dfrac{1}{3}\,\dfrac{e^{-2\tau}}{13}\,(-2\sin 3\tau - 3\cos 3\tau)\Big|_0^t$$

$$= \dfrac{1}{39}\,[3 - e^{-2t}(2\sin 3t + 3\cos 3t)].$$

5. We write $H(s) = \dfrac{1}{s^2(s + 3)}$ as the product $F(s)G(s)$, where

$F(s) = \dfrac{1}{s^2}$ and $G(s) = \dfrac{1}{s + 3}$. By Table 9.1, number 7,

$f(t) = \mathcal{L}^{-1}\{F(s)\} = \mathcal{L}^{-1}\left\{\dfrac{1}{s^2}\right\} = t$, and by number 2,

$g(t) = \mathcal{L}^{-1}\{G(s)\} = \mathcal{L}^{-1}\left\{\dfrac{1}{s + 3}\right\} = e^{-3t}$. Then

$$\mathcal{L}^{-1}\{H(s)\} = \mathcal{L}^{-1}\{F(s)G(s)\} = f(t)*g(t)$$

$$= \int_0^t f(\tau)g(t - \tau)d\tau$$

$$= \int_0^t \tau e^{-3(t-\tau)}d\tau.$$

or

$$\mathcal{L}^{-1}\{H(s)\} = \mathcal{L}^{-1}\{G(s)F(s)\} = g(t)*f(t)$$

$$= \int_0^t g(\tau)f(t - \tau)d\tau$$

$$= \int_0^t e^{-3\tau}(t - \tau)d\tau.$$

We evaluate the former of these two integral expressions.
We find

$$\mathcal{L}^{-1}\{H(s)\} = e^{-3t}\left[e^{3\tau}\,\dfrac{(3\tau - 1)}{9}\right]_0^t = \dfrac{-1 + 3t + e^{-3t}}{9}.$$

6. We write $H(s) = \dfrac{1}{(s + 2)(s^2 + 1)}$ as the product $F(s)G(s)$,

where $F(s) = \dfrac{1}{s + 2}$ and $G(s) = \dfrac{1}{s^2 + 1}$. By Table 9.1, numbers 2 and 3, respectively, $f(t) = \mathcal{L}^{-1}\{F(s)\}$

$$= \mathcal{L}^{-1}\left\{\frac{1}{s + 2}\right\} = e^{-2t} \text{ and } g(t) = \mathcal{L}^{-1}\{G(s)\} = \mathcal{L}^{-1}\left\{\frac{1}{s^2 + 1}\right\}$$

$= \sin t$. Then

$$\mathcal{L}^{-1}\{H(s)\} = \mathcal{L}^{-1}\{F(s)G(s)\} = f(t)*g(t)$$

$$= \int_0^t f(\tau)g(t - \tau)d\tau$$

$$= \int_0^t e^{-2\tau}\sin(t - \tau)d\tau$$

or

$$\mathcal{L}^{-1}\{H(s)\} = \mathcal{L}^{-1}\{G(s)F(s)\} = g(t)*f(t)$$

$$= \int_0^t g(\tau)f(t - \tau)d\tau$$

$$= \int_0^t [\sin \tau]e^{-2(t-\tau)}d\tau.$$

We evaluate the latter of these two integral expressions.
We find

$$\mathcal{L}^{-1}\{H(s)\} = e^{-2t} \cdot \frac{e^{2\tau}}{5} [2 \sin \tau - \cos \tau] \Big|_0^t$$

$$= e^{-2t}\left[\frac{e^{2t}}{5} (2 \sin t - \cos t) + \frac{1}{5}\right]$$

$$= \frac{2 \sin t - \cos t + e^{-2t}}{5}.$$

Section 9.3, Page 519.

3.   Step 1.   Taking the Laplace Transform of both sides of the D.E., we have

$$\mathcal{L}\{y'\} + 4\mathcal{L}\{y\} = 6\mathcal{L}\{e^{-t}\}. \tag{1}$$

Denoting $\mathcal{L}\{y\}$ by $Y(s)$ and applying Theorem 9.3, we have

$$\mathcal{L}\{y'\} = sY(s) - y(0).$$

Applying the I.C., this becomes

$$\mathcal{L}\{y'\} = sY(s) - 5.$$

Substituting this into the left member of (1) it becomes $sY(s) - 5 + 4Y(s)$. By Table 9.1, number 2, the right member becomes $\dfrac{6}{s + 1}$. Thus (1) becomes

$$(s + 4)Y(s) - 5 = \frac{6}{s + 1}.$$

Step 2.   Solving this for $Y(s)$, we obtain

$$Y(s) = \frac{5}{s + 4} + \frac{6}{(s + 1)(s + 4)},$$

that is

$$Y(s) = \frac{5s + 11}{(s + 1)(s + 4)}.$$

Step 3.   We must now determine

$$y = \mathcal{L}^{-1}\left\{\frac{5s + 11}{(s + 1)(s + 4)}\right\}.$$

We employ partial fractions. We have

$$\frac{5s + 11}{(s + 1)(s + 4)} = \frac{A}{s + 1} + \frac{B}{s + 4}.$$

Then $5s + 11 = A(s + 4) + B(s + 1)$. Letting $s = -1$, we find $A = 2$; and letting $s = -4$, we find $B = 3$. Thus

$$\frac{5s + 11}{(s + 1)(s + 4)} = \frac{2}{s + 1} + \frac{3}{s + 4}.$$

Thus

$$y = 2\mathcal{L}^{-1}\left\{\frac{1}{s + 1}\right\} + 3\mathcal{L}^{-1}\left\{\frac{1}{s + 4}\right\}.$$

Then using Table 9.1, number 2, with $a = -1$ and $a = -4$, respectively, we obtain

$$y = 2e^{-t} + 3e^{-4t}.$$

4. **Step 2.** Taking the Laplace Transform of both sides of the D.E., we have

$$\mathcal{L}\{y'\} + 2\mathcal{L}\{y\} = 16\mathcal{L}\{t^2\}. \tag{1}$$

Denoting $\mathcal{L}\{y\}$ by $Y(s)$ and applying Theorem 9.3, we have $\mathcal{L}\{y'\} = sY(s) - y(0)$. Applying the I.C., this becomes $\mathcal{L}\{y'\} = sY(s) - 7$. Substituting this into the left member of (1) it becomes $sY(s) - 7 + 2Y(s)$. By Table 9.1, number 7, the right member becomes $\frac{32}{s^3}$. Thus (1) becomes

$$(s + 2)Y(s) - 7 = \frac{32}{s^3}.$$

Step 2. Solving this for $Y(s)$, we obtain

$$Y(s) = \frac{7}{s + 2} + \frac{32}{s^3(s + 2)} = \frac{7s^3 + 32}{s^3(s + 2)}.$$

Step 3. We must now determine

$$y = \mathcal{L}^{-1}\left[\frac{7s^3 + 32}{s^3(s + 2)}\right].$$

We employ partial fractions. We have

$$\frac{7s^3 + 32}{s^3(s + 2)} = \frac{A}{s} + \frac{B}{s^2} + \frac{C}{s^3} + \frac{D}{s + 2}.$$

Then $7s^3 + 32 = As^2(s + 2) + Bs(s + 2) + C(s + 2) + Ds^3$
$= (A + D)s^3 + (2A + B)s^2 + (2B + C)s + 2C$. Letting $s = 0$
in this, we obtain $2C = 32$, so $C = 16$; letting $s = -2$, we
have $-8D = -24$, so $D = 3$. Then equating coefficients of
s^3, we have $A + D = 7$, so $A = 4$; and equating coefficients
of s^2, we have $2A + B = 0$, so $B = -8$. Thus we find

$$\frac{7s^3 + 32}{s^3(s + 2)} = \frac{4}{s} - \frac{8}{s^2} + \frac{16}{s^3} + \frac{3}{s + 2}.$$

Then

$$y = 4\mathcal{L}^{-1}\left\{\frac{1}{s}\right\} - 8\mathcal{L}^{-1}\left\{\frac{1}{s^2}\right\} + 8\mathcal{L}^{-1}\left\{\frac{2}{s^3}\right\}$$

$$+ 3\mathcal{L}^{-1}\left\{\frac{1}{s + 2}\right\}.$$

Finally, using Table 9.1, numbers 1, 7 (with n = 1), 7
(with n = 2), and 2, respectively, we find

$$y = 4 - 8t + 8t^2 + 3e^{-2t}.$$

5.   Step 1. Taking the Laplace Transform of both sides of the
     D.E., we have

$$\mathcal{L}\{y''\} - 5\mathcal{L}\{y'\} + 6\mathcal{L}\{y\} = \mathcal{L}\{0\}. \tag{1}$$

Denoting $\mathcal{L}\{y(t)\}$ by $Y(s)$ and applying Theorem 9.4, we have the following expressions for $\mathcal{L}\{y''\}$ and $\mathcal{L}\{y'\}$:

$$\mathcal{L}\{y''\} = s^2 Y(s) - sy(0) - y'(0),$$
$$\mathcal{L}\{y'\} = sY(s) - y(0).$$

Applying the I.C.'s to these, they become

$$\mathcal{L}\{y''\} = s^2 Y(s) - s - 2, \; \mathcal{L}\{y'\} = sY(s) - 1.$$

Substituting these expressions into the left member of (1) and using $\mathcal{L}(0) = 0$, (1) becomes $[s^2 Y(s) - s - 2]$ $- 5[sY(s) - 1] + 6Y(s) = 0$ or $(s^2 - 5s + 6)Y(s) - s + 3 = 0$.

Step 2. Solving the preceding for $Y(s)$, we have

$$Y(s) = \frac{s - 3}{s^2 - 5s + 6} = \frac{1}{s - 2}.$$

Step 3. We must now determine

$$y(t) = \mathcal{L}^{-1}\left\{\frac{1}{s - 2}\right\}.$$

By Table 9.1, number 2, we immediately find

$$y = e^{2t}.$$

7. Step 1. Taking the Laplace Transform of both sides of the D.E., we have

$$\mathcal{L}\{y''\} - 6\mathcal{L}\{y'\} + 9\mathcal{L}\{y\} = \mathcal{L}(0). \tag{1}$$

Denoting $\mathcal{L}\{y\}$ by $Y(s)$ and applying Theorem 9.4, we have $\mathcal{L}\{y''\} = s^2 Y(s) - sy(0) - y'(0)$, $\mathcal{L}\{y'\} = sY(s) - y(0)$. Applying the I.C.'s, these become $\mathcal{L}\{y''\} = s^2 Y(s) - 2s - 9$, $\mathcal{L}\{y'\} = sY(s) - 2$. Substituting into the left member of (1) and using $\mathcal{L}(0) = 0$, (1) becomes

$$s^2Y(s) - 2s - 9 - 6sY(s)$$
$$+ 12 + 9Y(s) = 0$$

or

$$(s^2 - 6s + 9)Y(s) - 2s + 3 = 0.$$

Step 2.   Solving for $Y(s)$, we obtain

$$Y(s) = \frac{2s - 3}{s^2 - 6s + 9} = \frac{2s - 3}{(s - 3)^2}.$$

Step 3.   We must now determine

$$y = \mathcal{L}^{-1}\left\{\frac{2s - 3}{(s - 3)^2}\right\} = \mathcal{L}^{-1}\left\{\frac{2(s - 3) + 3}{(s - 3)^2}\right\}$$

$$= 2\mathcal{L}^{-1}\left\{\frac{1}{s - 3}\right\} + 3\mathcal{L}^{-1}\left\{\frac{1}{(s - 3)^2}\right\}.$$

By Table 9.1, numbers 2 and 8, respectively,

$$\mathcal{L}^{-1}\left\{\frac{1}{s - 3}\right\} = e^{3t} \text{ and } \mathcal{L}^{-1}\left\{\frac{1}{(s - 3)^2}\right\} = te^{3t}. \text{ Thus}$$

$$y = 2e^{3t} + 3te^{3t} = (3t + 2)e^{3t}.$$

10.   Step 1.   Taking the Laplace Transform of both sides of the D.E., we have

$$\mathcal{L}\{y''\} + 9\mathcal{L}\{y\} = 36\mathcal{L}\{e^{-3t}\}. \tag{1}$$

Denoting $\mathcal{L}\{y\}$ by $Y(s)$ and applying Theorem 9.4, we have the following expression for $\mathcal{L}\{y''\}$:

$$\mathcal{L}\{y''\} = s^2Y(s) - sy(0) - y'(0).$$

Applying the I.C.'s to this, it becomes

$$\mathcal{L}\{y''\} = s^2 Y(s) - 2s - 3.$$

Thus the left member of (1) becomes $s^2 Y(s) - 2s - 3$ + 9Y(s). By Table 9.1, number 2, the right member of

(1) becomes $\dfrac{36}{s + 3}$. Thus equation (1) reduces to

$$(s^2 + 9)Y(s) - 2s - 3 = \frac{36}{s + 3}.$$

Step 2. Solving this for Y(s), we obtain

$$Y(s) = \frac{2s + 3}{s^2 + 9} + \frac{36}{(s + 3)(s^2 + 9)},$$

that is,

$$Y(s) = \frac{2s^2 + 9s + 45}{(s + 3)(s^2 + 9)}.$$

Step 3. We must now determine

$$y = \mathcal{L}^{-1}\left[\frac{2s^2 + 9s + 45}{(s + 3)(s^2 + 9)}\right].$$

We employ partial fractions. We have

$$\frac{2s^2 + 9s + 45}{(s + 3)(s^2 + 9)} = \frac{A}{s + 3} + \frac{Bs + C}{s^2 + 9}.$$

Then $2s^2 + 9s + 45 = A(s^2 + 9) + (Bs + C)(s + 3)$
= $(A + B)s^2 + (3B + C)s + (9A + 3C)$. From this,
$A + B = 2$, $3B + C = 9$, $9A + 3C = 45$; and hence $A = 2$,
$B = 0$, $C = 9$. Thus

$$\frac{2s^2 + 9s + 45}{(s + 3)(s^2 + 9)} = \frac{2}{s + 3} + \frac{9}{s^2 + 9}.$$

Therefore,

$$y = 2\mathcal{L}^{-1}\left\{\frac{1}{s + 3}\right\} + 3\mathcal{L}^{-1}\left\{\frac{3}{s^2 + 9}\right\}.$$

Then by Table 9.1, numbers 2 and 3, respectively, we find

$$y = 2e^{-3t} + 3 \sin 3t.$$

12. Step 1. Taking the Laplace Transform of both sides of the D.E., we have

$$2\mathcal{L}\{y''\} + \mathcal{L}\{y'\} = 5\mathcal{L}\{e^{2t}\}. \tag{1}$$

Denoting $\mathcal{L}\{y\}$ by $Y(s)$ and applying Theorem 9.4, we have

$$\mathcal{L}\{y''\} = s^2 Y(s) - sy(0) - y'(0),$$
$$\mathcal{L}\{y'\} = sY(s) - y(0).$$

Applying the I.C. to these, they become

$$\mathcal{L}\{y''\} = s^2 Y(s) - 2s,$$
$$\mathcal{L}\{y'\} = sY(s) - 2.$$

Thus the left member of (1) becomes $2s^2 Y(s) - 4s + sY(s) - 2$. By Table 9.1, number 2, the right member of (1) becomes $\frac{5}{s - 2}$. Thus equation (1) reduces to

$$(2s^2 + s)Y(s) - 4s - 2 = \frac{5}{s - 2}.$$

Step 2. Solving for $Y(s)$, we obtain

$$Y(s) = \frac{4s + 2}{2s^2 + s} + \frac{5}{(s - 2)(2s^2 + s)},$$

that is,

$$Y(s) = \frac{2}{s} + \frac{5}{s(s - 2)(2s + 1)},$$

or

$$Y(s) = \frac{4s^2 - 6s + 1}{s(s - 2)(2s + 1)}.$$

Step 3.   We must now determine

$$y = \mathcal{L}^{-1}\left\{\frac{4s^2 - 6s + 1}{s(s - 2)(2s + 1)}\right\}. \tag{2}$$

We first employ partial fractions.   We have

$$\frac{4s^2 - 6s + 1}{s(s - 2)(2s + 1)} = \frac{A}{s} + \frac{B}{s - 2} + \frac{C}{2s + 1}.$$

Then $4s^2 - 6s + 1 = A(s - 2)(2s + 1) + Bs(2s + 1)$
$+ Cs(s - 2)$.   Letting $s = 0$, we find $A = -1/2$; letting
$s = 2$, we find $B = 1/2$; and letting $s = -1/2$, we find
$C = 4$.   Thus

$$\frac{4s^2 - 6s + 1}{s(s - 2)(2s + 1)} = -\frac{1}{2}\left(\frac{1}{s}\right) + \frac{1}{2}\left(\frac{1}{s - 2}\right)$$
$$+ \frac{4}{2s + 1}.$$

Thus

$$y = -\frac{1}{2}\mathcal{L}^{-1}\left(\frac{1}{s}\right) + \frac{1}{2}\mathcal{L}^{-1}\left\{\frac{1}{s - 2}\right\}$$
$$+ 2\mathcal{L}^{-1}\left\{\frac{1}{s + 1/2}\right\}.$$

Then by Table 9.1, numbers 1, 2, and 2, respectively, we
find

$$y = -\frac{1}{2} + \frac{1}{2}e^{2t} + 2e^{-t/2}.$$

14.   Step 1.   Taking the Laplace Transform of both sides of the D.E., we have

$$\mathcal{L}\{y^{iv}\} - 2\mathcal{L}\{y''\} + \mathcal{L}\{y\} = \mathcal{L}\{0\}. \tag{1}$$

Denoting $\mathcal{L}\{y\}$ by $Y(s)$ and applying Theorem 9.4 with $n = 4$ and $n = 2$, respectively, we obtain

$$\mathcal{L}\{y^{iv}\} = s^4 Y(s) - s^3 y(0) - s^2 y'(0)$$
$$- sy''(0) - y'''(0),$$

$$\mathcal{L}\{y''\} = s^2 Y(s) - sy(0) - y'(0).$$

Applying the I.C.'s to these, they become

$$\mathcal{L}\{y^{iv}\} = s^4 Y(s) - 4s^2 - 8,$$

$$\mathcal{L}\{y''\} = s^2 Y(s) - 4.$$

Substituting into the left member of (1) and using $\mathcal{L}\{0\} = 0$, (1) becomes $s^4 Y(s) - 4s^2 - 8 - 2s^2 Y(s) + 8 + Y(s) = 0$ or

$$(s^4 - 2s^2 + 1)Y(s) - 4s^2 = 0.$$

Step 2.   Solving for $Y(s)$, we find

$$Y(s) = \frac{4s^2}{s^4 - 2s^2 + 1} = \frac{4s^2}{(s^2 - 1)^2}$$

$$= \frac{4s^2}{(s - 1)^2 (s + 1)^2}.$$

Step 3.   We must now find

$$\mathcal{L}^{-1}\left\{\frac{4s^2}{(s - 1)^2 (s + 1)^2}\right\}.$$

We employ partial fractions

$$\frac{4s^2}{(s - 1)^2(s + 1)^2} = \frac{A}{s - 1} + \frac{B}{(s - 1)^2}$$

$$+ \frac{C}{s + 1} + \frac{D}{(s + 1)^2}.$$

Then $4s^2 = A(s - 1)(s + 1)^2 + B(s + 1)^2$
$+ C(s + 1)(s - 1)^2 + D(s - 1)^2 = (A + C)s^3$
$+ (A + B - C + D)s^2 + (-A + 2B - C - 2D)s$
$+ (-A + B + C + D) = 0$. First, letting $s = 1$ in this
gives $4B = 4$, so $B = 1$. Then letting $s = -1$, we get
$4D = 4$, so $D = 1$. Next, equating coefficients of s^3 gives
$A + C = 0$, so $C = -A$. Equating coefficients of s^2 then
gives $A + B - C + D = 4$. Since $B = D = 1$ and $C = -A$, this
reduces to $2A + 2 = 4$, from which $A = 1$, and then $C = -1$.
Thus we find

$$\frac{4s^2}{(s - 1)^2(s + 1)^2} = \frac{1}{s - 1} + \frac{1}{(s - 1)^2}$$

$$- \frac{1}{s + 1} + \frac{1}{(s + 1)^2}.$$

So

$$y = \mathcal{L}^{-1}\left\{\frac{1}{s - 1}\right\} + \mathcal{L}^{-1}\left\{\frac{1}{(s - 1)^2}\right\}$$

$$- \mathcal{L}^{-1}\left\{\frac{1}{s + 1}\right\} + \mathcal{L}^{-1}\left\{\frac{1}{(s + 1)^2}\right\}.$$

Using Table 9.1, numbers 2, 8, 2, and 8, respectively, we
find

$$y = e^t + te^t - e^{-t} + te^{-t}$$

$$= (t + 1)e^t + (t - 1)e^{-t}.$$

15.   Step 1.   Taking the Laplace Transform of both sides of the
D.E., we have

$$\mathcal{L}\{y''\} - \mathcal{L}\{y'\} - 2\mathcal{L}\{y\} = \mathcal{L}\{18e^{-t}\sin 3t\}. \tag{1}$$

Denoting $\mathcal{L}\{y(t)\}$ by $Y(s)$ and applying Theorem 9.4, we have
the following expressions for $\mathcal{L}\{y''\}$ and $\mathcal{L}\{y'\}$:

$$\mathcal{L}\{y''\} = s^2Y(s) - sy(0) - y'(0),$$
$$\mathcal{L}\{y'\} = sY(s) - y(0).$$

Applying the I.C.'s to these, they become

$$\mathcal{L}\{y''\} = s^2Y(s) - 3,$$
$$\mathcal{L}\{y'\} = sY(s).$$

Substituting these expressions into the left member of
(1), this left member becomes $[s^2Y(s) - 3] - sY(s) - 2Y(s)$
or $(s^2 - s - 2)Y(s) - 3$.   By Table 9.1, number 11, the
right member of (1) becomes,

$$\frac{54}{(s + 1)^2 + 9}.$$

Thus (1) reduces to

$$(s^2 - s - 2)Y(s) - 3 = \frac{54}{(s + 1)^2 + 9}.$$

Step 2.   Solving the preceding for $Y(s)$, we have

$$Y(s) = \frac{3s^2 + 6s + 84}{(s + 1)(s - 2)[(s + 1)^2 + 9]}.$$

Step 3.   We must now determine

$$y(t) = \mathcal{L}^{-1}\left\{\frac{3s^2 + 6s + 84}{(s + 1)(s - 2)(s^2 + 2s + 10)}\right\}.$$

We employ partial fractions. We have

$$\frac{3s^2 + 6s + 84}{(s + 1)(s - 2)(s^2 + 2s + 10)} = \frac{A}{s + 1}$$

$$+ \frac{B}{s - 2} + \frac{Cs + D}{s^2 + 2s + 10}$$

or

$$3s^2 + 6s + 84 = A(s - 2)(s^2 + 2s + 10)$$
$$+ B(s + 1)(s^2 + 2s + 10)$$
$$+ (Cs + D)(s + 1)(s - 2) \tag{2}$$

or

$$3s^2 + 6s + 84 = (A + B + C)s^3$$
$$+ (3B - C + D)s^2$$
$$+ (6A + 12B - 2C - D)s$$
$$+ (-20A + 10B - 2D).$$

From this, we obtain

$$\begin{cases} A + B + C = 0, \quad 3B - C + D = 3, \\ 6A + 12B - 2C - D = 6, \ -20A + 10B - 2D = 84. \end{cases} \tag{3}$$

Letting $s = -1$ in (2), we find $-27A = 81$, so $A = -3$; and letting $s = 2$ in (2), we find $54B = 108$, so $B = 2$. Using the values for A and B, we find from (3), that $C = 1$, $D = -2$. Thus we have

$$\mathcal{L}^{-1}\left\{\frac{3s^2 + 6s + 84}{(s + 1)(s - 2)(s^2 + 2s + 10)}\right\}$$

$$= -3\mathcal{L}^{-1}\left\{\frac{1}{s + 1}\right\} + 2\mathcal{L}^{-1}\left\{\frac{1}{s - 2}\right\}$$

$$+ \mathcal{L}^{-1}\left\{\frac{s - 2}{s^2 + 2s + 10}\right\}$$

$$= -3\mathcal{L}^{-1}\left\{\frac{1}{s + 1}\right\} + 2\mathcal{L}^{-1}\left\{\frac{1}{s - 2}\right\}$$

$$+ \mathcal{L}^{-1}\left\{\frac{s + 1}{(s + 1)^2 + 9}\right\}$$

$$- \mathcal{L}^{-1}\left\{\frac{3}{(s + 1)^2 + 9}\right\}.$$

Then by Table 9.1, numbers 2, 2, 12, and 11, respectively, we find

$$y = -3e^{-t} + 2e^{2t} + e^{-t}\cos 3t - e^{-t}\sin 3t.$$

19. Step 1.   Taking the Laplace Transform of both sides of the D.E., we have

$$\mathcal{L}\{y''\} + 3\mathcal{L}\{y'\} + 2\mathcal{L}\{y\} = 10\mathcal{L}\{\cos t\}. \tag{1}$$

Denoting $\mathcal{L}\{y\}$ by $Y(s)$ and applying Theorem 9.4 we have

$$\mathcal{L}\{y''\} = s^2 Y(s) - sy(0) - y'(0),$$
$$\mathcal{L}\{y'\} = sY(s) - y(0).$$

Applying the I.C.'s to these, they become

$$\mathcal{L}\{y''\} = s^2 Y(s) - 7,$$
$$\mathcal{L}\{y'\} = sY(s).$$

Substituting into the left member of (1), it becomes $s^2 Y(s) - 7 + 3sY(s) + 2Y(s)$. By Table 9.1, number 4, the right member becomes $\dfrac{10s}{s^2 + 1}$. Thus (1) becomes

$$(s^2 + 3s + 2)Y(s) - 7 = \frac{10s}{s^2 + 1}.$$

Step 2. Solving for $Y(s)$, we obtain

$$Y(s) = \frac{7}{s^2 + 3s + 2} + \frac{10s}{(s^2 + 3s + 2)(s^2 + 1)}.$$

Step 3. We must now determine

$$y = \mathcal{L}^{-1}\left[\frac{7}{s^2 + 3s + 2} + \frac{10s}{(s^2 + 3s + 2)(s^2 + 1)}\right]$$

$$= \mathcal{L}^{-1}\left[\frac{7s^2 + 10s + 7}{(s + 1)(s + 2)(s^2 + 1)}\right].$$

We first employ partial fractions. We have

$$\frac{7s^2 + 10s + 7}{(s + 1)(s + 2)(s^2 + 1)} = \frac{A}{s + 1} + \frac{B}{s + 2}$$

$$+ \frac{Cs + D}{s^2 + 1}.$$

Then $7s^2 + 10s + 7 = A(s + 2)(s^2 + 1)$
$+ B(s + 1)(s^2 + 1) + (Cs + D)(s + 1)(s + 2)$. Letting
$s = -1$, we have $2A = 4$, so $A = 2$; and letting $s = -2$,
gives $-5B = 15$, so $B = -3$. Letting $s = 0$, we obtain
$2A + B + 2D = 7$. With $A = 2$, $B = -3$, this gives $D = 3$.
Finally, letting $s = 1$, we find $6A + 4B + 6(C + D) = 24$.
With $A = 2$, $B = -3$, $D = 3$, this gives $C = 1$. Thus we have

$$\frac{7s^2 + 10s + 7}{(s + 1)(s + 2)(s^2 + 1)} = \frac{2}{s + 1} - \frac{3}{s + 2} + \frac{s + 3}{s^2 + 1}.$$

Thus

$$y = 2\mathcal{L}^{-1}\left\{\frac{1}{s + 1}\right\} - 3\mathcal{L}^{-1}\left\{\frac{1}{s + 2}\right\} + \mathcal{L}^{-1}\left\{\frac{s}{s^2 + 1}\right\}$$

$$+ 3\mathcal{L}^{-1}\left\{\frac{1}{s^2 + 1}\right\}.$$

Then using Table 9.1, numbers 2, 2, 4, and 3, we have

$$y = 2e^{-t} - 3e^{-2t} + \cos t + 3\sin t.$$

20.   **Step 1.**   Taking the Laplace Transform of both sides of the D.E., we have

$$\mathcal{L}\{y''\} + 5\mathcal{L}\{y'\} + 4\mathcal{L}\{y\} = 6\mathcal{L}\{te^{-t}\}$$
$$+ 8\mathcal{L}\{e^{-t}\}. \tag{1}$$

Denoting $\mathcal{L}\{y\}$ by $Y(s)$ and applying Theorem 9.4 we have

$$\mathcal{L}\{y''\} = s^2 Y(s) - sy(0) - y'(0),$$
$$\mathcal{L}\{y'\} = sY(s) - y(0).$$

Applying the I.C.'s, these become

$$\mathcal{L}\{y''\} = s^2 Y(s) - s - 1,$$
$$\mathcal{L}\{y'\} = sY(s) - 1.$$

Substituting into the left member of (1), it becomes $s^2 Y(s) - s - 1 + 5sY(s) - 5 + 4Y(s)$ or $(s^2 + 5s + 4)Y(s) - s - 6$. Using Table 9.1, numbers 8 and 2, the right member of (1) becomes $\dfrac{6}{(s + 1)^2} + \dfrac{8}{s + 1}$. Thus (1) becomes

$$(s^2 + 5s + 4)Y(s) - s - 6 = \frac{6}{(s + 1)^2} + \frac{8}{s + 1}.$$

Step 2. Solving for $Y(s)$, we obtain

$$Y(s) = \frac{s + 6}{s^2 + 5s + 4} + \frac{8}{(s^2 + 5s + 4)(s + 1)}$$

$$+ \frac{6}{(s^2 + 5s + 4)(s + 1)^2}$$

Adding fractions and observing that $s^2 + 5s + 4$
$= (s + 1)(s + 4)$, we find

$$Y(s) = \frac{s^3 + 8s^2 + 21s + 20}{(s + 1)^3(s + 4)}.$$

Step 3. Thus we must determine

$$y = \mathcal{L}^{-1}\left[\frac{s^3 + 8s^2 + 21s + 20}{(s + 1)^3(s + 4)}\right]. \qquad (2)$$

We employ partial fractions, writing

$$\frac{s^3 + 8s^2 + 21s + 20}{(s + 1)^3(s + 4)} = \frac{A}{s + 1} + \frac{B}{(s + 1)^2}$$

$$+ \frac{C}{(s + 1)^3} + \frac{D}{s + 4}.$$

Then $s^3 + 8s^2 + 21s + 20 = A(s + 1)^2(s + 4)$
$+ B(s + 1)(s + 4) + C(s + 4) + D(s + 1)^3$. Letting $s = -1$,
we find $3C = 6$, so $C = 2$; and letting $s = -4$, we find
$D = 0$. Then letting $s = 0$, we obtain $4A + 4B + 4C + D$
$= 20$. With $C = 2$, $D = 0$, this gives $A + B = 3$. Finally,
letting $s = 1$, we obtain $20A + 10B + 5C + 8D = 50$, which,
with $C = 2$ and $D = 0$, reduces to $2A + B = 4$. From the two
equations in A and B, we find $A = 1$, $B = 2$. Thus

$$\frac{s^3 + 8s^2 + 21s + 20}{(s + 1)^3(s + 4)} = \frac{1}{s + 1} + \frac{2}{(s + 1)^2}$$

$$+ \frac{2}{(s + 1)^3}. \tag{3}$$

Alternatively, an astute observer might have noticed that $s^3 + 8s^2 + 21s + 20 = (s^2 + 4s + 5)(s + 4)$ and reduced the fraction in (2) to

$$\frac{s^2 + 4s + 5}{(s + 1)^3}.$$

Applying partial fractions to this, one writes

$$\frac{s^2 + 4s + 5}{(s + 1)^3} = \frac{A}{s + 1} + \frac{B}{(s + 1)^2} + \frac{C}{(s + 1)^3}.$$

Then $s^2 + 4s + 5 = A(s + 1)^2 + B(s + 1) + C$
$= As^2 + (2A + B)s + (A + B + C)$. From this, $A = 1$,
$2A + B = 4$, $A + B + C = 5$, and we find $A = 1$, $B = 2$,
$C = 2$. So we again obtain (3), and this time with an easier use of partial fractions. Therefore

$$y = \mathcal{L}^{-1}\left\{\frac{1}{s + 1}\right\} + 2\mathcal{L}^{-1}\left\{\frac{1}{(s + 1)^2}\right\} + \mathcal{L}\left\{\frac{2}{(s + 1)^3}\right\}.$$

By Table 9.1, numbers 2, 8, and 8, respectively, we find

$$y = e^{-t} + 2te^{-t} + t^2e^{-t} = (t + 1)^2e^{-t}.$$

21.　Step 1.　Taking the Laplace Transform of both sides of the D.E., we have

$$\mathcal{L}\{y'''\} - 5\mathcal{L}\{y''\} + 7\mathcal{L}\{y'\} - 3\mathcal{L}\{y\}$$
$$= \mathcal{L}\{20 \sin t\}. \tag{1}$$

Denoting $\mathcal{L}\{y(t)\}$ by $Y(s)$ and applying Theorem 9.4, we have the following expressions for $\mathcal{L}\{y'''\}$, $\mathcal{L}\{y''\}$, and $\mathcal{L}\{y'\}$:

$$\begin{cases} \mathcal{L}\{y'''\} = s^3Y(s) - s^2y(0) - sy'(0) - y''(0), \\ \mathcal{L}\{y''\} = s^2Y(s) - sy(0) - y'(0), \\ \mathcal{L}\{y'\} = sY(s) - y(0). \end{cases}$$

Applying the I.C.'s to these, they become

$$\mathcal{L}\{y'''\} = s^3Y(s) + 2,$$
$$\mathcal{L}\{y''\} = s^2Y(s),$$
$$\mathcal{L}\{y'\} = sY(s).$$

Substituting these expressions into the left member of (1), this left member becomes

$$s^3Y(s) + 2 - 5s^2Y(s) + 7sY(s) - 3Y(s)$$

or

$$(s^3 - 5s^2 + 7s - 3)Y(s) + 2.$$

By Table 9.1, number 3, the right member of (1) becomes $\dfrac{20}{s^2 + 1}$. Thus (1) reduces to

$$(s^3 - 5s^2 + 7s - 3)Y(s) + 2 = \frac{20}{s^2 + 1}.$$

Step 2. Solving the preceding for $Y(s)$, we have

$$Y(s) = \frac{18 - 2s^2}{(s^3 - 5s^2 + 7s - 3)(s^2 + 1)}$$

or

$$Y(s) = \frac{-2(s^2 - 9)}{(s - 1)^2(s - 3)(s^2 + 1)}$$

or finally

$$Y(s) = \frac{-2(s + 3)}{(s - 1)^2(s^2 + 1)}.$$

Step 3.   We must now determine

$$y(t) = \mathcal{L}^{-1}\left\{\frac{-2s - 6}{(s - 1)^2(s^2 + 1)}\right\}.$$

We employ partial fractions.   We have

$$\frac{-2s - 6}{(s - 1)^2(s^2 + 1)} = \frac{A}{s - 1} + \frac{B}{(s - 1)^2} + \frac{Cs + D}{s^2 + 1}$$

or $-2s - 6 = A(s - 1)(s^2 + 1) + B(s^2 + 1)$
$+ (Cs + D)(s - 1)^2$ or $-2s - 6 = (A + C)s^3$
$+ (-A + B - 2C + D)s^2 + (A + C - 2D)s + (-A + B + D)$.
From this, we obtain

$$\begin{cases} A + C = 0, & -A + B - 2C + D = 0, \\ A + C - 2D = -2, & -A + B + D = -6. \end{cases}$$

The first and third of these give $C = -A$, $D = 1$.   Then the
second and fourth reduce to $A + B = -1$, $-A + B = -7$,
respectively, from which $A = 3$, $B = -4$.   Hence $A = 3$,
$B = -4$, $C = -3$, $D = 1$; and we have

$$\mathcal{L}^{-1}\left\{\frac{-2s - 6}{(s - 1)^2(s^2 + 1)}\right\} = 3\mathcal{L}^{-1}\left\{\frac{1}{s - 1}\right\}$$

$$- 4\mathcal{L}^{-1}\left\{\frac{1}{(s - 1)^2}\right\} - 3\mathcal{L}^{-1}\left\{\frac{s}{s^2 + 1}\right\}$$

$$+ \mathcal{L}^{-1}\left\{\frac{1}{s^2 + 1}\right\}.$$

Then by Table 9.1, numbers 2, 8, 4, and 3, respectively, we find

$$y = 3e^t - 4te^t - 3 \cos t + \sin t.$$

Section 9.4A, Page 527.

1. The given function is $5u_6(t)$. Using formula (9.75), we have $\mathcal{L}\{5u_6(t)\} = \dfrac{5e^{-6s}}{s}$.

4. We may express the values of f in the form

$$f(t) = \begin{cases} 2 - 0, & 0 < t < 5, \\ 2 - 2, & t > 5. \end{cases}$$

Thus $f(t)$ can be expressed as $2 - 2u_5(t)$. Then using formulas (9.2) and (9.75) we find $\mathcal{L}\{f(t)\}$

$$= 2\mathcal{L}\{1\} - 2\mathcal{L}\{u_5(t)\} = \frac{2}{s} - \frac{2e^{-5s}}{s} = \frac{2(1 - e^{-5s})}{s}.$$

6. We may express the values of f in the form

$$f(t) = \begin{cases} 0 + 0, & 0 < t < 3, \\ -6 + 0, & 3 < t < 9, \\ -6 + 6, & t > 9. \end{cases}$$

Thus f can be expressed as

$$-6u_3(t) + 6u_9(t).$$

Then using formula (9.75), we find

$$\mathcal{L}\{f(t)\} = -6\mathcal{L}\{u_3(t)\} + 6\mathcal{L}\{u_9(t)\}$$

$$= \frac{-6e^{-3s}}{s} + \frac{6e^{-9s}}{s}$$

$$= \frac{6}{s}(e^{-9s} - e^{-3s}).$$

7.   We may express the values of f in the form

$$f(t) = \begin{cases} 1 + 0 + 0 - 0, & 0 < t < 2, \\ 1 + 1 + 0 - 0, & 2 < t < 4, \\ 1 + 1 + 1 - 0, & 4 < t < 6, \\ 1 + 1 + 1 - 3, & t > 6. \end{cases}$$

Thus f(t) can be expressed as

$$1 + u_2(t) + u_4(t) - 3u_6(t).$$

Then using formulas (9.2) and (9.75) we find

$$\mathcal{L}\{f(t)\} = \mathcal{L}\{1\} + \mathcal{L}\{u_2(t)\} + \mathcal{L}\{u_4(t)\} - 3\mathcal{L}\{u_6(t)\}$$

$$= \frac{1}{s} + \frac{e^{-2s}}{s} + \frac{e^{-4s}}{s} - \frac{3e^{-6s}}{s}$$

$$= \frac{1 + e^{-2s} + e^{-4s} - 3e^{-6s}}{s}.$$

9.   We may express the values of f in the form

$$f(t) = \begin{cases} 2 - 0 + 0, & 0 < t < 3, \\ 2 - 2 + 0, & 3 < t < 6, \\ 2 - 2 + 2, & t > 6. \end{cases}$$

Thus f can be expressed as

$$2 - 2u_3(t) + 2u_6(t).$$

Then using formulas (9.2) and (9.75), we find

$$L\{f(t)\} = 2\mathcal{L}\{1\} - 2\mathcal{L}\{u_3(t)\} + 2\mathcal{L}\{u_6(t)\}$$

$$= \frac{2}{s} - \frac{2e^{-3s}}{s} + \frac{2e^{-6s}}{s}$$

$$= \frac{2}{s} (1 - e^{-3s} + e^{-6s}).$$

11. We must first express $f(t)$ for $t > 2$ in terms of $t - 2$.
That is, we express t as $(t - 2) + 2$ and write

$$f(t) = \begin{cases} 0, & 0 < t < 2, \\ (t - 2) + 2, & t > 2. \end{cases}$$

This is the translated function defined by

$$u_2(t)\phi(t - 2) = \begin{cases} 0, & 0 < t < 2, \\ \phi(t - 2), & t > 2, \end{cases}$$

where $\phi(t) = t + 2$. By Theorem 9.9,
$$\mathcal{L}\{u_2(t)\phi(t - 2)\} = e^{-2s}\mathcal{L}\{\phi(t)\} = e^{-2s}\mathcal{L}\{t + 2\}$$

$$= e^{-2s}\left(\frac{1}{s^2} + \frac{2}{s}\right), \text{ where we have used (9.3) and (9.2).}$$

14. We express the values of f in the form

$$f(t) = \begin{cases} 2t - 0, & 0 < t < 5, \\ 2t - 2(t - 5), & t > 5. \end{cases}$$

Thus $f(t)$ can be expressed as $f(t) = 2t - u_5(t)\phi(t - 5)$,
where $\phi(t) = 2t$. By Formula (9.3), $\mathcal{L}\{2t\} = \frac{2}{s^2}$. By

Theorem 9.9, $\mathcal{L}\{u_5(t)\phi(t - 5)\} = e^{-5s}\mathcal{L}\{\phi(t)\} = e^{-5s}\mathcal{L}\{2t\}$

$$= \frac{2e^{-5s}}{s^2}. \quad \text{Thus } \mathcal{L}\{f(t)\} = \mathcal{L}\{2t\} - \mathcal{L}\{u_5(t)\phi(t - 5)\}$$

$$= \frac{2}{s^2} - \frac{2e^{-5s}}{s^2} = \frac{2(1 - e^{-5s})}{s^2}.$$

16. We express the values of f in the form

$$f(t) = \begin{cases} 0, & 0 < t < 2, \\ e^{-2}e^{-(t-2)}, & t > 2. \end{cases}$$

This is the translated function defined by

$$u_2(t)\phi(t - 2) = \begin{cases} 0, & 0 < t < 2, \\ \phi(t - 2), & t > 2, \end{cases}$$

where $\phi(t) = e^{-2}e^{-t}$. By Theorem 9.9,

$$\mathcal{L}\{u_2(t)\phi(t - 2)\} = e^{-2s}\mathcal{L}\{\phi(t)\}$$

$$= e^{-2s}\mathcal{L}\{e^{-2}e^{-t}\} = e^{-2s}e^{-2}\left(\frac{1}{s + 1}\right)$$

$$= \frac{e^{-2(s+1)}}{s + 1},$$

where we have used (9.4).

17. We express the values of f in the form

$$f(t) = \begin{cases} 0 - 0, & 0 < t < 4, \\ (t - 4) - 0, & 4 < t < 7, \\ (t - 4) - (t - 7), & t > 7. \end{cases}$$

Thus f(t) can be expressed as

$$f(t) = u_4(t)\phi(t - 4) - u_7(t)\psi(t - 7),$$

where $\phi(t) = t$ and $\psi(t) = t$. By formula (9.3),

$$L\{\phi(t)\} = L\{\psi(t)\} = L\{t\} = \frac{1}{s^2}. \quad \text{Then by Theorem 9.9,}$$

$$L\{f(t)\} = L\{u_4(t)\phi(t - 4)\} - L\{u_7(t)\psi(t - 7)\}$$

$$= e^{-4s}L\{\phi(t)\} - e^{-7s}L\{\psi(t)\}$$

$$= \frac{e^{-4s}}{s^2} - \frac{e^{-7s}}{s^2}$$

$$= \frac{e^{-4s} - e^{-7s}}{s^2}.$$

19.   We express the values of f in the form

$$f(t) = \begin{cases} 0 - 0, & 0 < t < 2\pi, \\ \sin(t - 2\pi) - 0, & 2\pi < t < 4\pi, \\ \sin(t - 2\pi) - \sin(t - 4\pi), & t > 4\pi. \end{cases}$$

Thus $f(t)$ can be expressed as

$$u_{2\pi}(t)\sin(t - 2\pi) - u_{4\pi}(t)\sin(t - 4\pi)$$

or

$$u_{2\pi}(t)\phi(t - 2\pi) - u_{4\pi}(t)\psi(t - 4\pi),$$

where $\phi(t) = \sin t$, $\psi(t) = \sin t$. By Theorem 9.9,

$$L\{f(t)\} = L\{u_{2\pi}(t)\phi(t - 2\pi)\} - L\{u_{4\pi}(t)\psi(t - 4\pi)\}$$

$$= e^{-2\pi s}L\{\phi(t)\} - e^{-4\pi s}L\{\psi(t)\}$$

$$= e^{-2\pi s}L\{\sin t\} - e^{-4\pi s}L\{\sin t\}$$

$$= e^{-2\pi s}\left[\frac{1}{s^2 + 1}\right] - e^{-4\pi s}\left[\frac{1}{s^2 + 1}\right]$$

$$= \frac{e^{-2\pi s} - e^{-4\pi s}}{s^2 + 1},$$

where we have used formula (9.5).

22. The period P = 2. By Theorem 9.10, formula (9.80),

$$\mathcal{L}\{f(t)\} = \frac{\displaystyle\int_0^2 e^{-st}f(t)dt}{1 - e^{-2s}}$$

$$= \frac{1}{1 - e^{-2s}}\left[\int_0^1 te^{-st}dt + \int_1^2 e^{-st}dt\right]$$

$$= \frac{1}{1 - e^{-2s}}\left[\frac{e^{-st}}{s^2}(-st - 1)\Big|_0^1 - \frac{1}{s}e^{-st}\Big|_1^2\right]$$

$$= \frac{1}{1 - e^{-2s}}\left[\frac{1}{s^2} - \frac{e^{-s}}{s^2} - \frac{e^{-2s}}{s}\right].$$

24. The period P = π. By Theorem 9.10, formula (9.80),

$$\mathcal{L}\{f(t)\} = \frac{\displaystyle\int_0^\pi e^{-st}f(t)dt}{1 - e^{-\pi s}}$$

$$= \frac{1}{1 - e^{-\pi s}}\left[\int_0^\pi e^{-st}\sin t\, dt\right]$$

$$= \frac{1}{1 - e^{-\pi s}}\left[\frac{e^{-st}}{s^2 + 1}(-s \sin t - \cos t)\right]\Big|_0^\pi$$

$$= \frac{1 + e^{-\pi s}}{(1 - e^{-\pi s})(s^2 + 1)}.$$

Section 9.4B, Page 530.

2. $F(s)$ is of the form $e^{-as}\Phi(s)$, where $a = 5$ and

$\Phi(s) = \dfrac{3s + 1}{(s - 2)^2}$. By formula (9.86) $\mathcal{L}^{-1}\{e^{-as}\Phi(s)\}$

$= u_a(t)\phi(t - a)$, where u_a is defined by (9.73) and

$\phi(t) = \mathcal{L}^{-1}\{\Phi(s)\}$. [See Theorem 9.9.] We must find $\phi(t)$.

Using Table 9.1, numbers 2 and 8, respectively,

$\phi(t) = \mathcal{L}^{-1}\left\{\dfrac{3s + 1}{(s - 2)^2}\right\} = 3\mathcal{L}^{-1}\left\{\dfrac{1}{(s - 2)}\right\} + 7\mathcal{L}^{-1}\left\{\dfrac{1}{(s - 2)^2}\right\}$

$= 3e^{2t} + 7te^{2t} = e^{2t}(7t + 3)$. Thus $\phi(t - 5)$

$= e^{2(t-5)}[7(t - 5) + 3] = e^{2(t-5)}(7t - 32)$. Then by

formula (9.86) [or Table 9.1, number 16],

$$\mathcal{L}^{-1}\left\{\dfrac{3s + 1}{(s - 2)^2}\, e^{-5s}\right\} = u_5(t)\phi(t - 5)$$

$$= \begin{cases} 0, & 0 < t < 5, \\ e^{2(t-5)}(7t - 32), & t > 5. \end{cases}$$

4. $F(s)$ is of the form $e^{-as}\Phi(s)$, where $a = 4$ and

$\Phi(s) = \dfrac{12}{s^2 + s - 2}$. By formula (9.86) $\mathcal{L}^{-1}\{e^{-as}\Phi(s)\}$

$= u_a(t)\phi(t - a)$, where u_a is defined by (9.73) and

$\phi(t) = \mathcal{L}^{-1}\{\Phi(s)\}$ [see Theorem 9.9]. We must find $\phi(t)$.
We first employ partial fractions. We write

$\dfrac{12}{s^2 + s - 12} = \dfrac{A}{s - 1} + \dfrac{B}{s + 2}$. Clearing fractions, we at

once find $A = 4$, $B = -4$. Thus, using Table 9.1, number 2, twice, we have $\phi(t) = \mathcal{L}^{-1}\left\{\dfrac{12}{s^2 + s - 12}\right\} = 4\mathcal{L}^{-1}\left\{\dfrac{1}{s - 1}\right\}$

$- 4\mathcal{L}^{-1}\left\{\dfrac{1}{s + 2}\right\} = 4e^t - 4e^{-2t}$. Thus $\phi(t - 4) = 4e^{t-4}$

$- 4e^{-2(t-4)}$. Then by formula (9.86) [or Table 9.1,

number 16],

$$\mathcal{L}^{-1}\left\{\dfrac{12}{s^2 + s - 2}\, e^{-4s}\right\} = u_4(t)\phi(t - 4)$$

$$= \begin{cases} 0, & 0 < t < 4, \\ 4e^{t-4} - 4e^{-2(t-4)}, & t > 4. \end{cases}$$

5. $F(s)$ is of the form $e^{-as}\Phi(s)$, where $a = \pi$ and

$\Phi(s) = \dfrac{5s + 6}{s^2 + 9}$. By formula (9.86), $\mathcal{L}^{-1}\{e^{-as}\Phi(s)\}$

$= u_a(t)\phi(t - a)$, where u_a is defined by (9.73) and

$\phi(t) = \mathcal{L}^{-1}\{\Phi(s)\}$. [See Theorem 9.9.] We must find $\phi(t)$.
Using Table 9.1, numbers 4 and 3, respectively, we find

$\phi(t) = \mathcal{L}^{-1}\{\Phi(s)\} = \mathcal{L}^{-1}\left\{\dfrac{5s + 6}{s^2 + 9}\right\} = 5\mathcal{L}^{-1}\left\{\dfrac{s}{s^2 + 9}\right\}$

$+ 2\mathcal{L}^{-1}\left\{\dfrac{3}{s^2 + 9}\right\} = 5\cos 3t + 2\sin 3t$. Thus

$\phi(t - a) = \phi(t - \pi) = 5\cos 3(t - \pi) + 2\sin 3(t - \pi)$

$= -5\cos 3t - 2\sin 3t$. Then by formula (9.86) [or Table
9.1, number 16], $\mathcal{L}^{-1}\{e^{-\pi s}\Phi(s)\} = u_\pi(t)\phi(t - \pi)$, that is,

$$\mathcal{L}^{-1}\{F(s)\} = u_\pi(t)(-5\cos 3t - 2\sin 3t)$$

$$= \begin{cases} 0, & 0 < t < \pi, \\ -5\cos 3t - 2\sin 3t, & t > \pi. \end{cases}$$

7. F(s) is of the form $e^{-as}\Phi(s)$, where a = $\pi/2$ and

$\Phi(s) = \dfrac{s + 8}{s^2 + 4s + 13}$. By formula (9.86),

$\mathcal{L}^{-1}\{e^{-as}\Phi(s)\} = u_a(t)\phi(t - a)$, where u_a is defined by

(9.73) and $\phi(t) = \mathcal{L}^{-1}\{\Phi(s)\}$ [see theorem 9.9]. We must find $\phi(t)$. We write

$$\frac{s + 8}{s^2 + 4s + 13} = \frac{s + 2}{(s + 2)^2 + 9} + \frac{6}{(s + 2)^2 + 9}$$

and proceed to use Table 9.1, numbers 12 and 11. By these respective entries, we find

$$\phi(t) = \mathcal{L}^{-1}\left\{\frac{s + 8}{s^2 + 4s + 13}\right\}$$

$$- \mathcal{L}^{-1}\left\{\frac{s + 2}{(s + 2)^2 + 3^2}\right\} + 2\mathcal{L}^{-1}\left\{\frac{3}{(s + 2)^2 + 3^2}\right\}$$

$$= e^{-2t}\cos 3t + 2e^{-2t}\sin 3t$$

$$= e^{-2t}(\cos 3t + 2\sin 3t).$$

Thus $\phi(t - \pi/2) = e^{-2(t-\pi/2)}[\cos 3(t - \pi/2)$
$+ 2\sin 3(t - \pi/2)] = e^{-2t+\pi}(2\cos 3t - \sin 3t)$. Then by formula (9.86) [or Table 9.1, number 16],

$$\mathcal{L}^{-1}\left\{\frac{s + 8}{s^2 + 4s + 13}e^{-(\pi s)/2}\right\} = u_{(\pi/2)}(t)\phi(t - \pi/2)$$

$$= \begin{cases} 0, & 0 < t < \pi/2, \\ e^{-2t+\pi}(2\cos 3t - \sin 3t), & t > \pi/2. \end{cases}$$

8.   F(s) is of the form $e^{-as}\Phi(s)$, where a = 3 and

$\Phi(s) = \dfrac{2s + 9}{s^2 + 4s + 13}$.   By formula (9.86), $\mathcal{L}^{-1}\{e^{-as}\Phi(s)\}$

$= u_a(t)\phi(t - a)$, where u_a is defined by (9.73) and

$\phi(t) = \mathcal{L}^{-1}\{\Phi(s)\}$.   We must find $\phi(t)$.   We have

$$\dfrac{2s + 9}{s^2 + 4s + 13} = \dfrac{2(s + 2)}{(s + 2)^2 + (3)^2} + \dfrac{5}{(s + 2)^2 + (3)^2}.$$

Using number 12 and 11, respectively, we find

$$\phi(t) = \mathcal{L}^{-1}\{\Phi(s)\} = \mathcal{L}^{-1}\left\{\dfrac{2s + 9}{s^2 + 4s + 13}\right\}$$

$$= 2\mathcal{L}^{-1}\left\{\dfrac{s + 2}{(s + 2)^2 + (3)^2}\right\} + \dfrac{5}{3}\mathcal{L}^{-1}\left\{\dfrac{3}{(s + 2)^2 + (3)^2}\right\}$$

$$= 2e^{-2t}\cos 3t + \left(\dfrac{5}{3}\right)e^{-2t}\sin 3t.$$

Thus

$\phi(t - a) = \phi(t - 3)$

$$= e^{-2(t-3)}\left[2\cos 3(t - 3) + \left(\dfrac{5}{3}\right)\sin 3(t - 3)\right]$$

Then by formula (9.86) [or Table 9.1, number 16],

$\mathcal{L}^{-1}\{e^{-3s}\Phi(s)\} = u_3(t)\phi(t - 3)$, that is,

$\mathcal{L}^{-1}\{F(s)\} = u_3(t)e^{-2(t-3)}[2\cos 3(t - 3)] + \left(\dfrac{5}{3}\right)\sin 3(t - 3)$

$$= \begin{cases} 0, & 0 < t < 3, \\ e^{-2(t-3)}\left[2\cos 3(t - 3) + \left(\dfrac{5}{3}\right)\sin 3(t - 3)\right], & t > 3. \end{cases}$$

10. We first write

$$\mathcal{L}^{-1}\left[\frac{e^{-3s} - e^{-8s}}{s^3}\right] = \mathcal{L}^{-1}\left[\frac{e^{-3s}}{s^3}\right] - \mathcal{L}^{-1}\left[\frac{e^{-8s}}{s^3}\right].$$

Each of the two inverse transforms on the right is of the form $\mathcal{L}^{-1}\{e^{-as}\Phi(s)\}$. By formula (9.86),
$\mathcal{L}^{-1}\{e^{-as}\Phi(s)\} = u_a(t)\phi(t - a)$, where u_a is defined by
(9.73) and $\phi(t) = \mathcal{L}^{-1}\{\Phi(s)\}$.

First consider $\mathcal{L}^{-1}\left\{\frac{e^{-3s}}{s^3}\right\}$. Here $a = 3$, $\Phi(s) = \frac{1}{s^3}$;

and from Table 9.1, number 7, $\phi(t) = \mathcal{L}^{-1}\{\Phi(s)\}$

$= \mathcal{L}^{-1}\left[\frac{1}{s^3}\right] = \frac{t^2}{2}$. Thus $\phi(t - 3) = \frac{(t - 3)^2}{2}$. By formula

(9.86),

$$\mathcal{L}^{-1}\left\{\frac{e^{-3s}}{s^3}\right\} = u_3(t)\phi(t - 3) = \begin{cases} 0, & 0 < t < 3, \\ \dfrac{(t - 3)^2}{2}, & t > 3. \end{cases}$$

Now consider $\mathcal{L}^{-1}\left\{\frac{e^{-8s}}{s^3}\right\}$. This is similar to the

preceding, with $a = 8$, $\Phi(s) = \frac{1}{s^3}$, and $\phi(t) = \frac{t^2}{2}$. Thus

$\phi(t - 8) = \frac{(t - 8)^2}{2}$, and

$$\mathcal{L}^{-1}\left\{\frac{e^{-8s}}{s^3}\right\} = u_8(t)\phi(t - 8) = \begin{cases} 0, & 0 < t < 8, \\ \dfrac{(t - 8)^2}{2}, & t > 8. \end{cases}$$

Thus

$$\mathcal{L}^{-1}\left\{\frac{e^{-3s} - e^{-8s}}{s^3}\right\} = u_3(t)\phi(t - 3) - u_8(t)\phi(t - 8)$$

$$= \begin{cases} 0 - 0, & 0 < t < 3, \\ \dfrac{(t - 3)^2}{2} - 0, & 3 < t < 8, \\ \dfrac{(t - 3)^2}{2} - \dfrac{(t - 8)^2}{2}, & t > 8, \end{cases}$$

$$= \begin{cases} 0, & 0 < t < 3, \\ \dfrac{1}{2}(t^2 - 6t + 9), & 3 < t < 8, \\ 5t - \dfrac{55}{2}, & t > 8. \end{cases}$$

13. We write

$$F(s) = \frac{2}{s^2 - 2s + 5} + \frac{2e^{-(\pi s)/2}}{s^2 - 2s + 5}, \qquad (1)$$

and first determine $\mathcal{L}^{-1}\left\{\dfrac{2}{s^2 - 2s + 5}\right\}$. By Table 9.1,

number 11, we have

$$\mathcal{L}^{-1}\left\{\frac{2}{s^2 - 2s + 5}\right\} = \mathcal{L}^{-1}\left\{\frac{2}{(s - 1)^2 + 4}\right\}$$

$$= e^t\sin 2t. \qquad (2)$$

Now letting $\Phi(s) = \dfrac{2}{s^2 - 2s + 5}$ and $\phi(t) = e^t\sin 2t$, we

see that (2) is $\mathcal{L}^{-1}\{\Phi(s)\} = \phi(t)$. Then by formula (9.86) [or Table 9.1, number 16],

$$\mathcal{L}^{-1}\left\{\frac{2e^{-(\pi s)/2}}{s^2 - 2s + 5}\right\} = \mathcal{L}^{-1}\{e^{-(\pi s)/2}\Phi(s)\}$$

$$= u_{\pi/2}(t)\phi\left(t - \frac{\pi}{2}\right) = \begin{cases} 0, & 0 < t < \frac{\pi}{2}, \\ \phi\left(t - \frac{\pi}{2}\right), & t > \frac{\pi}{2}, \end{cases}$$

$$= \begin{cases} 0, & 0 < t < \frac{\pi}{2}, \\ e^{t-\pi/2}\sin 2\left(t - \frac{\pi}{2}\right), & t > \frac{\pi}{2}. \end{cases}$$

That is,

$$\mathcal{L}^{-1}\left\{\frac{2e^{-(\pi s)/2}}{s^2 - 2s + 5}\right\} = \begin{cases} 0, & 0 < t < \frac{\pi}{2}, \\ -e^{t-\pi/2}\sin 2t, & t > \frac{\pi}{2}. \end{cases} \qquad (3)$$

Then, using (1), (2), and (3), we find

$$\mathcal{L}^{-1}\{F(s)\} = \begin{cases} e^t\sin 2t - 0, & 0 < t < \frac{\pi}{2}, \\ e^t\sin 2t - e^{t-\pi/2}\sin 2t. & t > \frac{\pi}{2}, \end{cases}$$

$$= \begin{cases} e^t\sin 2t, & 0 < t < \frac{\pi}{2}, \\ (1 - e^{-\pi/2})e^t\sin 2t, & t > \frac{\pi}{2}. \end{cases}$$

Section 9.4C, Page 533.

2. Step 1. Taking the Laplace Transform of both sides of the
   D.E., we have

$$3\mathcal{L}\{y'\} - 5\mathcal{L}\{y\} = \mathcal{L}\{h(t)\}. \qquad (1)$$

Denoting $\mathcal{L}\{y\}$ by $Y(s)$, applying Theorem 9.3, and then applying the I.C., we find that

$$\mathcal{L}\{y'\} = sY(s) - 4.$$

Substituting this into the left member of (1), it becomes $3[sY(s) - 4] - 5Y(s)$ or $(3s - 5)Y(s) - 12$. By the definition of the Laplace Transform

$$\mathcal{L}\{h(t)\} = \int_0^\infty e^{-st} h(t)dt = \int_6^\infty 10e^{-st}dt$$

$$= \lim_{R \to \infty}\left[-\frac{10}{s} e^{-st}\Big|_6^R\right] = \frac{10e^{-6s}}{s}.$$

Alternatively, we note that $h(t) = 10u_6(t)$; and then using Table 9.1, number 15, we have

$$\mathcal{L}\{h(t)\} = 10\mathcal{L}\{u_6(t)\} = \frac{10e^{-6s}}{s}.$$

Thus, (1) reduces to

$$(3s - 5)Y(s) - 12 = \frac{10e^{-6s}}{s}.$$

Step 2. Solving for $Y(s)$, we obtain

$$Y(s) = \frac{12}{3s - 5} + \frac{10e^{-6s}}{s(3s - 5)}.$$

Step 3. We must now determine

$$y = \mathcal{L}^{-1}\left\{\frac{12}{3s - 5} + \frac{10e^{-6s}}{s(3s - 5)}\right\}. \tag{2}$$

We first determine

$$\mathcal{L}^{-1}\left\{\frac{12}{3s - 5}\right\} = 4\mathcal{L}^{-1}\left\{\frac{1}{s - 5/3}\right\} = 4e^{(5/3)t}, \tag{3}$$

where we have used Table 9.1, number 2. Now consider

$\mathcal{L}^{-1}\left\{\dfrac{10e^{-6s}}{s(3s-5)}\right\}$. We first employ partial fractions to

$\dfrac{1}{s(s-5/3)}$, and write $\dfrac{1}{s(s-5/3)} = \dfrac{A}{s} + \dfrac{B}{s-5/3}$. Clearing

fractions, we at once find $A = -3/5$, $B = 3/5$. Then

$\dfrac{10}{s(3s-5)} = \dfrac{10}{3}\left[\dfrac{-3/5}{s} + \dfrac{3/5}{s-5/3}\right] = -\dfrac{2}{s} + \dfrac{2}{s-5/3}$. Then by

Table 9.1, numbers 1 and 2, respectively,

$$\mathcal{L}^{-1}\left\{\dfrac{10}{s(3s-5)}\right\} = -2\mathcal{L}^{-1}\left\{\dfrac{1}{s}\right\} + 2\mathcal{L}^{-1}\left\{\dfrac{1}{s-5/3}\right\}$$

$$= -2 + 2e^{(5/3)t}.$$

Now letting $F(s) = \dfrac{10}{s(3s-5)}$ and $f(t) = -2 + 2e^{(5/3)t}$ we

thus have $\mathcal{L}^{-1}\{F(s)\} = f(t)$. Then by Theorem 9.9,

$$\mathcal{L}^{-1}\{F(s)e^{-6s}\} = \mathcal{L}^{-1}\left\{\dfrac{10e^{-6s}}{s(3s-5)}\right\}$$

$$= u_6(t)f(t-6) = \begin{cases} 0, & 0 < t < 6, \\ f(t-6), & t > 6, \end{cases}$$

$$= \begin{cases} 0, & 0 < t < 6, \\ -2 + 2e^{(5/3)(t-6)}, & t > 6. \end{cases} \tag{4}$$

Then from (2), (3), and (4), we obtain

$$y = \begin{cases} 4e^{(5/3)t}, & 0 < t < 6, \\ 4e^{(5/3)t} - 2 + 2e^{(5/3)(t-6)}, & t > 6, \end{cases}$$

$$= \begin{cases} 4e^{(5/3)t}, & 0 < t < 6, \\ -2 + (4 + 2e^{-10})e^{(5/3)t}, & t > 6. \end{cases}$$

3.   Step 1.   Taking the Laplace Transform of both sides of the
D.E., we have

$$\mathcal{L}\{y''\} - 3\mathcal{L}\{y'\} + 2\mathcal{L}\{y\} = \mathcal{L}\{h(t)\}. \qquad (1)$$

Denoting $\mathcal{L}\{y(t)\}$ by $Y(s)$, applying Theorem 9.4 as in the
previous exercises, and then applying the I.C.'s we find
that

$$\mathcal{L}\{y''\} = s^2 Y(s) \quad \text{and} \quad \mathcal{L}\{y'\} = sY(s).$$

Substituting these expressions into the left member of
(1), this left member becomes

$$s^2 Y(s) - 3sY(s) + 2Y(s)$$

or $(s^2 - 3s + 2)Y(s)$.   By the definition of the Laplace
Transform,

$$\mathcal{L}\{h(t)\} = \int_0^\infty e^{-st} h(t)dt = \int_0^4 2e^{-st} dt$$

$$= \frac{2}{s} (1 - e^{-4s}).$$

Alternatively, writing

$$h(t) = \begin{cases} 2 - 0, & 0 < t < 4 \\ 2 - 2, & t > 4, \end{cases}$$

we have $h(t) = 2 - 2u_4(t)$.   Then using Table 9.1, numbers
1 and 15, respectively, we have

$$\mathcal{L}\{h(t)\} = \frac{2}{s} - \frac{2e^{-4s}}{s} = \frac{2}{s} (1 - e^{-4s}).$$

Thus (1) reduces to $(s^2 - 3s + 2)Y(s) = \frac{2}{s} (1 - e^{-4s})$.

Step 2. Solving the preceding for Y(s), we obtain

$$Y(s) = \frac{2(1 - e^{-4s})}{s(s - 1)(s - 2)}.$$

Step 3. We must now determine

$$y(t) = \mathcal{L}^{-1}\left\{\frac{2}{s(s - 1)(s - 2)}\right\}$$

$$- \mathcal{L}^{-1}\left[\frac{2e^{-4s}}{s(s - 1)(s - 2)}\right]. \tag{2}$$

We first apply partial fractions to the first term on the right of (2). We have

$$\frac{2}{s(s - 1)(s - 2)} = \frac{A}{s} + \frac{B}{s - 1} + \frac{C}{s - 2},$$

and from this we readily find $A = 1$, $B = -2$, $C = 1$. Thus

$$\mathcal{L}^{-1}\left\{\frac{2}{s(s - 1)(s - 2)}\right\} = \mathcal{L}^{-1}\left\{\frac{1}{s}\right\} - 2\mathcal{L}^{-1}\left\{\frac{1}{s - 1}\right\}$$

$$+ \mathcal{L}^{-1}\left\{\frac{1}{s - 2}\right\}.$$

By Table 9.1, numbers 1, 2, and 2, respectively, we find

$$\mathcal{L}^{-1}\left\{\frac{2}{s(s - 1)(s - 2)}\right\} = 1 - 2e^t + e^{2t}.$$

Letting $F(s) = \dfrac{2}{s(s - 1)(s - 2)}$ and $f(t) = 1 - 2e^t + e^{2t}$,

we thus have $\mathcal{L}^{-1}\{F(s)\} = f(t)$. We now consider

$$\mathcal{L}^{-1}\left\{\frac{2e^{-4s}}{s(s - 1)(s - 2)}\right\} = \mathcal{L}^{-1}\{F(s)e^{-4s}\}. \quad \text{By Theorem 9.9,}$$

$$\mathcal{L}^{-1}\{F(s)e^{-4s}\} = u_4(t)f(t - 4),$$

$$= \begin{cases} 0, & 0 < t < 4, \\ f(t - 4), & t > 4, \end{cases}$$

$$= \begin{cases} 0, & 0 < t < 4, \\ 1 - 2e^{t-4} + 2e^{2(t-4)}, & t > 4. \end{cases}$$

Thus from (2), we find $y = f(t) - u_4(t)f(t - 4)$

$$= \begin{cases} 1 - 2e^t + e^{2t}, & 0 < t < 4, \\ 1 - 2e^t + e^{2t} - [1 - 2e^{t-4} + e^{2(t-4)}], & t > 4. \end{cases}$$

Hence $y = \begin{cases} 1 - 2e^t + e^{2t}, & 0 < t < 4, \\ 2(e^{-4} - 1)e^t + (1 - e^{-8})e^{2t}, & t > 4. \end{cases}$

6. Step 1. Taking the Laplace Transform of both sides of the D.E., we have

$$\mathcal{L}\{y''\} + 6\mathcal{L}\{y'\} + 8\mathcal{L}\{y\} = \mathcal{L}\{h(t)\}. \tag{1}$$

Denoting $\mathcal{L}\{y(t)\}$ by $Y(s)$, applying Theorem 9.4 as in previous exercises, and then applying the I.C.'s, we find that

$$\mathcal{L}\{y''\} = s^2Y(s) - s + 1, \ \mathcal{L}\{y'\} = sY(s) - 1.$$

Substituting these expressions into the left member of (1), this left member becomes $[s^2Y(s) - s + 1]$ $+ 6[sY(s) - 1] + 8Y(s)$ or $(s^2 + 6s + 8)Y(s) - s - 5$. By the definition of the Laplace Transform,

$$\mathcal{L}\{h(t)\} = \int_0^\infty e^{-st}h(t)dt$$

$$= \int_0^{2\pi} 3e^{-st}dt = \frac{3}{s}(1 - e^{-2\pi s}).$$

Alternatively, writing

$$h(t) = \begin{cases} 3 - 0, & 0 < t < 2\pi, \\ 3 - 3, & t > 2\pi, \end{cases}$$

we have $h(t) = 3 - 3u_{2\pi}(t)$. Then using Table 9.1, numbers 1 and 15, respectively, we have

$$\mathcal{L}\{h(t)\} = \frac{3}{s} - \frac{3e^{-2\pi s}}{s} = \frac{3}{s}(1 - e^{-2\pi s}).$$

Thus (1) reduces to $(s^2 + 6s + 8)Y(s) - s - 5$
$= \frac{3}{s}(1 - e^{-2\pi s})$.

Step 2.   Solving the preceding for $Y(s)$, we obtain

$$Y(s) = \frac{s + 5}{s^2 + 6s + 8} + \frac{3(1 - e^{-2\pi s})}{s(s^2 + 6s + 8)}.$$

Step 3.   We must now determine

$$y = \mathcal{L}^{-1}\left\{\frac{s + 5}{(s + 2)(s + 4)}\right\}$$

$$+ \mathcal{L}^{-1}\left[\frac{3(1 - e^{-2\pi s})}{s(s + 2)(s + 4)}\right]. \qquad (2)$$

We first apply partial fractions to the first term on the right of (2). We have

$$\frac{s + 5}{(s + 2)(s + 4)} = \frac{A}{s + 2} + \frac{B}{s + 4},$$

and from this we readily find $A = 3/2$, $B = -1/2$.   Thus

$$\mathcal{L}^{-1}\left\{\frac{s + 5}{(s + 2)(s + 4)}\right\} = \frac{3}{2}\mathcal{L}^{-1}\left\{\frac{1}{s + 2}\right\}$$
$$- \frac{1}{2}\mathcal{L}^{-1}\left\{\frac{1}{s + 4}\right\}.$$

By Table 9.1, number 2, we find

$$\mathcal{L}^{-1}\left\{\frac{s + 5}{(s + 2)(s + 4)}\right\} = \frac{3}{2}e^{-2t} - \frac{1}{2}e^{-4t}. \tag{3}$$

We now consider $\mathcal{L}^{-1}\left\{\dfrac{3(1 - e^{-2\pi s})}{s(s + 2)(s + 4)}\right\}$.   We first apply

partial fractions to $\dfrac{3}{s(s + 2)(s + 4)}$.   We have

$$\frac{3}{s(s + 2)(s + 4)} = \frac{A}{s} + \frac{B}{s + 2} + \frac{C}{s + 4},$$

and from this we readily find $A = \dfrac{3}{8}$, $B = -\dfrac{3}{4}$, $C = \dfrac{3}{8}$.   Thus

$$\mathcal{L}^{-1}\left\{\frac{3}{s(s + 2)(s + 4)}\right\} = \frac{3}{8}\mathcal{L}^{-1}\left\{\frac{1}{s}\right\} - \frac{3}{4}\mathcal{L}^{-1}\left\{\frac{1}{s + 2}\right\}$$
$$+ \frac{3}{8}\mathcal{L}^{-1}\left\{\frac{1}{s + 4}\right\}.$$

By Table 9.1, numbers 1, 2, and 2, respectively, we find

$$\mathcal{L}^{-1}\left\{\frac{3}{s(s + 2)(s + 4)}\right\} = \frac{3}{8} - \frac{3}{4}e^{-2t} + \frac{3}{8}e^{-4t}.$$

Letting $F(s) = \dfrac{3}{s(s + 2)(s + 4)}$ and $f(t) = \dfrac{3}{8} - \dfrac{3}{4}e^{-2t}$
$+ \dfrac{3}{8}e^{-4t}$, we thus have $\mathcal{L}^{-1}\{F(s)\} = f(t)$.   We now consider

$$\mathcal{L}^{-1}\left\{\frac{3e^{-2\pi s}}{s(s + 2)(s + 4)}\right\} = \mathcal{L}^{-1}\{F(s)e^{-2\pi s}\}.   \text{ By Theorem 9.9,}$$

$$\mathcal{L}^{-1}\{F(s)e^{-2\pi s}\} = u_{2\pi}(t)f(t - 2\pi)$$

$$= \begin{cases} 0, & 0 < t < 2\pi \\ f(t - 2\pi), & t > 2\pi, \end{cases}$$

$$= \begin{cases} 0, & 0 < t < 2\pi, \\ \dfrac{3}{8} - \dfrac{3}{4}e^{-2(t-2\pi)} + \dfrac{3}{8}e^{-4(t-2\pi)}, & t > 2\pi. \end{cases}$$

Thus

$$\mathcal{L}^{-1}\left\{\frac{3(1 - e^{-2\pi s})}{s(s + 2)(s + 4)}\right\}$$

$$= \begin{cases} \dfrac{3}{8} - \dfrac{3}{4}e^{-2t} + \dfrac{3}{8}e^{-4t}, & 0 < t < 2\pi, \\ \dfrac{3}{8} - \dfrac{3}{4}e^{-2t} + \dfrac{3}{8}e^{-4t} - \dfrac{3}{8} + \dfrac{3}{4}e^{-2(t-2\pi)} - \dfrac{3}{8}e^{-4(t-2\pi)}, \\ \qquad\qquad\qquad\qquad\qquad\qquad\qquad\qquad t > 2\pi, \end{cases}$$

$$= \begin{cases} \dfrac{3}{8} - \dfrac{3}{4}e^{-2t} + \dfrac{3}{8}e^{-4t}, & 0 < t < 2\pi, \\ \dfrac{3}{4}(e^{4\pi} - 1)e^{-2t} + \dfrac{3}{8}(1 - e^{8\pi})e^{-4t}, & t > 2\pi. \end{cases} \qquad (4)$$

Hence, using (3) and (4), (2) becomes

$$y = \begin{cases} \dfrac{3}{8} + \dfrac{3}{4}e^{-2t} - \dfrac{1}{8}e^{-4t}, & 0 < t < 2\pi, \\ \dfrac{3}{4}(e^{4\pi} + 1)e^{-2t} - \dfrac{1}{8}(1 + 3e^{8\pi})e^{-4t}, & t > 2\pi. \end{cases}$$

8. Step 1. Taking the Laplace Transform of both sides of the
D.E., we have

$$\mathcal{L}\{y''\} + \mathcal{L}\{y\} = \mathcal{L}\{h(t)\}. \qquad (1)$$

Denoting $\mathcal{L}\{y(t)\}$ by $Y(s)$, applying Theorem 9.4 as in
previous exercises, and then applying the I.C.'s, we find
that

$$\mathcal{L}\{y''\} = s^2Y(s) - 2s - 3.$$

Substituting this expression into the left member of (1), this left member becomes $(s^2 + 1)Y(s) - 2s - 3$. We now find $\mathcal{L}\{h(t)\}$. We have

$$h(t) = \begin{cases} t, & 0 < t < \pi, \\ \pi, & t > \pi, \end{cases}$$

$$= \begin{cases} t - 0, & 0 < t < \pi, \\ t - (t - \pi), & t > \pi, \end{cases}$$

and hence $h(t) = t - u_\pi(t)f(t - \pi)$, where $f(t) = t$. Thus, using Theorem 9.9 and Table 9.1, number 7, we find

$$\mathcal{L}\{h(t)\} = \mathcal{L}\{t\} - \mathcal{L}\{u_\pi(t)f(t - \pi)\} = \frac{1}{s^2} - e^{-\pi s}\left(\frac{1}{s^2}\right)$$

$$= \frac{1 - e^{-\pi s}}{s^2}. \quad \text{Thus (1) reduces to } (s^2 + 1)Y(s) - 2s - 3$$

$$= \frac{1 - e^{-\pi s}}{s^2}.$$

Step 2. Solving the preceding for $Y(s)$, we obtain

$$Y(s) = \frac{2s + 3}{s^2 + 1} + \frac{1 - e^{-\pi s}}{s^2(s^2 + 1)}.$$

Step 3. We must now determine

$$y = \mathcal{L}^{-1}\left\{\frac{2s + 3}{s^2 + 1}\right\} + \mathcal{L}^{-1}\left\{\frac{1}{s^2(s^2 + 1)}\right\}$$

$$- \mathcal{L}^{-1}\left\{\frac{e^{-\pi s}}{s^2(s^2 + 1)}\right\}. \tag{2}$$

Using Table 9.1, numbers 4 and 3, we at once find

$$\mathcal{L}^{-1}\left\{\frac{2s + 3}{s^2 + 1}\right\} = 2\cos t + 3\sin t. \tag{3}$$

We now apply partial fractions to the second term in the right member of (2). We have

$$\frac{1}{s^2(s^2 + 1)} = \frac{A}{s} + \frac{B}{s^2} + \frac{Cs + D}{s^2 + 1}$$

and from this we readily find $A = 0$, $B = 1$, $C = 0$, $D = -1$. Thus

$$\mathcal{L}^{-1}\left\{\frac{1}{s^2(s^2 + 1)}\right\} = \mathcal{L}^{-1}\left\{\frac{1}{s^2}\right\} - \mathcal{L}^{-1}\left\{\frac{1}{s^2 + 1}\right\}.$$

Then using Table 9.1, numbers 7 and 3, we find

$$\mathcal{L}^{-1}\left\{\frac{1}{s^2(s^2 + 1)}\right\} = t - \sin t. \tag{4}$$

Now letting $F(s) = \dfrac{1}{s^2(s^2 + 1)}$ and $f(t) = t - \sin t$,

(4) states that $\mathcal{L}^{-1}\{F(s)\} = f(t)$. Using Theorem 9.9, we now find

$$\mathcal{L}^{-1}\left\{\frac{e^{-\pi s}}{s^2(s^2 + 1)}\right\} = \mathcal{L}^{-1}\{F(s)e^{-\pi s}\}$$

$$= u_\pi(t)f(t - \pi)$$

$$= \begin{cases} 0, & 0 < t < \pi, \\ f(t - \pi), & t > \pi, \end{cases}$$

$$= \begin{cases} 0, & 0 < t < \pi, \\ t - \pi - \sin(t - \pi), & t > \pi. \end{cases}$$

That is,

$$\mathcal{L}^{-1}\left\{\frac{e^{-\pi s}}{s^2(s^2 + 1)}\right\} = \begin{cases} 0, & 0 < t < \pi, \\ t - \pi + \sin t, & t > \pi, \end{cases} \tag{5}$$

Thus, using (3), (4), and (5), (2) becomes

$$y = \begin{cases} (2 \cos t + 3 \sin t) + (t - \sin t) - 0, & 0 < t < \pi, \\ (2 \cos t + 3 \sin t) + (t - \sin t) - (t - \pi + \sin t), & t > \pi \end{cases}$$

$$= \begin{cases} 2 \sin t + 2 \cos t + t, & 0 < t < \pi, \\ \sin t + 2 \cos t + \pi, & t > \pi. \end{cases}$$

10.   Step 1.   Taking the Laplace Transform of both sides of the D.E., we have

$$\mathcal{L}\{y''\} + 6\mathcal{L}\{y'\} + 25\mathcal{L}\{y\} = 25[\mathcal{L}\{u_2(t)\} - \mathcal{L}\{u_4(t)\}]. \tag{1}$$

Denoting $\mathcal{L}\{y\}$ by $Y(s)$, applying Theorem 9.4, and then applying the I.C.'s, we have

$$\mathcal{L}\{y''\} = s^2Y(s) - 3s + 9, \quad \mathcal{L}\{y'\} = sY(s) - 3.$$

Substituting these into the left member of (1), it becomes $s^2Y(s) - 3s + 9 + 6sY(s) - 18 + 25Y(s)$ or $(s^2 + 6s + 25)Y(s) - 3s - 9$.   By Table 9.1, number 15;

twice, the right member becomes $25\left[\dfrac{e^{-2s}}{s} - \dfrac{e^{-4s}}{s}\right]$.   Thus (1)

becomes

$$(s^2 + 6s + 25)Y(s) - 3s - 9 = \frac{25}{s}[e^{-2s} - e^{-4s}].$$

Step 2.   We solve this for $Y(s)$ to obtain

$$Y(s) = \frac{3s + 9}{s^2 + 6s + 25} + \frac{25[e^{-2s} - e^{-4s}]}{s(s^2 + 6s + 25)}.$$

Step 3. We must find

$$y = \mathcal{L}^{-1}\left\{\frac{3s + 9}{s^2 + 6s + 25} + \frac{25[e^{-2s} - e^{-4s}]}{s(s^2 + 6s + 25)}\right\} \qquad (2)$$

Using Table 9.1, number 12, we first find

$$\mathcal{L}^{-1}\left\{\frac{3s + 9}{s^2 + 6s + 25}\right\} = 3\mathcal{L}^{-1}\left\{\frac{s + 3}{(s + 3)^2 + 4^2}\right\}$$

$$= 3e^{-3t}\cos 4t. \qquad (3)$$

Now consider $\mathcal{L}^{-1}\left\{\dfrac{25}{s(s^2 + 6s + 25)}\right\}$ and employ partial fractions. We have

$$\frac{25}{s(s^2 + 6s + 25)} = \frac{A}{s} + \frac{Bs + C}{s^2 + 6s + 25}$$

Clearing fractions, we find $25 = A(s^2 + 6s + 25)$ $+ s(Bs + C) = (A + B)s^2 + (6A + C)s + 25A$. From this, $A = 1$, $B = -1$, $C = -6$. Thus

$$\mathcal{L}^{-1}\left\{\frac{25}{s(s^2 + 6s + 25)}\right\} = \mathcal{L}^{-1}\left\{\frac{1}{s}\right\} - \mathcal{L}^{-1}\left\{\frac{s + 6}{s^2 + 6s + 25}\right\}$$

$$= \mathcal{L}^{-1}\left\{\frac{1}{s}\right\} - \mathcal{L}^{-1}\left\{\frac{s + 3}{(s + 3)^2 + 4^2}\right\}$$

$$- \frac{3}{4}\mathcal{L}^{-1}\left\{\frac{4}{(s + 3)^2 + 4^2}\right\}$$

$$= 1 - e^{-3t}\cos 4t - \frac{3}{4}e^{-3t}\sin 4t,$$

where we have used Table 9.1, numbers 1, 12, and 11, respectively.

Letting $F(s) = \dfrac{25}{s(s^2 + 6s + 25)}$ and

$f(t) = 1 - e^{-3t}\cos 4t - \dfrac{3}{4} e^{-3t}\sin 4t$, we thus have

$\mathcal{L}^{-1}\{F(s)\} = f(t)$. Then by Table 9.1, number 16, we find

$$\mathcal{L}^{-1}\left\{\frac{25e^{-2s}}{s(s^2 + 6s + 25)}\right\} = \mathcal{L}^{-1}\{F(s)e^{-2s}\}$$

$$= u_2(t)f(t - 2) = \begin{cases} 0, & 0 < t < 2, \\ f(t - 2), & t > 2, \end{cases}$$

$$= \begin{cases} 0, & 0 < t < 2, \\ 1 - e^{-3(t-2)}\cos 4(t - 2) - \dfrac{3}{4} e^{-3(t-2)}\sin 4(t - 2), \\ \quad t > 2. \end{cases} \tag{4}$$

Similarly, we find

$$\mathcal{L}^{-1}\left\{\frac{25e^{-4s}}{s(s^2 + 6s + 25)}\right\} = \begin{cases} 0, & 0 < t < 4, \\ f(t - 4), & t > 4, \end{cases}$$

$$= \begin{cases} 0, & 0 < t < 4, \\ 1 - e^{-3(t-4)}\cos 4(t-4) - \dfrac{3}{4} e^{-3(t-4)}\sin 4(t-4), & t > 4. \end{cases} \tag{5}$$

Finally, using (2), (3), (4), and (5), we find

$$y = \begin{cases} 3e^{-3t}\cos 4t, & 0 < t < 2, \\[4pt] 3e^{-3t}\cos 4t + 1 - e^{-3(t-2)}\cos 4(t-2) - \dfrac{3}{4} e^{-3(t-2)}\sin 4(t-2), \\ \quad 2 < t < 4, \\[4pt] 3e^{-3t}\cos 4t - e^{-3(t-2)}\cos 4(t-2) - \dfrac{3}{4} e^{-3(t-2)}\sin 4(t-2) \\ \quad + e^{-3(t-4)}\cos 4(t-4) + \dfrac{3}{4} e^{-3(t-4)}\sin 4(t-4), & t > 4. \end{cases}$$

12. Step 1. Taking the Laplace Transform of both sides of the
D.E., we have

$$\mathcal{L}\{y''\} - \mathcal{L}\{y'\} = \mathcal{L}\{h(t)\}. \tag{1}$$

Denoting $\mathcal{L}\{y\}$ by $Y(s)$, applying Theorem 9.4, and then applying the I.C.'s, we have

$$\mathcal{L}\{y''\} = s^2 Y(s) - s + 2, \quad \mathcal{L}\{y'\} = sY(s) - 1.$$

Substituting these into the left member of (1) and simplifying, it becomes $(s^2 - s)Y(s) - s + 3$. Now observe that

$$h(t) = \begin{cases} 4 - 2t, & 0 \le t \le 2, \\ 0, & t > 2, \end{cases}$$

$$= \begin{cases} 4 - 2t + 0, & 0 \le t \le 2, \\ 4 - 2t + 2(t - 2), & t > 2, \end{cases}$$

$$= 4 - 2t + 2u_2(t)(t - 2)$$

$$= 4 - 2t + 2u_2(t)f(t - 2),$$

where $f(t - 2) = t - 2$. Then by Table 9.1, numbers 1, 7, and 16, respectively, we have

$$\mathcal{L}\{h(t)\} = \frac{4}{s} - \frac{2}{s^2} + 2e^{-2s}F(s),$$

where $F(s) = \mathcal{L}\{f(t)\} = \mathcal{L}\{t\} = \frac{1}{s^2}$, using number 7 again.

Thus the right member of (1) becomes $\frac{4}{s} - \frac{2}{s^2} + \frac{2e^{-2s}}{s^2}$, and (1) becomes

$$(s^2 - s)Y(s) - s + 3 = \frac{4}{s} - \frac{2}{s^2} + \frac{2e^{-2s}}{s^2}.$$

Step 2. We solve this for Y(s), obtaining

$$Y(s) = \frac{s - 3}{s(s - 1)} + \frac{4}{s^2(s - 1)} - \frac{2}{s^3(s - 1)}$$

$$+ \frac{2e^{-2s}}{s^3(s - 1)}.$$

Step 3. We must find

$$y = \mathcal{L}^{-1}\left\{\frac{s - 3}{s(s - 1)} + \frac{4}{s^2(s - 1)} - \frac{2}{s^3(s - 1)}\right\}$$

$$+ \mathcal{L}^{-1}\left\{\frac{2e^{-2s}}{s^3(s - 1)}\right\}. \tag{2}$$

We first find the first of these two inverse transforms.
We write the expression in the first pair of braces over a
lowest common denominator and then employ partial
fractions. We have

$$\frac{s^3 - 3s^2 + 4s - 2}{s^3(s - 1)} = \frac{A}{s} + \frac{B}{s^2} + \frac{C}{s^3} + \frac{D}{s - 1}.$$

Clearing fractions, we obtain

$$s^3 - 3s^2 + 4s - 2 = As^2(s - 1) + Bs(s - 1)$$
$$+ C(s - 1) + Ds^3 = (A + D)s^3 + (B - A)s^2$$
$$+ (C - B)s - C.$$

From this we easily find A = 1, B = -2, C = 2, D = 0.
Hence we seek

$$\mathcal{L}^{-1}\left\{\frac{1}{s}\right\} - 2\mathcal{L}^{-1}\left\{\frac{1}{s^2}\right\} + \mathcal{L}^{-1}\left\{\frac{2}{s^3}\right\};$$

and using Table 9.1, numbers 1, 7, and 7, respectively, we
obtain $1 - 2t + t^2$. Thus

$$\mathcal{L}^{-1}\left\{\frac{s-3}{s(s-1)} + \frac{4}{s^2(s-1)} - \frac{2}{s^3(s-1)}\right\}$$

$$= 1 - 2t + t^2. \tag{3}$$

We now return to (2) and proceed to find

$\mathcal{L}^{-1}\left\{\dfrac{2e^{-2s}}{s^3(s-1)}\right\}$. We consider $\dfrac{2}{s^3(s-1)}$, and again employ

partial fractions. We have

$$\frac{2}{s^3(s-1)} = \frac{A}{s} + \frac{B}{s^2} + \frac{C}{s^3} + \frac{D}{s-1}.$$

Clearing fractions and simplifying, we obtain

$$2 = (A + D)s^3 + (B - A)s^2 + (C - B)s - C,$$

from which $A = B = C = -2$ and $D = 2$. Thus

$$\mathcal{L}^{-1}\left\{\frac{2}{s^3(s-1)}\right\} = -2\mathcal{L}^{-1}\left\{\frac{1}{s}\right\} - 2\mathcal{L}^{-1}\left\{\frac{1}{s^2}\right\}$$

$$- \mathcal{L}^{-1}\left\{\frac{2}{s^3}\right\} + 2\mathcal{L}^{-1}\left\{\frac{1}{s-1}\right\}.$$

Then by Table 9.1, numbers 1, 7, 7, and 2, respectively, we obtain

$$\mathcal{L}^{-1}\left\{\frac{2}{s^3(s-1)}\right\} = -2 - 2t - t^2 + 2e^t.$$

Letting $F(s) = \dfrac{2}{s^3(s-1)}$ and $f(t) = -2 - 2t - t^2$

$+ 2e^t$, we thus have $\mathcal{L}^{-1}\{F(s)\} = f(t)$. Then by Table 9.1, number 16, we find

$$\mathcal{L}^{-1}\left\{\frac{2e^{-2s}}{s^3(s-1)}\right\} = \mathcal{L}^{-1}\{F(s)e^{-2s}\}$$

$$= u_2(t)f(t-2)$$

$$= \begin{cases} 0, & 0 < t < 2, \\ f(t-2), & t > 2, \end{cases}$$

$$= \begin{cases} 0, & 0 < t < 2, \\ -2 - 2(t-2) - (t-2)^2 + 2e^{t-2}, & t > 2, \end{cases}$$

$$= \begin{cases} 0, & 0 < t < 2, \\ -2 + 2t - t^2 + 2e^{t-2}, & t > 2. \end{cases} \tag{4}$$

Then, using (2), (3), and (4), we obtain

$$y = \begin{cases} (t-1)^2, & 0 < t < 2, \\ -1 + 2e^{t-2}, & t > 2. \end{cases}$$

Section 9.4D, Page 539.

1. Step 1. We take the Laplace Transform of both sides of
    the D.E. to obtain

    $$\mathcal{L}\{y'\} - 4\mathcal{L}\{y\} = \mathcal{L}\{\delta(t-2)\}. \tag{1}$$

    We denote $\mathcal{L}\{y\}$ by $Y(s)$ and apply Theorem 9.3 and then the
    I.C., to obtain

    $$\mathcal{L}\{y'\} = sY(s) - y(0) = sY(s) - 3.$$

    Then the left member of equation (1) becomes
    $(s-4)Y(s) - 3.$ By formula (9.103), $\mathcal{L}\{\delta(t-2)\} = e^{-2s}.$
    Thus (1) reduces to

    $$(s-4)Y(s) - 3 = e^{-2s}.$$

Step 2. We solve this for $Y(s)$, obtaining

$$Y(s) = \frac{3}{s-4} + \frac{e^{-2s}}{s-4}.$$

Step 3. We must now determine

$$y = 3\mathcal{L}^{-1}\left\{\frac{1}{s-4}\right\} + \mathcal{L}^{-1}\left\{\frac{e^{-2s}}{s-4}\right\}.$$

Using Table 9.1, numbers 2 and 16, respectively, we find
$y = 3e^{4t} + u_2(t)f(t-2)$, where $f(t) = e^{4t}$. Thus

$$y = 3e^{4t} + u_2(t)e^{4(t-2)}$$

$$= \begin{cases} 3e^{4t}, & 0 < t < 2, \\ 3e^{4t} + e^{4(t-2)}, & t > 2. \end{cases}$$

3. Step 1. We take the Laplace Transform of both sides of
the D.E. to obtain

$$\mathcal{L}\{y''\} + \mathcal{L}\{y\} = \mathcal{L}\{\delta(t - \pi)\}. \tag{1}$$

We denote $\mathcal{L}\{y\}$ by $Y(s)$ and apply Theorem 9.3 and then the
I.C., to obtain

$$\mathcal{L}\{y''\} = s^2Y(s) - sy(0) - y'(0)$$

$$= s^2Y(s) - 1.$$

Then the left member of equation (1) becomes
$(s^2 + 1)Y(s) - 1$. By formula (9.103), $\mathcal{L}\{\delta(t - \pi)\} = e^{-\pi s}$.
Thus (1) reduces to

$$(s^2 + 1)Y(s) - 1 = e^{-\pi s}.$$

Step 2.   We solve this for Y(s), obtaining

$$Y(s) = \frac{1}{s^2 + 1} + \frac{e^{-\pi s}}{s^2 + 1}.$$

Step 3.   We must now determine

$$y = \mathcal{L}^{-1}\left\{\frac{1}{s^2 + 1}\right\} + \mathcal{L}^{-1}\left\{\frac{e^{-\pi s}}{s^2 + 1}\right\}.$$

Using Table 9.1, numbers 3 and 16, respectively, we find
y = sin t + u$_\pi$(t)f(t − π), where f(t) = sin t.   Thus

$$y = \sin t + u_\pi(t)\sin(t - \pi)$$

$$= \begin{cases} \sin t, & 0 < t < \pi, \\ \sin t - \sin t, & t > \pi, \end{cases}$$

$$= \begin{cases} \sin t, & 0 < t < \pi, \\ 0, & t > \pi. \end{cases}$$

4.   Step 1.   We take the Laplace Transform of both sides of
the D.E. to obtain

$$\mathcal{L}\{y''\} + 3\mathcal{L}\{y'\} + 2\mathcal{L}\{y\} = \delta(t - 4). \tag{1}$$

We denote $\mathcal{L}\{y\}$ by Y(s) and apply Theorem 9.4 and then the
I.C.'s, to obtain

$$\mathcal{L}\{y''\} = s^2Y(s) - sy(0) - y'(0) = s^2Y(s) - 2s + 6,$$
$$\mathcal{L}\{y'\} = sY(s) - y(0) = sY(s) - 2.$$

Substituting these into the left member of (1) that left
member becomes

$$s^2Y(s) - 2s + 6 + 3sY(s) - 6 + 2Y(s)$$

or

$$(s^2 + 3s + 2)Y(s) - 2s.$$

By formula (9.103), $\mathcal{L}\{\delta(t - 4)\} = e^{-4s}$. Thus (1) reduces to

$$(s^2 + 3s + 2)Y(s) - 2s = e^{-4s}.$$

Step 2. We solve this for $Y(s)$, obtaining

$$Y(s) = \frac{2s}{s^2 + 3s + 2} + \frac{e^{-4s}}{s^2 + 3s + 2}$$

Step 3. We must now determine

$$y = \mathcal{L}^{-1}\left\{\frac{2s}{(s + 1)(s + 2)}\right\}$$

$$+ \mathcal{L}^{-1}\left\{\frac{e^{-4s}}{(s + 1)(s + 2)}\right\}. \tag{2}$$

We employ partial fractions to find the first of the two inverse transforms needed. We write

$$\frac{2s}{(s + 1)(s + 2)} = \frac{A}{s + 1} + \frac{B}{s + 2},$$ and upon clearing fractions, at once find $A = -2$, $B = 4$. Thus

$$\mathcal{L}^{-1}\left\{\frac{2s}{(s + 1)(s + 2)}\right\} = -2\mathcal{L}^{-1}\left\{\frac{1}{s + 1}\right\} + 4\mathcal{L}^{-1}\left\{\frac{1}{s + 2}\right\}.$$

Then using Table 9.1, number 2, twice, we find

$$\mathcal{L}^{-1}\left\{\frac{2s}{(s + 1)(s + 2)}\right\} = -2e^{-t} + 4e^{-2t}. \tag{3}$$

Now we proceed to determine $\mathcal{L}^{-1}\left\{\frac{e^{-4s}}{(s + 1)(s + 2)}\right\}.$

Again using partial fractions, we find

$$\frac{1}{(s + 1)(s + 2)} = \frac{1}{s + 1} - \frac{1}{s + 2}.$$

Thus by Table 9.1, number 2, $\mathcal{L}^{-1}\left\{\frac{1}{(s + 1)(s + 2)}\right\}$

$= e^{-t} - e^{-2t}.$ Then by Table 9.1, number 16,

$$\mathcal{L}^{-1}\left\{\frac{e^{-4s}}{(s + 1)(s + 2)}\right\} = u_4(t)f(t - 4), \tag{4}$$

where $f(t) = e^{-t} - e^{-2t}.$ Hence using (2), (3), and (4), we find

$$y = \begin{cases} -2e^{-t} + 4e^{-2t}, & 0 < t < 4, \\ -2e^{-t} + 4e^{-2t} + e^{-(t-4)} - e^{-2(t-4)}, & t > 4, \end{cases}$$

$$= \begin{cases} -2e^{-t} + 4e^{-2t}, & 0 < t < 4, \\ (-2 + e^4)e^{-t} + (4 - e^8)e^{-2t}, & t > 4. \end{cases}$$

Section 9.5, Page 542.

3.   Step 1.   Taking the Laplace Transform of both sides of each D.E. of the system, we have

$$\begin{cases} \mathcal{L}\{x'\} - 5\mathcal{L}\{x\} + 2\mathcal{L}\{y\} = 3\mathcal{L}\{e^{4t}\}, \\ \mathcal{L}\{y'\} - 4\mathcal{L}\{x\} + \mathcal{L}\{y\} = \mathcal{L}\{0\}. \end{cases} \tag{1}$$

We denote $\mathcal{L}\{x\}$ by $X(s)$ and $\mathcal{L}\{y\}$ by $Y(s)$.

Applying Theorem 9.3 and the given I.C.'s, we express $\mathcal{L}\{x'\}$ and $\mathcal{L}\{y'\}$ in terms of $X(s)$ and $Y(s)$, respectively, as follows

$$\begin{cases} \mathcal{L}\{x'\} = sX(s) - x(0) = sX(s) - 3, \\ \mathcal{L}\{y'\} = sY(s) - y(0) = sY(s). \end{cases} \quad (2)$$

By Table 9.1, number 2, $\mathcal{L}\{e^{4t}\} = \dfrac{1}{s - 4}$. Thus, from this and (2), we see that (1) becomes

$$\begin{cases} sX(s) - 3 - 5X(s) + 2Y(s) = \dfrac{3}{s - 4}, \\ sY(s) - 4X(s) + Y(s) = 0, \end{cases}$$

or

$$\begin{cases} (s - 5)X(s) + 2Y(s) = \dfrac{3s - 9}{s - 4}, \\ -4X(s) + (s + 1)Y(s) = 0. \end{cases}$$

Step 2. We solve this system for the two unknowns $X(s)$ and $Y(s)$. We have

$$\begin{cases} (s + 1)(s - 5)X(s) + 2(s + 1)Y(s) = \dfrac{(3s - 9)(s + 1)}{s - 4}, \\ -8X(s) + 2(s + 1)Y(s) = 0. \end{cases}$$

Subtracting, we obtain

$$(s^2 - 4s + 3)X(s) = \frac{(3s - 9)(s + 1)}{s - 4},$$

from which we find

$$X(s) = \frac{3(s - 3)(s + 1)}{(s - 1)(s - 3)(s - 4)} = \frac{3(s + 1)}{(s - 1)(s - 4)}.$$

In like manner, we find

$$Y(s) = \frac{12}{(s - 1)(s - 4)}.$$

Step 3.   We must now determine

$$x = \mathcal{L}^{-1}\{X(s)\} = \mathcal{L}^{-1}\left\{\frac{3(s + 1)}{(s - 1)(s - 4)}\right\}.$$

and

$$y = \mathcal{L}^{-1}\{Y(s)\} = \mathcal{L}^{-1}\left\{\frac{12}{(s - 1)(s - 4)}\right\}.$$

We first find x. We employ partial fractions. We have

$\dfrac{3(s + 1)}{(s - 1)(s - 4)} = \dfrac{A}{s - 1} + \dfrac{B}{s - 4}$, from which we readily find

$A = -2$, $B = 5$.   Thus

$$x = -2\mathcal{L}^{-1}\left\{\frac{1}{s - 1}\right\} + 5\mathcal{L}^{-1}\left\{\frac{1}{s - 4}\right\},$$

and using Table 9.1, number 2, we obtain

$$x = -2e^{t} + 5e^{4t}.$$

Similarly, we find y. Again employing partial fractions,
we obtain

$$y = -4\mathcal{L}^{-1}\left\{\frac{1}{s - 1}\right\} + 4\mathcal{L}^{-1}\left\{\frac{1}{s - 4}\right\}.$$

Again using Table 9.1, number 2, we obtain

$$y = -4e^{t} + 4e^{4t}.$$

4.   Step 1.   Taking the Laplace Transform of both sides of
each D.E. of the system, we have

$$\begin{cases} \mathcal{L}\{x'\} - 2\mathcal{L}\{x\} - 3\mathcal{L}\{y\} = 0, \\ \mathcal{L}\{y'\} + \mathcal{L}\{x\} + 2\mathcal{L}\{y\} = \mathcal{L}\{t\}. \end{cases} \tag{1}$$

Denote $\mathcal{L}\{x\}$ by $X(s)$ and $\mathcal{L}\{y\}$ by $Y(s)$. Applying Theorem
9.3 and the given I.C.'s, we express $\mathcal{L}\{x\}$ in terms of $X(s)$
and $\mathcal{L}\{y\}$ in terms of $Y(s)$ as follows:

$$\begin{cases} \mathcal{L}\{x'\} = sX(s) - x(0) = sX(s) + 1, \\ \mathcal{L}\{y'\} = sY(s) - y(0) = sY(s). \end{cases} \tag{2}$$

By Table 9.1, number 7, $\mathcal{L}\{t\} = \dfrac{1}{s^2}$. Substituting this and the expressions for $\mathcal{L}\{x'\}$ and $\mathcal{L}\{y'\}$ given by (2) into (1), we see that (1) becomes

$$\begin{cases} sX(s) + 1 - 2X(s) - 3Y(s) = 0, \\ sY(s) + X(s) + 2Y(s) = 1/s^2. \end{cases}$$

or

$$\begin{cases} (s - 2)X(s) - 3Y(s) = -1, \\ X(s) + (s + 2)Y(s) = 1/s^2. \end{cases}$$

Step 2. We solve this system for the two unknowns $X(s)$ and $Y(s)$. We have

$$\begin{cases} (s - 2)(s + 2)X(s) - 3(s + 2)Y(s) = -(s + 2), \\ 3X(s) + 3(s + 2)Y(s) = 3/s^2. \end{cases}$$

Adding, we obtain

$$(s^2 - 1)X(s) = -(s + 2) + 3/s^2,$$

from which we find

$$X(s) = -\frac{s + 2}{s^2 - 1} + \frac{3}{s^2(s^2 - 1)} = \frac{-s^3 - 2s^2 + 3}{s^2(s^2 - 1)}.$$

In like manner, we find

$$Y(s) = \frac{1}{s^2 - 1} + \frac{s - 2}{s^2(s^2 - 1)} = \frac{s^2 + s - 2}{s^2(s^2 - 1)}.$$

Step 3.   We must now determine

$$x = \mathcal{L}^{-1}\left\{\frac{-s^3 - 2s^2 + 3}{s^2(s^2 - 1)}\right\}.$$

$$y = \mathcal{L}^{-1}\left\{\frac{s^2 + s - 2}{s^2(s^2 - 1)}\right\}.$$

We first determine x.   Using partial fractions, we write

$$\frac{-s^3 - 2s^2 + 3}{s^2(s - 1)(s + 1)} = \frac{A}{s} + \frac{B}{s^2} + \frac{C}{s - 1} + \frac{D}{s + 1}.$$ Clearing

fractions, we obtain $-s^3 - 2s^2 + 3 = As(s - 1)(s + 1)$
$+ B(s - 1)(s + 1) + Cs^2(s + 1) + Ds^2(s - 1)$
$= (A + C + D)s^3 + (B + C - D)s^2 - As - B$.   From this, we
find A = 0, B = -3, C = 0, D = -1.   Thus

$$x = -3\mathcal{L}^{-1}\left\{\frac{1}{s^2}\right\} - \mathcal{L}^{-1}\left\{\frac{1}{s + 1}\right\};$$

and using Table 9.1, numbers 7 and 2, respectively, we
find

$$x = -3t + e^{-t}.$$

In like manner, using partial fractions, we obtain

$$y = -\mathcal{L}^{-1}\left\{\frac{1}{s}\right\} + 2\mathcal{L}^{-1}\left\{\frac{1}{s^2}\right\} + \mathcal{L}^{-1}\left\{\frac{1}{s + 1}\right\};$$

and using Table 9.1, numbers 1, 7, and 2, respectively, we
find

$$y = -1 + 2t + e^{-t}.$$

7.   Step 1.   Taking the Laplace Transform of both sides of
each D.E. of the system, we have

$$\begin{cases} 2L\{x'\} + L\{y'\} - L\{x\} - L\{y\} = L\{e^{-t}\} \\ L\{x'\} + L\{y'\} + 2L\{x\} + L\{y\} = L\{e^{t}\}. \end{cases} \quad (1)$$

We denote $L\{x\}$ by $X(s)$ and $L\{y\}$ by $Y(s)$. Applying Theorem
9.3 and the given I.C.'s, we express $L\{x'\}$ and $L\{y'\}$ in
terms of $X(s)$ and $Y(s)$, respectively, as follows:

$$L\{x'\} = sX(s) - x(0) = sX(s) - 2,$$
$$L\{y'\} = sY(s) - y(0) = sY(s) - 1. \quad (2)$$

By Table 9.1, number 2, $L\{e^{-t}\} = \dfrac{1}{s+1}$, $L\{e^{t}\} = \dfrac{1}{s-1}$.
Thus, from this and (2), we see that (1) becomes

$$\begin{cases} 2sX(s) - 4 + sY(s) - 1 - X(s) - Y(s) = \dfrac{1}{s+1}, \\ sX(s) - 2 + sY(s) - 1 + 2X(s) + Y(s) = \dfrac{1}{s-1}, \end{cases}$$

or

$$\begin{cases} (2s - 1)X(s) + (s - 1)Y(s) = \dfrac{5s+6}{s+1}, \\ (s + 2)X(s) + (s + 1)Y(s) = \dfrac{3s-2}{s-1}. \end{cases}$$

Step 2.   We solve this system for the two unknowns $X(s)$
and $Y(s)$.   We have

$$\begin{cases} (s + 1)(2s - 1)X(s) + (s + 1)(s - 1)Y(s) = 5s + 6, \\ (s - 1)(s + 2)X(s) + (s + 1)(s - 1)Y(s) = 3s - 2. \end{cases}$$

Subtracting, we obtain $(s^2 + 1)X(s) = 2s + 8$, from which
we find

$$X(s) = \frac{2s + 8}{s^2 + 1}.$$

In like manner, we find

$$Y(s) = \frac{s^3 - 12s^2 - s + 14}{(s + 1)(s - 1)(s^2 + 1)}.$$

Step 3.   We must now determine

$$x = \mathcal{L}^{-1}\{X(s)\} = \mathcal{L}^{-1}\left\{\frac{2s + 8}{s^2 + 1}\right\}$$

and

$$y = \mathcal{L}^{-1}\{Y(s)\} = \mathcal{L}^{-1}\left\{\frac{s^3 - 12s^2 - s + 14}{(s + 1)(s - 1)(s^2 + 1)}\right\}.$$

We first obtain x. Using Table 9.1, numbers 4 and 3, we find

$$x = 2\mathcal{L}^{-1}\left\{\frac{s}{s^2 + 1}\right\} + 8\mathcal{L}^{-1}\left\{\frac{1}{s^2 + 1}\right\}$$

$$= 2\cos t + 8\sin t.$$

We proceed to find y. We first employ partial fractions. We have $\dfrac{s^3 - 12s^2 - s + 14}{(s + 1)(s - 1)(s^2 + 1)} = \dfrac{A}{s - 1} + \dfrac{B}{s + 1} + \dfrac{Cs + D}{s^2 + 1}$ or

$$s^3 - 12s^2 - s + 14 = A(s + 1)(s^2 + 1)$$

$$+ B(s - 1)(s^2 + 1) + (Cs + D)(s - 1)(s + 1). \quad (3)$$

or

$$s^3 - 12s^2 - s + 14 = (A + B + C)s^3$$
$$+ (A - B + D)s^2 + (A + B - C)s$$
$$+ (A - B - D).$$

From this, we have

$$\begin{cases} A + B + C = 1, \quad A - B + D = -12, \\ A + B - C = -1, \quad A - B - D = 14. \end{cases} \qquad (4)$$

Letting $s = 1$ in (3), we find $A = \frac{1}{2}$; and letting $s = -1$ in (3), we find $B = -\frac{1}{2}$. Using these values and (4), we find $C = 1$, $D = -13$. Thus we have

$$y = \frac{1}{2} \mathcal{L}^{-1}\left\{\frac{1}{s - 1}\right\} - \frac{1}{2} \mathcal{L}^{-1}\left\{\frac{1}{s + 1}\right\}$$
$$+ \mathcal{L}^{-1}\left\{\frac{s}{s^2 + 1}\right\} - 13\mathcal{L}^{-1}\left\{\frac{1}{s^2 + 1}\right\}.$$

Using Table 9.1, numbers 2, 2, 4, and 3, respectively, we find

$$y = \frac{1}{2} e^t - \frac{1}{2} e^{-t} + \cos t - 13 \sin t.$$

8. Step 1. Taking the Laplace Transform of both sides of each D.E. of the system, we have

$$\begin{cases} 2\mathcal{L}\{x'\} + \mathcal{L}\{y'\} + \mathcal{L}\{x\} + 5\mathcal{L}\{y\} = 4\mathcal{L}\{t\}, \\ \mathcal{L}\{x'\} + \mathcal{L}\{y'\} + 2\mathcal{L}\{x\} + 2\mathcal{L}\{y\} = 2\mathcal{L}\{1\}. \end{cases} \qquad (1)$$

Denote $\mathcal{L}\{x\}$ by $X(s)$ and $\mathcal{L}\{y\}$ by $Y(s)$. Applying Theorem 9.3 and the given I.C.'s, we express $\mathcal{L}\{x\}$ in terms of $X(s)$ and $\mathcal{L}\{y\}$ in terms of $Y(s)$ as follows:

$$
\begin{cases}
\mathcal{L}\{x'\} = sX(s) - x(0) = sX(s) - 3, \\
\mathcal{L}\{y'\} = sY(s) - y(0) = sY(s) + 4.
\end{cases} \qquad (2)
$$

By Table 9.1, numbers 7 and 2, respectively, we find
$\mathcal{L}\{t\} = 1/s^2$ and $\mathcal{L}\{1\} = 1/s$. Substituting these and the
expressions for $\mathcal{L}\{x'\}$ and $\mathcal{L}\{y'\}$ given by (2) into (1), we
see that (1) becomes

$$
\begin{cases}
2sX(s) - 6 + sY(s) + 4 + X(s) + 5Y(s) = 4/s^2, \\
sX(s) - 3 + sY(s) + 4 + 2X(s) + 2Y(s) = 2/s.
\end{cases}
$$

or

$$
\begin{cases}
(2s + 1)X(s) + (s + 5)Y(s) = 2 + 4/s^2, \\
(s + 2)X(s) + (s + 2)Y(s) = -1 + 2/s.
\end{cases}
$$

Step 2. We solve this system for the two unknowns $X(s)$
and $Y(s)$. We have

$$
\begin{cases}
(s+2)(2s+1)X(s) + (s+2)(s+5)Y(s) = (s+2)(2+4/s^2), \\
(s+2)(s+5)X(s) + (s+2)(s+5)Y(s) = (s+5)(-1+2/s).
\end{cases}
$$

Subtracting, we obtain

$$
(s^2 - 2s - 8)X(s) = 3s + 7 - \frac{6}{s} + \frac{8}{s^2},
$$

from which we find

$$
X(s) = \frac{3s + 7}{s^2 - 2s - 8} - \frac{6}{s(s^2 - 2s - 8)}
$$

$$
+ \frac{8}{s^2(s^2 - 2s - 8)}
$$

$$
= \frac{3s^3 + 7s^2 - 6s + 8}{s^2(s - 4)(s + 2)}.
$$

In like manner, we find

$$Y(s) = \frac{-4s - 1}{s^2 - 2s - 8} - \frac{2}{s(s^2 - 2s - 8)}$$

$$- \frac{8}{s^2(s^2 - 2s - 8)}$$

$$= \frac{-4s^3 - s^2 - 2s - 8}{s^2(s - 4)(s + 2)}.$$

Step 3. We must now determine

$$x = \mathcal{L}^{-1}\left[\frac{3s^3 + 7s^2 - 6s + 8}{s^2(s - 4)(s + 2)}\right],$$

$$y = \mathcal{L}^{-1}\left[\frac{-4s^3 - s^2 - 2s - 8}{s^2(s - 4)(s + 2)}\right].$$

We first determine x. Using partial fractions, we write

$$\frac{3s^3 + 7s^2 - 6s + 8}{s^2(s - 4)(s + 2)} = \frac{A}{s} + \frac{B}{s^2} + \frac{C}{s - 4} + \frac{D}{s + 2}.$$ Clearing
fractions, we have $3s^3 + 7s^2 - 6s + 8 = As(s - 4)(s + 2)$
$+ B(s - 4)(s + 2) + Cs^2(s + 2) + Ds^2(s - 4)$
$= (A + C + D)s^3 + (-2A + B + 2C - 4D)s^2 + (-8A - 2B)s$
$- 8B$. From this, we find $A = 1$, $B = -1$, $C = 3$, $D = -1$.
Thus

$$x = \mathcal{L}^{-1}\left\{\frac{1}{s}\right\} - \mathcal{L}^{-1}\left\{\frac{1}{s^2}\right\} + 3\mathcal{L}^{-1}\left\{\frac{1}{s - 4}\right\} - \mathcal{L}^{-1}\left\{\frac{1}{s + 2}\right\}.$$

Then, using Table 9.1, numbers 1, 7, 2, and 2,
respectively, we find

$$x = 1 - t + 3e^{4t} - e^{-2t}.$$

In like manner, using partial fractions, we obtain

$$y = \mathcal{L}^{-1}\left\{\frac{1}{s^2}\right\} - 3\mathcal{L}^{-1}\left\{\frac{1}{s-4}\right\} - \mathcal{L}^{-1}\left\{\frac{1}{s+2}\right\}.$$

Then, using Table 9.1, numbers 7, 2, and 2, respectively, we find

$$y = t - 3e^{4t} - e^{-2t}.$$

10. **Step 1.** Taking the Laplace Transform of both sides of each D.E. of the system, we have

$$\begin{cases} \mathcal{L}\{x''\} - 3\mathcal{L}\{x'\} + \mathcal{L}\{y'\} + 2\mathcal{L}\{x\} - \mathcal{L}\{y\} = 0, \\ \\ \mathcal{L}\{x'\} + \mathcal{L}\{y'\} - 2\mathcal{L}\{x\} + \mathcal{L}\{y\} = 0. \end{cases} \quad (1)$$

We denote $\mathcal{L}\{x\}$ by $X(s)$ and $\mathcal{L}\{y\}$ by $Y(s)$. Applying Theorem 9.4 and the given I.C.'s, we express $\mathcal{L}\{x''\}$ and $\mathcal{L}\{x'\}$ in terms of $X(s)$ and $\mathcal{L}\{y'\}$ in terms of $Y(s)$ as follows:

$$\mathcal{L}\{x''\} = s^2 X(s), \quad \mathcal{L}\{x'\} = sX(s),$$
$$\mathcal{L}\{y'\} = sY(s) + 1.$$

Substituting these expressions into (1), we see that (1) becomes

$$\begin{cases} s^2 X(s) - 3sX(s) + sY(s) + 1 + 2X(s) - Y(s) = 0, \\ sX(s) + sY(s) + 1 - 2X(s) + Y(s) = 0, \end{cases}$$

or

$$\begin{cases} (s^2 - 3s + 2)X(s) + (s - 1)Y(s) = -1, \\ (s - 2)X(s) + (s + 1)Y(s) = -1. \end{cases}$$

Step 2. We solve this system for the two unknowns $X(s)$ and $Y(s)$. We have

$$\begin{cases} (s^2 - 3s + 2)(s + 1)X(s) + (s - 1)(s + 1)Y(s) = -(s + 1), \\ (s - 2)(s - 1)X(s) + (s + 1)(s - 1)Y(s) = -(s - 1). \end{cases}$$

Subtracting, we obtain $(s^3 - 3s^2 + 2s)X(s) = -2$ from which we find

$$X(s) = -\frac{2}{s(s - 1)(s - 2)}.$$

In like manner, we find

$$Y(s) = -\frac{s - 2}{s(s - 1)}.$$

Step 3.   We must now determine

$$x = \mathcal{L}^{-1}\{X(s)\} = \mathcal{L}^{-1}\left\{\frac{-2}{s(s - 1)(s - 2)}\right\}, \quad \text{and}$$

$$y = \mathcal{L}^{-1}\{Y(s)\} = \mathcal{L}^{-1}\left\{\frac{-s + 2}{s(s - 1)}\right\}.$$

We first find x using partial fractions.   We have

$$\frac{-2}{s(s - 1)(s - 2)} = \frac{A}{s} + \frac{B}{s - 1} + \frac{C}{s - 2} \quad \text{or}$$

$-2 = A(s - 1)(s - 2) + Bs(s - 2) + Cs(s - 1)$.   Letting $s = 0$, we find $A = -1$; letting $s = 1$, we find $B = 2$; and letting $s = 2$, we find $C = -1$.   Thus we have

$$x = -\mathcal{L}^{-1}\left\{\frac{1}{s}\right\} + 2\mathcal{L}^{-1}\left\{\frac{1}{s - 1}\right\} - \mathcal{L}^{-1}\left\{\frac{1}{s - 2}\right\}.$$

Now using Table 9.1, numbers 1, 2, and 2, respectively, we find $x(t) = -1 + 2e^t - e^{2t}$.

We now find y, again using partial fractions.   We have $\frac{-s + 2}{s(s - 1)} = \frac{A}{s} + \frac{B}{s - 1}$.   We readily find $A = -2$, $B = 1$; and hence

$$y = -2\mathcal{L}^{-1}\left\{\frac{1}{s}\right\} + \mathcal{L}^{-1}\left\{\frac{1}{s - 1}\right\}.$$

Then from Table 9.1, numbers 1 and 2, we find

$$y = -2 + e^t.$$

Answers to Even–Numbered Exercises

Section 1.1, Page 5.

2. ordinary; third; linear.

4. ordinary; first; nonlinear.

6. partial; fourth; linear.

8. ordinary; second; nonlinear.

10. ordinary; second; nonlinear.

Section 1.3, Page 21.

2. (a) $y = (x^2 + 2)e^{-x}$; (b) $y = (x^2 + 3e^{-1})e^{-x}$.

Section 2.1, Page 36.

2. $xy^2 + 3x - 4y = c$. 4. not exact.

6. $(\theta^2 + 1)\sin r = c$. 8. not exact.

10. not exact.

12. $x^3y^2 - y^3x + x^2 + y + 1 = 0$.

14. $e^x y + xy^2 + 2e^x = 8$.

16. $2x^{1/3} y^{-1/3} + 4x^{4/3} y^{1/3} = 9$.

18. (a) $A = 3$; $x^3 y + 2xy^2 = c$.

    (b) $A = -2$; $x^{-2} y - x^{-1} y = c$.

20. (a) $xy^4 + 2x^3 y^2 + \phi(x)$;   (b) $e^x y^2 + e^{3x} y^3 + \phi(x)$.

22. (b) $n = -2$;   (c) $x + x^2 y^{-1} = c$.

24. $4 \arctan \dfrac{x}{y} + (x^2 + y^2)^2 = c$.

Section 2.2, Page 46.

2. $x(x + 2)(y + 2)^2 = c$.   4. $\sin x - \cos y = c$.

6. $(\sin u + 1)(e^v + 1) = c$.   8. $y = x \ln|cx|$.

10. $v^2 = u^2(\ln v^2 + c)$.

12. $t^3 - 3t^2 s - 3ts^2 - 2s^3 = c$.

14. $x\left(\sqrt{\dfrac{x^2 - y^2}{x^2}} + 1\right) = c$.   16. $4x - 2\sin 2x + \tan y = \dfrac{\pi}{3}$.

18. $x^2 + y^2 = 5x^3$.   20. $2(3x^2 + 3xy + y^2)^2 = 9x^5$.

22. (a) $x^2 + 4xy - y^2 = c$;   26. (a) $y = x \ln|cx|$;

    (b) $3x^2 - 2xy - y^2 = c$.   (b) $y^2 + xy = cx^3$.

Section 2.3, Page 56.

2. $x^2y + \dfrac{1}{x} = c.$

4. $y = 2 + ce^{-2x^2}.$

6. $(u^2 + 1)^2v = \dfrac{3u^4}{4} + \dfrac{3u^4}{2} + c.$

8. $3(x + 2)y = x - 1 + c(x - 1)^{-2}.$

10. $xy = 1 + ce^{-y^2/2}.$

12. $2r = (\theta + \sin\theta\cos\theta + c)\cos\theta$

14. $(1 + \sin^2 x)y = \sin x + c.$ 16. $x^3y^3(2x^3 + c) = 1.$

18. $x^2 = 2 + ct^{-1}e^{-t}.$ 20. $3y = 1 + 5e^{-x^3}.$

22. $y = -2x^2 - 3.$

24. $x = \dfrac{1}{5}(2e^t - \sin 2t - 2\cos 2t).$

26. $\dfrac{1}{\sqrt{xy}} = -\dfrac{1}{2}x + 1.$

28. $\begin{cases} y = 5 + e^{-x} & , \; 0 \leq x < 10; \\ y = 1 + (4e^{10} + 1)e^{-x} & , \quad x \geq 10. \end{cases}$

30. $\begin{cases} y = \dfrac{x^2 + 1}{2(x + 1)} & , \; 0 \leq x < 3, \\ y = \dfrac{3x - 4}{x + 1} & , \quad x \geq 3. \end{cases}$

36. (b) $y = ce^{-x} + \sum_{k=1}^{5} \dfrac{\sin kx - k \cos kx}{1 + k^2}$

40. $\dfrac{e^{-x^2/2}}{y - x} = \displaystyle\int e^{-x^2/2}\, dx + c.$

Miscellaneous Review Exercises, Page 59.

2. $x^2 y^3 - xy = c.$

4. $y = \dfrac{x^2}{4} + \dfrac{c}{x^2}.$

6. $(e^{2x} - 2)y^2 = c.$

8. $\left(\dfrac{y + x}{y + 2x}\right)^2 = \left|\dfrac{c}{x}\right|.$

10. $y = e^{-x}[-1 + c(x + 1)].$

12. $y^2 = \dfrac{1}{(1 + cx^2)}.$

14. $(y - 2x)(y + x) = cx.$

16. $(x + 1)\sqrt{y^2 + 4} = 4(x - 1).$

18. $x^3 + x^2 y^2 + 2y^3 = 21.$

20. $5(2x + y)^2 = 16(x + 2y).$

22. $y = \begin{cases} 1 - e^{-x} & , \ 0 \leq x < 2, \\ (e^2 - 1)e^{-x} & , \ \quad x > 2. \end{cases}$

24. $y = \dfrac{\sqrt{2x}}{\sqrt[4]{x^4 + 1}}$

Section 2.4, Page 67.

2. $x^2 \cos y + x \sin y = c.$

4. $x^2 + xy^{-1} + y^2 = c.$

6. $x^{2/3} y^{5/3}(x^2 - y) = c.$

8. $x - 2y + \ln|3x - y - 2| = c.$

10. $(2x + y + 3)^3 (x - y + 1)^2 = c.$

12. $\ln[3(x - 1)^2 + (y + 3)^2] + \dfrac{2}{\sqrt{3}} \arctan \dfrac{y + 3}{\sqrt{3}(x - 1)}$

$\qquad = \ln 4 + \dfrac{\pi}{3\sqrt{3}}.$

14. $\ln|2x + y - 1| + \dfrac{x - 2}{2x + y - 1} = 1.$

Section 3.1, Page 78.

2. $y^2 + 2x^2 = k^2.$

4. $y^2 \left(\ln|y| - \dfrac{1}{2} \right) = -x^2 + k.$

6. $x^2 + y^2 - 2x + \ln(x + 2)^4 = k.$

8. $x^2 y = \dfrac{x^4}{4} + k.$

10. $y = k(x^2 + 3y^2).$

12. $x^2 + y^2 = ky.$

14. $n = 3.$

16. $\ln(x^2 + y^2) + 2 \arctan(y/x) = k.$

18. $\ln|3x^2 + 3xy + 4y^2| - (2/\sqrt{39}) \arctan[(3x + 8y)/\sqrt{39}x] = k.$

Section 3.2, Page 88.

2.   (a) v = 9 ft./sec.;   x = 539.16 ft.;
     (b) v = 9 ft./sec.

4.   (a) $v = \frac{25}{2} (1 - e^{-t/250})$;   (b) 12.5 ft./sec.;
     (c) 402 sec.

6.   (a) 33.99 cm/sec.;   8.   (a) 7.16 ft./sec.;
     (b) 88.82 cm/sec.   (b) 4.95 sec.

10.   (a) v = 100 tan (arc tan 10 - 0.32t);
      (b) 4.60 sec.

12.   3.06 ft./sec.   14.   17.5 ft./sec.

16.   0.25   18. $v = \left[\frac{2gR^2}{x} + v_0^2 - 2gR\right]^{1/2}$.

Section 3.3, Page 102.

2.   (a) 34.5%;   4.   (a) 19.8%;
     (b) t = 43.4 min.   (b) 9 hrs., 58 min.

6.   (a) 10.22 oz.;   (b) 23 hrs., 34 min.

8.   18.91 min.

10.   (a) 151.41⁰;
      (b) Between 22 min., 31 sec., after 10 A.M., and 30 min.,
          12 sec., after.

12. 40,833.

14. (a) $x = (4219)(10^6)e^{0.02(t-1978)}$.

(b) 2410 million.   (c) 6551 million.

(d) 886 million.   (e) 48,405 million.

16. (a) $x = \dfrac{kx_0}{\lambda x_0 + (k - \lambda x_0)e^{-k(t-t_0)}}$

18. 94,742

20. (a) 34.66%;   (b) 0.85 years

22. (a) 7.81 lbs; concentration is 0.0539 lb./gal.

(b) 0.0217 lb./gal.

24. (a) 466.12 lb.;   (b) 199.99 lb.

26. 4,119.65 gm.   28. 292.96 $(ft.)^3$/min.

30. 2016.   32. 7.14 grams.

Section 4.1B, Page 122.

2. $y = 0$ for all real x.

4. (b) and (c). Theorem 4.2.

8. (b) $y = c_1 e^x + c_2 x e^x$.

(c) $y = e^x + 3x e^x$; $-\infty < x < \infty$.

10.   (b) $y = c_1 x^2 + \dfrac{c_2}{x^2}$.

   (c) $y = \dfrac{1}{4} x^2 + \dfrac{8}{x^2}$; Theorem 4.1; $0 < x < \infty$.

12. $y = c_1 e^{-x} + c_2 e^{3x} + c_3 e^{4x}$.

Section 4.1D, Page 132.

2. $y = c_1 (x + 1) + c_2 (x + 1)^3$.

4. $y = c_1 x + c_2 (x - 1) e^x$.

6. $y = xe^x$; $y = c_1 x^2 + c_2 xe^x$.

8. $y = c_1 e^x + c_2 x^{-1}$.

10.   (b) $y = c_1 e^x + c_2 e^{2x}$.

   (d) $y = c_1 e^x + c_2 e^{2x} + 2x^2 + 6x + 7$.

Section 4.2, Page 143.

2. $y = c_2 e^{3x} + c_2 e^{-x}$.   4. $y = c_1 e^{5x} + c_2 e^{-x/3}$.

6. $y = c_1 e^{x/2} + c_2 e^{-2x}$.   8. $y = (c_1 + c_2 x) e^{2x}$.

10. $y = e^{-x}\left(c_1 \sin \dfrac{3x}{4} + c_2 \cos \dfrac{3x}{4} \right)$.

12. $y = c_1 e^{-x} + c_2 e^{3x} + c_3 e^{4x}$.

14. $y = c_1 e^{-2x} + (c_2 + c_3 x) e^{x/2}$.

16. $y = c_1 e^{-3x} + e^{-x/2} \left(c_2 \sin \frac{\sqrt{7}}{2} x + c_3 \cos \frac{\sqrt{7}}{x} x \right)$.

18. $y = (c_1 + c_2 x) e^{-x/2}$.

20. $y = e^{-3x} (c_2 \sin 4x + c_2 \cos 4x)$.

22. $7 = c_1 \sin \frac{x}{2} + c_2 \cos \frac{x}{2}$.

24. $y = (c_1 + c_2 x + c_3 x^2) e^{-x/2}$.

26. $y = c_1 e^{x} + c_2 e^{-x} + c_3 \sin x + c_4 \cos x$.

28. $y = c_1 e^{x} + c_2 e^{2x} + (c_3 + c_4 x) e^{-x}$.

30. $y = (c_1 + c_2 x) e^{-2x} + e^{-x} (c_3 \sin \sqrt{2} x + c_4 \cos \sqrt{2} x)$.

32. $y = (c_1 + c_2 x + c_3 x^2 + c_4 x^3 + c_5 x^4) e^{-x}$.

34. $y = (c_1 + c_2 x) e^{x}$

$\qquad + e^{-x/2} \left[(c_3 + c_4 x) \sin \frac{\sqrt{3}x}{2} + (c_5 + c_6 x) \cos \frac{\sqrt{3}x}{2} \right]$

36. $y = c_1 \sin 2x + c_2 \cos 2x + e^{\sqrt{3}x}(c_3 \sin x + c_4 \cos x)$

$+ e^{-\sqrt{3}x}(c_5 \sin x + c_6 \cos x)$.

38. $y = -6e^{-2x} + 2e^{-5x}$.   40. $y = 2e^{-2x}$.

42. $y = 3xe^{3x/2} + 4e^{3x/2}$.   44. $y = (3 - 2x)e^{x/3}$.

46. $y = e^{-3x}\left(\dfrac{2}{7} \sin 7x - \cos 7x\right)$.

48. $y = e^{-x}(4 \sin 2x + 2 \cos 2x)$.

50. $y = e^{-x/2}(-\sin 3x + 2 \cos 3x)$.

52. $y = e^{2x} - \sin 2x + \cos 2x$.

54. $y = e^{x} + e^{2x}(2 \sin x - \cos x)$.

56. $y = e^{-2x} - \sin 2x$.

58. $y = e^{-x/2}\left[(c_1 + c_2 x)\sin \dfrac{\sqrt{7}x}{2} + (c_3 + c_4 x)\cos \dfrac{\sqrt{7}x}{2}\right]$.

60. $y = (c_1 + c_2 x + c_3 x^2 + c_4 x^3 + c_5 x^4 + c_6 x^5)e^{2x}$

$+ e^{3x}[(c_7 + c_8 x + c_9 x^2)\sin 4x$

$+ (c_{10} + c_{11}x + c_{12}x^2)\cos 4x]$.

62. $y = c_1 e^{-2x} + c_2 e^{-3x} + e^{x}(c_3 \sin 2x + c_4 \cos 2x)$.

Section 4.3, Page 159.

2. $y = c_1 e^{4x} + c_2 e^{-2x} - \frac{1}{2} e^{2x} - 3e^{-3x}$.

4. $y = e^{-x}(c_1 \sin x + c_2 \cos x) - \frac{7}{13} \sin 4x - \frac{4}{13} \cos 4x$.

6. $y = c_1 e^{4x} + c_2 e^{-x} + 2e^{2x} - 4x + 3$.

8. $y = e^{-x}(c_1 \sin 3x + c_2 \cos 3x) + \frac{1}{2} xe^{-2x} + \frac{1}{10} e^{-2x}$.

10. $y = c_1 e^{-2x} + c_2 e^{-4x} + c_2 e^{-4x} + \frac{1}{4} xe^{2x} - \frac{5}{48} e^{2x}$

$\qquad + x^2 - \frac{3}{2} x + \frac{7}{8}$.

12. $y = c_1 e^{2x} + c_2 e^{-2x} + 2x^2 e^{2x} - xe^{2x}$.

14. $y = (c_1 + c_2 x)e^{3x} + 3x^2 e^{3x} + 5xe^{4x} - 10e^{4x}$.

16. $y = c_1 e^{2x} + e^{-2x}(c_2 \sin x + c_3 \cos x) - 2xe^{-2x} - \frac{1}{2} e^{-2x}$.

18. $y = c_1 e^{-x} + c_2 e^{x/2} + c_3 e^{3x/2} + x^3 + 5x^2 + 22x + 42$.

20. $y = c_1 e^{x} + c_2 e^{-2x} + xe^{x} - 2xe^{-2x} + 2x^2 + 2x + 3$.

22. $y = e^{2x}(c_1 \sin x + c_2 \cos x) + 3xe^{2x} \sin x$.

24. $y = c_1 e^{x} + c_2 e^{-x} + c_3 e^{2x} + 3xe^{2x} - e^{3x}$.

26. $y = c_1 + c_2 x + c_3 e^x + c_4 e^{2x} + \frac{1}{2} e^{-x} + \frac{3}{2} x e^{2x}$

$- \frac{1}{2} x^3 - \frac{9}{4} x^2.$

28. $y = (c_1 + c_2 x) e^x + c_3 e^{2x} - \frac{1}{4} x^4 e^x + \frac{1}{2} x^2 e^x.$

30. $y = c_1 + c_2 e^{2x} + c_3 e^{-2x} + 2x^2 e^{2x} - 3x e^{2x} + 2x^3 + 3x.$

32. $y = c_1 \sin 2x + c_2 \cos 2x + 3x^2 - \frac{3}{2} - 2x^2 \sin 2x - x \cos 2x.$

34. $y = c_1 e^{2x} + c_2 e^{3x} + c_3 \sin x + c_4 \cos x + \frac{1}{13} \sin 2x$

$+ \frac{5}{13} \cos 2x + \frac{1}{4} x \sin x - \frac{1}{4} x \cos x.$

36. $y = 3e^{-x} + 2e^x + 4x - 5.$

38. $y = e^{-2x} - 2x e^{-3x} - e^{-3x}.$

40. $y = (3x - 5) e^{-3x} + 3e^{-6x}.$

42. $y = 2e^{5x} (2 \sin 2x - \cos 2x + 1).$

44. $y = e^{-2x} + 4e^{3x} - 2e^{2x} - x e^{3x}.$

46. $y = \frac{9e^x}{8} - \frac{e^{-x}}{8} + \frac{x^3 e^x}{2} - \frac{3x^2 e^x}{4} + \frac{3x e^x}{4}.$

48. $y = 5 \sin 2x + 6 \cos 2x - 2x \cos 2x.$

50. $y = \left(\frac{122}{9} - \frac{4}{3} x \right) e^x - \frac{5}{9} e^{4x} - 2x^2 - 9x - 15 + 3e^{2x}.$

52. $y_p = Ae^{3x} + Be^{-3x} + Ce^{3x}\sin 3x + De^{3x}\cos 3x.$

54. $y_p = Ax^4e^x + Bx^3e^x + Cx^2e^x + Dxe^x + Ee^x + Fx^3e^{2x}$

$\qquad + Gx^2e^{2x} + Hxe^{2x} + Ie^{2x} + Jx^4e^{3x} + Kx^3e^{3x} + Lx^2e^{3x}.$

56. $y_p = Ax^3e^x + Bx^2e^x + Cxe^x + Dx^2e^{2x} + Exe^{2x} + Fx^3$

$\qquad + Gx^2 + Hx.$

58. $y_p = Ax^4e^{-x} + Bx^3e^{-x} + Cx^2e^{-x} + Dxe^{-x/2}\sin\frac{\sqrt{3}}{2}x$

$\qquad + Exe^{-x/2}\cos\frac{\sqrt{3}}{2}x.$

60. $y_p = Ax^7 + Bx^6 + Cx^5 + Dx^4 + Ex^2e^{-x} + Fxe^{-x} + Ge^{-x}$

$\qquad + Hxe^{-x}\sin 2x + Ixe^{-x}\cos 2x.$

62. $y_p = Ax^2e^{\sqrt{2}x}\sin\sqrt{2}x + Bx^2e^{\sqrt{2}x}\cos\sqrt{2}x + Cxe^{\sqrt{2}x}\sin\sqrt{2}x$

$\qquad + Dxe^{\sqrt{2}x}\cos\sqrt{2}x + Exe^{-\sqrt{2}x}\sin\sqrt{2}x + Fxe^{-\sqrt{2}x}\cos\sqrt{2}x.$

64. $y_p = Ax\sin x\sin 2x + Bx\sin x\cos 2x + Cx\cos x\sin 2x$

$\qquad + Dx\cos x\cos 2x.$

Section 4.4, Page 169.

2. $y = c_1\sin x + c_2\cos x + (\sin x)[\ln|\sec x + \tan x|] - 2.$

4. $y = c_1\sin x + c_2\cos x + \sin x\tan x - \dfrac{\sec x}{2}.$

$$OR \ y = c_1 \sin x + c_2 \cos x + \frac{(\sin x \tan x)}{2}.$$

6. $y = c_1 \sin x + c_2 \cos x - \sin x \ln|\cos x| + x \cos x.$

8. $y = e^x(c_1 \sin 2x + c_2 \cos 2x)$

$$- \frac{1}{4} e^x \cos 2x[\ln|\sec 2x + \tan x|].$$

10. $y = c_1 e^x + c_2 x e^x - \frac{5x^3 e^x}{36} + \frac{x^3 e^x \ln|x|}{6}.$

12. $y = c_1 \sin x + c_2 \cos x + \frac{\tan x}{2} + \frac{3}{2} \cos x[\ln|\sec x + \tan x|].$

14. $y = c_1 e^{-x} + c_2 e^{-2x} + e^{-x} \arctan e^x - \frac{1}{2} e^{-2x} \ln(1 + e^{2x}).$

16. $y = c_1 e^x + c_2 x e^x + \frac{e^x \sin^{-1} x}{4} + \frac{x^2 e^x \sin^{-1} x}{2} + \frac{3x e^x \sqrt{1 - x^2}}{4}.$

18. $y = (c_1 + c_2 x)e^x - e^x \left[\int \frac{x^2 \ln x}{e^x} \, dx \right] + x e^x \left[\int \frac{x \ln x}{e^x} \, dx \right].$

20. $y = c_1(x + 1) + c_2(x + 1)^2 + \frac{1}{2}.$

22. $y = c_1 x + c_2 x e^x - x^2 - x.$

24. $y = c_1 x + c_2 (x + 1)^{-1} + x^2 - \frac{(2x^3 + 3x^2)(x + 1)^{-1}}{6}.$

26. $y = c_1 e^x + c_2 x^2 e^x + \frac{2}{3} e^x x^{-1}.$

28. $y = c_1 e^x + c_2 e^{-x} + c_3 e^{3x} - \dfrac{x^3 e^x}{12} - \dfrac{x e^x}{8}.$

Section 4.5, Page 176.

2. $y = c_1 x^2 + \dfrac{c_2}{x^2}.$

4. $y = c_1 x^2 + c_2 x^2 \ln x.$

6. $y = x^2 [c_1 \sin(\ln x^3) + c_2 \cos(\ln x^3)].$

8. $y = c_1 \sin(\ln x^3) + c_2 \cos(\ln x^3).$

10. $y = x^3 [c_1 \sin(\ln x) + c_2 \cos(\ln x)].$

12. $y = \dfrac{c_1}{x} + \dfrac{c_2}{x^2} + c_3 x^4.$

14. $y = c_1 x + c_2 x^{-1} + c_3 x^2 + c_4 x^4.$

16. $y = c_1 x^2 + c_2 x^4 - 2x^3.$

18. $y = c_1 \sin(\ln x^2) + c_2 \cos(\ln x^2) + \dfrac{x \ln x^2}{5} - \dfrac{4x}{25}.$

20. $y = x^2 [c_1 \sin(\ln x) + c_2 \cos(\ln x)] + 5x^2.$

22. $y = (c_1 + c_2 \ln x)x + c_3 x^2 + \dfrac{x^3}{4}.$

24. $y = -2x^2 + x^3$

26. $y = x^{-1} + x^2 - 2x + 4.$

28. $y = -x^2 + 2x^{-3} + 2x^2 \ln x$.

30. $y = \frac{1}{6}\left(\frac{x^{-2}}{2} + \frac{x^3}{3} - \ln x + \frac{1}{6}\right)$.

32. $y = c_1(2x - 3) + c_2(2x - 3)^3$.

Section 4.6, Page 184.

1.   (a) $f_1(x) = 2e^x - e^{2x}$, $f_2(x) = -e^x + e^{2x}$.

   (b) $5f_1(x) + 7f_2(x)$.

Section 5.2, Page 197.

2.   (a) $x = \frac{1}{4} \sin 8t + \frac{1}{3} \cos 8t$.

   (b) $x = -\frac{1}{4} \sin 8t + \frac{1}{3} \cos 8t$.

   (c) $x = \frac{1}{4} \sin 8t - \frac{1}{3} \cos 8t$.

4. $\sqrt{442}/7$ (cm); $\pi/7$ (sec.); $7/\pi$ (cycles/sec.).

6.   (a) $\frac{1}{2}$ (ft.); $2\sqrt{3}$ ft./sec.

   (b) $\frac{\pi}{3} + \frac{2n\pi}{4}$ sec.; $2\sqrt{3}$ ft./sec.

8. $k = 4$; $A = \frac{\sqrt{71}}{4}$.

10.   (a) $\theta = c_1 \sin\sqrt{\frac{g}{\ell}}\, t + c_2 \cos\sqrt{\frac{g}{\ell}}\, t$; $A = \theta_0$; period $= 2\pi\sqrt{\frac{\ell}{g}}$.

      (b) $\frac{d\theta}{dt} = \pm\sqrt{\frac{2g}{\ell}}\, \sqrt{\cos\theta - \cos\theta_0}$.

Section 5.3, Page 208.

2. $x = \frac{\sqrt{3}}{3} e^{-6t} \sin 2\sqrt{3}t$.

4. $x = -\frac{1}{12} e^{-16t} + \frac{1}{3} e^{-4t}$.

6. $x = \frac{e^{-4t}}{100} (\sqrt{6} \sin 2\sqrt{6}t + 2 \cos 2\sqrt{6}t)$.

8.   (a) $x = e^{-4t}(\sin 4t + \cos 4t)$; $x = \sqrt{2}e^{-4t}\cos\left(4t - \frac{\pi}{4}\right)$;

      (b) $\pi/2$; $\sqrt{2}e^{-4t}$;

      (c) $\pi/4$; 0.04(ft.)

10.   (a) $x = \left(\frac{1}{4} + 2t\right)e^{-8t}$.

      (b) $x = e^{-4t}\left(\frac{\sqrt{3}}{12} \sin 4\sqrt{3}t + \frac{1}{4} \cos 4\sqrt{3}t\right)$.

      (c) $x = \left(\frac{3 + 2\sqrt{3}}{24}\right)e^{(-16+8\sqrt{3})t} + \left(\frac{3 - 2\sqrt{3}}{24}\right)e^{(-16-8\sqrt{3})t}$

12.   (a) $8\sqrt{6}$;

      (b) $8\sqrt{2}$

Section 5.4, Page 217.

2.   (a) $x = e^{-4t}\left(-\frac{21}{74} \sin 8t + \frac{11}{37} \cos 8t\right)$

$+ \frac{8}{37} \sin 16t - \frac{11}{37} \cos 16t$.

4. (a) $x = e^{-8t}\left(\dfrac{11}{20}\sin 8t + \dfrac{9}{10}\cos 8t\right) + \dfrac{7}{10}\sin 4t - \dfrac{2}{5}\cos 4t.$

   (b) amplitude: $\dfrac{\sqrt{65}}{10}.$

6. $x = e^{-t}(c_1\sin 3t + c_2\cos 3t) + \dfrac{9}{85}\sin t - \dfrac{2}{85}\cos t$

   $\qquad + \dfrac{3}{104}\sin 2t - \dfrac{1}{52}\cos 2t + \dfrac{1}{333}\sin 3t - \dfrac{6}{333}\cos 3t,$

   where $\quad c_1 = -54{,}854/1{,}471{,}860, \qquad c_2 = 29{,}819/490{,}620.$

10. (b) $x = \dfrac{\sqrt{37}}{8}\,e^{-2t}[\cos(4t - \phi)] + \dfrac{\sqrt{5}}{4}\,[\cos(2t - \theta)],$

   where $\phi = \cos^{-1}\dfrac{6}{\sqrt{37}}\quad$ and $\quad \theta = \cos^{-1}\left(-\dfrac{1}{\sqrt{5}}\right)$

   or $x = e^{-2t}\left[\dfrac{3}{4}\cos 4t + \dfrac{1}{8}\sin 4t\right] + \dfrac{1}{2}\sin 2t - \dfrac{1}{4}\cos 2t.$

Section 5.5, Page 224.

2. $a = 5\sqrt{2 - \dfrac{\pi^2}{32}}.$

Section 5.6, Page 232.

2. $i = \dfrac{15}{17}\left(\cos 200t + 4\sin 200t - e^{-50t}\right).$

4. $q = \left(\dfrac{1}{200} + \dfrac{t}{2}\right)e^{-100t} - \dfrac{1}{200}\cos 100t.$

6. $i = e^{-200t}(1.0311 \sin 979.8t + 0.1031 \cos 979.8t)$

$\qquad - 0.1031\, e^{-100t}.$

8. (d) $\omega = 50$; ampl.: $\dfrac{100}{\sqrt{(-187.5)^2 + (20)^2}} \approx \dfrac{100}{190}.$

Section 6.1, Page 249.

2. $y = c_0(1 + 2x^2 - 2x^4 + \cdots) + c_1\left(x - \dfrac{2}{3}x^3 + \dfrac{2}{3}x^5 + \cdots\right).$

4. $y = c_0\left(1 - \dfrac{1}{4}x^4 + \dfrac{1}{20}x^5 - \cdots\right)$

$\qquad + c_1\left(x - \dfrac{1}{2}x^2 + \dfrac{1}{6}x^3 - \dfrac{1}{24}x^4 - \dfrac{17}{120}x^5 + \cdots\right).$

6. $y = c_0\left(1 + 2x^2 + \dfrac{1}{4}x^4 + \cdots\right)$

$\qquad + c_1\left(x + \dfrac{1}{2}x^3 - \dfrac{1}{20}x^5 + \cdots\right).$

8. $y = c_0\left(1 + x^2 - \dfrac{1}{2}x^3 + \cdots\right) + c_1\left(x + \dfrac{1}{2}x^3 - \dfrac{1}{4}x^4 + \cdots\right).$

10. $y = c_0\left(1 - \dfrac{1}{2}x^2 - \dfrac{1}{24}x^4 - \dfrac{1}{20}x^5 - \dfrac{1}{240}x^6 + \cdots\right)$

$\qquad + c_1\left(x + \dfrac{1}{12}x^4 + \dfrac{1}{120}x^6 + \cdots\right).$

12. $y = c_0\left(1 + \dfrac{1}{3}x^3 + \dfrac{1}{35}x^4 + \cdots\right) + c_1\left(x + x^2 + \dfrac{1}{2}x^3 + \cdots\right).$

14. $y = c_0\left(1 - \dfrac{1}{6}x^2 + \dfrac{1}{18}x^3 - \cdots\right)$

$\qquad + c_1\left(x - \dfrac{1}{3}x^2 + \dfrac{1}{36}x^4 + \cdots\right).$

16. $y = x + \dfrac{1}{6} x^3 - \dfrac{1}{120} x^5 + \cdots$.

18. $y = 2 + 3x - \dfrac{7x^3}{6} - \dfrac{x^4}{2} + \dfrac{21x^5}{40} + \cdots$.

20. $y = \displaystyle\sum_{n=0}^{\infty} (-1)^n x^n$.

22. $y = c_0\left[1 + \dfrac{1}{2}(x-1)^2 - \dfrac{5}{6}(x-1)^3 + \cdots\right]$

$\qquad + c_1\left[(x-1) - \dfrac{3}{2}(x-1)^2 + \dfrac{13}{6}(x-1)^3 - \cdots\right]$.

24.   (a) $y = c_0\left[1 - \dfrac{n(n+1)}{2!} x^2\right.$

$\qquad\qquad + \dfrac{n(n-2)(n+1)(n+3)}{4!} x^4 - \cdots\Bigg]$

$\qquad + c_1\left[x - \dfrac{(n-1)(n+2)}{3!} x^3\right.$

$\qquad\qquad + \dfrac{(n-1)(n-3)(n+2)(n+4)}{5!} x^5 - \cdots\Bigg]$.

Section 6.2, Page 269.

2. $x = 0$ and $x = -1$ are regular singular pts.

4. $x = 0$ is an irregular singular pt.;
$\quad$ $x = -3$ and $x = 2$ are regular singular pts.

6. $y = C_1 x^{3/2}\left(1 - \dfrac{1}{9} x^2 + \dfrac{1}{234} x^4 - \cdots\right)$

$\qquad + C_2 x^{-1}\left(1 + x^2 - \dfrac{1}{6} x^4 + \cdots\right)$.

8. $y = C_1 x^{5/3}\left(1 - \dfrac{3}{10}\, x^2 + \dfrac{9}{320}\, x^4 + \cdots\right)$

$\qquad + C_2 x^{1/3}\left(1 - \dfrac{3}{2}\, x^2 + \dfrac{9}{32}\, x^4 + \cdots\right).$

10. $y = C_1 x^{1/2}\left(1 - \dfrac{2}{3}\, x + \dfrac{2}{15}\, x^2 - \cdots\right)$

$\qquad + C_2\left(1 - 2x + \dfrac{2}{3}\, x^2 - \dfrac{4}{45}\, x^3 + \cdots\right).$

12. $y = C_1 x^{1/2}\left(1 - \dfrac{2x}{7} + \dfrac{2x^2}{63} - \dfrac{4x^3}{2079} + \cdots\right)$

$\qquad + C_2 x^{-2}\left(1 + \dfrac{2x}{3} + \dfrac{2x^2}{3} - \dfrac{4x^3}{9} + \cdots\right).$

14. $y = C_1 x^{5/2}\left(1 - \dfrac{25x}{44} + \dfrac{1225x^2}{4576} - \cdots\right) + C_2 x^{-2}\left(1 + \dfrac{4x}{7} + \dfrac{2x^2}{35}\right).$

16. $y = C_1 x^{1/2}\left(1 - \dfrac{x^2}{6} + \dfrac{x^4}{120} - \cdots\right)$

$\qquad + C_2 x^{-1/2}\left(1 - \dfrac{x^2}{2} + \dfrac{x^4}{24} - \cdots\right).$

18. $y = C_1\left[1 + \displaystyle\sum_{n=1}^{\infty} \dfrac{x^{2n}}{2^n n!}\right]$

$\qquad + C_2 x^3\left[1 + \displaystyle\sum_{n=1}^{\infty} \dfrac{x^{2n}}{[5 \cdot 7 \cdot 9 \cdots (2n + 3)]}\right]$

$\quad = C_1\left(1 + \dfrac{x^2}{2} + \dfrac{x^4}{8} + \cdots\right) + C_2 x^3\left(1 + \dfrac{x^2}{5} + \dfrac{x^4}{35} + \cdots\right).$

20. $y = C_1 x^{1/2} + C_2 x^{3/2}.$

22. $y = C_1 x^{-1} \sum_{n=0}^{\infty} \frac{(-1)^n x^n}{(n + 2)!} + C_2 x^{-3}(1 - x).$

24. $y = C_1 x^{5/2}\left(1 - \frac{x^2}{3} + \frac{x^4}{12} - \cdots\right) + C_2 x^{-3/2}(1 - x^2).$

26. $y = C_1 x^3\left(1 - \frac{1}{4} x^2 + \frac{5}{128} x^4 - \cdots\right)$

  $+ C_2\left[x^{-1}\left(1 - \frac{1}{4} x + \frac{1}{192} x^5 - \cdots\right) + \frac{1}{16} y_1(x)\ln|x|\right],$

where $y_1(x)$ denotes the solution of which C_1 is the

coefficient.

28. $y = C_1 x^{3/2}\left[1 + \sum_{n=1}^{\infty} \frac{(-1)^n [3 \cdot 5 \cdot 7 \cdots (2n + 1)]}{2^{n-1} n!(n + 2)!} x^n\right]$

  $+ C_2\left[x^{-1/2}\left(- \frac{1}{2} + \frac{1}{4} x - \frac{5}{64} x^2 + \cdots\right)\right.$

  $\left. - \frac{1}{16} y_1(x)\ln|x|\right],$ where $y_1(x)$ denotes the solution

of which C_1 is the coefficient.

30. $y = C_1\left[1 + \sum_{n=1}^{\infty} \frac{(-1)^n x^n}{2^{n-1} n!(n + 2)!}\right]$

  $+ C_2\left[x^{-2}\left(- \frac{1}{2} - \frac{1}{4} x + \frac{29}{576} x^2 + \cdots\right)\right.$

  $\left. + \frac{1}{16} y_1(x)\ln|x|\right],$ where $y_1(x)$ denotes the solution

of which C_1 is the coefficient.

32. $y = C_1 x^3 \left[1 + \sum_{n=1}^{\infty} \frac{(-1)^n x^{2n}}{2^{n-1} n! (n + 2)!} \right]$

$\qquad + C_2 \left[x^{-1} \left(-\frac{1}{4} - \frac{1}{16} x^2 + \frac{29}{4608} x^4 + \cdots \right) \right.$

$\qquad \left. + \frac{1}{64} y_1(x) \ln|x| \right]$, where $y_1(x)$ denotes the solution

of which C_1 is the coefficient.

Section 7.1, Page 296.

2. $\begin{cases} x = ce^{-t} - 2, \\[2mm] y = -2ce^{-t} - t^2 - 2t + 4. \end{cases}$

4. $\begin{cases} x = c_1 e^t + c_2 e^{3t}, \\[2mm] y = -\dfrac{c_1}{3} e^t - c_2 e^{3t} - 3e^{2t}. \end{cases}$

6. $x = c_1 e^t - \dfrac{\sin t}{2}, \quad y = -\dfrac{c_1 e^t}{3} + \dfrac{\sin t}{2}.$

8. $\begin{cases} x = c_1 e^{-t} + t - 1 - \dfrac{1}{2} e^t, \\[2mm] y = c_2 e^t - \dfrac{5}{2} c_1 e^{-t} - \dfrac{1}{2} t e^t - 4t + 1. \end{cases}$

10. $\begin{cases} x = c_1 e^{\sqrt{3}t} + c_2 e^{-\sqrt{3}t} + 3t - 3, \\[2mm] y = \dfrac{\sqrt{3}}{3} c_1 e^{\sqrt{3}t} - \dfrac{\sqrt{3}}{3} c_2 e^{-\sqrt{3}t} - 2t + \dfrac{4}{3}. \end{cases}$

12.
$$\begin{cases} x = c_1 \sin t + c_2 \cos t - t - 3, \\ y = (c_2 - c_1)\sin t - (c_1 + c_2)\cos t - 1. \end{cases}$$

14.
$$\begin{cases} x = 3e^t, \\ y = -2e^t - \frac{1}{2} e^{2t}. \end{cases}$$

16.
$$\begin{cases} x = c_1 + c_2 e^{-2t} + 2t^2 + t, \\ y = (1 - c_1) - 3c_2 e^{-2t} - t^2 - 3t. \end{cases}$$

18.
$$\begin{cases} x = c_1 e^{2t} + \frac{1}{2} t^2 + \frac{3}{2} t + \frac{3}{4}, \\ y = -3c_1 e^{2t} + c_2 e^t - \frac{1}{2} t^2 + \frac{3}{2} t + \frac{15}{4}. \end{cases}$$

20.
$$\begin{cases} x = e^{-t}(c_1 \sin 2t + c_2 \cos 2t) + 2, \\ y = e^{-t}[2(-c_1 + c_2)\sin 2t - 2(c_1 + c_2)\cos 2t] \\ \qquad + t^2 + 2t - 14. \end{cases}$$

22.
$$\begin{cases} x = c_1 + c_2 e^t + c_3 t e^t + \frac{1}{2} e^{2t}, \\ y = -c_1 - c_2 e^t - c_3(t + 1)e^t - \frac{1}{2} e^{2t}. \end{cases}$$

24.
$$\begin{cases} x = c_1 + c_2 e^{2t} + c_3 e^{-2t} - \frac{2}{3} e^t, \\ y = -4c_1 + 2c_2 e^{2t} - 2c_3 e^{-2t} - \frac{5}{3} e^t. \end{cases}$$

26.
$$\begin{cases} x = c_1 e^{-t} + e^{-2t}(c_2\sin t + c_3\cos t) - \dfrac{4}{51} e^{2t}, \\[2mm] y = \dfrac{1}{4} c_1 e^{-t} + e^{-2t}\left(\dfrac{2}{5} c_2 + \dfrac{1}{5} c_3\right)\sin t \end{cases}$$

$$+ \left(-\dfrac{1}{5} c_2 + \dfrac{2}{5} c_3\right)\cos t + \dfrac{5}{51} e^{2t}.$$

28.
$$\begin{cases} \dfrac{dx_1}{dt} = x_2, \\[3mm] \dfrac{dx_2}{dt} = x_3, \\[3mm] \dfrac{dx_3}{dt} = 2x_1 + x_2 - 2x_3 + e^{3t}. \end{cases}$$

30.
$$\begin{cases} \dfrac{dx_1}{dt} = x_2, \\[3mm] \dfrac{dx_2}{dt} = x_3, \\[3mm] \dfrac{dx_3}{dt} = x_4, \\[3mm] \dfrac{dx_4}{dt} = -2tx_1 + t^2 x_3 + \cos t. \end{cases}$$

Section 7.2, Page 307.

2. (b)
$$\begin{cases} x = c_1 + c_2 e^{-kt/m}, \\[2mm] y = c_3 + c_4 e^{-kt/m} - \dfrac{mg}{k} t. \end{cases}$$

6.
$$\begin{cases} x = 10 e^{-t/5} + 20, \\[2mm] y = -20 e^{-t/5} + 20. \end{cases}$$

Section 7.3, Page 317.

2.   (b)
$$\begin{cases} x = 3c_1 e^{7t} + c_2 e^{-t}, \\ \\ y = 2c_1 e^{7t} - 2c_2 e^{-t}. \end{cases}$$

(c)
$$\begin{cases} x = 3e^{7t} - 3e^{-t}, \\ \\ y = 2e^{7t} + 6e^{-t}. \end{cases}$$

Section 7.4, Page 328.

2.
$$\begin{cases} x = c_1 e^{4t} + c_2 e^{2t}, \\ \\ y = c_1 e^{4t} + 3c_2 e^{2t}. \end{cases}$$

4.
$$\begin{cases} x = c_1 e^{t} + c_2 e^{-t}, \\ \\ y = -3c_1 e^{t} - c_2 e^{-t}. \end{cases}$$

6.
$$\begin{cases} x = c_1 e^{5t} + c_2 e^{3t}, \\ \\ y = c_1 e^{5t} + 3c_2 e^{3t}. \end{cases}$$

8.
$$\begin{cases} x = e^{2t}(c_1 \cos 3t + c_2 \sin 3t), \\ \\ y = e^{2t}(3c_1 \sin 3t - 3c_2 \cos 3t). \end{cases}$$

10.
$$\begin{cases} x = c_1 e^{5t} + c_2 e^{-3t}, \\ y = c_1 e^{5t} - 3c_2 e^{-3t}. \end{cases}$$

12.
$$\begin{cases} x = e^{2t}(c_1 \cos 3t - c_2 \sin 3t), \\ y = e^{2t}(c_1 \cos 3t + c_2 \sin 3t). \end{cases}$$

14.
$$\begin{cases} x = 2e^{3t}(c_1 \cos 2t + c_2 \sin 2t), \\ y = e^{3t}[c_1(\cos 2t + \sin 2t) + c_2(\sin 2t - \cos 2t)]. \end{cases}$$

16.
$$\begin{cases} x = 5(c_1 \cos 3t + c_2 \sin 3t), \\ y = -c_1(\cos 3t + 15 \sin 3t) + c_2(3 \cos 3t - 5 \sin 3t). \end{cases}$$

18.
$$\begin{cases} x = -5e^{4t}(c_1 \cos t + c_2 \sin t), \\ y = e^{4t}[(c_2 - 2c_1)\cos t - (c_1 + 2c_2 \sin t]. \end{cases}$$

20.
$$\begin{cases} x = 2c_1 e^{5t} + c_2(2t + 1)e^{5t}, \\ y = -c_1 e^{5t} - c_2 t e^{5t}. \end{cases}$$

22.
$$\begin{cases} x = c_1 e^{-t} + c_2(2t + 1)e^{-t}, \\ y = c_1 e^{-t} + 2c_2 t e^{-t}. \end{cases}$$

24.
$$\begin{cases} x = 2c_1 e^{6t} + c_2(2t + 1)e^{6t}, \\ y = c_1 e^{6t} + c_2 t e^{6t}. \end{cases}$$

26.
$$\begin{cases} x = 2c_1 + c_2(2t + 1), \\ y = c_1 + c_2 t. \end{cases}$$

28.
$$\begin{cases} x = e^{5t} + 5e^{-3t}, \\ y = 7e^{5t} - 5e^{-3t}. \end{cases}$$

30.
$$\begin{cases} x = e^{4t}\left(5 \cos 3t - \dfrac{10}{3} \sin 3t\right), \\ y = -e^{4t}\left(\cos 3t + \dfrac{11}{3} \sin 3t\right). \end{cases}$$

32.
$$\begin{cases} x = e^{5t} - te^{5t}, \\ y = 3e^{5t} - 2te^{5t}. \end{cases}$$

34.
$$\begin{cases} x = (8t + 6)e^{-3t}, \\ y = (16t + 8)e^{-3t}. \end{cases}$$

36.
$$\begin{cases} x = c_1 t^2 + c_2 t^4, \\ y = c_1 t^2 + 3c_2 t^4. \end{cases}$$

Section 7.5A, Page 340.

2. (a) $\begin{pmatrix} 3 & 6 \\ 21 & -9 \end{pmatrix}$

(b) $\begin{pmatrix} -4 & 12 & -20 \\ -24 & 8 & 0 \\ 12 & -4 & -8 \end{pmatrix}$

(c) $\begin{pmatrix} -15 & 3 & -6 \\ -12 & 9 & 6 \\ 0 & -9 & 18 \end{pmatrix}$

4. (a) $\begin{pmatrix} 2x_1 + x_2 - 4x_3 \\ 5x_1 - 2x_2 + 3x_3 \\ x_1 - 3x_2 + 2x_3 \end{pmatrix}$ (b) $\begin{pmatrix} -35 \\ 10 \\ -7 \end{pmatrix}$

(c) $\begin{pmatrix} x_1 - 2x_2 + 3x_3 \\ -3x_1 - 4x_2 - 4x_3 \\ -2x_1 + x_2 - 2x_3 \end{pmatrix}$

6. (a) (i.) $\begin{pmatrix} 10t \\ -18t^2 + 2t \\ 4t - 5 \end{pmatrix}$ (ii.) $\begin{pmatrix} \dfrac{5t^3}{3} \\ -\dfrac{3t^4}{2} + \dfrac{t^3}{3} \\ \dfrac{2t^3}{3} - \dfrac{5t^2}{2} \end{pmatrix}$

(b) (i.) $\begin{pmatrix} 3e^{3t} \\ (6t + 11)e^{3t} \\ (3t^2 + 2t)e^{3t} \end{pmatrix}$

(ii.) $\begin{pmatrix} \dfrac{(e^{3t} - 1)}{3} \\ \dfrac{[(6t + 7)e^{3t} - 7]}{9} \\ \dfrac{[(9t^2 - 6t + 2)e^{3t} - 2]}{27} \end{pmatrix}$

(c) (i.)
$$\begin{pmatrix} 3\cos 3t \\ -3\sin 3t \\ 3t\cos 3t + \sin 3t \\ -3t\sin 3t + \cos 3t \end{pmatrix}$$

(ii.)
$$\begin{pmatrix} \dfrac{(1 - \cos 3t)}{3} \\ \dfrac{1}{3}\sin 3t \\ \dfrac{1}{9}\sin 3t - \dfrac{t}{3}\cos 3t \\ \dfrac{1}{9}\cos 3t + \dfrac{t}{3}\sin 3t - \dfrac{1}{9} \end{pmatrix}$$

Section 7.5B, Page 351.

2. $AB = \begin{pmatrix} -9 & 50 \\ 2 & 17 \end{pmatrix}$; $BA = \begin{pmatrix} 27 & 26 \\ -10 & -19 \end{pmatrix}$.

4. $AB = \begin{pmatrix} 6 & 20 & 11 & 21 \\ 5 & 11 & 12 & 26 \end{pmatrix}$; BA not defined.

6. $AB = \begin{pmatrix} -5 & -6 & -3 \\ -4 & -6 & 9 \\ -11 & -14 & 1 \end{pmatrix}$; $BA = \begin{pmatrix} -9 & -9 & -9 \\ -13 & -8 & -3 \\ 23 & 15 & 7 \end{pmatrix}$.

8. $AB = \begin{pmatrix} 7 & 14 & 21 \\ 5 & -4 & 6 \\ 5 & 17 & 20 \end{pmatrix}$; $BA = \begin{pmatrix} 11 & 7 & -3 \\ 29 & 19 & -8 \\ -3 & -4 & -7 \end{pmatrix}$.

10. $AB = \begin{pmatrix} 12 & 8 & -16 \\ 21 & 13 & 19 \\ -6 & 6 & -22 \end{pmatrix}$; $BA = \begin{pmatrix} 20 & 8 & 20 \\ 2 & 7 & -4 \\ 12 & -2 & -24 \end{pmatrix}$.

12. $\begin{pmatrix} 6 & 6 & 18 \\ 0 & 2 & 4 \\ -16 & -10 & -8 \end{pmatrix}.$

14. $\begin{pmatrix} 4 & -\dfrac{5}{2} \\ 1 & -\dfrac{1}{2} \end{pmatrix}.$

16. $\begin{pmatrix} \dfrac{5}{3} & 2 \\ \dfrac{2}{3} & 1 \end{pmatrix}.$

18. $\begin{pmatrix} -\dfrac{1}{3} & \dfrac{7}{6} & -\dfrac{1}{2} \\ -1 & \dfrac{3}{2} & -\dfrac{1}{2} \\ \dfrac{2}{3} & -\dfrac{5}{6} & \dfrac{1}{2} \end{pmatrix}.$

20. $\begin{pmatrix} \dfrac{3}{5} & -\dfrac{2}{5} & 1 \\ -\dfrac{2}{5} & \dfrac{3}{5} & -1 \\ -\dfrac{6}{5} & \dfrac{4}{5} & -1 \end{pmatrix}.$

22. $\begin{pmatrix} 9 & -3 & 1 \\ -4 & \dfrac{3}{2} & -\dfrac{1}{2} \\ 1 & -\dfrac{1}{2} & \dfrac{1}{2} \end{pmatrix}.$

24. $\begin{pmatrix} \dfrac{5}{2} & 3 & -\dfrac{3}{2} \\ -\dfrac{1}{12} & \dfrac{1}{6} & \dfrac{1}{12} \\ \dfrac{25}{12} & \dfrac{17}{6} & -\dfrac{13}{12} \end{pmatrix}.$

Section 7.5D, Page 367.

2. Characteristic values: 5 and -3; Respective corresponding characteristic vectors:

$$\begin{pmatrix} k \\ k \end{pmatrix} \quad \text{and} \quad \begin{pmatrix} k \\ -3k \end{pmatrix},$$

where in each vector k is an arbitrary nonzero real number.

4. Characteristic values: 5 and −5; Respective corresponding characteristic vectors:

$$\begin{pmatrix} k \\ k \end{pmatrix} \quad \text{and} \quad \begin{pmatrix} 7k \\ -3k \end{pmatrix},$$

where in each vector k is an arbitrary nonzero real number.

6. Characteristic values: 7 and −2; Respective corresponding characteristic vectors:

$$\begin{pmatrix} 5k \\ -4k \end{pmatrix} \quad \text{and} \quad \begin{pmatrix} k \\ k \end{pmatrix},$$

where in each vector k is an arbitrary nonzero real number.

8. Characteristic values: 2, 3, and 5; Respective corresponding characteristic vectors:

$$\begin{pmatrix} 10k \\ -7k \\ -3k \end{pmatrix}, \quad \begin{pmatrix} k \\ -k \\ -k \end{pmatrix}, \quad \text{and} \quad \begin{pmatrix} k \\ -k \\ -3k \end{pmatrix},$$

where in each case k is an arbitrary nonzero real number.

10. Characteristic values: 1, 2, and −1; Respective corresponding characteristic vectors:

$$\begin{pmatrix} k \\ 0 \\ -k \end{pmatrix}, \quad \begin{pmatrix} k \\ k \\ k \end{pmatrix}, \quad \text{and} \quad \begin{pmatrix} k \\ -2k \\ k \end{pmatrix},$$

where in each case k is an arbitrary nonzero real number.

12. Characteristic values: 1, 3, and 4; Respective corresponding characteristic vectors:

$$\begin{pmatrix} k \\ 0 \\ k \end{pmatrix}, \quad \begin{pmatrix} 0 \\ k \\ 2k \end{pmatrix}, \quad \text{and} \quad \begin{pmatrix} 2k \\ -k \\ k \end{pmatrix},$$

where in each vector k is an arbitrary nonzero real number.

14. Characteristic values: 1, -1, and 4; Respective corresponding characteristic vectors:

$$\begin{pmatrix} 2k \\ k \\ 0 \end{pmatrix}, \quad \begin{pmatrix} 0 \\ 3k \\ k \end{pmatrix}, \quad \text{and} \quad \begin{pmatrix} k \\ -2k \\ -k \end{pmatrix},$$

where in each vector k is an arbitrary nonzero real number.

Section 7.6, Page 377.

2. $\begin{cases} x_1 = c_1 e^{4t} + c_2 e^{2t}, \\ x_2 = c_1 c^{4t} + 3c_2 e^{2t}. \end{cases}$

4. $\begin{cases} x_1 = c_1 e^{t} + c_2 e^{-t}, \\ x_2 = -3c_1 e^{t} - c_2 e^{-t}. \end{cases}$

6. $\begin{cases} x_1 = c_1 e^{5t} + c_2 e^{3t}, \\ x_2 = c_1 e^{5t} + 3c_2 e^{3t}. \end{cases}$

8. $\begin{cases} x_1 = c_1 e^{5t} + c_2 e^{-3t}, \\ x_2 = c_1 e^{5t} - 3c_2 e^{-3t}. \end{cases}$

10.
$$\begin{cases} x_1 = e^{2t}(c_1\cos 3t - c_2\sin 3t), \\[2mm] x_2 = e^{2t}(c_1\cos 3t + c_2\sin 3t). \end{cases}$$

12.
$$\begin{cases} x_1 = 2e^{3t}(c_1\cos 2t + c_2\sin 2t), \\[2mm] x_2 = e^{3t}[c_1(\cos 2t + \sin 2t) + c_2(\sin 2t - \cos 2t)]. \end{cases}$$

14.
$$\begin{cases} x_1 = 5e^{5t}(c_1\cos 2t + c_2\sin 2t), \\[2mm] x_2 = -e^{5t}[c_1(\cos 2t - 2\sin 2t) + c_2(2\cos 2t + \sin 2t)]. \end{cases}$$

16.
$$\begin{cases} x_1 = -2c_1 e^{5t} - c_2(2t + 1)e^{5t}, \\[2mm] x_2 = c_1 e^{5t} + c_2 t\, e^{5t}. \end{cases}$$

18.
$$\begin{cases} x_1 = c_1 e^{-t} + c_2(2t + 1)e^{-t}, \\[2mm] x_2 = c_1 e^{-t} + 2c_2 t e^{-t}. \end{cases}$$

20.
$$\begin{cases} x_1 = c_1 e^{5t} + c_2(t + 1)e^{5t}, \\[2mm] x_2 = 2c_1 e^{5t} + c_2(2t + 1)e^{5t}. \end{cases}$$

Section 7.7, Page 400.

2.
$$
\begin{cases}
x_1 = 10c_1 e^{2t} + c_2 e^{3t} + c_3 e^{5t}, \\
x_2 = -7c_1 e^{2t} - c_2 e^{3t} - c_3 e^{5t}, \\
x_3 = -3c_1 e^{2t} - c_2 e^{3t} - 3c_3 e^{5t}.
\end{cases}
$$

4.
$$
\begin{cases}
x_1 = c_1 e^{t} + c_3 e^{4t}, \\
x_2 = 2c_2 e^{3t} - c_3 e^{4t}, \\
x_3 = c_2 e^{3t} - c_3 e^{4t}.
\end{cases}
$$

6.
$$
\begin{cases}
x_1 = c_1 e^{t} + c_2 e^{2t} + c_3 e^{-t}, \\
x_2 = c_2 e^{2t} - 2c_3 e^{-t}, \\
x_3 = -c_1 e^{t} + c_2 e^{2t} + c_3 e^{-t}.
\end{cases}
$$

8.
$$
\begin{cases}
x_1 = c_1 e^{2t} + c_2 e^{-t} + c_3 e^{-4t}, \\
x_2 = 2c_1 e^{2t} + c_2 e^{-t} + c_3 e^{-4t}, \\
x_3 = c_1 e^{2t} + c_2 e^{-t} + 2c_3 e^{-4t}.
\end{cases}
$$

10.
$$
\begin{cases}
x_1 = 17c_1 e^{3t} + 2c_2 \sin 2t - 2c_3 \cos 2t, \\
x_2 = 3c_1 e^{3t} + c_2 \cos 2t + c_3 \sin 2t, \\
x_3 = 7c_1 e^{3t} + c_2 (\cos 2t + 2 \sin 2t) + c_3 (\sin 2t - 2\cos 2t).
\end{cases}
$$

12.
$$
\begin{cases}
x_1 = c_1 e^{2t} + c_2 e^{t}, \\
x_2 = 2c_1 e^{2t} - c_2 e^{t} + c_3 e^{t}, \\
x_3 = c_1 e^{2t} + 2c_3 e^{t}.
\end{cases}
$$

14.
$$\begin{cases} x_1 = c_1 e^t + c_2 e^{3t}, \\ x_2 = -c_1 e^t + 2c_3 e^{3t}, \\ x_3 = 3c_1 e^t + c_3 e^{3t}. \end{cases}$$

16.
$$\begin{cases} x_1 = c_1 e^{28t} + c_2 e^t, \\ x_2 = 2c_1 e^{28t} + c_3 e^t, \\ x_3 = c_1 e^{28t} - c_3 e^t. \end{cases}$$

18.
$$\begin{cases} x_1 = 5c_1 e^{-t} + 5c_2 e^{2t} + c_3 e^{3t}, \\ x_2 = 3c_1 e^{-t} + 5c_2 e^{2t} + c_3 e^{3t}, \\ x_3 = -2c_1 e^{-t} - c_2 e^{2t}. \end{cases}$$

20.
$$\begin{cases} x_1 = 3c_1 e^{-t} + c_2 e^{3t} + c_3 t e^{3t}, \\ x_2 = -8c_1 e^{-t} + 2c_3 e^{3t}, \\ x_3 = c_1 e^{-t} - c_2 e^{3t} - c_3(t + 3)e^{3t}. \end{cases}$$

22.
$$\begin{cases} x_1 = c_1 e^{2t} + 3c_2 + c_3(3t + 2), \\ x_2 = 4c_2 + 4c_2 t, \\ x_3 = c_1 e^{2t} + c_2 + c_3(t + 3). \end{cases}$$

24.
$$\begin{cases} x_1 = c_1 e^{-t} + 3c_2 e^t + c_3\left(3t - \frac{1}{5}\right)e^t, \\ x_2 = 2c_1 e^{-t} + 5c_2 e^t + 5c_3 t e^t, \\ x_3 = -2c_1 e^{-t} - 4c_2 e^t + c_3\left(-4t + \frac{3}{5}\right)e^t. \end{cases}$$

26. $\begin{cases} x_1 = c_1 e^{2t} + c_2 e^{2t} + c_3(2t + 1)e^{2t}, \\ x_2 = 2c_1 e^{2t} + 2c_3 t e^{2t}, \\ x_3 = 2c_2 e^{2t} + 2c_3 t e^{2t}. \end{cases}$

28. $\begin{cases} x_1 = 2c_1 e^{2t} + c_2(2t + 1)e^{2t} + c_3(t^2 + t + 1)e^{2t}, \\ x_2 = -c_1 e^{2t} - c_2 t e^{2t} - \frac{1}{2} c_3 t^2 e^{2t}, \\ x_3 = -2c_1 e^{2t} - 2c_2 t e^{2t} + c_3(-t^2 + 1)e^{2t}. \end{cases}$

Section 8.1A, Page 421.

2.

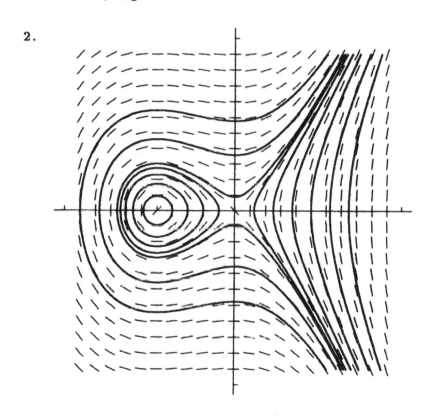

4.

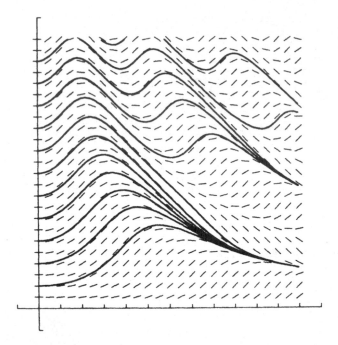

6.

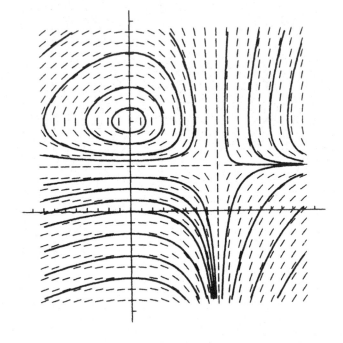

8.

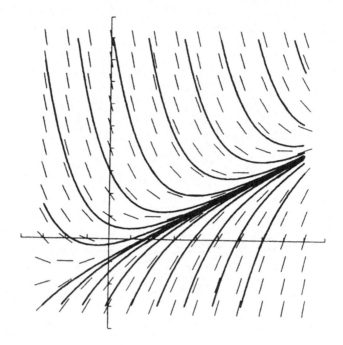

10.

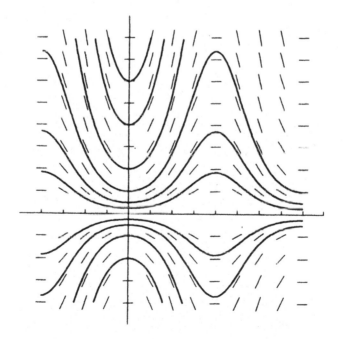

12.   (a) 4;   (b) 1;   (c) 6;   (d) 2;   (e) 5;   (f) 3.

Section 8,.1B, Page 427

2.

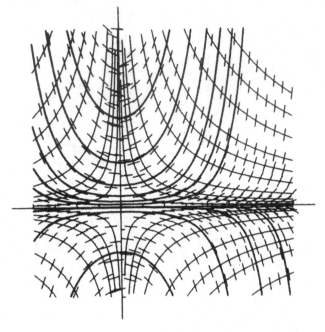

4.

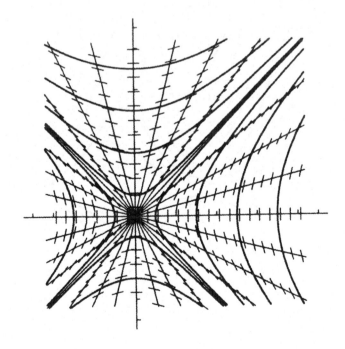

6.

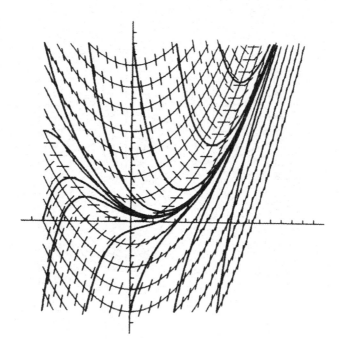

8.

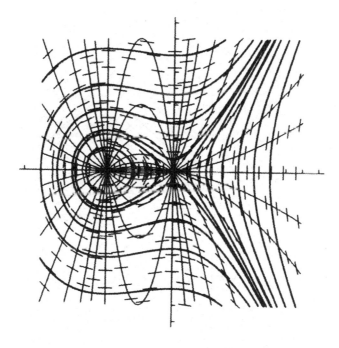

10.

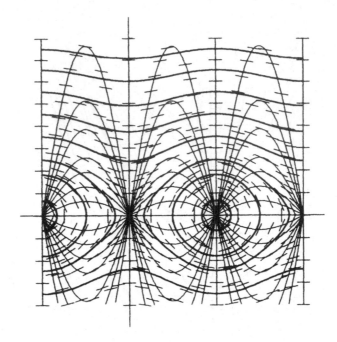

12.

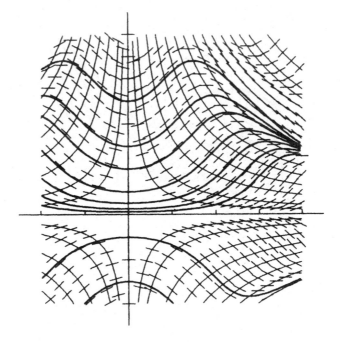

Section 8.2, Page 434.

2. $y = 4 + 32x + 256x^2 + \dfrac{6145}{3} x^3 + \cdots$.

4. $y = 3 + 27x + \dfrac{729}{2} x^2 + \dfrac{10,935}{2} x^3 + \cdots$.

6. $y = x + \dfrac{1}{3} x^3 + \dfrac{1}{15} x^5 + \cdots$.

8. $y = 1 + x + 2x^2 + \dfrac{14}{3} x^3 + \dfrac{35}{3} x^4 + \cdots$.

10. $y = 1 + 2(x - 1) + \dfrac{7(x - 1)^2}{2} + \dfrac{14(x - 1)^3}{3}$

$\qquad + \dfrac{73(x - 1)^4}{12} + \cdots$.

12. $y = \pi + \dfrac{(x - 1)^2}{2} + \dfrac{(x - 1)^5}{40} + \cdots$.

14. $y = 2 + (e + 2)(x - 1) + \left(\dfrac{3}{2} e + 1\right)(x - 1)^2$

$\qquad + \left(e + \dfrac{1}{3}\right)(x - 1)^3 + \left(\dfrac{5}{12} e + \dfrac{1}{12}\right)(x - 1)^4 + \cdots$.

Section 8.3, Page 438.

2. $\phi_1(x) = 1 + x + \dfrac{x^2}{2}$, $\phi_2(x) = 1 + x + x^2 + \dfrac{x^3}{6}$,

$\qquad \phi_3(x) = 1 + x + x^2 + \dfrac{x^3}{3} + \dfrac{x^4}{24}$.

4. $\phi_1(x) = x$, $\phi_2(x) = x + \dfrac{x^4}{4}$, $\phi_3(x) = x + \dfrac{x^4}{4} + \dfrac{x^7}{14} + \dfrac{x^{10}}{160}$.

6. $\phi_1(x) = 1 - \cos x,$

   $\phi_2(x) = 1 + \dfrac{3}{2} x - \cos x - 2 \sin x + \dfrac{1}{4} \sin 2x.$

8. $\phi_1(x) = x,\ \phi_2(x) = x + x^6,$

   $\phi_3(x) = x + x^6 + \dfrac{24}{11} x^{11} + \dfrac{9}{4} x^{16} + \dfrac{8}{7} x^{21} + \dfrac{6}{25} x^{25}.$

Section 8.4

	x_n	Exact Sol'n	Euler Approx.	Error	% Rel Errors
2.	0.25	0.026633	0.000000	0.026633	100.0
	1.50	0.512447	0.503906	0.008541	1.666615
4.	−0.80	1.268869	1.200000	0.068869	5.427554
	0.00	5.291792	3.783680	1.508112	28.499080
6.	1.50	2.598076	2.625000	0.026924	1.036297
	3.00	4.242641	4.304352	0.061711	1.454539
8.	0.20	0.510067	0.500000	0.010067	1.973606
	1.00	0.791798	0.713110	0.078687	9.937829
10.	1.20	0.692821	1.200000	0.507179	73.204864
	2.00	1.743560	2.000000	0.256440	14.707844
12.	0.20	1.019739	1.000000	0.019739	1.935654
	2.00	1.957624	1.935192	0.022432	1.145899

Section 8.5

	x_n	Exact Sol'n	Improved Euler	Error	% Rel Errors
2.	0.25	0.026633	0.031250	0.004617	17.337112
	1.50	0.512447	0.514901	0.002454	0.478956
4.	−0.80	1.268869	1.260000	0.008869	0.698932
	0.00	5.291792	5.075616	0.216176	4.085122
6.	1.50	2.598076	2.602679	0.004602	0.177145
	3.00	4.242641	4.251370	0.008730	0.205757
8.	0.20	0.510067	0.509933	0.000133	0.026123
	1.00	0.791798	0.789224	0.002574	0.325092
10.	1.20	0.692821	0.800000	0.107179	15.469910
	2.00	1.743560	1.789042	0.045482	2.608548
12.	0.20	1.019739	1.019867	0.000128	0.012583
	2.00	1.957624	1.955473	0.002152	0.109918

Section 8.6

	x_n	Exact Sol'n	Runge- Kutta	Error	% Rel Errors
2.	0.25	0.026633	0.026693	0.000060	0.225450
	1.50	0.512447	0.512476	0.000030	0.005776
4.	−0.80	1.268869	1.268800	0.000069	0.005400
	0.00	5.291792	5.290095	0.001697	0.032064

6.	1.50	2.598076	2.598127	0.000051	0.001961
	3.00	4.242641	4.242717	0.000077	0.001805
8.	0.20	0.510067	0.510067	0.000000	0.000000
	1.00	0.791798	0.791796	0.000001	0.000139
10.	1.20	0.692821	0.725782	0.032960	4.757407
	2.00	1.743560	1.756940	0.013380	0.767424
12.	0.20	1.019739	1.019739	0.000001	0.000064
	2.00	1.957624	1.957627	0.000003	0.000137

Section 8.7

	x_n	Exact Sol'n	ABAM approx.	Error	% Rel Errors
2.	1.00	0.283834	0.283619	0.000215	0.075706
	1.50	0.512447	0.512201	0.000245	0.047869
4.	-0.20	3.564774	3.563793	0.000981	0.027531
	0.00	5.291792	5.289874	0.001918	0.036247
6.	3.00	4.242641	4.242756	0.000116	0.002724
8.	0.80	0.677156	0.677153	0.000003	0.000440
	1.00	0.791798	0.791787	0.000011	0.001363
10.	1.80	1.509967	1.539989	0.030021	1.988212
	2.00	1.743560	1.769706	0.026146	1.499599
12.	0.80	1.267512	1.267597	0.000085	0.006677
	2.00	1.957624	1.957674	0.000050	0.002540

Section 8.8 using the Euler method

t_n	Exact $\begin{cases} x(t_n) \\ y(t_n) \end{cases}$	Euler $\begin{cases} x_n \\ y_n \end{cases}$	Error	% Rel Errors
2. 0.100	2.69972	2.60000	0.09972	3.694
	2.69972	2.60000	0.09972	3.694
0.500	8.96338	7.42586	1.53752	17.15
	8.96338	7.42586	1.53752	17.15
4. 0.100	0.90484	0.90000	0.00484	0.535
	-0.90484	-0.90000	0.00484	0.535
0.500	0.60653	0.59049	0.01604	2.645
	-0.60653	-0.59049	0.01604	2.645
6. 0.100	3.11518	3.10000	0.01518	0.487
	2.01001	2.00000	0.01001	0.498
0.500	3.90397	3.81151	0.09246	2.368
	2.25525	2.20100	0.05425	2.406
8. 0.100	2.44281	2.40000	0.04281	1.752
	3.66421	3.60000	0.06421	1.752
0.500	5.43656	5.03860	0.39797	7.320
	8.15485	7.43811	0.71673	8.789
10. 0.100	1.34264	1.30000	0.04264	3.176
	3.87874	3.80000	0.07874	2.030
0.500	3.69453	3.45744	0.23709	6.417
	7.38906	8.45152	1.06246	14.38
12. 0.100	8.84140	8.80000	0.04140	0.468
	8.84281	8.80000	0.04281	0.484
0.500	13.21828	12.89295	0.32533	2.461
	13.43656	13.03010	0.40646	3.025

Section 8.8 using the Runge-Kutta method

	t_n	Exact $\begin{cases} x(t_n) \\ y(t_n) \end{cases}$	R-K $\begin{cases} x_n \\ y_n \end{cases}$	Error	% Rel Errors
2.	0.100	2.6997176 2.6997176	2.6996750 2.6996750	0.0000426 0.0000426	0.00158 0.00158
	0.500	8.9633781 8.9633781	8.9626707 8.9626707	0.0007074 0.0007074	0.00789 0.00789
4.	0.100	0.9048374 −0.9048374	0.9048375 −0.9048375	0.0000001 0.0000001	0.00001 0.00001
	0.500	0.6065307 −0.6065307	0.6065309 −0.6065309	0.0000003 0.0000003	0.00005 0.00005
6.	0.100	3.1151793 2.0100083	3.1151792 2.0100083	0.0000001 0.0000000	0.00000 0.00000
	0.500	3.9039732 2.2552519	3.9039722 2.2552516	0.0000010 0.0000004	0.00003 0.00002
8.	0.100	2.4428055 3.6642083	2.4428285 3.6642073	0.0000230 0.0000010	0.00094 0.00003
	0.500	5.4365637 8.1548455	5.4371347 8.1550023	0.0005710 0.0001568	0.01050 0.00192
10.	0.100	1.3426422 3.8787442	1.3426667 3.8789333	0.0000244 0.0001891	0.00182 0.00488
	0.500	3.6945280 7.3890561	3.6960378 7.3973576	0.0015098 0.0083015	0.04087 0.11235
12.	0.100	8.8414028 8.8428055	8.8414025 8.8428058	0.0000003 0.0000003	0.00000 0.00000
	0.500	13.2182818 13.4365637	13.2182719 13.4365489	0.0000100 0.0000148	0.00008 0.00011

Section 9.1A, Page 488

2. $\dfrac{1}{s^2 - 1}$.

4. $\dfrac{2}{s}\left(2 - e^{-3s}\right)$.

6. $\left(e^{-s} - e^{-2s}\right)\left(\dfrac{1}{s} + \dfrac{1}{s^2}\right)$.

8. $\dfrac{2}{s^2}\left(1 - e^{-s} - e^{-3s}\right)$.

Section 9.1B, Page 496.

2. $\dfrac{2abs^2}{[s^2 + (a - b)^2][s^2 + (a + b)^2]}$.

4. $\dfrac{s^3 + 7a^2 s}{(s^2 + a^2)(s^2 + 9a^2)}$; $\dfrac{as^2 + 3a^3}{(s^2 + a^2)(s^2 + 9a^2)}$.

6. $\dfrac{24}{s^5}$.

8. $\dfrac{3s + 11}{s^2 + 4s - 8}$.

10. $\dfrac{-s^3 + s + 1}{s^4 - s^2}$.

12. $\dfrac{12s - 2}{3s^2 - 5s + 7} + \dfrac{2}{(3s^2 - 5s + 7)(s^2 + 4)}$.

14. $\dfrac{2b^2}{(s - a)[(s - a)^2 + 4b^2]}$.

16. $\dfrac{24bs(s^2 - b^2)}{(s^2 + b^2)^4}$.

18. $\dfrac{24}{(s - a)^5}$.

Section 9.2A, Page 504.

2. $4e^{-2t} + 7$.

4. $2 \cos 3t$.

6. $5t + 3t^2$.

8. $e^{-4t}(3 \sin 2t + \cos 2t)$.

10. $2 \cosh 2t + \frac{3}{2} \sinh 2t$.

12. $-\frac{1}{2} e^{-t/2} + \frac{3}{4} e^{3t/4}$.

14. $\frac{1}{2} - \frac{1}{2} \cos\sqrt{2}t + \frac{1}{\sqrt{2}} \sin\sqrt{2}t$.

16. $t^2 e^{-3t}\left(1 + \frac{t}{6}\right)$.

18. $te^{-t/2}\left(1 + \frac{1}{4} t\right)$.

20. $t(\cos 2t - \sin 2t)$.

22. $e^{-2t}\left(5 \cos 3t + \frac{7}{3} \sin 3t\right)$.

24. $\frac{5}{2} e^{t/2} - 2e^{-t}$.

26. $-\frac{13}{4} + \frac{5}{2} t + 4e^{-t} - \frac{3}{4} e^{-2t}$.

28. $2t^2 + 3te^{2t}$.

30. $\sin 3t - 3t \sin 3t + 2t \cos 3t$.

Section 9.2B, Page 509.

2. $\dfrac{(e^t - e^{-4t})}{5}$.

4. $\dfrac{[3 - e^{-2t}(2 \sin 3t + 3 \cos 3t)]}{39}$.

6. $\dfrac{(2 \sin t - \cos t + e^{-2t})}{5}$.

Section 9.3, Page 519.

2. $y = \sin t - \cos t$.

4. $y = 8t^2 - 8t + 4 + 3e^{-2t}$.

6. $y = \dfrac{(15e^{3t} + 13e^{-4t})}{7}$.

8. $y = e^{-t}(3 \sin 2t + 2 \cos 2t)$.

10. $y = 3 \sin 3t + 2e^{-3t}$.

12. $y = -\dfrac{1}{2} + 2e^{-t/2} + \dfrac{1}{2} e^{2t}$.

14. $y = (t + 1)e^t + (t - 1)e^{-t}$.

16. $y = -e^{-t} + 2te^{-t} + 2e^{-2t} + te^{-2t}$.

18. $y = 3e^{3t} - 2e^{5t} + 4e^{2t} + 3te^{2t}$.

20. $y = (t + 1)^2 e^{-t}$.

22. $y = -4e^t + 14e^{3t} + 6te^{4t} - 11e^{4t}$.

Section 9.4A, Page 527.

2. $\dfrac{-3e^{-10s}}{s}$.

4. $\dfrac{2}{s} (1 - e^{-5s})$.

6. $\dfrac{6}{s} (e^{-9s} - e^{-3s})$.

8. $\dfrac{3}{s} (3 - e^{-5s} - e^{-10s} - e^{-15s})$.

10. $\dfrac{1}{s} (4 - 4e^{-5s} + 3e^{-10s})$.

12. $3e^{-4s}\left(\dfrac{1}{s^2} + \dfrac{4}{s}\right)$.

14. $\dfrac{2}{s^2} (1 - e^{-5s})$.

16. $\dfrac{e^{-2(s+1)}}{s + 1}$.

18. $\dfrac{6}{s} - \dfrac{2e^{-s}}{s^2} + \dfrac{2e^{-3s}}{s^2}$.

20. $\dfrac{e^{-3\pi s/2} + e^{-9\pi s/2}}{s^2 + 1}$.

22. $\dfrac{1 - e^{-s} - se^{-2s}}{s^2(1 - e^{-2s})}$.

24. $\dfrac{1}{1 - e^{-\pi s}} \left[\dfrac{e^{-\pi s} + 1}{s^2 + 1} \right]$.

Section 9.4B, Page 530.

2. $\begin{cases} 0, & 0 < t < 5, \\ e^{2(t-5)}(7t - 32), & t > 5. \end{cases}$

4. $\begin{cases} 0, & 0 < t < 4, \\ 4e^{t-4} - 4e^{-2(t-4)}, & t > 4. \end{cases}$

6. $\begin{cases} 0, & 0 < t < 2, \\ -e^{-4(t-2)} + 2e^{2(t-2)}, & t > 2. \end{cases}$

8. $\begin{cases} 0, & 0 < t < 3, \\ e^{-2(t-3)}\left[2 \cos 3(t - 3) + \dfrac{5}{3} \sin 3(t - 3)\right], & t > 3. \end{cases}$

10. $\begin{cases} 0, & 0 < t < 3, \\ \dfrac{(t - 3)^2}{2}, & 3 < t < 8, \\ 5t - \dfrac{55}{2}, & t > 8. \end{cases}$

12.
$$
\begin{cases}
\dfrac{2(\sin 3t)}{3}, & 0 < t < 3, \\[4mm]
\dfrac{2(\sin 3t)}{3} - \sin 3(t - 3), & t > 3.
\end{cases}
$$

14.
$$
\begin{cases}
\cos 2t - 1, & 0 < t < 2, \\[2mm]
\cos 2t - \cos 2(t - 2), & t > 2.
\end{cases}
$$

Section 9.4C, Page 533.

2. $y = \begin{cases} 4e^{5t/3}, & 0 < t < 6, \\[2mm] -2 + (4 + 2e^{-10})e^{5t/3}, & t > 6. \end{cases}$

4. $y = \begin{cases} 1 - 3e^{-2t} + 2e^{-3t}, & 0 < t < 2. \\[2mm] 3(e^4 - 1)e^{-2t} + 2(1 - e^6)e^{-3t}, & t > 2. \end{cases}$

6. $y = \begin{cases} \dfrac{3}{8} + \dfrac{3}{4} e^{-2t} - \dfrac{1}{8} e^{-4t}, & 0 < t < 2\pi, \\[3mm] \dfrac{3}{4} e^{-2t}(1 + e^{4\pi}) - \dfrac{1}{8} e^{-4t}(1 + 3e^{8\pi}), & t > 2\pi. \end{cases}$

8. $y = \begin{cases} 2 \sin t + 2 \cos t + t, & 0 < t < \pi. \\[2mm] \sin t + 2 \cos t + \pi, & t > \pi. \end{cases}$

$$
10. \quad y = \begin{cases}
3e^{-3t}\cos 4t, & 0 < t < 2, \\[2mm]
3e^{-3t}\cos 4t + 1 - e^{-3(t-2)}\cos 4(t - 2) \\[1mm]
\quad - \dfrac{3}{4} e^{-3(t-2)}\sin 4(t - 2), & 2 < t < 4, \\[3mm]
3e^{-3t}\cos 4t - e^{-3(t-2)}\cos 4(t - 2) \\[1mm]
\quad - \dfrac{3}{4} e^{-3(t-2)}\sin 4(t - 2) + e^{-3(t-4)}\cos 4(t - 2) \\[1mm]
\quad + \dfrac{3}{4} e^{-3(t-4)}\sin 4(t - 2), & t > 4.
\end{cases}
$$

$$
12. \quad y = \begin{cases}
(t - 1)^2, & 0 < t < 2, \\[2mm]
-1 + 2e^{t-2}, & t > 2.
\end{cases}
$$

Section 9.4D, Page 539.

$$
2. \quad y = \begin{cases}
0, & 0 < t < 2\pi, \\[2mm]
e^{-2(t-2\pi)}\sin t, & t > 2\pi.
\end{cases}
$$

$$
4. \quad y = \begin{cases}
-2e^{-t} + 4e^{-2t}, & 0 < t < 4, \\[2mm]
(e^4 - 2)e^{-t} - (e^8 - 4)e^{-2t}, & t > 4.
\end{cases}
$$

$$
6. \quad y = \begin{cases}
e^{-2t}\cos t, & 0 < t < \pi, \\[2mm]
e^{-2t}(\cos t - e^{2\pi}\sin t), & t > \pi.
\end{cases}
$$

Section 9.5, Page 542.

2. $\begin{cases} x = 2 + 2e^t - e^{2t}, \\ y = e^t - e^{2t}. \end{cases}$
4. $\begin{cases} x = -3t - e^{-t}, \\ y = 2t - 1 + e^{-t}. \end{cases}$

6. $\begin{cases} x = e^{2t} + 2\cos 2t - \sin 2t, \\ y = 2e^{2t} + 5\sin 2t. \end{cases}$

8. $\begin{cases} x = 1 - t + 3e^{4t} - e^{-2t}, \\ y = t - 3e^{4t} - e^{-2t}. \end{cases}$

10. $\begin{cases} x = 2e^t - e^{2t} - 1, \\ y = e^t - 2. \end{cases}$

CPSIA information can be obtained at www.ICGtesting.com
Printed in the USA
BVOW08s0413280515

401910BV00006BA/57/P